The Secret World of Roots

This first appeared in the *New Scientist*, London, the weekly review of science and technology.

The Rhizosphere

WILEY SERIES IN
ECOLOGICAL AND APPLIED MICROBIOLOGY

Edited by Ralph Mitchell

MICROBIAL LECTINS AND AGGLUTININS: Properties and Biological Activity
David Mirelman Editor

THERMOPHILES: General, Molecular, and Applied Microbiology
Thomas D. Brock Editor

INNOVATIVE APPROACHES TO PLANT DISEASE CONTROL
Ilan Chet Editor

PHAGE ECOLOGY
Sagar M. Goyal, Charles P. Gerba, Gabriel Bitton Editors

BIOLOGY OF ANAEROBIC MICROORGANISMS
Alexander J. B. Zehnder Editor

BIOFILMS
William G. Characklis, Kevin C. Marshall Editors

The Rhizosphere

Edited by

J. M. LYNCH

AFRC Institute of Horticultural Research
Littlehampton
West Sussex, BN17 6LP
UK

WILEY

A WILEY-INTERSCIENCE PUBLICATION

JOHN WILEY & SONS

Chichester · New York · Brisbane · Toronto · Singapore

Other Wiley Editorial Offices

John Wiley & Sons, Inc., 605 Third Avenue,
New York, NY 10158-0012, USA

Jacaranda Wiley Ltd, G.P.O. Box 859, Brisbane,
Queensland 4001, Australia

John Wiley & Sons (Canada) Ltd, 22 Worcester Road,
Rexdale, Ontario M9W 1L1, Canada

John Wiley & Sons (SEA) Pte Ltd, 37 Jalan Pemimpin 05-04,
Block B, Union Industrial Building, Singapore 2057

Library of Congress Cataloging-in-Publication Data:

The Rhizosphere / edited by J.M. Lynch.
p. cm.—(Wiley series in ecological and applied microbiology)
"A Wiley-Interscience publication."
Includes bibliographical references.
ISBN 0 471 92548 9
1. Rhizosphere—Microbiology. 2. Root-tubercles. I. Lynch, J. M. (James Michael) II. Series.
QR111.R47 1990
574.5′26404—dc20 89-24819
CIP

British Library Cataloguing in Publication Data:

The Rhizosphere.
1. Plants. Roots
I. Lynch, J. M. (James Michael)
581.1′0428

ISBN 0 471 92548 9

Phototypeset by Dobbie Typesetting Limited, Plymouth, Devon
Printed in Great Britain by Courier International Ltd., Tiptree, Essex

TO MUM AND DAD

O fortunatos nimium,
sua si bona norint,
Agricolas

Virgil 70-19 BC
ii 458

CONTRIBUTORS

M. J. BAZIN, Microbial Physiology Research Group, King's College (University of London), Campden Hill Road, London W8 7AH, UK

R. CAMPBELL, Department of Botany, University of Bristol, Woodland Road, Bristol BS8 1UG, UK

R. J. COOK, USDA-ARS, Johnson Hall, Washington State University, Pullman, WA 99164-6421, USA

E. A. CURL, Department of Plant Pathology, 139 Funchess Hall, Auburn University, Alabama 36849, USA

M. C. DREW, Department of Horticultural Sciences, Texas A & M University, College Station, Texas 77843-2133, USA

L. F. ELLIOTT, USDA-ARS Cotton Research Station, 17053 Shafter Avenue, Shafter, CA 93263, USA

C. A. GILLIGAN, Botany School, University of Cambridge, Downing Street, Cambridge CB2 3EA, UK

M. P. GREAVES, AFRC Institute of Arable Crops Research, Long Ashton Research Station, Bristol BS18 9AF, UK

G. E. HARMAN, Department of Horticultural Sciences, New York State Agricultural Experiment Station, Cornell University, Sturtevant Hall, Geneva, NY 14456, USA

J. D. HARPER, Department of Entomology, Box 7613, North Carolina State University, Raleigh, NC 27695, USA

R. W. HEDGES*, Rijksuniversiteit Gent, Laboratorium Genetika, Ledeganckstraat 35, B-9000 Gent, Belgium

*Deceased

D. HORNBY, AFRC Institute of Arable Crops Research, Rothamsted Experimental Station, Harpenden, Herts AL5 2JQ, UK

R. D. LUMSDEN, Biocontrol of Plant Diseases Laboratory, Plant Protection Institute, USDA–ARS, Beltsville, MD 20705, USA

J. M. LYNCH, AFRC Institute of Horticultural Research, Littlehampton, West Sussex BN17 6LP, UK

P. MARKHAM, Microbial Physiology Research Group, King's College (University of London), Campden Hill Road, Kensington, London W8 7AH, UK

E. MESSENS, Rijksuniversiteit Gent, Laboratorium Genetika, Ledeganckstraat 35, B-9000 Gent, Belgium

B. METTING, Enviros Corporation, 600 Skyline Tower, 10900 NE 4th Street, Bellevue, WA 98004, USA

C. P. P. REID, Department of Forestry, 118 Newins-Ziegler Hall, University of Florida, Gainesville, FL 32611, USA

A. D. ROVIRA, CSIRO Division of Soils, Private Bag No. 2, Glen Osmond, South Australia 5064

E. SCOTT, Microbial Physiology Research Group, King's College (University of London), Campden Hill Road, Kensington, London W8 7AH, UK

J. M. WHIPPS, AFRC Institute of Horticultural Research, Littlehampton, West Sussex BN17 6LP, UK

D. A. ZUBERER, Department of Soil & Crop Sciences, Texas A & M University, College Station, Texas 77843-2474, USA

SERIES PREFACE

The Ecological and Applied Microbiology series of monographs and edited volumes is being produced to facilitate the exchange of information relating to the microbiology of specific habitats, biochemical processes of importance in microbial ecology, and evolutionary microbiology. The series will also publish texts in applied microbiology, including biotechnology, medicine, and engineering.

During the past decade we have seen dramatic advances in the study of microbial ecology. It is gratifying that today's microbial ecologists not only cooperate with colleagues in other disciplines but also study the comparative biology of different habitats. Modern microbial ecologists, investigating ecosystems, gain insights into previously unknown biochemical processes, comparative ecology, and evolutionary theory. They also isolate new microorganisms with application to medicine, industry and agriculture.

Applied microbiology has also undergone a revolution in the past decade. The field of industrial microbiology has been transformed by new techniques in molecular genetics. Because of these advances, we now have the potential to utilize microorganisms for industrial processes in ways microbiologists could not have imagined 20 years ago. At the same time, we face the challenge of determining the consequences of releasing genetically engineered microorganisms into the natural environment.

New concepts and methods to study this extraordinary range of exciting problems in microbiology are now available. Young microbiologists are increasingly being trained in ecological theory, mathematics, biochemistry, and genetics. The barriers between the disciplines essential to the study of modern microbiology are disappearing. It is my hope that this series in Ecological and Applied Microbiology will facilitate the reintegration of microbiology and stimulate research in the tradition of Louis Pasteur.

Cambridge, Massachusetts
October 1989

RALPH MITCHELL

PREFACE

Since Ralph Mitchell first suggested to me during the International Congress of Microbiology in 1986 that I edit a volume on the rhizosphere, it has been a great pleasure to see such a vast expansion of interest in the subject in so many countries. This is resulting in public awareness that the productivity and quality of flowers, fruit, leaves and seeds are determined by the activities of root systems and their associated physical and biological environment. In a recent issue of *New Scientist* which carried an article on the rhizosphere (Lynch, 1988) the artist Stuart Robertson depicted a lettuce plant with genies (instead of the intended genes) working hard around roots! This is reproduced at the beginning of the book in black and white with the permission of the artist.

The rhizosphere has become an important area to test and evaluate new opportunities being developed in biotechnology. This is, of course, not a new concept for the rhizosphere, as rhizobial inoculants have been used to enhance nitrogen fixation and growth of legumes since the turn of the century. However, the novelty in recent years is the recognition that biological processes are regulated and can be manipulated genetically, opening enormous opportunities in the future for optimization of plant productivity in managed and natural ecosystems, while minimizing the potential of environmental damage by misuse of agrochemicals so feared by Rachel Carson in *Silent Spring*.

It is becoming increasingly clear that molecular biology will have a major impact on rhizosphere studies. An early recruit into the new breed of molecular biologists applying their attentions to the rhizosphere was Bob Hedges. He went to Gent in 1983 to join the research group of Marc van Montagu and Jeff Schell who had already made such an important impact on the rhizosphere by characterizing the Ti plasmid as being responsible for the induction of root tumours by *Agrobacterium*. Bob had read botany in Oxford and had been greatly influenced by G. W. Beadle in discussing microbial genetics. After working at Caltec, Trinity College Dublin and the Institut Pasteur in Paris, he obtained a Ph.D. at Leicester University. At the suggestion of Sir Ernst Chain, he then joined Professor Naomi Datta at the Royal Postgraduate Medical School in London. Together they collected and classified plasmids determining antibiotic resistance. They noticed the unexpected frequency of recombination between unrelated plasmids in which one acquired a resistance gene from the other. Bob suggested that DNA sequences, including a resistance gene, were "jumping" from one DNA replicon to another. In 1974, with A. E. Jacob, he found the evidence to confirm this and he coined the word "transposon" for the transposing sequence. So emerged a major new area of molecular biology and one which is already having an enormous impact on rhizosphere methodology because *Tn-5* insertions have proved to be reliable and extremely sensitive sequences to track organisms introduced into soil.

I first met Bob in 1983 when he went to Gent and in the five succeeding years enjoyed a stimulating dialogue with him by correspondence, on novel approaches to the rhizosphere, especially when I persuaded him to write the genetics chapter in this book. He became the first author to deliver his manuscript but tragically died shortly afterwards at the age of 50. As some work was still needed on the chapter his colleague and friend in Gent, Eric Messens, agreed to complete it and I hope that this chapter will be a fitting final tribute to Bob.

This book has been a pleasure to edit, in large part due to the stimulating cooperation of the contributors. The chapters arrived over a period of about 9 months. For example, John Whipps produced his chapter shortly after Bob Hedges and, of course, in fast-moving fields it can be irritating for authors to have their manuscripts delayed even for months. There were also some difficult editorial decisions in determining where to allow overlap between chapters. Where this happens, cross-referencing is made and views expressed are not always coincident. It seemed to me more sensible to leave the reader to judge the evidence, for example the role of nitrogen fertilizer form in take-all infection (Chapters 9 and 15). The editorial effort involved required considerable understanding and constructive comment by my colleagues at the AFRC Institute of Horticultural Research, Littlehampton, as well as my colleagues at King's College London, where I am Visiting Professor. I am particularly grateful to my secretary, Sue Bewsey, who not only produced my manuscript but helped me "chase" contributors. I am also grateful for the excellent support from Michael Dixon, Patricia Sharp and Lesley Winchester at the Wiley offices in Chichester. Most of my time on the book was spent "at home" and I am very grateful to my wife Mary, and children Luke, Dominic, Cordelia and Francesca for their forbearance.

Littlehampton, West Sussex JIM LYNCH
April 1989

REFERENCE

Lynch, J. M. (1988). Microbes are rooting for better crops. *New Scientist,* **118**:45–49.

CONTENTS

1

INTRODUCTION: SOME CONSEQUENCES OF MICROBIAL RHIZOSPHERE COMPETENCE FOR PLANT AND SOIL

J. M. LYNCH
AFRC Institute of Horticultural Research, Littlehampton, West Sussex BN17 6LP, UK

> **rhīz′o-**comb. form of Greek **rhiza**, a root.
>
> **sphēre** *n*. one's field of action, influence, or existence; one's natural surroundings.

When the term "rhizosphere" was first used by Hiltner (1904) he used it to describe specifically the interaction between bacteria and legume roots. That was too narrow as a definition and the dictionary definition of the derived word is more appropriate. Today we should also recognize the component regions of the various cell layers of the root itself (*endorhizosphere*) where microorganisms also colonize, the area surrounding the root (*ectorhizosphere*) and the root surface (*rhizoplane*) (Fig. 1). The ectorhizosphere can extend a substantial distance from the root with the development of mycorrhizal fungal associations (Chapter 11), and it is sometimes termed the *mycorrhizosphere*. Another boundary which is difficult to fix is that between the rhizosphere and the *spermosphere* (the region around the germinating seed). This latter

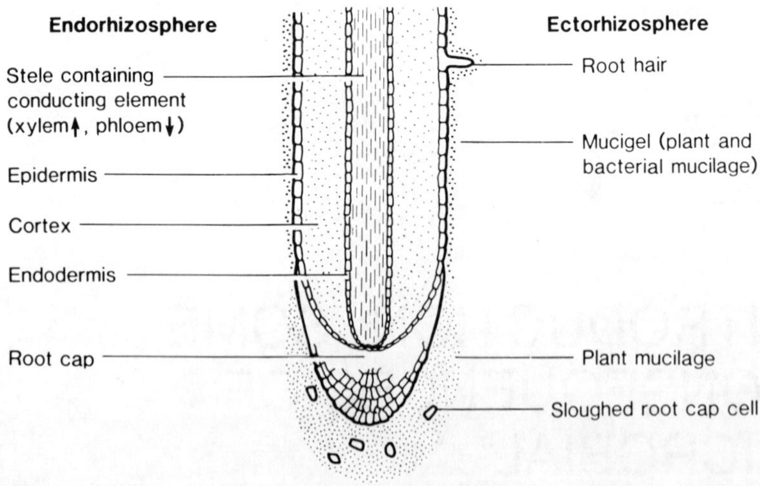

Fig. 1. Root region. (From Lynch, 1983, with permission.)

region (Fig. 2) has received far less attention than the rhizosphere and yet its colonization by microorganisms, usually from the embryo-end, is generally the predisposing determinant of *rhizosphere colonization*. Depending on the plant species, microbial colonization by airborne microorganisms can occur during seed formation and these can become primary colonists of the spermosphere and rhizosphere.

Although the colonization of roots by microorganisms has long been recognized, few attempts to quantify this were made in the earlier literature (e.g. Timonin, 1940). Bowen (1979) suggested that the initial rapid growth of microorganisms on roots could be regarded as an ''*intensity*'' *factor*, indicating the richness of the substrate available for microbial growth, and a slower phase could be regarded as a ''*capacity*'' *factor*, in which substrate supply balances maintenance requirements of the microorganisms. The term *colonization potential* was preferred by Bennett and Lynch (1981a) to describe this latter phase, in terms of biomass or number of bacterial cells per unit length or weight of root. This was characteristic of the type of microorganism and plant but did not depend on inoculum size with gnotobiotic cereal plants. It did, however, depend on competition between organisms around roots, a factor important to the development of biocontrol strategy (Bennett and Lynch, 1981b). On potato pieces in non-sterile soil, the colonization potential depended on inoculum size (Loper et al., 1985); this pattern of colonization could be analysed both temporally and spatially and depended on soil type and irrigation (Bahme and Schroth, 1987).

It has been proposed that the colonization proceeds in two phases (Howie et al., 1987). In *phase 1*, bacteria become distributed by passive carriage

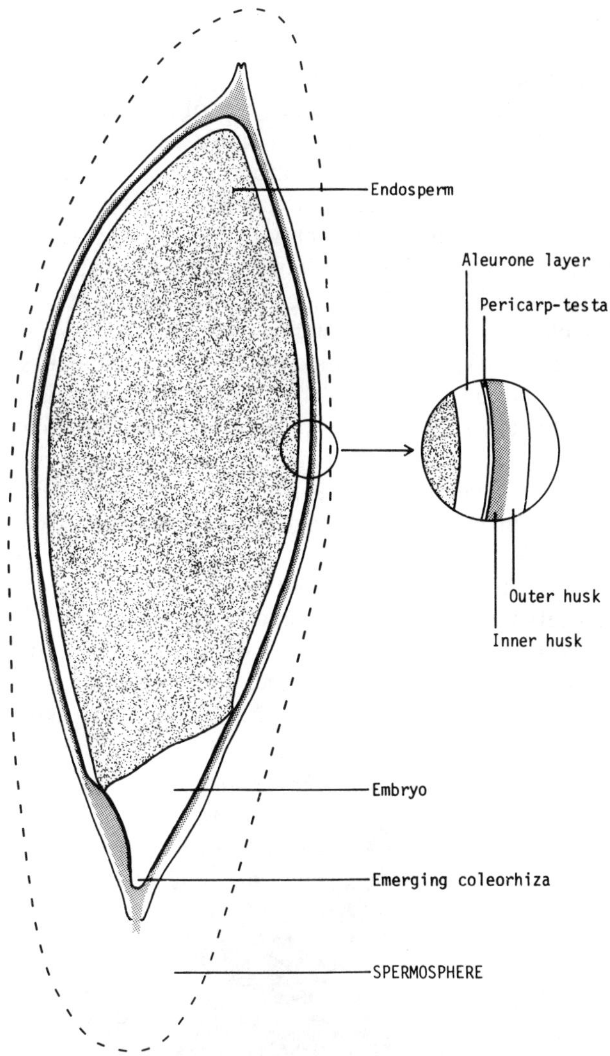

Fig. 2. The germinating barley seed and its associated spermosphere. (From Lynch, 1983, with permission.)

downward with root extension through soil, thereby accounting for progressively lower populations on the roots at increasing distances away from the seed. *Phase 2* is the multiplication and survival phase and occurs during and after phase 1, whereby the population increases to the limits of the ecological niche. Such considerations apply equally well to fungi as bacteria, and Ahmad and Baker (1987) proposed the term *rhizosphere competence* to describe the number of fungal propagules colonizing roots. This unfortunately can give rise to problems of interpretation because, unlike bacterial colonization, we have shown that there is no positive correlation between propagule number and fungal biomass on roots (Lumsden et al., 1989) and rhizosphere competence more usefully should be described in biomass terms.

In the rhizosphere the quantities and types of substrates are different from those in the bulk soil and this leads to colonization by different populations of bacteria, fungi, protozoa and nematodes. Other physicochemical factors which can be different in this region are acidity, moisture and nutrient status, electrical conductivity and redox potential. The total rhizosphere environment is determined by an interacting trinity of the soil, the plant and the organisms associated with the roots (Fig. 3). As such it can only satisfactorily be interpreted with interdisciplinary approaches including soil chemists/physicists, agronomists, plant physiologists/geneticists, pathologists and microbiologists (Lynch, 1987).

A major justification of studying rhizosphere microorganisms must be their influence on plants. The association between organisms and roots can be beneficial, harmful or neutral, but often the effects depend on soil conditions and must therefore be regarded as variable (Fig. 4). Clearly the goal in manipulating the rhizosphere must be to increase the balance of beneficial over harmful effects. With the recognition that the biology of all systems in nature is under genetic control, there are now major opportunities to control these effects by modifying the plant or its associated organisms. This has brought a new impetus for rhizosphere studies, especially with the rapid advancement of techniques for molecular biology and more quantitative approaches to population biology and ecophysiology. There are new opportunities to control

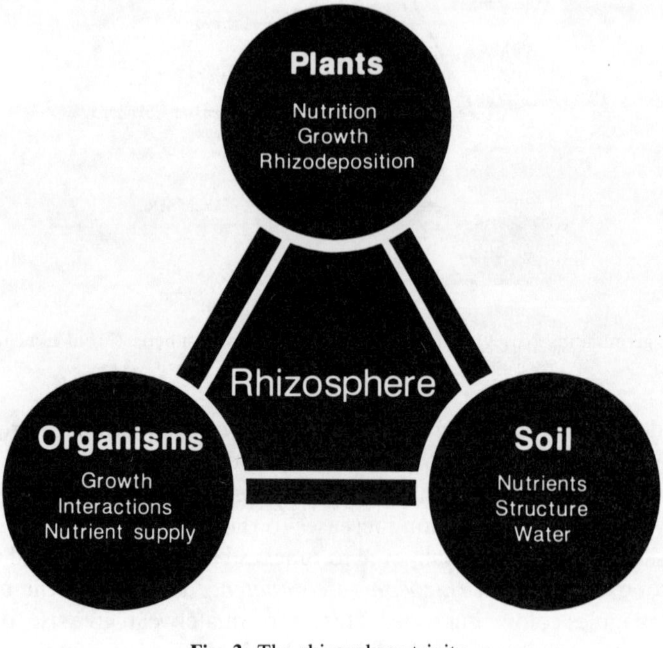

Fig. 3. The rhizosphere trinity.

the rhizosphere by inoculation with genetically modified organisms, or to boost the populations of useful indigenous strains. Alternatively it should be possible to modify the nature and quantity of rhizodeposition products by genetically modifying the plant.

Earlier volumes related to the rhizosphere (e.g. Russell, 1977; Dommergues and Krupa, 1978; Harley and Russell, 1979; Cook and Baker, 1983; Lynch, 1983; Curl and Truelove, 1986; Wild, 1988) were written prior to the recognition of some of the new biological opportunities in rhizosphere studies. Microbial ecology as a discipline is advancing conceptually (Lynch and Hobbie, 1988) and it seemed that a new volume on the rhizosphere would be timely. However, even since planning the present volume in 1987, important new areas are emerging.

One such illustration is the recognition that viruses could be transmitted in the rhizosphere (Brunt and Shikata, 1986). The disease rhizomania (or "root madness") is caused by the vector fungus *Polymyxa betae* transmitting beet necrotic yellow vein virus (BNYVV) to sugar beet (*Beta vulgaris* var. *saccarifera*) and Swiss chard (*B. vulgaris* var. *cyla oleracea*). The significance of viruses in

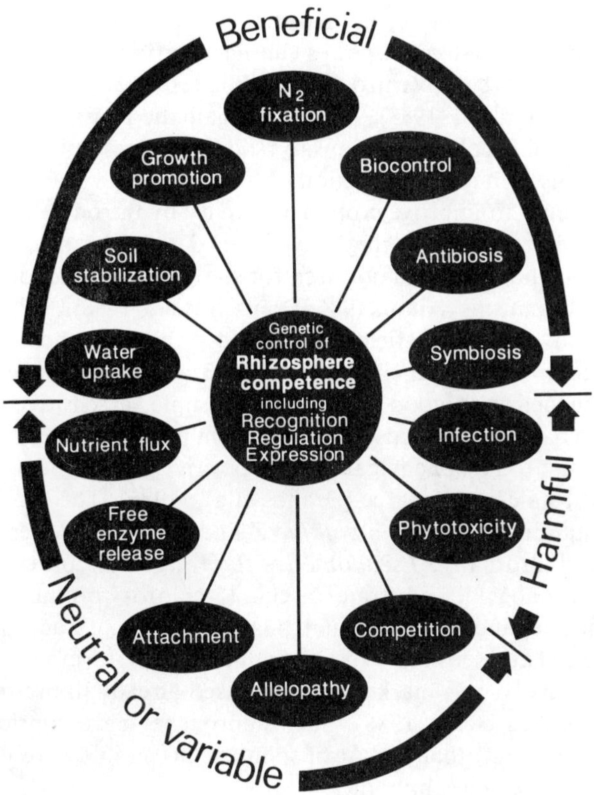

Fig. 4. Process values of rhizosphere organisms.

the rhizosphere has received relatively little attention hitherto but must warrant further study. Where vectors are involved in virus transmission, opportunities might arise for biocontrol of the vector as opposed to the pathogen itself.

Another area which is likely to contribute to our understanding of rhizosphere processes is the study of electrical currents generated by plant roots and the consequence of these for rhizosphere colonization. It is now possible to measure these fields around roots with a vibrating electrode (Miller et al., 1988). Protons circulate in species of five taxonomically diverse families (Gramineae, Leguminosae, Rosaceae, Solanaceae and Pinaceae) in the same way; positive electrical current flows to the root tip where cells are dividing, making the rhizosphere in this region relatively alkaline in pH because protons are taken up from the soil (Miller and Gow, 1989). The current down the root may be involved in root extension or message transmission to determine tip direction. The root tip and the root region immediately behind it are critical sites for determining rhizosphere colonization. The physicochemical gradients so generated could be involved in chemotaxis, electrotaxis and pH taxis of fungi and bacteria to roots, and the understanding of these phenomena could be crucial in the rational design of new biological and chemical crop protection agents. However, it should also be recognized that for roots growing in soil, the full rhizospheres of different crop species can have various acidic or alkaline pH values and this can be governed by mineral (especially nitrogen) nutrition (Marschner and Römheld, 1983). Therefore, again the significance of interesting laboratory physiological observations to field conditions can only be evaluated with full interdisciplinary cooperation.

The opportunity to improve crop productivity by introducing organisms to the rhizosphere, especially if they have been derived by recombinant DNA technology, is highlighting a major need for study of the detection and tracking of introduced organisms (Anon, 1986; Domsch et al., 1988; Levin et al., 1988). This is a rapidly advancing field and should provide methods for assessing rhizosphere competence in the field. Various strategies for bacterial detection have been pursued with good effect. For example, a sensitive and selective marker system has been developed for fluorescent pseudomonads, natural strains of which have no *o*-nitrophenylgalactoside positive phenotype or the ability to utilize lactose as a carbon source (Drahos et al., 1986). The broad host-range plasmids containing *Escherichia coli lac Z* and *lac Y* genes were constructed and introduced into native *Pseudomonas fluorescens* to confer the ability to cleave the chromogenic substrate X-Gal (5-chloro-4-bromo-3-indolyl-β-D-galactopyranoside) and grow on minimal lactose medium. Lac^+ transformants could be detected at a sensitivity of less than 10 colony-forming units per gram of soil. Additionally, this marker has been used already in pre-release testing procedures (Drahos et al., 1988). Such approaches have considerably more precision and sensitivity than the use of spontaneous antibiotic-resistant mutants or the use of polyclonal antibodies.

Another useful strategy, however, has been the use of transposable elements. Transposon Tn5 insertions produce well-characterized mutations on the bacterial chromosome, multiple insertions occur at low frequencies, insertions occur at a large number of sites to increase the likelihood of obtaining a mutant which is not biochemically impaired in respect of the parent, and a region of DNA homology for DNA hybridization is available. This technique has been used successfully for *Pseudomonas putida* and *Rhizobium* spp. using transposon-associated kanamycin resistance and coupled with genotypic (DNA probe) analysis specific for the Tn5 (Fredrickson et al., 1988). The technique was successful in enumerating populations as low as 10 (*P. putida*) to 100 (*Rhizobium*) cells per gram of soil. The technique was also applied to tracking the fate of *Azospirillum* transposon mutants in microcosms of soil (Bentjen et al., 1989). Using most probable number DNA hybridization, *A. lipoferum* Tn5 mutant populations varied from undetectable to 10^6 per gram of dry root.

The luminescence technique is another which is showing promise as a tracking technique (Meiklejohn et al., 1989). It is based on the phenomenon of natural bioluminescence. It has been tested with *Escherichia coli* containing a plasmid-borne *lux* cassette which codes for the enzyme luciferase, with the output of light measured by luminometry; it detected 10 cells per gram of soil.

Fluorescence techniques have traditionally been difficult to use in the quantitative determination of microorganisms in soils. However, by tagging cells with proteins for which fluorescent probes exist or relying on natural fluorescence, flow cytometry and fluorescence-activated cell sorting can be used for tracking by discriminating between and quantifying different types of prokaryotic cells (Watson, 1987). The technique has already been exploited in marine systems (Burkill, 1987) and it seems likely to make an impact on rhizosphere studies.

The above studies and the use of other plasmid markers (Elsas et al., 1988) will all be useful in developing model and field experimental systems to track the fate of introduced organisms and their genetic information in the soil and rhizosphere and to monitor the population dynamics (Chapter 5). Whereas the discussion is related only to bacteria at present, it is confidently predicted that DNA probe techniques will become available shortly for fungi (Bainbridge, personal communication). In the meantime monoclonal antibody techniques combined with ELISA (enzyme-linked immunosorbent assay) (Dewey, 1988) or polyclonals against specific cell proteins (Carter and Lynch, unpublished) coupled to ELISA and slot-blot procedures are providing useful.

Even with the advance of molecular techniques to characterize and manipulate the rhizosphere, a firm understanding of the ecophysiology of this region is necessary to justify the inputs of the molecular methodologies. Implicit in this is the need for a better understanding of the community interactions between microorganisms, fauna and the plant itself. In an earlier review (Whipps and Lynch, 1986) we indicated that the scheme identifying the role of bacteria, fungi and protozoa in root nitrogen uptake, as proposed by Clarholm (1983) (Fig. 5), was worthy of further evaluation. Following their earlier studies with communities in chemostats (Coleman et al., 1978) (see Chapter 5), Ingham et al. (1985) studied the interactions between blue grama grass (*Bouteloua gracilis*)

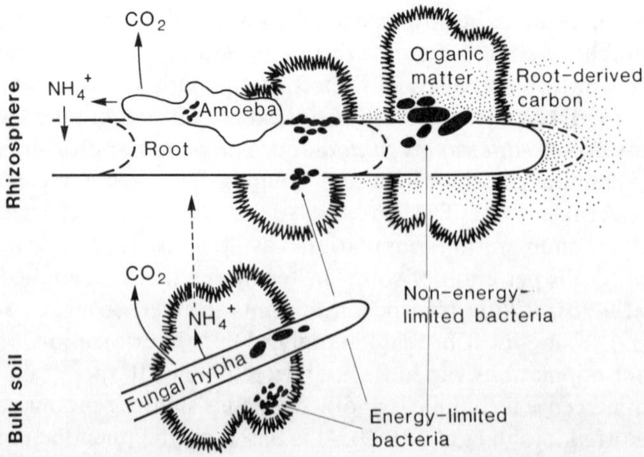

Fig. 5. Model of proposed interactions in the rhizosphere and in the bulk soil. A root is growing in the soil from left to right. Under the influence of root-derived carbon (dots) bacteria utilizing organic matter are temporarily not energy-limited and start to mineralize nitrogen from the organic matter, which will be immediately immobilized in an increased bacterial biomass. The pulse of carbon is soon depleted and the bacteria production will be consumed by naked amoebae that are attracted to the site. When digesting the bacteria, the protozoa release part of the bacterial nitrogen as ammonium on the root surface, where it can be taken up by the root. Below the root, in the bulk soil, a fungal hypha is decomposing organic material. Ammonium will be released as a waste product and it can diffuse toward the root as ammonium or, after nitrification, as NO_3^-. (From Clarholm 1983, with permission.)

in gnotobiotic microcosms containing sandy loam soil low in inorganic N. The soil was inoculated with a bacterium (*Pseudomonas paucimobilis* or *P. stutzeri*) or a fungus (*Fusarium oxysporum*), with half the bacterial microcosms inoculated with bacterial-feeding nematodes (*Pelodera* sp. or *Acrobeloides* sp.) and half the fungal microcosms inoculated with fungal-feeding nematodes (*Aphelenchus avenae*). All the inoculants increased their populations in rhizosphere versus non-rhizosphere soil. Plants growing in soil with bacteria and bacterial-feeding nematodes grew faster and initially took up more N than plants in soil with only bacteria, because of increased N mineralization by bacteria, NH_4^+-N excretion by nematodes, and greater initial exploitation of soil by plant roots. This was not achieved by fungal-feeding nematodes, which excreted less NH_4^+-N than those feeding on bacteria. It is difficult to extrapolate these studies to natural field soils but it is likely that the additional mineralization resulting from the activities of microbial grazers could be of benefit in natural or managed ecosystems when mineralization by microorganisms alone is inadequate to meet the plant's need. The use of ^{15}N tracers and microbial tracking procedures would facilitate the evaluation of the field significance of these concepts. Similar evaluations could be made using ^{32}P tracers to evaluate the phosphorus cycle in the rhizosphere. The quantification of the competence of specific populations in natural and manipulated rhizospheres will hopefully open new horizons in approaches to optimized crop productivity while minimizing inputs of chemicals and energy. The chapters which follow range from the analyses of the rhizosphere's *raison d'être* (the physiology of the root) through to some examples of the consequences of the rhizosphere in specific crop production systems. Hopefully the reader will find that many of the concepts will apply in a range of agricultural and forestry systems.

REFERENCES

Ahmad, J. S. and R. Baker (1987). Rhizosphere competence of *Trichoderma harzianum*. *Phytopathology*, **77**:182–189.

Anon. (1986). *Recombinant DNA Safety Considerations*. Organisation for Economic Co-operation and Development, Paris, 69 pp.

Bahme, J. B. and M. N. Schroth (1987). Spatial-temporal colonization patterns of a rhizobacterium on the underground organs of potato. *Phytopathology*, **77**:1093–1100.

Bennett, R. A. and J. M. Lynch (1981a). Colonization potential of bacteria in the rhizosphere. *Curr. Microbiol.*, **6**:137–138.

Bennett, R. A. and J. M. Lynch (1981b). Bacterial growth and development in the rhizosphere of gnotobiotic cereal plants. *J. Gen. Microbiol.*, **125**:95–102.

Bentjen, S. A., J. K. Fredrickson, P. van Voris, and S. W. Li (1989). Microcosm evaluation of the fate of and ecological impacts from the release of *Azospirillum lipoferum* transposon mutant. *Appl. Environ. Microbiol.*, **55**:198–202.

Bowen, G. D. (1979). Integrated and experimental approaches to the study of growth of organisms around roots. In *Soil-Borne Plant Pathogens* (B. Schippers and W. Gams, eds.). Academic Press, London, pp. 207–227.

Brunt, A. A. and E. Shikata (1986). Fungus-transmitted and similar labile rod-shaped viruses. In *The Plant Viruses, Vol. 2* (M. H. V. Regenmortel and H. Fraenkel-Conrat, eds.). Plenum Publishing Corporation, New York, pp. 305–335.

Burkill, P. H. (1987). Analytical flow cytometry and its application to marine microbial ecology. In *Microbes in the Sea* (M. A. Sleigh, ed.). Ellis Horwood, Chichester, pp. 139–166.

Clarholm, M. (1983). *Dynamics of Soil Bacteria in Relation to Plants, Protozoa and Inorganic Nitrogen*, Report No. 17, Department of Microbiology, Swedish University of Agricultural Science, Uppsala.

Coleman, D. C., C. V. Cole, H. W. Hunt, and D. A. Klein (1978). Trophic interactions as they affect energy and nutrient dynamics. I. Introduction. *Microb. Ecol.*, **4**:345–349.

Cook, R. J. and K. F. Baker (1983). *The Nature and Practice of Biological Control of Plant Pathogens*. American Phytopathological Society, St. Paul, 539 pp.

Curl, E. A. and B. Truelove (1986). *The Rhizosphere*. Springer-Verlag, Berlin, 288 pp.

Dewey, F. M. (1988). Development of immunological diagnostic assays for fungal plant pathogens. *Brighton Crop Protection Conference—Pests and Diseases*, **2**:777–786.

Dommergues, Y. R. and S. V. Krupa (eds.) (1978). *Interactions Between Non-Pathogenic Soil Micro-organisms and Plants*. Elsevier, Amsterdam, 475 pp.

Domsch, K. H., A. J. Driesel, W. Goebel, W. Andersch, W. Lindenmaier, W. Lotz, H. Reber, and F. Schmidt (1988). DECHEMA Working Party "Safety in Biotechnology" Report: Considerations on Release of Gene-Technologically Engineered Micro-organisms into the Environment. *FEMS Microbiology Ecology*, **53**:261–272.

Drahos, D. J., B. C. Hemming, and S. McPherson (1986). Tracking recombinant organisms in the environment: β-galactosidase as a selectable non-antibiotic marker for fluorescent pseudomonads. *Bio/Technology*, **4**:439–444.

Drahos, D. J., G. F. Barry, B. C. Hemming, E. J. Brandt, H. D. Skipper, E. L. Kline, D. A. Klepfel, T. A. Hughes, and D. T. Gooden (1988). Pre-release testing procedures: US field test of a lac ZY-engineered soil bacterium. In *Release of Genetically Engineered Micro-organisms* (M. Sussman, C. H. Collins, F. A. Skinner, and D. E. Stewart-Tull, eds.). Academic Press, London, pp. 181–191.

Elsas, J. D. van, J. T. Trevors, and M. E. Staroclub (1988). Plasmid transfer in the soil and rhizosphere. In *Risk Assessment for Deliberate Releases* (W. Klingmüller, ed.). Springer-Verlag, Berlin, pp. 89–99.

Fredrickson, J. K., D. F. Bezdicek, F. J. Brockman, and S. W. Li (1988). Enumeration of Tn5 mutant bacteria in soil using a most-probable-number-DNA hybridization procedure and antibiotic resistance. *Appl. Environ. Microbiol.*, **54**:446–453.

Harley, J. L. and R. S. Russell (eds.) (1979). *The Soil–Root Interface.* Academic Press, London, 448 pp.

Hiltner, L. (1904). Uber neuere Erfahrungen und Problem auf dem Gebeit der Bodenbakteriologie und unter besonderer Berucksichtigung der Grundungung und Brache. *Arb Dtsch. Landwirt. Ges.*, **98**:59–78.

Howie, W. J., R. J. Cook, and D. M. Weller (1987). Effects of soil matric potential and cell motility on wheat root colonization by fluorescent pseudomonads suppressive to take-all. *Phytopathology*, **77**:286–292.

Ingham, R. E., J. A. Trofymow, E. R. Ingham, and D. C. Coleman (1985). Interactions of bacteria, fungi, and their nematode grazers: effects on nutrient cycling and plant growth. *Ecol. Monogr.*, **55**:119–140.

Levin, M. A., R. Seidler, A. W. Borquin, J. R. Fowle, and T. Barkay (1987). EPA developing methods to assess environmental release. *Bio/Technology*, **5**:38–45.

Loper, J. E., C. Haack, and M. N. Schroth (1985). Population dynamics of soil pseudomonads in the rhizosphere of potato (*Solanum tuberosum* L.). *Appl. Environ. Microbiol.*, **49**:416–422.

Lumsden, R. D., J. P. Carter, J. M. Whipps, and J. M. Lynch (1989). Comparison of biomass and viable propagule measurements in the antagonism of *Trichoderma harzianum* against *Pythium ultimum. Soil Biol. Biochem.* (in press).

Lynch, J. M. (1983). *Soil Biotechnology. Microbiological Factors in Crop Productivity.* Blackwell Scientific Publications, Oxford, 191 pp.

Lynch, J. M. (1987). Soil biology—accomplishments and potential. *Soil Science Soc. Amer. J.* **51**:1409–1412.

Lynch, J. M. and J. B. Hobbie (eds.) (1988). *Micro-organisms in Action: Concepts and Applications in Microbial Ecology.* Blackwell Scientific Publications, Oxford, 363 pp.

Marschner, H. and V. Römheld (1983). *In vivo* measurement of root-induced pH changes at the soil–root interface: effect of plant species and nitrogen source. *Z. Pflanzenphysiol.*, **111**:241–251.

Meiklejohn, E. M., A. Meikle, L. A. Glover, K. Killham, and J. I. Prosser (1989). A novel detection system for genetically manipulated microbial inoculations in soil. *J. Sci. Fd. Agric.* (in press).

Miller, A. L. and N. A. P. Gow (1989). Correlation between profile of ion-current circulation and root development. *Physiol. Plant*, **75**:102–108.

Miller, A. L., E. Shand, and N. A. R. Gow (1988). Ion currents associated with root tips, emerging laterals and induced wound sites in *Nicotiana tabacum*: spatial relationship proposed between resulting electrical fields and phytophthoran zoospore infection. *Pl. Cell Environ.* **11**:21–25.

Russell, R. S. (1977). *Plant Root Systems. Their Function and Interaction with the Soil.* McGraw-Hill, London, 298 pp.

Timonin, M. I. (1940). The interaction of higher plants and soil micro-organisms. I. Microbial population of rhizosphere of seedlings of certain cultivated plants. *Can. J. Res.*, **18**:307–317.

Watson, J. V. (1987). Flow cytometry in biomedical science. *Nature*, **325**:714–742.

Whipps, J. M. and J. M. Lynch (1986). The influence of the rhizosphere on crop productivity. *Adv. Microb. Ecol.*, **6**:187–244.

Wild, A. (ed.) (1988). *Russell's Soil Conditions and Plant Growth*, 11th edn. Longman, London, 991 pp.

2

ANATOMY AND COMMUNITY STRUCTURE OF THE RHIZOSPHERE

R. CAMPBELL

Department of Botany, University of Bristol, Woodland Road, Bristol BS8 1UG, UK

M. P. GREAVES

AFRC Institute of Arable Crops Research, Long Ashton Research Station, Bristol BS18 9AF, UK

1. REVIEW OF CURRENT METHODOLOGY

1.1. The Root

Root anatomy has been studied classically by light microscopy using whole mounts and sections stained after various fixation and dehydration procedures (O'Brien and McCully, 1981). Free-hand sections of fresh material are valuable in presenting live material without fixation and dehydration artefacts (McCully, 1987). There are various more recent techniques, such as epifluorescence microscopy (see below), which have allowed additional components to be identified and have given information on the distribution of particular chemicals, for example in cell walls, especially when fluorescence is combined with immunological techniques. Scanning and transmission electron microscopy have added to the information on ultrastructure of different sorts of root cells and have provided general illustrations of structure, without altering basic concepts.

The growth and development of roots and their distribution in soil, can be studied by the use of boxes with glass or perspex sides to enable the roots to be observed with the minimum of disturbance. Refinements of this system include using a cover glass as part of the box so that microscopic observations can be made (Polonenko and Mayfield, 1979), staining roots or microorganisms *in situ*, and the use of fibre-optic systems to observe roots through tubes in field situations (Rush et al., 1984).

1.2. Microorganisms On and Near Roots

Methods for studying soil microorganisms are widely reviewed, most recently by Poindexter and Leadbetter (1986), Burns and Slater (1982) and Dhingra and Sinclair (1985). This section will concentrate on those methods or modifications of standard procedures which relate to microorganisms on or near roots.

There are three main ways of analysing microbial populations. Their activity can be assayed by measuring various features of metabolism (CO_2 evolution, substrate utilization and the production of enzymes). Attempts may be made to culture organisms to estimate the numbers of viable propagules and subsequently to identify them. Thirdly, microbes can be looked for, with a microscope, and counted or measured directly.

Measurements of metabolic activity and of the pool sizes of important metabolic intermediates and essential components like DNA, have been used to assess the activity of microorganisms (Karl, 1986). The simplest measure, at first sight anyway, is to determine respiration by CO_2 evolution or O_2 uptake, and methods for this are available for both field and laboratory studies. Enzymes such as dehydrogenases and urease may be extracted and measured, though care needs to be taken that these enzymes are not persistent in the soil rather than in live microorganisms (Burns, 1978). ATP has been measured as an indicator of microbial activity (Tate and Jenkinson, 1982) and can be detected in very small quantities by systems based on light emission by luciferase. There are conversion factors to estimate biomass from average ATP levels, though there can be problems with extracting ATP from soil. Perhaps of more interest is the relationship between ATP, ADP and AMP (adenylate energy charge, AEC) as a measure of metabolic turnover. Adenylate energy charge is defined as:

$$\mathrm{AEC} = \frac{[\mathrm{ATP}] + 0.5\,[\mathrm{ADP}]}{[\mathrm{ATP}] + [\mathrm{ADP}] + [\mathrm{AMP}]}$$

In general an $\mathrm{AEC} > 0.8$ indicates actively growing microbes and 0.5 to 0.7 indicates dormancy. Anything less than 0.5 suggests that most microbes are dead, although some deeply dormant spores may have an AEC of < 0.1 (Brookes, 1986; Brookes et al., 1987).

All the metabolic methods suffer from one serious problem: the separation of the root effects from the microorganisms. How much of the activity is due to microorganisms and how much to the root itself? It is impossible to remove very fine roots from soil. Even if this were possible, soil samples without roots, or plant roots without microbes (gnotobiotic plants) have little relevance to rhizosphere studies. Some distinctions can be made by the use of various antibiotics to inhibit bacteria, or fungi or both (e.g. Domsch et al., 1979), but it is not clear what effects they may have on the roots and the distinction between microbes is never clear-cut because of the antibiotic resistance of some organisms. There is no system of inhibitors which claims to distinguish between effects on roots and rhizosphere organisms.

Metabolic activity can also be measured by the use of radioactive isotopes, either to detect respiratory products in quantitative assays or, by autoradiography of whole roots or root sections, to determine the destination and microbial use of plant products. The plant is usually supplied with radioactive CO_2 and allowed to photosynthesize and translocate the products, which later appear in the rhizosphere and the associated microorganisms if they are actively metabolizing. Usually ^{14}C is used but Minchin and McNaughton (1984) used ^{11}C to study exudation in non-sterile systems treated with *p*-chloromercuribenzene sulphuric acid to inhibit microorganisms' membrane transport systems and so prevent microbial uptake of exudates. For detailed autoradiography, at the resolution of individual bacterial cells, low-energy radioactive emitters are required. Thus ^{32}P is unsuitable, as silver grains may be affected a long way from the emitting microorganisms; ^{14}C is more suitable but still lacks sufficient resolution for detailed studies. The preferred isotope is ^{3}H so that individual bacteria, that are, for example, incorporating tritiated nucleotides, can be located. Autoradiography has been recommended by Brock (1987) but, admittedly, mostly for aquatic microbiology.

Culture methods (dilution plates, most probable number, etc.), despite widespread admission of their faults and problems, are still much used (Brock, 1987). They depend on a range of media which are inevitably selective to some extent, and it is usual to recover only 1–10% of the organisms known to be present. Rhizosphere organisms can be recovered by gently shaking roots and associated soil in a suitable diluent. Organisms growing on the root surface can be removed by more vigorous agitation with glass beads or by sonication, and those present in the root tissues can be brought into suspension by macerating the roots by grinding or stomaching (Greaves et al., 1978). Culture methods may be useful to recover and enumerate known organisms, including those which have been genetically marked or selected for antibiotic resistance, where the ability to grow on selected media is well characterized. Many of the other techniques used to analyse microbial populations in natural environments, such as principal component analysis (Rosswall and Kvillner, 1979), depend on obtaining a culture and so, automatically, have the limitation of cultural methods.

Direct observation is usually the method of choice for the study of community structure. The scanning and transmission electron microscopes can give quantitative data as well as useful qualitative information (Foster, 1988). Distortions, especially of soft tissues and mucilage (see below), can be caused by dehydration procedures (Costerton et al., 1986) and pose severe problems. In the scanning electron microscope distortion can be reduced by using frozen hydrated specimens (Campbell and Porter, 1982), though it is then more difficult to locate bacteria and fungi within the aqueous phase.

Phase-contrast or interference microscopy of fresh material is possible, but associated soil and the root structure itself can give a confusing background against which to observe microorganisms. The size of microorganisms, especially bacteria, means that high magnification objectives with short working distance and depth of field are necessary. Stains, such as phenol acetic aniline blue, differentiate between microorganisms and root tissue (Rovira et al., 1974), and there are more specific visible-light stains, e.g. trypan blue, for vesicular arbuscular mycorrhizas. The greatest advance in recent years has been the development of fluorochromes for use with epifluorescence microscopy. The most commonly used fluorochrome is acridine orange (Newell et al., 1986) but there may be serious problems with background fluorescence and non-specific staining. This may be overcome to some extent by counter staining and washing sequences (van Vuurde and Elenbaas, 1978). Living cells may be differentiated from dead ones with fluorescein diacetate (FDA), fluorescein isothiocyanate (FITC) and ethidium bromide, though these may give artificially low values because of problems with uptake of the dye.

A greater staining specificity is possible by the use of immunofluorescence techniques using antibody prepared against specific organisms—or a part, such as the walls (Bohlool and Schmidt, 1980; Chantler and McIllmurray, 1987). There are major problems with specificity of the antibody and with non-specific binding to, and autofluorescence of, soil components. In order to have a realistic chance of detecting organisms in the necessarily thin soil smears, the method depends on having quite high numbers (10^6–10^7 g^{-1}) of the target organism present (McCully, 1987). The most successful uses of this technique have involved identification of strains after culturing or when concentrated in root nodules, rather than examination of roots and soil *per se*.

Individual strains of bacteria can also be identified by markers introduced by genetic engineering techniques. Thus Drahos et al. (1986) have introduced the *lac* Z gene into some pseudomonads so that when plated onto media containing suitable chromogenic substrates they will release a dye and identify that colony (see Chapter 1).

In general, the methodology for rhizosphere studies is difficult and, in many ways, unsatisfactory. However, by using a combination of methods it is possible to obtain useful information, though only through the expenditure of considerable time and effort.

2. ROOT ANATOMY

2.1. Introduction

The basic structure of roots has been known for many years and is described in many plant anatomy texts (e.g. Esau, 1977; see also Chapter 3). Young roots of terrestrial plants have a protective root cap over the apical meristem, behind which is a zone where cells elongate and the stele differentiates and then a region where root hairs are produced. In annual plants, there may be some secondary development, involving lignification or thickening of the walls of the cortex (McCully, 1987), but generally the cortex dies and the surface of the root may be the exposed endodermis (see discussion below). In perennial plants a cambium develops, usually in the pericycle below the endodermis, and produces xylem and phloem to give a woody root with bark. The anatomy of root xylem and phloem is frequently different from that in the stem. Almost nothing is known about the microbiology of woody roots and their surfaces except for comments made, in passing, when describing tree diseases.

Root anatomy does, of course, vary with plant species, with the physiological state of the plant, and with the presence of microbial infections such as mycorrhizas (Harley and Smith, 1983; see Chapter 11) and root nodules (Chapter 12), and with the cultural conditions. In this chapter consideration is limited to those tissues outside the endodermis. Microorganisms in the stele are almost invariably pathogens, though saprophytes are known (Gagné et al., 1987).

2.2. The Rhizosphere

The term "rhizosphere" was introduced by Hiltner (1904) but has since been modified and redefined. At its simplest, it is that volume of soil influenced by the root. The extent may vary with soil type, plant species, age and many other factors (Curl and Truelove, 1986), but it is usually assumed to extend from the root surface (rhizoplane) out into the soil for up to a few millimetres, or possibly a few centimetres in extreme cases in some desert and sand dune plants. The root surface is not usually a well-defined entity and its nature may change during the life of the root. Furthermore, the cortex may be invaded by apparently harmless bacteria and fungi (Greaves and Darbyshire, 1972) and cells of the cortex and epidermis may die or be damaged while the root as a whole remains alive and healthy. The rhizosphere can therefore be considered as a microbial continuum stretching out into the soil from the root endodermis (Old and Nicholson, 1978). It is frequently divided into the ectorhizosphere (i.e. the soil region) and the endorhizosphere, which includes the rhizoplane (if it can be distinguished) and the epidermis and cortical cells of the root that are invaded by microbes.

An exception to this lack of physically defined limits to the rhizosphere is found in sheathed roots. These occur in xerophytic plants, especially grasses (Wullstein and Pratt, 1981; Buckley, 1982), in mesophytic agricultural grasses (Duell and Peacock, 1985) and in crop plants such as maize (*Zea mays*) (Vermeer and McCully, 1981). Sheathed roots are coated with a well-defined layer of sand or soil which remains attached when the plants are gently removed from the soil, being held in place by mucilage (mostly from the root cap) and, to a lesser extent, by persistent root hairs. These so-called rhizosheaths are extensively

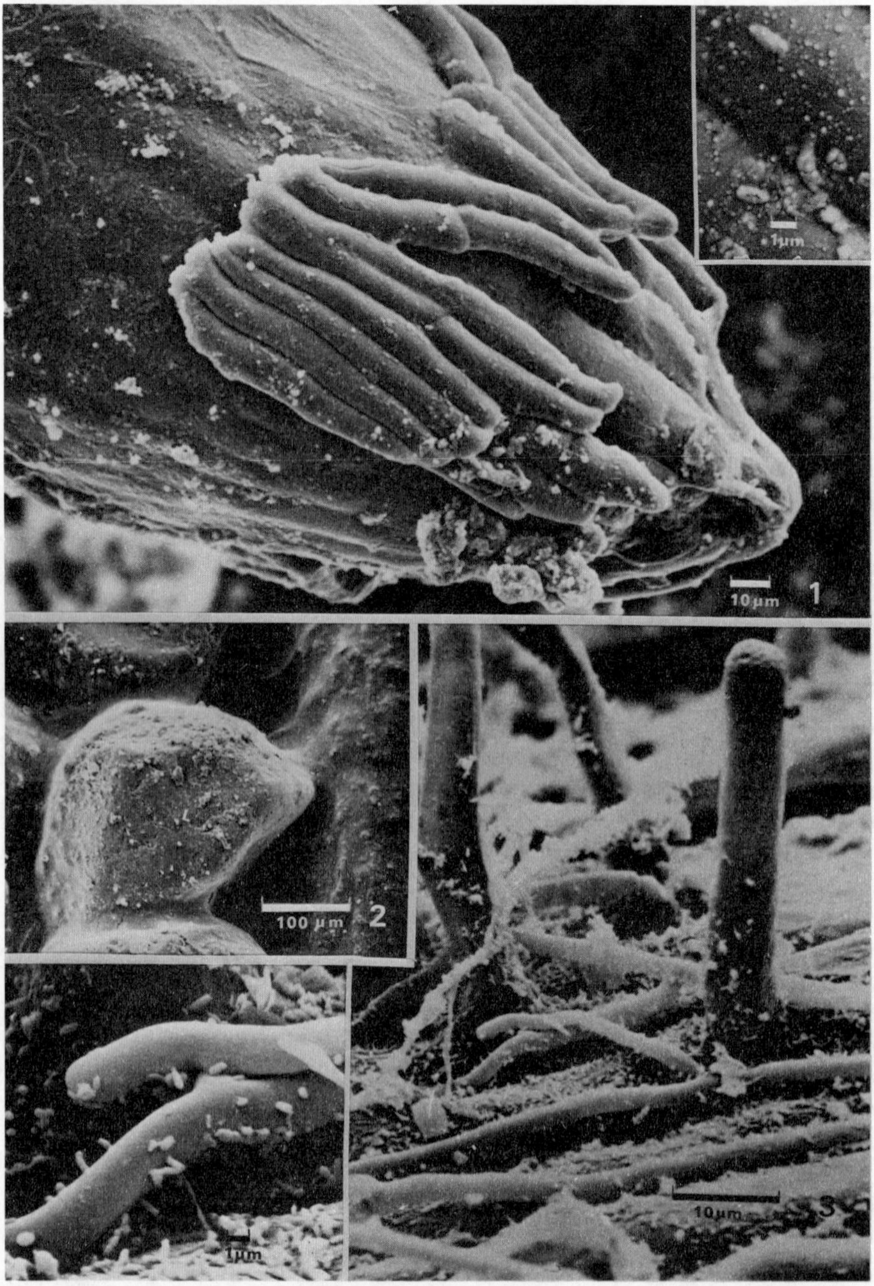

Fig. 1. The root tip of wheat (low temperature scanning electron microscopy (LTSEM); partially freeze-dried (PFD)). The root cap cells are being sloughed off as the tip grows. Inset: detail showing bacteria on root cap cells.

Fig. 2. The root and soil particles (mostly sand grains). The surface of the mucilage layer over the root (on right) is viewed from a soil void: the mucilage is continuous with the water films around the sand grains. LTSEM, fully hydrated. (From Campbell and Porter, 1982.)

Fig. 3. The microbial colonization of the root hair region, showing fungal hyphae and a dense colonization by bacteria. Scanning electron microscopy (SEM), critical point dried. Inset: detail of colonization of hyphae by bacteria.

colonized by bacteria and fungi and, at least in maize, the root cortex and epidermis remain alive and active (McCully, 1987).

2.3. The Root Cap

The root cap arises from a group of cells which may be part of the main root meristem or be separate from it. The cells of the cap are continuously produced and transitory, passing across the cap in a few days and being shed or abraded from the surface (Esau, 1977). The production of cap cells can be considerable, ranging from 400 to 600 per root cap per day for *Vicia faba* to a maximum of nearly 21 000 per root cap per day for maize (Rougier, 1981). These cap cells (Fig. 1) are left behind as the root tip grows through the soil and they may be found, still alive, in the soil surrounding the root (Vermeer and McCully, 1981). The surfaces of root cap cells may be colonized by bacteria (Fig. 1 inset) and the walls can also be penetrated (Vermeer and McCully, 1981).

The root cap produces mucilage (Fig. 1), which is usually considered to be a lubricant to help the tip push through the soil. It may also protect against desiccation, especially if it contains arabinogalactans which bind soil particles and help to ensure continuity of the water film (Fig. 2) between the soil and the root (Lucas, 1987). The mucilage is left behind in the soil round the root, where it may play a major role in adhesion of soil particles and microbes and act as a nutrient source for the latter. There is a great deal known about the ultrastructure and chemistry of slime production by the root cap cells (Rougier, 1981). The mucilage originates in the Golgi apparatus and is transported across the plasma membrane by the fusion of vesicles. The amount produced, and probably its chemical nature, varies with species. Wheat (*Triticum aestivum*) and, especially, maize roots produce so much that droplets form on the root cap, but plants such as peas (*Pisum* spp.) and other dicotyledenous species produce much less and may barely form a slime layer over the cap surface. The chemical analyses, which have largely been done on maize mucilage because of the quantities available, show that the polysaccharide is mostly a β 1–4 glucan with hydrophilic polymers of galactose and fucose and lesser amounts of xylose, arabinose, galacturonic acids, some phenolics (like ferulic acid) and also proteins (Rougier, 1981). Fucose occurs in particular layers of maize root cap cell walls and in the mucilage, where it persists for long periods on field-grown roots and is important in host-pathogen-symbiont recognition systems (e.g. *Azospirillum lipoferum*) (Vermeer and McCully, 1981).

2.4. Epidermal Cells and Root Hairs

Root hairs form behind the zone of elongation by outgrowth of an epidermal cell (Fig. 3). They grow apically and are usually twisted and distorted by the need to pass around and between soil crumbs, with which they form close

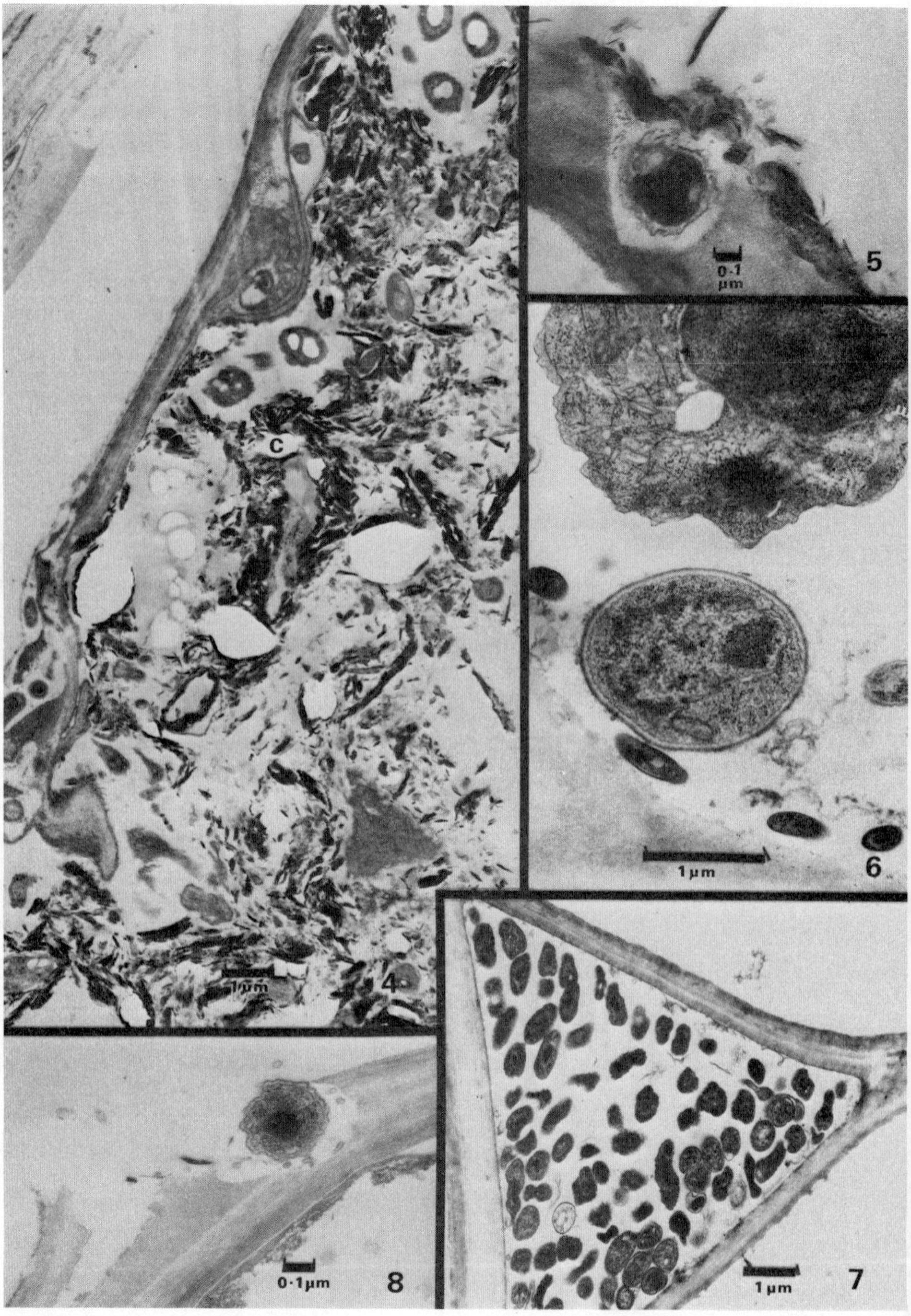
c
1μm
4
0·1 μm
5
1μm
6
1μm
7
0·1μm
8

attachments, especially in root sheaths where they persist for the whole growing season (Vermeer and McCully, 1982; McCully, 1987). Normally, however, they are quite short-lived and collapse and die, as do most of the epidermal cells (McCully, 1987). The broken epidermal cells and dead root hairs may be a common site for the invasion of the cortex (Fig. 4), though there can be direct, presumably enzymic, degradation of the intact walls (Fig. 5). Root hairs may be colonized by bacteria in general (Fig. 3), but they are also known to be the main entry points for *Rhizobium* (and possibly *Frankia*), with which strain-specific recognition systems are known to occur (Chapters 6 and 12). Epidermal cells, including root hairs, secrete a mucilage similar to pectin (polygalacturonic acid) (Leppard and Colvin, 1971) and this may bind soil particles and micro-organisms to the root surface (Greaves and Darbyshire, 1972). The whole root hair region can be extensively colonized by bacteria and fungi (Fig. 3).

2.5. The Cortex

The cells between the endodermis and the epidermis are usually unspecialized and thin-walled, with large vacuoles and intercellular spaces. They occur in young roots, but may be sloughed off as the annual roots age. In biennial or perennial roots the cortex is not present. When plants are grown in normal soil conditions some cells of the cortex are invariably invaded after only a few days and the outer layers of the cortex may die (Fig. 4). As soon as there are deal cells, microbial invasion can become considerable, with fungi and protozoa as well as bacteria (Fig. 6). An exception must again be made for sheathed roots, where the cortex and the epidermis remain alive and may differentiate to form thickened, lignified cell walls (McCully, 1987). Intercellular spaces in normal roots (Fig. 7) are also colonized by bacteria, which may later penetrate into neighbouring cells (Fig. 8).

There is confusion in the literature as to when and how cortical cells die. Is microbial invasion the cause or the effect? Clearly, pathogens can penetrate

---

Fig. 4. The root surface and associated bacteria and soil. Clays (c) are present as electron opaque lamellae, and organic matter is the more amorphous material. Bacteria of various morphologies, with and without storage materials and often with capsules, are scattered through the soil as individual cells and microcolonies. Bacteria have also invaded the endorhizosphere (the outer cells of the root, which are dead) through breaks in the wall (bottom left) and apparently by penetrating the wall. Transmission electron microscope (TEM) photograph, by kind permission of Dr A. Beckett, University of Bristol.

Fig. 5. A bacterium penetrating an epidermal cell wall, covered by clay particles. TEM.

Fig. 6. An amoeba, a bacterium and a fungal hypha in the endorhizosphere. The cortical cell within which they occur is dead. TEM.

Fig. 7. An intercellular space in the endorhizosphere, extensively colonized by bacteria. TEM. (From Faull and Campbell, 1979, with permission.)

Fig. 8. Penetration of a cortical cell wall by a bacterium; the frayed ends of the cell wall fibres are visible. TEM. (From Faull and Campbell, 1979, with permission.)

root cells actively and cause the death of the cells, saprotrophic colonization of the dead cell following quickly. Darbyshire and Greaves (1971) have shown that *Pseudomonas fluorescens*, normally regarded as a saprophyte, can invade roots of the pea. There is also the possibility that cortical cells might die quite quickly without the intervention of pathogens. Henry and Deacon (1981) reported the death of uninvaded cortical cells of wheat and barley (*Hordeum vulgare*), as judged by the absence of nuclear staining with acridine orange; in wheat roots only 16 days old, five out of the six cortical cell layers in 52–89% of roots had apparently died. Barley roots also had many fewer cells without stained nuclei. Brown and Hornby (1987) observed less than 27% of the cortical cells with stainable nuclei after only 4 days, and by 11 days almost all the epidermis and outer cortical layers were apparently anucleate. However, McCully (1987), cautions that the stain may not penetrate, at least in maize, suggesting that there may be cells with unstained nuclei that could be alive and healthy. Electron microscopy of field-grown wheat roots up to several months old, usually shows many cortical cells to have normal, apparently live, cytoplasm with only some cells dead and invaded by microbes, such as in Fig. 6. If cells do die as quickly as nuclear staining suggests, they must be a source of nutrients for rhizosphere organisms and possibly provide easy access to the root.

So it is normal, even in apparently healthy roots, to find some dead cells invaded by saprotrophs, but whether this is natural death followed by colonization, or death caused by microorganisms is not always clear. Except in very young roots (1 or 2 days), roots grown in special conditions—such as water culture (see below)—and gnotobiotic plants, it is rare to find intact an epidermis or cortex.

Lateral roots are usually initiated in the pericycle and force their way out through the cortex and epidermis, the breakage of which may release nutrients and encourage microbial growth (Foster et al., 1983), though the exit point soon forms a "graft-like union" and seals the edges of the hole to the new root (McCully, 1987). The cortex and epidermis of lateral roots of *Lolium perenne* have a higher proportion of nucleate cells than does the main axis of equivalent age (D. Rice, personal communication).

3. EXUDATES AND RHIZODEPOSITION

Apart from sloughing off cells and producing mucilage, roots also exude a wide variety of soluble materials, especially sugars and amino acids. This has been extensively reviewed (e.g. Newman, 1985; Whipps and Lynch, 1985; Curl and Truelove, 1986; Vancura, 1988; see Chapter 4). Exudates vary with plant species, physiological conditions (age, nutrient status) and abiotic conditions (temperature, soil structure, aeration, water content). They are readily available as nutrients for microorganisms and are usually considered one of the main reasons for increased microbial numbers and activity in the rhizosphere. Total rhizodeposition may amount to a considerable proportion of the plant's photosynthate; 12–40% is common (Whipps and Lynch, 1986) although values of up to 70–80% of the material transported to the root have been reported (Whipps and Lynch, 1985); often, however, the conditions from which these data are derived are not known. Values of the order of 20% of photosynthates are most commonly reported (Whipps, 1985). Newman (1985) has pointed out that estimates vary greatly and the methodology is fraught with difficulty. Furthermore, the total amount of exudate and root cap plus mucilage (10–100 mg g^{-1} root dry weight and 20–50 mg g^{-1} dry weight respectively) is only 3–15% of the dry weight of the root, which eventually dies and is available to the soil organisms.

4. THE MUCIGEL AND THE SOIL–ROOT INTERFACE

4.1. Soil-grown Plants

This section is concerned with the structure of the boundary between soil and the mature root having an epidermis and cortex (alive or not). This boundary has already been illustrated (Figs. 3 and 4), and is the sum of the effects of most of the phenomena which have already been discussed (root cap mucilage and cells, exudates, cortical and epidermal cells) together with the components of the soil. Care is needed in interpreting observations of this region because most methods of microscopy involve some dehydration, which may seriously damage or change the structure of mucilage from a gel to a fibrous matrix, or even to a thin discontinuous sheet after complete drying as for some scanning electron microscope studies.

The epidermal cells secrete mucilage. Its distribution is somewhat variable depending on plant species, age and other factors (Foster, 1986). It may be bounded by a distinct pellicle, membrane or cuticle (Greaves and Darbyshire, 1972; Foster, 1981, 1982, 1986), but it may just diminish with distance away from the root (Greaves and Darbyshire, 1972; Foster, 1982). Foster et al. (1983) showed that the outer layers of the epidermal cell walls may become gelatinized and form part of the mucilage, possibly as a discrete gelatinous layer (Fig. 9) (Campbell and Porter, 1982). Material, which may be fibrillar, derived from the outer layers of the cell wall may extend out into the soil. There is also the remains of mucilage from the root cap and, in addition, the microorganisms themselves may produce mucilage (Rovira and Campbell, 1974). This combination of at least four sorts of mucilage, probably varying in amount and importance with different plant species, and with physiological age and edaphic factors, has been called mucigel by Jenny and Grossenbacher (1963). Its chemistry is complex, as might be imagined with such diverse origins, but essentially it is a polysaccharide composed mostly of hexose and pentose sugars and uronic acids (Oades, 1978).

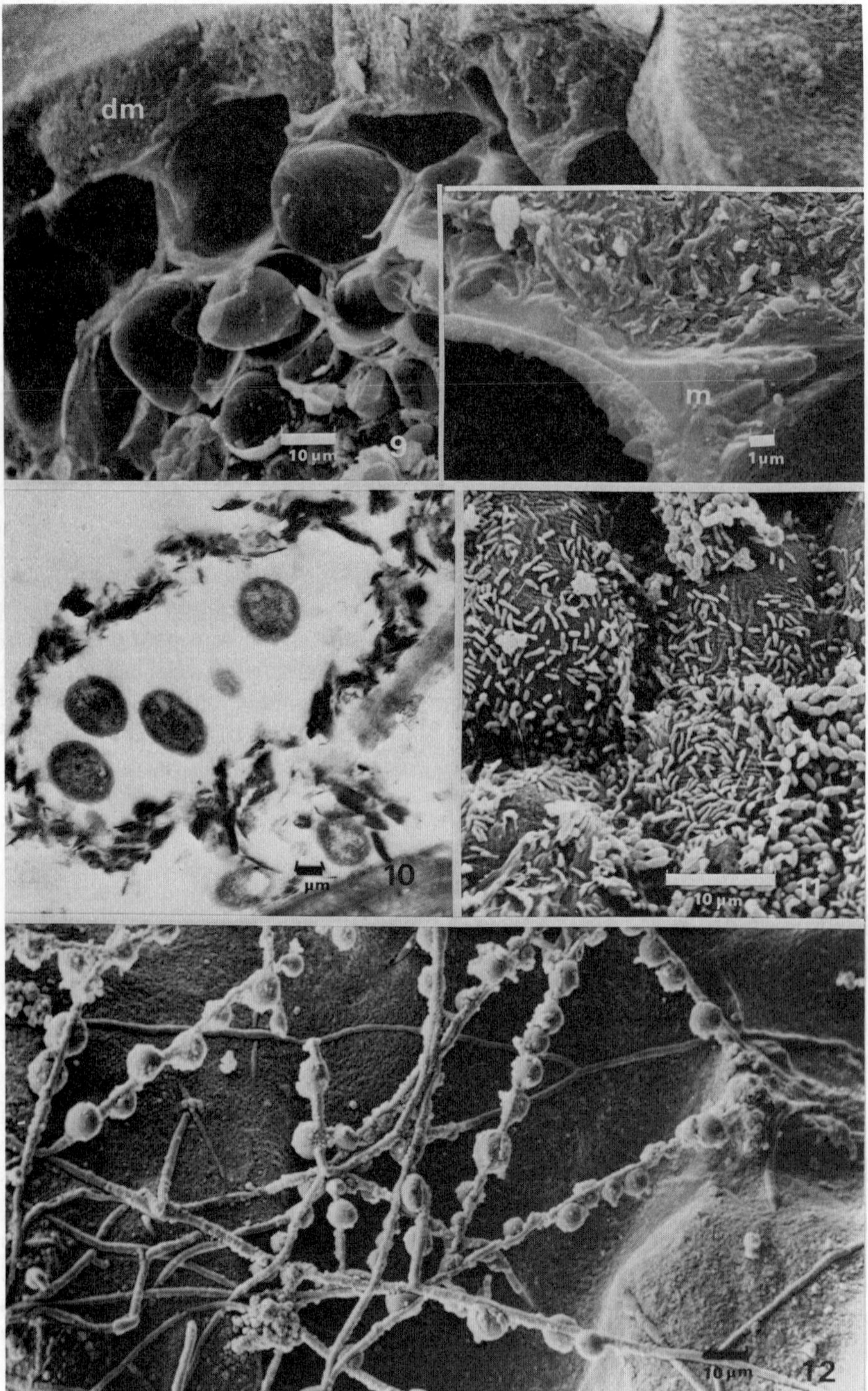
dm
10 µm
9
m
1 µm
µm
10
10 µm
11
10 µm
12

The mucilage binds together clay particles, other minerals, organic matter, individual bacteria and microcolonies into a layer around the root epidermis or the remains of dead cell layers (Figs. 4 and 9) and many of the bacterial microcolonies are surrounded by clays present in the soil (Figs. 4 and 10). The surface of the mucilage layer around the root is often discrete (Figs. 2 and 9) (Campbell, 1983).

4.2. Plants Grown in Soil-free Systems

The root surfaces of plants grown in solution culture or in soil are very different. Provided that aeration in solution culture is adequate, the epidermis and cortical cells remain alive and apparently active throughout the life of the root (Campbell and Drew, 1983). The epidermal cells have convex outer surfaces with deep invaginations between the cells, compared with cells from normal growth conditions (Fig. 9 compared with Fig. 11). There may be very dense microbial cover (Fig. 11) with the microorganisms spread in large, dense colonies, rather than in microcolonies separated by clay and other soil constituents. Since there is no abrasion, as there is in soil, the root cap may be very large and produce considerable quantities of mucilage, and be densely colonized by microbes. It is not clear how such changes in the root surface and its associated microflora affect the physiology of the root and the dynamics of nutrient uptake, but there is a *prima-facie* case for exercising some caution when extrapolating data from liquid culture to field conditions.

5. THE COMMUNITY STRUCTURE OF THE RHIZOSPHERE

There are numerous reviews of microbial occurrence and activity near roots (Curl and Truelove, 1986; Vancura and Kunc, 1988) and of the effects of microorganisms on nutrient turnover and supply to the plant (Clarholm, 1985;

Fig. 9. A cross-fractured root (LTSEM, fully hydrated), with the centre of the root to the bottom of the photograph showing (inset) the gelatinous inner layer of mucilage (m) on the epidermal cell walls, over which is the layer of clay particles, bacteria, etc. held together by diffuse mucilage (dm) (main photograph). (From Campbell and Porter, 1982, with permission.)

Fig. 10. A bacterial colony on the root surface surrounded by mucilage (unstained) and adsorbed clay particles. TEM. (From Faull and Campbell, 1979, with permission.)

Fig. 11. The root surface of a solution-grown plant with massive colonization by bacteria and yeasts. SEM, critical point dried.

Fig. 12. The surface of soil particles in the rhizosphere (the root lies behind the picture) where hyphae cross a soil void. Most hyphae have water droplets on them (soil at field capacity). There are hyphae growing on and within the water films around the soil particles, sometimes dipping in and out of the surface (e.g. at left). LTSEM, fully hydrated. (From Campbell, 1983, with permission.)

Coleman, 1986; Whipps and Lynch, 1986). We are here concerned with information on the spatial organization and morphology of the organisms.

There is usually no distinct boundary between the rhizosphere and the bulk soil. Fungal hyphae, especially, may cross the voids out into the bulk soil and, in soils at field capacity, usually carry water drops along them (Fig. 12). Where hyphae re-enter the mucigel they dip in and out of the surface (Figs. 12 and 13), which will lead to great changes in their surface properties, affecting adsorption of nutrients and soil colloids. This in turn can affect the growth of microorganisms and the interactions between them. Where hyphae enter narrow voids they may distort to fill the volume available (Foster, 1988). Bacteria occur mostly within the mucilage, although occasionally they are on the gas side of the interface (Campbell, 1983).

Bacteria are the organisms seen most often by direct observations of the rhizosphere. It is possible to distinguish various morphological types (Fig. 4) (Foster et al., 1983; Foster, 1988) and to measure their sizes. This shows a clear rhizosphere effect over the distance of a few micrometres, with an increase in numbers and diversity of morphologies near the root. Furthermore, it seems that rhizosphere bacteria are, in general, larger than those in the bulk soil, possibly reflecting their better nutrition (Foster, 1986). Most of the bacterial individuals or colonies in the rhizosphere are surrounded by capsules (Figs. 4 and 10) and may attach to roots by pili and/or fimbriae. Many bacteria contain conspicuous storage materials, probably polyhydroxybutyrate. Fungi and protozoa are seen more rarely (Fig. 6) and are not generally identifiable. They are so much larger than bacteria that even though their total biomass in the rhizosphere may be as great, they occur less frequently. They are frequently associated with bacteria (Fig. 3), which may cause death and lysis of the hyphae (Fig. 14).

The rate of root cap cell sloughing (except in solution-grown plants), and the rate of cell extension in the zone behind the cap, is such that colonization of these regions by microorganisms is low. In the root hair region (Fig. 3) microbial populations may be very high, partly at least in response to exudates. By the time epidermal and cortical cell death is occurring (Figs. 4 and 6) both the ecto- and the endorhizosphere are colonized and there are data showing not only changes in microbial numbers with root age, but also changes in the types of bacteria present, which presumably reflects the different substrates available (Curl and Truelove, 1986; Vancura and Kunc, 1988).

6. EFFECTS OF MICROORGANISMS ON ROOT STRUCTURE AND FUNCTION

6.1. Introduction

It has long been recognized that symbiotic microorganisms such as *Rhizobium* spp. and the mycorrhizal fungi markedly affect both root morphology and root

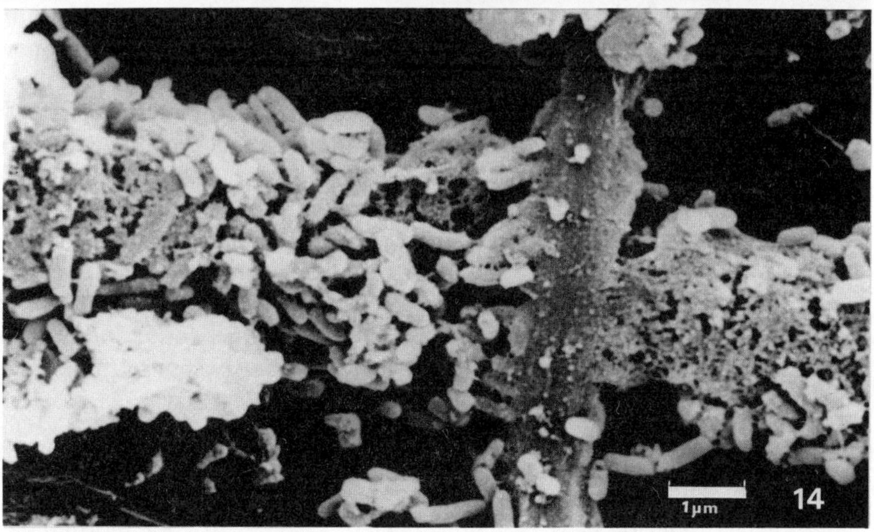

Fig. 13. Hyphae growing in the mucigel. The true root surface is at the top left with a mucilage droplet. The root is mostly covered with mucilage, the surface of which is viewed in this picture. The hyphae dip in and out of the surface film, and the meniscus is sometimes pulled down as though some hyphal surfaces were hydrophobic. LTSEM, fully hydrated. (From Campbell, 1983, with permission.)

Fig. 14. A hypha running from left to right which has been densely colonized by bacteria; these have lysed the wall so that almost nothing remains. SEM, critical point dried.

function. The effects of non-symbiotic organisms on the morphology of roots were also recognized many decades ago but, in contrast to the symbionts, these microorganisms received little attention until comparatively recently. In the 1950s in particular, reports from Russia of striking increases in plant growth and yield following inoculation with phosphate-dissolving or nitrogen-fixing species (Cooper, 1959) caused an upsurge in interest. The subsequent demonstration that such effects were unreliable, resulted in a decline in attention. The subject has been reviewed by Burr and Caesar (1984). Most recently, the demonstration of plant growth promoting rhizobacteria (PGPR) has resulted in another surge of interest in the effects of microorganisms on plant roots.

Inevitably, the "biotechnological" potential suggested by this recent research has caused most emphasis to be placed on the effects of selected species inoculated onto plant roots, via either seed coating or root or soil drenching. More fundamental studies of the effects of the indigenous flora on roots have been neglected. A principal reason for this is the undoubted complexity of the required experimentation and the inevitable difficulties in accurate interpretation of the data. While comparisons of roots grown in soil under gnotobiotic conditions with those grown in normal unsterilized soil are possible, the effects of soil sterilization on soil nutrient content and on production of phytotoxic substances in soil cannot be quantified or eliminated (Cawse, 1975).

Although the effects of plant inoculation result from the exposure of plants to massive inocula, they serve to demonstrate that roots do respond both positively and negatively to the presence of microorganisms. Such responses must inevitably occur in roots growing naturally, although they may be localized to the site of colonization or may be transient, owing to the small natural inoculum size and relatively low concentrations of microbial metabolites involved. In many cases in natural systems, effects may be cancelled by equal and opposite effects of other types of organism growing nearby. Overall, however, it is safe to say that form and function is a reflection of not only plant, edaphic and climatic factors, but also of the size and activity of rhizosphere microbial populations. Unfortunately, there is insufficient information in the literature to allow any generalizations about the way in which the various factors interact with microorganisms in determining plant root response.

6.2. Gross Root Morphology

The most well-known, and best documented, changes in root morphology associated with microorganisms are those responses to colonization of roots by symbionts. The curling of root hairs in response to *Rhizobium* spp. (and *Frankia* spp.) and the subsequent formation of nodules on the roots have been the subject of many classical studies, as have the stunting and forking of roots infected by certain ectomycorrhizas (Chapters 11 and 12).

There is considerably less information about the changing morphology of roots in the presence of non-symbiotic microorganisms; and most of that information is from inoculation experiments, some using hydroponically grown plants. The most typical responses described can be summarized as: no effect; temporary or permanent inhibition; or stimulation of root growth (Campbell et al., 1986; Alström, 1987). This admittedly vague summary is a consequence of the extreme variability of response which is recorded following inoculation of roots with organisms isolated from soils. This variability arises from differences in plant age, plant species or variety, soil conditions, environmental factors and from the changeability of microorganisms themselves. Microorganisms can show marked changes in physiological properties following subculture or storage in the laboratory. Indeed, in a repeated series of experiments, inoculating roots of wheat seedlings with fluorescent *Pseudomonas* spp. from the rhizosphere of field-grown wheat, it was shown that some isolates that initially stimulated root growth later became inhibitory and *vice versa* (Greaves, unpublished data). Regardless of these particular changes, the overall results obtained from the batch of more than 150 isolates studied were remarkably constant. Approximately 40% stimulated root growth, 40% inhibited it and 20% had no effect. Although no data on the levels of activity of such organisms *in situ* in soils were obtained, these results might suggest that there would be no measurable direct effect on roots, all the effects cancelling each other out. There is a further complication in that the effects of root inoculation with the same bacterial isolate may vary with the host plant cultivar (Åström and Gerhardson, 1988). These experiments only studied effects on shoots, but it is likely that root development was also affected.

The microbial effects detected in recent work on root responses to inoculation are measured as changes in length of seedling roots, or in root weight. It must be borne in mind, however, that changes in the former do not necessarily entail changes in the latter. Thus, Darbyshire and Greaves (1971) showed that although the roots of *Pisum sativum* were markedly stunted following inoculation with *Pseudomonas fluorescens*, they were also thicker and root weight was not significantly affected. Root branching can also be affected, presumably as a response to effects on apical dominance. This has been reported principally where plants have been treated with xenobiotics such as herbicides (see below). To date no serious attempt has been made to quantify any changes in root branching which might result from microbial action, even though such changes could have significant consequences for plant growth and vigour. Similarly, there is little, if any, information on the effects of saprophytic organisms on root hair proliferation and longevity. There is circumstantial evidence that, as root hairs may provide readily available infection points for microorganisms to invade the epidermis, the presence of microbes may reduce root hair longevity (see Section 4 above).

6.3. Root Microscopic Morphology

Microbial colonization of roots is frequently associated with visible changes in the morphology of root cells. Apart from the damage to, and invasion of, root hairs, epidermal cell walls are seen to be eroded in the vicinity of microorganisms (Fig. 8), the wall apparently fraying such that cellulose microfibrils are detached and contribute to the mucigel (cf. Foster et al., 1983). Such cells show loss of turgor pressure and are often collapsed. Where these cells contain visible protoplasm it is often possible to recognize that organelles and membranes are disorganized and often degraded (Fig. 8).

Perhaps the commonest response to the presence of microorganisms on roots is that the mucigel layer is frequently much thicker than in the absence of microbes (Greaves and Darbyshire, 1972). While this is in part due to the physical presence of the microbial biomass submerged in the mucilage, it is also a reflection of the contribution made by microbial extracellular products, such as capsules (Fig. 4). The presence of microorganisms in the ''mucigel'' may also be the cause of a detectable thin ''membrane'' layer on the outer surface of the mucilage (Greaves and Darbyshire, 1972). This is most noticeable with *Rhizobium* on legume roots. The consequences of such an outer ''skin'' on the mucigel can so far only be guessed.

7. EFFECTS OF XENOBIOTICS ON ROOT–MICROFLORA INTERACTIONS

The impact of xenobiotics, such as pesticides, on the soil microflora is detailed in many hundreds of papers in the last decade or so (see reviews by Greaves et al., 1976; Anderson, 1978; Papavizas and Lewis, 1979; Greaves, 1982). Despite this there is relatively little information regarding effects on the rhizosphere microflora. Most, if not all, the recorded side-effects on the general soil microflora will occur on rhizosphere organisms also. However, the rhizosphere microflora is also susceptible to possibly more profound effects which arise from xenobiotic effects on the plant. In the case of symbiotic microorganisms those effects might be expected to assume the greatest practical significance. Indeed, pesticide effects on the *Rhizobium*–legume symbiosis must be determined as part of the registration requirements of several countries. While there are many suggested ways of doing this (acetylene-ethylene reduction assays of nitrogenase, nodule counting, etc.) the preferred options recognize the symbiotic association with the plant and concentrate on analysis of plant performance (Somerville et al., 1985). So far little attention has been paid to vesicular arbuscular mycorrhizas in this regard. Perhaps the most detailed analysis of xenobiotic effects on rhizosphere microflora has been done in the case of herbicides. This is not surprising, as these chemicals are designed to be phytotoxic, and may affect roots. This toxicity can occur in the crops in which the herbicides are used, though the effects are limited, generally to changes in physiology and root morphology which do not affect final yield. Thus, several chemical classes of herbicides are known to affect plants in these ways (Curry and Teem, 1976).

Amongst the herbicides the phenoxy group has received the most attention both for this root damage (Friesen et al., 1964; Maas, 1968) and for their effect on root–microflora interactions. This latter emphasis is, undoubtedly, due to early reports that the phenoxy-acid herbicide mecoprop increased cereal root infection by the take-all fungus (Nilsson, 1973a, 1973b; Tottman and Davies, 1978). Recent investigations (Greaves and Sargent, 1986; Greaves et al., 1986; Whipps and Greaves, 1986) have revealed the complexities of the interactions between herbicide (mecoprop), root and microflora which may typify many of the effects produced by other xenobiotics.

Mecoprop caused decreases in shoot and root growth in wheat, and affected plants by causing the production of large numbers of short, swollen, atypically branched lateral roots before recovering to normal growth. The herbicide treatment was also associated with an immediate and lasting decrease in photosynthetic $^{14}CO_2$-fixation, and an initial reduction followed by an increase in the proportion of fixed carbon transported to the roots (Whipps and Greaves, 1986). Examination of the roots in the electron microscope revealed cytoplasmic responses to the herbicide (Greaves and Sargent, 1986; Greaves et al., 1986). These changes involve, mainly, the endomembrane system: rough endoplasmic reticulum increases; actively secreting dictyosomes proliferate; the plasma membrane becomes crenellated, indicating rapid discharge of vesicle contents into the cell wall; osmiophilic granules accumulate within the endoplasmic reticulum and just outside the plasmalemma; and starch granules are formed in the cytoplasm with unusual frequency. All these symptoms point to increased synthesis in the cells and increased discharge of metabolites from the cells to the cell walls, probably leading to increased exudation from the root. This conclusion is supported by the apparent degradation of cell walls and middle lamella which is also seen.

In normal roots microbial invasion is restricted to areas of epidermis damaged by lateral root emergence or some other trauma. In roots affected by mecoprop—and many other herbicides (Greaves, unpublished data)—invasion and colonization of root tissues is much more extensive. The principal colonizers are bacteria of many different morphological types (Fig. 4); fungi are rarely seen. Of the bacteria, *Pseudomonas* spp. are particularly numerous, as shown by plate counts. Myxobacteria (*Cytophaga* spp.) are also common (Greaves, unpublished data).

Bacterial colonization of the root tissue appears to take place, principally along the interfaces of cells where the middle lamella has been degraded. The organisms almost certainly contribute to this degradation, as they do to that of the cell walls (Greaves and Sargent, 1986). Cell walls are eventually ruptured, cell lumens colonized and their contents digested.

Greaves and Sargent (1986) concluded, from the observed sequence of events, that mecoprop induces a series of physiological changes in affected cells which result in increased exudation. This stimulates bacterial population growth on the root and makes the root cells susceptible to invasion. Invasion itself results in further root damage, possibly including the observed root morphological

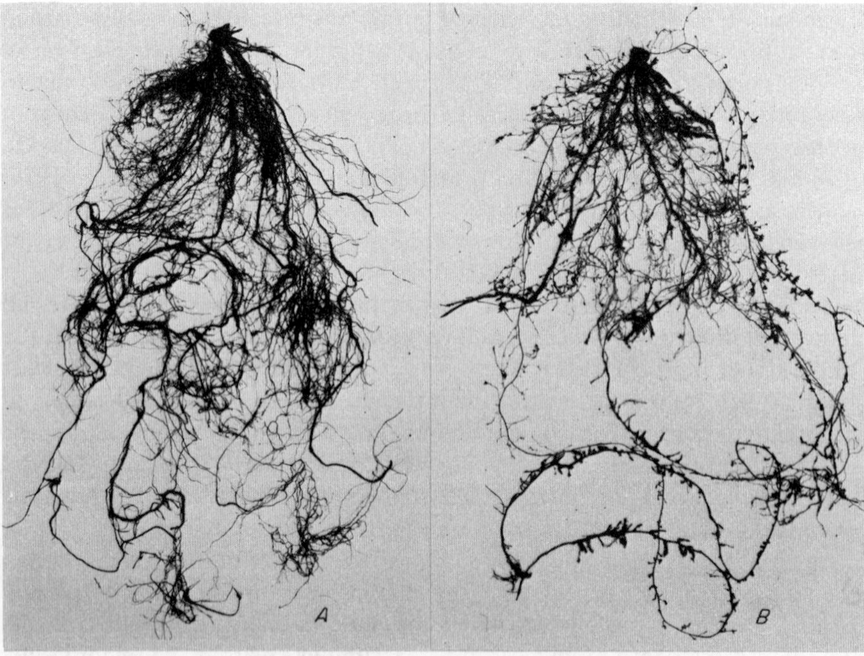

Fig. 15. Roots of wheat plants treated with the herbicide mecroprop at 2.4 kg a.i. ha^{-1}. (*A*) Plants grown in soil amended with 1% (w/w) lucerne meal. (*B*) soil not amended.

changes, which might reflect the effects of plant growth substances produced by the microorganisms (Burr and Caesar, 1984). The relationship between increased exudation and increased bacterial growth on the roots, and invasion of roots, may involve the development of a gradient of decreasing available carbon concentration as one moves away from affected tissue into the soil. This gradient may induce a chemotactic attraction of organisms to the root surface and into the tissues. Interestingly, the microbial response to mecoprop-treated roots is eliminated if the plants are grown in soil supplemented with an available carbon source (Fig. 15), for example 1% (w/w) lucerne meal (Greaves, unpublished data). This seems to support the idea of a chemical gradient effect, as the amendment would swamp such a gradient.

This example uses a xenobiotic designed to be phytotoxic, albeit in this case, to a plant which is supposedly tolerant to it. Where tolerance to a non-herbicidal xenobiotic is not present, effects at the cellular level can also be expected and many of the attendant consequences as described here, could occur. This is a fruitful and fascinating area of research which is worthy of considerable further study and which may well give valuable insight into means by which inocula of desirable microbes may be maintained on plant roots in an active and proliferating state. It exemplifies, also, the great benefits to be obtained in rhizosphere microbiological research from employing the skills of several disciplines, especially plant physiology.

8. CONCLUSION

This chapter has introduced many subjects that will be considered in more detail later. The importance of a basic understanding of root anatomy and growth cannot be over-emphasized for those microbiologists, physiologists and biochemists who study the rhizosphere. The problems of methodology, although well-known, limit the development of concepts and models of rhizosphere functioning. We are far from understanding microbial communities and the physics and chemistry of this dynamic interface between the root and the soil, although the increased interest in the practical applications, such as the modification of plant growth, improvement in plant health and the interactions with xenobiotics, has stimulated much of the recent research reported in this volume.

REFERENCES

Alström, S. (1987). Influence of root-zone inhabiting bacteria on growth of plants and soil-borne fungal pathogens. Dissertation, Dept. of Plant & Forest Protection, Swedish Univ., Ag. Sci., Uppsala.

Anderson, J. R. (1978). Pesticide effects on non-target soil micro-organisms. In *Pesticide Microbiology* (I. R. Hill and S. J. L. Wright, eds.). Academic Press, London, pp. 313–533.

Åström, B. and B. Gerhardson (1988). Differential reactions of wheat and pea genotypes to root inoculation with growth-affecting rhizobacteria. *Plant and Soil*, **109**:263–269.

Bohlool, B. B. and E. L. Schmidt (1980). The immunofluorescence approach to microbial ecology. *Adv. Microbial Ecology*, **4**:203–241.

Brock, T. D. (1987). The study of microorganisms *in situ*: progress and problems. In *Ecology of Microbial Communities* (M. Fletcher, T. R. G. Gray and J. G. Jones, eds.). SGM Symposium 41, Cambridge University Press, Cambridge, pp. 1–17.

Brookes, P. C. (1986). The adenylate energy charge ratio and ATP concentration of soil microbial biomass. In *Perspectives in Microbial Ecology* (F. Meğuser and M. Gantar, eds.). Slovene Society for Microbiology, Ljubljana, pp. 673–678.

Brookes, P. C., A. D. Newcombe, and D. S. Jenkinson (1987). Adenylate energy charge measurements in soil. *Soil Biol. Biochem.*, **19**:211–217.

Brown, M. E. and D. Hornby (1987). Effects of nitrate and ammonium on wheat roots in gnotobiotic culture: amino acids, cortical cell death and take-all (*Gaeumannomyces graminis* var. *tritici*). *Soil Biol. Biochem.*, **19**:567–574.

Buckley, R. (1982). Sand rhizosheath of an arid zone grass. *Plant and Soil*, **66**:417–421.

Burns, R. G. (1978). *Soil Enzymes*. Academic Press, London, 380 pp.

Burns, R. G. and J. H. Slater (1982). *Experimental Microbial Ecology*. Blackwell Scientific Publications, Oxford, 683 pp.

Burr, T. J. and A. Caesar (1984). Beneficial plant bacteria. *CRC Crit. Rev. Pl. Sci.*, **2**:1–20.

Campbell, R. (1983). Ultrastructural studies of *G. graminis* in waterfilms on wheat roots and the effect of clay on the interaction between this fungus and antagonistic bacteria. *Can. J. Microbiol.* **29**:39–45.

Campbell, R. and M. C. Drew (1983). Electron microscopy of gas space (aerenchyma) formation in adventitious roots of *Zea mays* L. subjected to oxygen shortage. *Planta*, **157**:350–357.

Campbell, R. and R. Porter (1982). Low-temperature scanning electron microscopy of micro-organisms in soil. *Soil Biol. Biochem*, **14**:241–245.

Campbell, R., A. Renwick and S. K. A. M. Coe (1986). Antagonism and siderophore production by biocontrol agents, plant growth promoting organisms and the general rhizosphere population. In *Iron, Siderophores and Plant Diseases, NATO ASI Series A, Life Sciences, Vol. 117* (T. R. Swinburne, ed.). Plenum Press, London, pp. 179–188.

Cawse, P. A. (1975). Microbiology and biochemistry of radiated soils. In *Soil Biochemistry Vol. 3* (E. A. Paul and A. D. McLaren, eds.). Marcel Dekker, New York, pp. 214–267.

Chantler, S. M. and M. B. McIllmurray (1987). Labelled-antibody methods for detection and identification of microorganisms. In *Methods in Microbiology. Vol. 19* (R. R. Colwell and R. Grigorova, eds.). Academic Press, New York, pp. 273–332.

Clarholm, M. (1985). Possible roles for roots, bacteria, protozoa and fungi in supplying nitrogen to plants. In *Ecological Interactions in Soil, Plants, Microbes and Animals* (A. H. Fitter, ed.). Special Publication 4, British Ecological Society, Blackwell, Oxford, pp. 355–365.

Coleman, D. C. (1986). The role of microfloral and faunal interactions in affecting soil processes. In *Microfloral and Faunal Interaction in Natural and Agro-Ecosystems* (M. J. Mitchell and J. P. Nakas, eds.). Martinus Nijhoff/Dr W. Junk Publishers, Dordrecht, pp. 317–348.

Cooper, R. (1959). Bacterial fertilizers in the Soviet Union. *Soils & Ferts.*, **22**:327–333.

Costerton, J. W., J. C. Nickel, and T. I. Ladd (1986). Suitable methods for the comparative study of free-living and surface-associated bacterial populations. In *Bacteria in Nature, Vol. 2, Methods and Special Applications in Bacterial Ecology* (J. S. Poindexter and E. R. Leadbetter, eds.). Plenum Press, New York, pp. 49–84.

Curl, E. A. and B. T. Truelove (1986). *The Rhizosphere*. Springer-Verlag, Berlin, 288 pp.

Curry, W. L. and D. H. Teem (1976). Herbicides and root development. *Soil Crop Sci. Soc. Fl. Proc.*, **36**:23–28.

Darbyshire, J. F. and M. P. Greaves (1971). The invasion of pea roots, *Pisum sativum* L. by soil microorganisms, *Acanthamoeba palestinensis* (Reich) and *Pseudomonas* sp. *Soil Biol. Biochem.*, **3**:151–155.

Dhingra, O. D. and J. B. Sinclair (1985). *Basic Plant Pathology Methods*. CRC Press, Boca Raton, 355 pp.

Domsch, K. H., T. Beck, J. P. E. Anderson, B. Söderström, D. Parkinson, and G. Trolldenier (1979). A comparison of methods for soil microbial population and biomass studies. *Z. Pflanzenernaehr Dung. Bodenk.*, **142**:520–533.

Drahos, D. J., B. C. Hemming, and S. McPherson (1986). Tracking recombinant organisms in the environment: β-galactosidase as a selectable non-antibiotic marker for pseudomonads. *Biotechnology*, **4**:439–444.

Duell, R. W. and G. R. Peacock (1985). Rhizosheaths on mesophytic grasses. *Crop Science*, **25**:880–883.

Esau, K. (1977). *Anatomy of Seed Plants*, 2nd edn. John Wiley & Sons, New York, 550 pp.

Faull, J. L. and R. Campbell (1979). Ultrastructure of the interaction between the take-all fungus and antagonistic bacteria. *Can. J. Bot.*, **57**:1800–1808.

Foster, R. C. (1981). The ultrastructure and histochemistry of the rhizosphere. *New Phytol.*, **89**:263–273.

Foster, R. C. (1982). The fine structure of epidermal cell mucilages of roots. *New Phytol.*, **91**: 727–740.

Foster, R. C. (1986). The ultrastructure of the rhizoplane and rhizosphere. *A. Rev. Phytopath.*, **24**:211–234.

Foster, R. C. (1988). Microenvironments of soil microorganisms. *Biol. Fertil. Soil*, **6**:189–203.

Foster, R. C., A. D. Rovira, and T. W. Cock (1983). *Ultrastructure of the Root–Soil Interface*. American Phytopathological Society, St. Paul, Minn., 157 pp.

Friesen, H. A., H. Baenziger, and C. H. Keys (1964). Morphological and cytological effects of dicamba on wheat and barley. *Can. J. Plant Sci.*, **44**:288–294.

Gagné, S., C. Richard, H. Rousseau, and H. Antoun (1987). Xylem-residing bacteria in alfalfa roots. *Can. J. Microbiol.*, **33**:996–1000.

Greaves, M. P. (1982). Effects of pesticides on soil microorganisms. In *Experimental Microbial Ecology* (R. G. Burns and J. H. Slater, eds.). Blackwell Scientific Publications, Oxford, pp. 613–630.

Greaves, M. P. and J. F. Darbyshire (1972). The ultrastructure of the mucilagenous layer on plants' roots. *Soil Biol. Biochem.*, **4**:443–449.

Greaves, M. P. and J. A. Sargent (1986). Herbicide-induced microbial invasion of plant roots. *Weed Sci.* (Suppl. 1), **34**:50–53.

Greaves, M. P., H. A. Davies, J. A. P. Marsh, and G. I. Wingfield (1976). Herbicides and soil microorganisms. *CRC Crit. Rev. Microbiol.*, **5**:1–38.

Greaves, M. P., S. L. Cooper, H. A. Davies, J. A. P. Marsh, and G. I. Wingfield (1978). *Methods of Analysis for Determining the Effects of Herbicides on Soil Microorganisms and their Activities.* Technical Report 45, ARC Weed Research Organization, 53 pp.

Greaves, M. P., J. A. Sargent, and J. M. Whipps (1986). Herbicide-induced interactions between cereal roots and fluorescent *Pseudomonas* spp. In *Iron, Siderophores and Plant Disease, NATO ASI Series A, Life Sciences, Vol. 117* (T. R. Swinburne, ed.). Plenum Press, London, pp. 189–201.

Harley, J. L. and S. E. Smith (1983). *Mycorrhizal Symbiosis.* Academic Press, London, 483 pp.

Henry, C. M. and J. W. Deacon (1981). Natural (non-pathogenic) death of the cortex of wheat and barley seminal roots, as evidenced by nuclear staining with acridine orange. *Plant and Soil*, **60**:255–274.

Hiltner, L. (1904) Uber neuere Erfahrungen und Probleme auf dem Gebeit der Bodenbakteriologie und unter besonderer Berucksichtigung der Grundungung und Brache. *Arb. Dsch. Landwirt. Ges.*, **98**:59–78.

Jenny, H. and K. Grossenbacher (1963). Root–soil boundary zones as seen in the electron microscope. *Soil Sci. Soc. Am. Proc.*, **27**:273–277.

Karl, D. M. (1986). Determination of *in situ* microbial biomass, viability, metabolism and growth. In *Bacteria in Nature, Vol. 2, Methods and Special Applications in Bacterial Ecology* (J. S. Poindexter and E. R. Leadbetter, eds.). Plenum Press, New York, pp. 85–176.

Leppard, G. G. and J. R. Colvin (1971). Fibrillar lignin or fibrillar pectin? *J. Polymer Sci.*, **C36**:321–326.

Lucas, W. J. (1987). Functional aspects of cells in root apices. In *Root Development and Function.* (P. J. Gregory, J. V. Lake, and D. A. Rose, eds.). Cambridge University Press, Cambridge, pp. 27–52.

Maas, G. (1968). Damage to the anchorage roots (crown roots) of cereals after herbicide treatment. *Z. PflKrankh. PflPath. PflSchutz.*, **75** (Suppl. 4):139–144.

McCully, M. E. (1987). Selected aspects of the structure and development of field-grown roots with special reference to maize. In *Root Development and Function* (P. J. Gregory, J. V. Lake, and D. A. Rose, eds.). Cambridge University Press, Cambridge, pp. 53–70.

Minchin, P. E. H. and G. S. McNaughton (1984). Exudation of recently fixed carbon by non-sterile roots. *J. Exp. Bot.*, **35**:74–82.

Newell, S. Y., R. D. Fallon, and P. S. Tabor (1986). Direct microscopy of natural assemblages. In *Bacteria in Nature, Vol. 2, Methods and Special Applications in Bacterial Ecology.* (J. S. Poindexter and E. R. Leadbetter, eds.). Plenum Press, New York, pp. 1–48.

Newman, E. I. (1985). The rhizosphere: carbon sources and microbial populations. In *Ecological Interactions in Soil, Plants, Microbes and Animals. Special Publication 4, British Ecological Society* (A. H. Fitter, ed.). Blackwell Scientific Publications, Oxford, pp. 107–121.

Nilsson, H. E. (1973a). Influence of the herbicide mecoprop on *Gaeumannomyces graminis* and take-all disease of spring wheat. *Swed. J. Agric. Res.*, **3**:105–113.

Nilsson, H. E. (1973b). Influence of herbicides on take-all and eye-spot disease of winter wheat in a field trial. *Swed. J. Agric. Res.*, **3**:115–118.

Oades, J. M. (1978). Mucilages at the root surface. *J. Soil Sc.*, **29**:1–16.

O'Brien, T. P. and M. E. McCully (1981). *The Study of Plant Structure: Principles and Selected Methods*. Termarcarphi Ptg. Ltd., Melbourne. Distributed by Blackwell, Oxford.

Old, K. M. and T. H. Nicholson (1978). The root cortex as part of a microbial continuum. In *Microbial Ecology* (M. W. Loutit and J. A. R. Miles, eds.). Springer-Verlag, Berlin, pp. 291–294.

Papavizas, G. C. and J. A. Lewis (1979). Side-effects of pesticides on soil-borne plant pathogens. In *Soil-Borne Plant Pathogens* (B. Schippers and W. Gams, eds.). Academic Press, London, pp. 483–505.

Poindexter, J. S. and E. R. Leadbetter (1986). *Bacteria in Nature, Vol. 2, Methods and Special Applications in Bacterial Ecology*. Plenum Press, New York, 385 pp.

Polonenko, D. R. and C. I. Mayfield (1979). A direct observation technique of studies on rhizoplane and rhizosphere colonization. *Plant and Soil*, **51**:405–420.

Rosswall, T. and E. Kvillner (1979). Principal component analysis for description of microbial populations. *Adv. Microb. Ecol.*, **2**:1–48.

Rougier, M. (1981). Secretory activity of the root cap. In *Plant Carbohydrates II, Extracellular Carbohydrates: Encyclopedia of Plant Physiology NS13B* (W. Tanner and F. A. Leowus, eds.). Springer-Verlag, Berlin, pp. 542–574.

Rovira, A. D. and R. Campbell (1974). Scanning electron microscopy of microorganisms on the roots of wheat. *Microb. Ecol.*, **1**:15–23.

Rovira, A. D., E. I. Newman, H. J. Bowen, and R. Campbell (1974). Quantitative assessment of the rhizoplane microflora by direct microscopy. *Soil Biol. Biochem.*, **6**:211–216.

Rush, C. M., D. R. Upchurch, and T. J. Gerik (1984). *In situ* observations of *Phymatotrichum omnivorum* with a borescope mini-rhizotron system. *Phytopathology*, **74**:104–105.

Somerville, L., M. P. Greaves, K. H. Domsch, W. Verstraete, N. J. Poole, H. van Dijk, and J. P. E. Anderson (1985). *Recommended Laboratory Tests for Assessing the Side-Effects of Pesticides on the Soil Microflora. Proc. 3rd Int. Workshop, Cambridge.* Long Ashton Research Station (Publ.), Bristol, 29 pp.

Tate, K. R. and D. S. Jenkinson (1982). Adenosine triphosphate measurement in soil: an improved method. *Soil Biol. Biochem.*, **14**:331–335.

Tottman, D. R. and E. L. P. Davies (1978). The effect of herbicides on the root system of wheat plants. *Ann. Appl. Biol.*, **90**:93–99.

Vancura, V. (1988). Plant metabolites in soil. In *Soil Microbial Associations* (V. Vancura and F. Kunc, eds.). Elsevier, Amsterdam, pp. 57–144.

Vancura, V. and F. Kunc (1988). *Soil Microbial Associations*. Elsevier, Amsterdam, 498 pp.

van Vuurde, J. W. L. and P. F. M. Elenbaas (1978). Use of fluorochromes for direct observation of microorganisms associated with roots. *Can. J. Microbiol.*, **24**:1272–1275.

Vermeer, J. and M. E. McCully (1981). Fucose in the surface deposits of axenic and field grown roots of *Zea mays* L. *Protoplasma*, **109**:233–248.

Vermeer, J. and M. E. McCully (1982). The rhizosphere of *Zea*: new insight into structure and development. *Planta*, **156**:45–61.

Whipps, J. M. (1985). Effect of CO_2 concentration on growth, carbon distribution and loss of carbon from the roots of maize. *J. Exp. Bot.*, **36**:644–651.

Whipps, J. M. and M. P. Greaves (1986). Effect of mecoprop on plant growth and distribution of photosynthate in wheat (*Triticum aestivum* L.) seedlings. *Weed Res.*, **26**:227–232.

Whipps, J. M. and J. M. Lynch (1985). Energy losses by the plant in rhizodeposition. *Plant Products and the New Technology, Ann. Proc. Phytochem. Soc. of Europe*, **26**: (K. W. Fuller and J. R. Gallon, eds.). Clarendon Press, Oxford, pp. 59–71.

Whipps, J. M. and J. M. Lynch (1986). The influence of the rhizosphere in crop productivity. *Adv. Microb. Ecol.*, **9**:187–244.

Wullstein, L. H. and S. A. Pratt (1981). Scanning electron microscopy of rhizosheaths of *Oryzopsis hymenoides*. *Amer. J. Bot.*, **68**:408–419.

3

ROOT FUNCTION, DEVELOPMENT, GROWTH AND MINERAL NUTRITION

M. C. DREW
Department of Horticultural Sciences, Texas A & M University, College Station, Texas 77843-2133, USA

1. INTRODUCTION: ROOT GROWTH AND ROOT FUNCTION

From the moment the root emerges from the germinating seed and extends into the soil, its future development and function will depend on a complex interplay between genotype and environment that illustrates very clearly the ''plasticity'' in phenotype found in many biological systems. The major ''architectural'' features of the developing root system are genetically determined: most dicotyledons and gymnosperms form a tap root comprising a single main axis with lateral roots; most monocotyledons form a fibrous root system in which the embryonic root axis is supplemented by numerous adventitious axes originating at the stem base (Esau, 1965). Genetics also determines whether a primary root becomes woody with increasing age, with production of secondary xylem and phloem, and development of a periderm (bark) and lenticels. But beyond these anatomical generalizations, it is not possible to predict the total length, weight or depth of roots that a plant will generate in its lifetime except in highly uniform (and usually experimental) environments. Nor can the precise physiological function of its parts and their longevity be predicted.

As the young root axis (and its branches) continuously extend at the apex into fresh soil volumes, the growing cells at the apex experience both temporal and positional variation in the environment: the surface soil may quickly desiccate during drought; natural or agriculturally-produced pans may block extension into the subsoil; diurnal temperature gradients with wide amplitude at the soil surface (contrasting with a more stable deep-soil temperature) may be lethal. Each of these components can profoundly modify the growth rate of roots, their structure at the tissue, cellular and subcellular levels, and their function in relation to the growth of the whole plant. However, while the root system is modified by the soil environment, the root itself changes the properties of the soil: fluxes of water, ions and gases that are essential for the growth of shoots and roots take place at the soil/root interface, and thereby locally change the concentration or free energy of these metabolites in the soil. The implication for the micro-habitat that constitutes the rhizosphere is obvious.

In this chapter, some of the mechanisms by which the soil environment modifies the properties of plant root systems and, in turn, the ability of the root surface to provide a habitat for soil microorganisms, are discussed.

2. ENVIRONMENTAL FACTORS CONTROLLING ROOT GROWTH AND DEVELOPMENTAL PATTERN IN THE SOIL

2.1. General Considerations

The root systems of all land plants, despite wide variation in morphology and size, retain certain functions in common. The transport of water and dissolved mineral (nutrient) ions from the soil to the plant is an essential and unique property of roots, as is the production of metabolites, including phytohormones, whose synthesis is responsive to the environment.

The root system in return receives virtually its entire carbon source from the leaves, but the effectiveness with which the root system uses that carbon in growth, respiration and transport processes, or the "costs" versus "benefits" to the shoot, are not well documented.

The primary growth of roots of all woody and herbaceous plants takes place by cell division in the apical meristem, with subsequent expansion (predominantly in the longitudinal direction) in the subapical zone (Barlow, 1987). A cell, newly produced in the meristem, may complete the process in only a few hours under favourable conditions. A cell of a maize root that is extending at the rate of 40 mm day^{-1} will have completed cell elongation by the time it reaches a position 10 mm from the tip, equivalent to 6 h since its initiation in the meristem. As soon as cell expansion is complete, many species produce root hairs as protuberances from the epidermal cells. The hairs increase the coefficient of friction and thereby help anchor the subapical zone in the soil so that the root tip can exert a physical force on soil particles during extension through resistant soil. The hairs may also contribute to water and ion uptake (Nye and Tinker, 1977), although the duration of their viability is variable and probably dependent on the local microbiological population in any zone along the root as well as on soil physical conditions. The hairs also help bind soil particles and aggregates together in the form of rhizosheaths (Wullstein and Prat, 1981; Buckley, 1982; Goodchild and Myers, 1987).

With time, a number of genetically determined developmental changes to the structure and morphology of the primary roots take place. These changes are also sensitive to soil conditions, and they greatly influence the movement of materials at the soil/root interface. Lateral root primordia are initiated, usually in the pericycle (McCully, 1987). In barley this occurs within 10 mm of the root tip, but the primordia remain microscopic for several days before they begin rapid division and expansion to first emerge as lateral roots (see Chapter 2) some 50–100 mm behind the tip. The lateral roots in cereals and other members of the Gramineae make up the major part of the root system, whether in terms of weight, length or surface area (Hackett, 1968).

At about the same position along the root where laterals are first emerging, the endodermis and sometimes the hypodermis (see Fig. 1 for root structure) lay down continuous layers of suberin in the wall interior to the primary wall. Suberin is a complex material composed of ω-hydroxy fatty acids of 16–24 carbon chain length, the corresponding dicarboxylic acids, and long chain alcohols, together with waxes (Kolattukudy, 1980). Its deposition coincides with

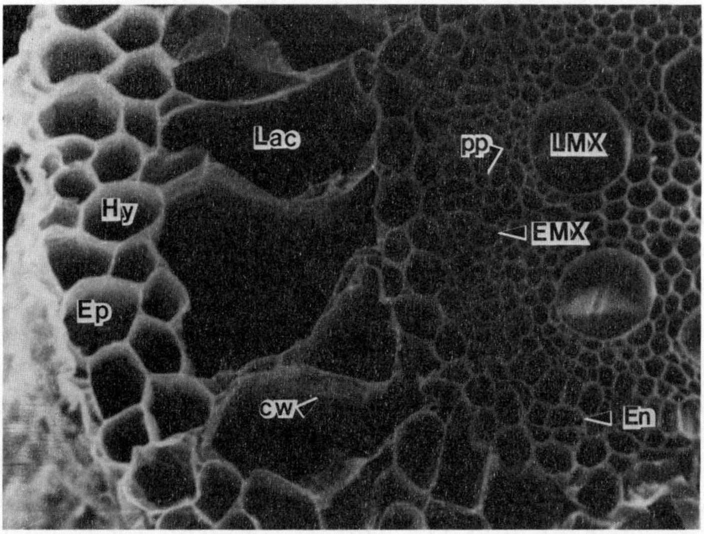

Fig. 1. Part of a transverse section of an aerenchymatous nodal (adventitious) root in wheat growing under hypoxic (partial oxygen deficient) conditions. Note the prominent lacunae (Lac) formed by lysis of cells of the cortex. Oxygenated roots of comparable age do not normally form such cavities in the cortex, so that the lacunae are occupied by intact cells that link the epidermis to the xylem or phloem. Ep = epidermis; Hy = hypodermis; cw = cell wall material, remaining after cell lysis; En = endodermis; pp = protophloem; EMX = early maturing metaxylem; LMX = late maturing metaxylem. SEM.

several changes in root properties, especially in relation to ion and water uptake. Suberization of the endodermis comprises continuous lamellae over the inner surface of the primary cell wall (state II of development), except for pits where plasmodesmata link adjacent cells. State II development coincides with a marked decline in radial transport of water, Ca^{2+} and Mg^{2+} to the xylem (Clarkson, 1974, 1984). Other ions that are believed to move symplastically across the root, that is from cell to cell via plasmodesmata, are unaffected by suberization. Thus, root tips become the zones where fluxes of water and divalent cations through the rhizosphere are greatest, while fluxes of potassium and phosphate, and probably also NO_3^- and NH_4^+, are more uniform along the root under favourable environmental conditions. Suberization of the hypodermis is more variable in its effects, apparently depending on the circumstances at the time of its deposition (see later sections).

Woody plants, because of production of new cells by a cambium, continue to increase root diameter and tensile strength. This probably terminates the direct involvement of the root surface in selective ion uptake and water absorption and leakage of soluble carbon compounds, although the same portion of root allows axial conduction of materials in xylem and phloem, to and from younger zones. Some mineral ions and water may enter through lenticels and through splits in the bark, and along living ray parenchyma cells to the xylem (Kramer and Bullock, 1966), but the quantitative significance is unknown. Mathematical modelling of fluxes of ions and water in the rhizosphere (Nye and Tinker, 1977; de Willigen and van Nordwick, 1987) is thus hampered by variability in geometry and absorptive competence of the root surfaces that are produced.

The total amount of carbon allocated to the structure of the root system per annum can be appreciable for herbaceous and woody species (Chapter 4). In deep soil, cereal root systems will extend to depths of 1–2 m, the total root length at flowering making 20 km per m^2 of land area, and the dry weight (mostly carbon) comprising 1–2 t ha^{-1} (Welbank et al., 1974). In the topsoil, especially in permanent grassland, root length densities of 36–340 cm cm^{-3} can occur (see references cited in Andrews and Newman, 1970), which must mean that any volume of the soil is within 1 mm of a root surface! With forest trees, root length densities of fine root members are much less (0.5–1.0 cm cm^{-3}) but there is a great turnover of roots. Short, small-diameter "feeder roots" that last from a week to several months, fail to undergo secondary thickening, frequently become mycorrhizal, and are shed. It is estimated that annual productivity of such roots is 1–2 t ha^{-1} for *Pinus sylvestris* and *Betula verrucosa* (Head, 1973). With *Picea excelsa*, about 2 t ha^{-1} of transient roots died each year in a 25-year-old stand. Cold desert shrubs like *Atriplex confertifolia* show appreciable annual turnover of the fine root members, amounting to 24%, and equivalent to an input of 1.2 t ha^{-1} of carbon to the rhizosphere soil (Caldwell and Camp, 1974).

2.2. Mechanical Impedance

Root growth in soil invariably necessitates that the tip is pushed through the soil by the expansion of cells in the zone of elongation. The tip is thus forced to displace some particles and compress others, and this is done by cylindrical enlargement of pores or channels. By reaction, the root tip is subject to stress from the soil. This stress can be estimated approximately by a fine-tipped steel penetrometer that is forced into the soil while the resistance to its movement is measured (Graecen et al., 1968). In detail, however, the processes of penetration are different: the root cap cells are compressible, unlike steel; the coefficients of friction are lowered by cap cells detaching from the tip and by mucigel production; and the root tip, unlike a rigid steel probe, is not forced to take a preset direction through the soil, but follows lines of least resistance.

Because of the slowing of root extension in resistant media, random contact with a region of low resistance ensures that a root will follow lines of weakness, especially channels left by worms, or occupied by straw, decayed roots and partially decomposed organic matter (Hamblin, 1985). Once in a channel or crack, root tips have difficulty penetrating the soil matrix again, because contacting the ped surface at a glancing angle tends to deflect the root (like the internal reflection of light in a "light pipe") back into the channel.

Studies in artificial media (glass spheres), in which roots were able to grow only by forcing incompressible particles apart against a controlled pressure, suggest that extension is highly sensitive to stress (Goss, 1977). Extension of barley roots was halved by a confining pressure of only 0.2 bar. This is small relative to the turgor (hydrostatic pressure) commonly measured in fully hydrated cells (5–10 bar) that should be available to drive cell elongation. However, growth-induced water potentials in the zone of cell expansion depress turgor (Westgate and Boyer, 1984) so that little hydrostatic pressure in excess of the yield threshold may remain to drive expansion growth. Additionally, physical stresses experienced at the root cap are probably transmitted to the dividing initials in the meristem thereby slowing cell division rates, and hence root elongation, by an effect other than one involving expansion of cells.

The slowing of root extension must be considered separately from the *maximum* pressures that confined plant organs can exert. Pressures of the order of 60 bar have been recorded in axial or radial directions for confined roots, and these would be sufficient to expand rigid pores in the soil. The enlargement of such pores is possible during secondary growth in woody plants when the cambium becomes active. For radial thickening, a rapid growth rate is of less concern than for a rapidly extending root axis (when plant survival may depend on its ability to reach water deep in the soil profile).

Because roots follow lines of least resistance, root distribution in a natural soil profile tends to follow the principal structural units when there are well-developed peds, but in uncompacted soils of coarse texture, in well-structured loams and in the plough layer, roots have more opportunity to follow directions that are determined by their geotropic response (Barley and Graecen, 1967; Hamblin, 1985). Where extension is transiently slowed by a resistant soil zone, lateral roots proliferate along the axis, close to the apex, and root hairs are more frequent at the epidermis (Goss and Russell, 1980), both indicative of a decline of "apical dominance" of the stressed root apex.

Because of the sensitivity of root extension to relatively small radial stresses when roots are grown in packings of glass spheres (described above), there is interest in whether changes in production of phytohormones in the root tip control root growth under mechanical impedance (Goss and Russell, 1980; Masle and Passioura, 1987). There is a temporary absence of growth inhibition in roots with the root cap removed—the cap is a possible site of abscisic acid (ABA) production (Pilet and Barlow, 1987)—and ABA may act as a growth inhibitor under these conditions. Furthermore, ethylene, another phytohormone produced by plant tissue, including roots, inhibits root growth in a way that resembles some effects of mechanical impedance. Mechanical impedance, like the effect of exogenous ethylene, causes a reorientation of cell expansion from longitudinal to lateral, coinciding with changes in the predominant orientation of cellulose microfibrils in cell walls. Mechanical impedance also stimulates ethylene biosynthesis by confined roots (Kays et al., 1974; Moss et al., 1988). Ethylene, whether endogenous or exogenous, can evoke profound changes in metabolism of responsive cells, affecting protein synthesis and gene expression (Nichols and Laties, 1984; Tucker and Laties, 1984; Cassab et al., 1988), and is involved in lysis of cells to form aerenchyma (see Section 2.5 below). The hormone may also be involved in plant–microorganism interactions at the cell surface that influence cell wall chemistry (Roby et al., 1985).

The possibility that increased ethylene biosynthesis accounts for sensitivity to mechanical impedance is attractive, particularly because endogenous ethylene production under unstressed conditions can be sufficient to slightly depress cell extension rates (Whalen and Feldmann, 1988) so that its augmentation could readily account for growth inhibition during physical stress. However, roots of maize seedlings failed to show a greater tolerance of mechanical impedance when extending through packings of glass spheres in the presence of inhibitors of ethylene production or ethylene action (Moss et al., 1988). Additionally, mechanical impedance did not influence the root content of another potential hormonal inhibitor of growth, ABA. These results suggest that additional production of ethylene may be incidental to inhibition by physical impedance. Perhaps the significance of additional ethylene is to increase root diameter to help exert a greater force at the root tip (pressure × area) and in soil with well-defined peds or cracking patterns, to expand planes of weakness (tensile rupture) ahead of the root tip (Barley and Graecen, 1967).

Impeded roots develop smaller root caps and have smaller volumes of mucigel at the apex, suggesting a greater degree of wearing away of cells and mucigel at the tip (Wilson and Robards, 1979). There was less mucigel produced by Golgi apparatus in the outer cap cells, indicating less deposition to the rhizosphere. Because mechanically impeded roots extend more slowly, while maturation of the endodermis is less affected, suberization of the endodermis occurs closer to the root tip than in unimpeded roots (Wilson and Robards, 1978). Structural changes alone suggest that fluxes of water, Ca^{2+} and Mg^{2+} would be restricted to a zone closer to the tip (Clarkson, 1984). However, little research has been done on ion transport in physically impeded roots, although it seems significant that Ca^{2+} content of shoots and especially roots was reduced in plants with impeded roots in packings of glass spheres (Lindberg and Petterson, 1985).

Leaves of plants with mechanically impeded roots show different stomatal conductance and photosynthetic rates compared with unstressed controls (Herold and McNeil, 1979; Masle and Passioura, 1987; Masle and Farquar, 1988). In wheat seedlings, net CO_2 fixation per unit leaf area increased with mechanical impedance (Masle and Farquar, 1988). There was a greater proportion of carbon translocated to the roots, rather like the response of water-deficient plants. This change in distribution of carbon may help account for continued maturation of impeded roots (Wilson and Robards, 1978), their greater dry weight/fresh weight ratio (Lindberg and Petterson, 1985) and increased deposition of soluble assimilates to the rhizosphere (Barber and Gunn, 1974).

2.3. Temperature

Cell division and expansion are temperature-dependent processes that control the rate of root growth (Barlow, 1987). Above and below the temperature range to which a plant is adapted, root growth is inhibited or even irreversibly arrested (Kramer, 1983), depending on the magnitude of the temperature change, the rate of change, and the ability of cells to acclimate. Many aspects of cell metabolism are adversely affected by temperatures that depart from optimal; these include cell division, protein synthesis, respiration and ion transport (Levitt, 1980).

Unfavourable temperatures, particularly when suddenly imposed, cause leakage of ions and metabolites from the root to the surrounding medium. Additionally, sudden changes of temperature cause a lowering of conductivity of roots to water, and inhibition of ion transport (reviewed by Lyons, 1973; Levitt, 1980; Kramer, 1983). This can indirectly affect the carbon economy of the whole plant by leading to more negative leaf water potentials, partial stomatal closure, and hence slowing of net CO_2 fixation. The fluidity of various cellular membranes may be a fundamental aspect that accounts for this range of responses. Measurement of the physical processes of natural or synthetic membranes indicates that distinct changes take place at temperatures coinciding with chilling or freezing injury to plants. There is a greater degree of ordering, perhaps a transition from semi-crystalline to gel, and this greatly alters the permeability properties of the lipid bilayer matrix and of the transport proteins and enzymes in the membrane. Contrary to earlier reports, it seems unlikely that such phase changes will occur abruptly (Bagnall and Wolfe, 1978; Martin, 1986), since the lipid components have a range of freezing points, but rather that localized zones of ordered structure develop as temperatures decline.

Roots display a considerable ability to acclimate to suboptimal temperatures. With rye or barley acclimated by previous exposure to 8 °C, net transport of K^+, Ca^{2+} and $H_2PO_4^-$ (per unit weight of root) into the xylem sap of detached roots was enhanced by factors of 2 to 3 compared with controls maintained at 20 °C (Clarkson, 1976; White et al., 1987), with a three-fold increase in flux of water to the xylem. These changes virtually compensated for the smaller size of the root system developed at cooler temperatures. Interestingly, low-temperature acclimated roots contained 9–15 times more soluble carbohydrates (mostly sucrose and fructose) than controls at 20 °C (Clarkson et al., 1974). The carbon efflux to the rhizosphere under such conditions, therefore, might be considerably enhanced if permeability to these carbohydrates was unaffected by temperature.

Part of the process of low-temperature acclimation in chill-sensitive and chill-insensitive plant cells involves a greater degree of desaturation of membrane lipids (Lyons, 1973; Yoshida, 1984; Williams et al., 1988). Such changes are thought to lower the temperature at which apparent phase changes begin (Lyons, 1973; Vigh et al., 1987). In roots, there was desaturation of acyl lipids during acclimation to cool temperatures (Clarkson et al., 1980; Osmond et al., 1982) with an increase in synthesis of phospholipids (Kinney et al., 1987). However, when cooled roots were warmed to 20 °C, ion transport returned in a few hours to the same rate as in controls kept at 20 °C throughout, while the turnover of acyl lipids in biomembranes was much slower (Clarkson et al., 1980). Such observations on de-acclimation raise doubts about the essentiality of membrane composition changes to all the phenomena that are found during adjustment to low temperature.

Freezing injury occurs when ice crystals penetrate the plasma membrane (and possibly subcellular organelles) and cause mechanical rupture (Brown et al., 1974). The effect is irreversible and, upon thawing, leakage of solutes and metabolites from the cytosol is concomitant with cell death. The release of carbon compounds to the microbiological population in the surrounding soil and the rhizosphere of surviving roots must be considerable. Whether ice-nucleating bacteria have a role in frost damage to roots of frost-sensitive species, as they do in leaves (Lindow et al., 1982), is unknown.

Low temperature (like mechanical impedance) tends to slow root extension more than cell maturation, so that the endodermis (and hypodermis) become suberized (state II) and thickened (state III) closer to the tip, and so effectively block apoplastic (cell wall) movement of water and solutes. In roots of maize that had been acclimated to cool temperatures, suberization did not decrease water absorption by roots (Clarkson et al., 1987), perhaps because the radial pathway for water was symplastic and therefore unaffected by suberization.

The effects of suberization on transport of carbohydrates in roots and the leakage of carbon compounds from roots to the rhizosphere have received scant attention. There is evidence that in *Pisum sativum* and maize, sucrose is transferred from the phloem to the surrounding xylem parenchyma cells (living cells that border the xylem) through the symplast (Dick and ap Rees, 1975; Giaquinta et al., 1983). Although the apoplastic pathway is blocked by suberin at the endodermis and hypodermis, carbon could still move outwards in the symplasm, but this would take place against a current of water and ions travelling in the opposite direction towards the xylem. This structure–function dilemma is badly in need of resolution.

In the surface layers of the soil, insolation can cause temperatures to rise to 50–60 °C even in temperate climates, a range that would be lethal for the cells of the leaf. The thermal response of mature root zones (in contrast to apical meristems) has received little attention, although it is probable that plasma membrane leakage of ions and metabolites will occur at temperatures that are close to the lethal level. The ability of mature root zones to function in transport processes in the phloem and xylem is probably much less sensitive to heat damage than are processes involving growth at the apical meristem: cell division ceases at 35 °C in barley and 40 °C in onion (Barlow, 1987).

High-temperature injury is associated closely with membrane disorder arising from excess fluidity, and its effect on the properties of membrane-associated proteins (Bjorkman et al., 1980). Additionally, at lethal temperatures, proteins denature with loss of enzyme activity (Raison et al., 1980). Plants acclimate to these conditions by increased saturation of membrane lipids (making membranes less fluid) and with changes in the thermal stability of enzymes (Bjorkman et al., 1980). The induction of the heat shock proteins (HSP) in roots and shoots is an area of current interest because of the high degree of conservation of the structural genes of some HSP between bacteria, animals and plants (Sachs and Ho, 1986), implying a number of common mechanisms of heat tolerance (and regulation of gene induction). Furthermore, some HSP are inducted by water deficit or salt stress (Heikkila et al., 1984), although the precise role of the HSP at the cellular level is unknown.

2.4. Water Deficit

In uniformly wet but well-drained soil, the greatest density (in cm root per cm^3 soil) is in the upper 20 cm (Welbank et al., 1974). Rooting density in a mature root system often decreases almost exponentially with depth under favourable soil conditions, but that pattern is almost always disrupted by changes in the distribution of water in the soil profile. Surface desiccation enhances rooting depth, while shallow, frequent irrigations encourage proliferation of roots at the soil surface, with inhibition of deeper rooting, which increases susceptibility to drought.

When roots extend into dry soil, the apical zone may develop a water deficit and lose turgor, so that cell expansion rates are slowed (Sharp and Davies, 1979; Westgate and Boyer, 1985). Osmotic adjustment takes place over a period of several days during acclimation to water deficits that are gradually imposed, so that the root and shoot can maintain a low (negative) water potential (ψ_w), and thereby maintain the flow of water from the rhizosphere to the root without simultaneous loss of turgor. However, roots seem less sensitive to water deficit than shoots at the same ψ_w; although both shoots and roots osmotically adjust to about the same extent, the roots can continue to extend if the medium does not become mechanically more resistant (Westgate and Boyer, 1985). This difference in response to water deficit may result from the viscoelastic properties of the cell walls, with less resistance to extension in root cells subjected to water deficit. In the field, soils generally develop a markedly greater resistance to deformation at low water (matric) potentials so that root extension is inhibited by mechanical resistance (Taylor and Ratliff, 1969; Mirreh and Ketcheson, 1972, 1973; Hamblin, 1985; Almaras et al., 1988).

From the viewpoint of modelling the rhizosphere it would be advantageous to define a critical ψw for the root or soil at which root growth ceases, or at which fluxes of solutes into or out of the rhizosphere are inhibited. At present, this is not possible with any precision because of the different water requirements and ψw values of growing and mature zones (Westgate and Boyer, 1985), and the different responses encountered among different plant species during acclimation of roots to desiccation and rehydration (Drew, 1987). Growing cells in the root tip zone readily lose water to a drier environment, and in maize irreversible damage to cells occurs when water loss exceeds 70%. With less severe dehydration, disorganization of plasma membranes, plastids, mitochondria and nuclei takes place, but these structures can reassemble when water is again available (Nir et al., 1969) and viability is maintained.

The water potential of root cells at which ability to transport ions and water ceases has received little study, despite the fluctuations in soil drying that occur in nature. Similarly, the effect of wetting–drying cycles on leakage of carbon compounds from roots to the rhizosphere has received little direct measurement, although desiccation is known to change physical and chemical properties of plant cell membranes (Seneratna et al., 1987). In mature zones of the root of barley, desiccation caused degeneration of the epidermis and outer cortex (Shone and Flood, 1983) implying a release of appreciable amounts of carbon compounds to the rhizosphere. There was also prominent thickening of the walls of the endodermis and stele. Following rehydration, transport of ^{32}P-phosphate resumed despite the greatly modified structure (Shone and Flood, 1983).

In roots that undergo hypodermal suberization, the properties of the suberin may depend on the physical environment during its deposition. In maize, suberin laid down in well-moistened roots offered no barrier to ion transport or water flow (Clarkson, 1984; Clarkson et al., 1987). This contrasts with suberin deposited in root zones exposed to air, which were highly impermeable (Ferguson and Clarkson, 1976; Clarkson, 1984). Suberization of the hypodermis, like the periderm on woody roots, may thus sometimes retard root desiccation in a drying soil, but allow subsequent inward movement of water when rehydrated. Such behaviour accounts for the ability of superficial roots to rapidly absorb water from light rainfall after extended drought. In some species of pine, suberin extends from the endodermis or hypodermis over the root tip when it has ceased growth in drying soil. Such "metacutinized" roots remain dormant for months, and quickly resume extension on rehydration (Leshem, 1970), which implies that water can readily penetrate such roots.

Some plants are able to transfer water internally through the root system at night from deeper or adjacent wet zones and release it to superficial dry zones, a process that has been called "hydraulic lift". Water movement within roots is more rapid than through the soil when it is dry. This process is known to occur in barley, in Bermuda grass (*Cynodon dactylon*) and in sagebrush (*Artemisia tridentata*) (Shone and Flood, 1980; Van Bavel and Baker, 1985; Richards and Caldwell, 1987) as well as in some forest tree species. The advantage to the plant may be in solubilization of soil nutrients in the topsoil, so that ions can be taken up by roots, and in sustaining mycorrhizas and free-living rhizosphere organisms that would otherwise become desiccated during extended periods of drought.

Recent evidence indicates that root tip cells synthesize ABA (and possibly some other unidentified solutes) when they desiccate and lose turgor (Cornish and Zeevaart, 1985; Robertson et al., 1985; Zhang et al., 1987; Zhang and Davies, 1988; Munns and King, 1988). The ABA moves in the transpiration stream to the leaves and can hasten stomatal closure before the leaves suffer water deficits. Thus the slowing of root growth in drying soil may provide plants with an "early warning system" of impending water deficit stresses. Apart from ABA's role in stomatal closure, mounting evidence implicates ABA as a regulator of gene expression in water-deficient plants (Bensen et al., 1988). New transcripts and proteins are produced in plant cells during water deficit or by elevated ABA (Heikkila et al., 1984; Bozarth et al., 1987; Vartanian et al., 1987; Bensen et al., 1988; Guerrero and Mullet, 1988). Some of these are likely to be involved with the synthesis of osmolites in osmoregulation as well as changes in the composition of the cell wall that affect its extensibility.

2.5. Oxygen Deficiency

In well-drained soil, roots obtain oxygen directly from their surroundings. When the gas-filled porous system in the soil becomes partially or completely blocked by water, and temperatures are warm (>20 °C), oxygen consumption by roots and soil microorganisms is more rapid than the rate of replenishment from the atmosphere, and dissolved oxygen concentrations in the soil water and in roots decline (Russell, 1973). Dissolved oxygen can disappear in as little as 24 h from water-filled peds or aggregates when soil temperatures are warm (Currie, 1961; Russell, 1973) even though the major pores between peds are air-filled. Under these conditions, roots and soil aerobes are in direct competition for oxygen—microorganisms account for about half the O_2 consumption of soil with densely rooted arable crops (Russell, 1973). When O_2 has disappeared within peds, a combination of chemical and biological transformations (Gambrell and Patrick, 1978) result in release to the soil solution of a sequence of reduced substances—NO_2^-, then Mn^{2+}, Fe^{2+}, H_2S—as well as microbiological end-products such as acetic and butyric acid (see also Chapter 7). Some of these can reach concentrations in the rhizosphere that are injurious to roots, or phytotoxic if accumulated in leaves (Drew and Lynch, 1980), the rate of production depending on the supplies of organic substrates for microbiological activity, and on the chemistry of the soil (presence of Fe^{3+}, Mn^{4+} and SO_4^{2-} for reduction).

Lack of oxygen in roots is alone sufficient to induce injury (Trought and Drew, 1980). Cells, beginning in the root interior, quickly experience a decline in aerobic respiration if O_2 is not replenished (Saglio et al., 1984). Ethanolic fermentation is detectable (Berry and Norris, 1949; Bertani and Brambilla, 1982), and energy status (estimated by various criteria: ATP concentration, adenylate energy charge [AEC] and the free energy of hydrolysis of ATP) declines (Saglio et al., 1980; Roberts et al., 1984; Saglio et al., 1988). Maintenance of plasma membrane structure depends on energy status, and decline of the latter causes leakage of organic and inorganic solutes to the medium (Hiatt and Low, 1967), inhibition of ion uptake and transport to shoots (Cheeseman and Hanson, 1979; Trought and Drew, 1980) and low conductivity to water (Everard and Drew, 1987, 1989). Cell death in unacclimated cells of maize takes place after about 20 h of anoxia (Roberts et al., 1984) through a gradual leakage of H^+ from the vacuole (pH 5.8) to the cytoplasm (pH 7.4), presumably because of inhibition of tonoplast H^+-ATPases that normally "prime" the vacuole with H^+. By contrast, cells gradually exposed to oxygen shortage acclimate to anoxia and this extends viability greatly (Saglio et al., 1988). Acclimation is associated with induction of about 20 proteins (Sachs et al., 1980), including alcohol dehydrogenase and pyruvate decarboxylase (essential for fermentation) and some of the enzymes of the glycolytic pathway. Ethanolic fermentation gives a net yield of ATP, and ethanol (being membrane permeable) leaks readily to the surrounding medium (Bertani et al., 1980).

The critical oxygen pressure (COP) in the rhizosphere, below which oxygen consumption by roots is first inhibited, depends on temperature and respiration rate (Berry and Norris, 1949). For maize root tips at 25 °C, the COP is close to that in ambient air (21% oxygen) (Saglio et al., 1984), so that small restrictions to oxygen diffusion through the soil to root surfaces result in the beginning of an "anaerobic core".

When the soil is oxygen deficient, diffusion and convection of oxygen occur through the intercellular spaces between cells. These spaces form a continuous porous system to the leaves or, in woody plants, to stem and root lenticels that open to the air (Armstrong, 1979). Resistance to oxygen movement along this pathway is large, and flux is sufficient to provide O_2 for aerobic respiration for roots of maize only over a length of about 100 mm (Saglio et al., 1983). Many crop and native species develop aerenchymatous roots in poorly aerated soil (Justin and Armstrong, 1987) in which lysis or separation of cells gives relatively large cavities (lacunae) that interconnect to leaves or other organs in contact with air (Fig. 1). The smaller resistance to diffusion of O_2 allows such roots to extend into anaerobic media, and the internal oxygen status is improved (Drew et al., 1985). The formation of aerenchyma in roots appears to be hormonally controlled. In maize roots, ethylene biosynthesis is stimulated by hypoxia (Drew et al., 1979; Jackson et al., 1985), and the additional ethylene causes lysis of cells in the mid cortex to give a lysigenous aerenchyma. Soil microorganisms also produce ethylene under hypoxia or anoxia (Drew and Lynch, 1980), and this may contribute further to aerenchyma formation. However, N or P deficiency can also cause aerenchyma to form in maize under well-aerated conditions without enhanced ethylene production. Here, sensitivity to the usual low concentrations of this gaseous hormone that are present in root tissue is enhanced (Drew et al., 1989). Maize roots, like those of other dryland species, are not especially adapted to conserve O_2 moving within their lacunae, and roots may not receive sufficient O_2 at the root apex over distances of 250 mm from the leaves (Drew et al., 1985). By contrast, wetland species conserve internal O_2, probably because of suberized and lignified layers in the hypodermis and outer cortex (Armstrong and Armstrong, 1988). These stop radial leakage outward, allowing more O_2 to reach the apical meristem. Leakage of O_2 occurs at the tip, but it serves to oxidize and therefore detoxify potentially harmful reducing substances in the rhizosphere (Armstrong, 1979; McKee et al., 1988).

2.6. Nutrient Ions and pH in the Rhizosphere

Nutrient ions move in the soil towards plant roots by both mass flow with the soil water, and by diffusion (Barber, 1962). Mass flow is caused by bulk, convective movement of soil solution towards roots, mainly in response to transpiration-driven water absorption, and ions moving this way are drawn radially towards the root over distances of several centimetres per day when soil is moist. Diffusion occurs in response to a concentration gradient for a particular ion resulting from its absorption by the root and depletion from soil immediately next to the root surface. For typical agricultural soils, mass flow supplies Ca^{2+}, Mg^{2+}, Cl^-, SO_4^{2-} and usually NO_3^-, but diffusion predominates for K^+ and $H_2PO_4^-$ because of the very low concentrations of these ions in the free soil solution (Nye and Tinker, 1977; Barber, 1984). Where mass flow brings in ions at a much faster rate than absorption by a root surface, ions accumulate in the rhizosphere (Barber, 1962; Barber and Ozanne, 1970), the concentration depending on how rapidly diffusion away from the root towards the soil takes place. Concentrations of Na^+ in saline soil can thus increase to injurious levels in the rhizosphere (Sinha and Singh, 1974), and Ca^{2+} is sometimes precipitated as carbonates to form pedotubules around roots. The relative amounts of a nutrient supplied by mass flow compared with diffusion can also change during the season, and with soil depth. For example, a high concentration of NO_3^- in the upper soil layers is supplied by mass flow initially but when it is depleted the supply to roots in deeper soil is mainly by diffusion (Renger and Strebel, 1979; Strebel et al., 1983).

Diffusion coefficients for movement of ions through soil are 10 to 10 000 times less than in free solution (Nye and Tinker, 1977), depending on the extent to which the diffusing ions react with the surfaces of soil particles along the pathway. Typical diffusion coefficients for K^+ (10^{-7} $cm^2 s^{-1}$) and phosphate (10^{-9} $cm^2 s^{-1}$) indicate that these ions would move, respectively, only 10 to 11 mm per day in the bulk soil. Roots would thus extract these ions principally from the rhizosphere. To account for measured rates of absorption by roots from soil, calculation of the ion concentration in soil at the root surface suggests that a very intense depletion would be necessary for $H_2PO_4^-$, but with relatively small depletion of K^+ (Drew et al., 1969; Drew and Nye, 1970; Bhat and Nye, 1974). Direct measurements of depletion profiles in the rhizosphere of *Brassica napus* (rape) were made by quantitative autoradiography, the labile soil phosphate having been previously labelled with ^{33}P (Bhat and Nye, 1973). Unexpectedly, these revealed a much wider zone of depletion than predicted from diffusion theory, whether or not the root hairs were assumed to be functional in P uptake (Fig. 2). Clearly, the presence of the root had modified the mobility of $H_2PO_4^-$ in the soil. Further study (Grinstead et al., 1982; Hedley et al., 1982) indicated that a major factor in the enhanced mobility of P in the rhizosphere of rape was a shift to a lower rhizosphere pH at the time that plant cation absorption was in excess of anion absorption. This pH change had the effect of raising the concentration of $H_2PO_4^-$ in the soil water and thereby raising its diffusion coefficient. Although these results have not yet been confirmed in comparable studies with other species, the known ability of roots

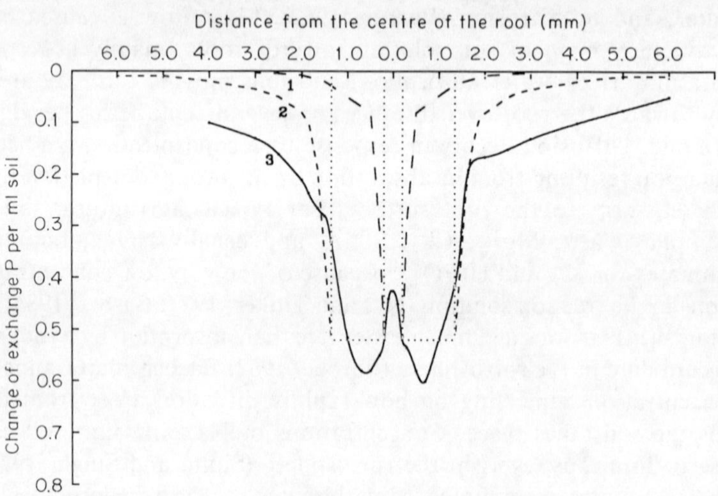

Fig. 2. Quantitative depletion of soil phosphate around root of *Brassica napus* (rape) seedling. Curve 1: calculated for P diffusion to root surface assuming root hairs inactive. Curve 2: calculated for P diffusion to surface of an imaginary cylinder, represented by the tips of the root hairs: this implies intense depletion of P from within the soil volume occupied by root hairs. Curve 3: experimentally measured change in phosphate concentration, from autoradiography of ^{33}P-labelled soil phosphate. Fine vertical lines indicate limits of root surface and root hair cylinder to which phosphate diffuses in curves 1 and 2, respectively. (Redrawn from Bhat and Nye, 1973.)

to change soil pH (see below) makes it probable as a generalized mechanism for acquisition of P from the soil.

Pronounced changes in pH in the rhizosphere, by as much as one or two units, occur under agricultural conditions because of the major influence of the form of nitrogen on cation/anion balance of plants. With NH_4^+-based fertilizer there is a marked solubilization of phosphate in calcareous soil when banded with NH_4^+ because of a net release of H^+ by roots (Riley and Barber, 1971; Soon and Miller, 1977). On the other hand, NO_3^- in fertilizer leads to more alkaline conditions (efflux of OH^-) and tends to release phosphate from the chemical forms in which it is held in an acid soil. The rhizosphere of legumes fixing N_2 also becomes markedly more acid. A model for calculation of rhizosphere pH changes in plants supplied with NH_4^+-N or NO_3^--N has been proposed by Nye (1981). Such changes in pH in the rhizosphere have been clearly demonstrated with pH-sensitive dyes with a range of monocot and dicot crop species (Marschner and Römheld, 1983). These techniques show surprisingly large differences in rhizosphere pH between species growing in the same soil, and pH differences exceeding 2 pH units at different points along the root of an individual plant (Marschner and Römheld, 1983; Römheld and Marschner, 1983). In general, dicots tend to take up excess cations compared with anions, so that acidification of the rhizosphere is usually greater than for monocots, which show a cation/anion ratio close to unity.

Iron and manganese nutrition of plants is especially sensitive to changes in the chemistry of the soil in the rhizosphere. Under conditions of iron deficiency, non-specific acidification of the rhizosphere (through cation/anion balance) helps raise the solubility of Fe^{3+} which is mainly controlled by the solubility of the Fe (III) hydrous oxides. At pH 6, the solubility of Fe^{3+} is only 10^{-15} M, but at pH 4 it rises to 10^{-9} M. Some plants specifically respond to iron deficiency by rhizosphere acidification, as part of an apparent overall strategy to mobilize iron (reviewed by Marschner, 1986; see also Chapter 7). Iron deficiency-induced acidification is a typical response of many dicots (strategy I), which also form transfer-cell-like structures at the root epidermis thereby increasing the plasma membrane surface in contact with the free space solution. These plants also release phenolics to chelate Fe^{3+} and raise its concentration in the rhizosphere, and are able to reduce Fe^{3+} to Fe^{2+} at the plasma membrane. Reduction of Mn^{4+} is simultaneously stimulated in response to iron deficiency. By contrast, most monocots (strategy II) rely heavily on an ability to release to the rhizosphere phytosiderophores—non-proteinogenic amino acids like avenic acid (chemically related to nicotinamide) that complex Fe (III) and can even mobilize Fe from Fe O OH at high pH (Fushiya et al., 1982). Fe (III) is transported into root cells as the chelate and Fe^{3+} released for translocation within the xylem as ferric citrate.

Roots modify the concentration of ions in the rhizosphere through their absorptive activities, but at the same time the concentrations around the roots regulate root morphology and ion transport characteristics. In barley, low concentrations of NO_3^- or $H_2PO_4^-$ depress initiation of lateral root primordia so that there is little branching (Drew et al., 1973; Drew and Saker, 1978). Zones along the root that develop in contact with high concentrations, by contrast, show intense proliferation of 1st and 2nd order lateral roots (in terms of both number per centimetre of the main axis, and extension rate). This results in intense root growth (carbon allocation) and uptake of nutrients from locally enriched zones in the soil (Drew and Saker, 1975, 1978). Extension of root hairs from the root epidermis is stimulated by phosphate deficiency (Foehse and Jungk, 1983). This response is in keeping with the proposed role of root hairs in enhancing the capture by the root of ions that slowly diffuse in the soil as discussed above.

The rate at which root surfaces remove nutrient ions from the rhizosphere depends on the growth rate of the plant (hence the physical environment of shoot and root) and also on the previous nutritional history of the plant. Brief periods of nutrient starvation enhance transport of K^+, $H_2PO_4^-$ or NO_3^- when these are again accessible to roots (Drew et al., 1984; Drew and Saker, 1975, 1978, 1984; Lee and Drew, 1986), implying a more intense depletion from the rhizosphere of relatively immobile ions. It thus seems that the geometry of the absorbing root system as well as the mechanism of ion transport to the root xylem are responsive to the nutritional status of the soil and plant. Such relations imply mutual regulation of shoot and root activities, but the mechanisms of coordination are poorly understood at present.

3. CONCLUSIONS

This chapter has examined separately the influence of each factor in the root environment. In practice, several of these factors interact in a complex manner; mechanical impedance, soil aeration and soil water content are almost always dependent on each other. Under wet conditions, soils are more easily deformed by farming operations, and particles or aggregates are compressed, with elimination of large pores (Hamblin, 1985; Almaras et al., 1988). Such compacted soil becomes mechanically resistant to root growth when it dries (Mirreh and Ketcheson, 1973). Furthermore, its "plant-available" water content is smaller and plants suffer water deficits sooner. Conversely, under wet conditions, small soil pores fill with water and the compacted soil becomes prone to waterlogging and aeration problems (Hopkins and Patrick, 1969). Water deficit is often associated with high temperatures, especially for leaves and superficial roots. Freezing tolerance in some tissues involves slow dehydration, and is thus similar in part to mechanisms of tolerance to water deficit-induced desiccation. Furthermore, at the molecular level the same stress proteins (and the mRNAs that encode them) are induced by very different environmental factors, such as heat shock, wounding or water deficit (Heikkila et al., 1984). This raises the possibility of cross protection against one stress by another. Plant hormones such as ABA and ethylene are especially interesting because they may act as messengers to alter gene expression, in response to a range of environmental factors.

Progress in understanding the precise responses of plant roots to environmental conditions necessitates experiments being conducted in artificial, controlled conditions. In few cases do these conditions simulate well the soil environment, so that important interactions between plant roots, their physicochemical environment and the rhizosphere are not usually reproduced. Although there is qualitative information on the growth and structure of roots, and how their release of carbon to the rhizosphere is dependent on developmental (genetic) and environmental factors, quantification of root-induced changes in the rhizosphere is poorly documented. In part this arises because gnotobiotic techniques are tedious to apply with whole plants; but their use is essential if we are to understand better the extent to which controlled manipulation of the root environment can be used as a tool in the regulation of plant performance.

REFERENCES

Almaras, R. R., J. M. Kraft, and P. E. Miller (1988). Effects of soil compaction and incorporated crop residues on root health. *Ann. Rev. Phytopathol.*, **26**:219–243.

Andrews, R. E. and E. I. Newman (1970). Root density and competition for nutrients. *Oecologia Plant.*, **5**:320–334.

Armstrong, J. and W. Armstrong (1988). *Phragmites australis*—a preliminary study of soil-oxidising sites and internal gas transport pathways. *New Phytol.*, **108**:373–382.

Armstrong, W. (1979). Aeration in higher plants. *Adv. Botan. Res.*, **7**:225–331.

Bagnall, D. J. and J. A. Wolfe (1978). Chilling sensitivity in plants: do the activation energies of growth processes show an abrupt change at a critical temperature? *J. Exp. Bot.*, **29**:1231–1242.

Barber, D. A. and K. B. Gunn (1974). The effect of mechanical forces on the exudation of organic substances by the roots of cereal plants grown under sterile conditions. *New Phytol.*, **73**:39–45.

Barber, S. A. (1962). A diffusion and massflow concept of soil nutrient availability. *Soil Sci.*, **93**:39–49.

Barber, S. A. (1984). *Soil Nutrient Bioavailability: A Mechanistic Approach*. John Wiley, New York.

Barber, S. A. and P. G. Ozanne (1970). Autoradiographic evidence for the differential effect of four plant species in altering the calcium content of the rhizosphere soil. *Soil Sci. Soc. Ann. Proc.*, **34**:635–637.

Barley, K. P. and E. L. Graecen (1967). Mechanical resistance as a soil factor influencing the growth of roots and underground shoots. *Adv. Agron.*, **19**:1–43.

Barlow, P. W. (1987). The cellular organization of roots and its response to the physical environment. In *Root Development and Function* (P. J. Gregory, J. V. Lake, and D. A. Rose, eds.). Cambridge University Press, Cambridge, pp. 1–26.

Bensen, R. J., J. S. Boyer, and J. H. Mullet (1988). Water deficit-induced changes in abscisic acid, growth, polysomes, and translatable RNA in soybean hypocotyls. *Plant Physiol.*, **88**:289–294.

Berry, L. J. and W. E. Norris (1949). Studies of onion root respiration. I. Velocity of oxygen consumption in different segments of root at different temperatures as a function of partial pressure of oxygen. *Biochem. Biophys. Acta*, **3**:593–606.

Bertani, A. and I. Brambilla (1982). Effects of decreasing oxygen concentrations on wheat roots: growth and induction of anaerobic metabolism. *Z. Pflanzenphysiol.*, **108**:283–288.

Bertani, A., I. Brambilla, and F. Menegus (1980). Effect of anaerobiosis on rice seedlings: growth, metabolic rate, and fate of fermentation products. *J. Exp. Bot.*, **31**:325–331.

Bhat, K. K. S. and P. H. Nye (1973). Diffusion of phosphate to plant roots in soil. I. Quantitative autoradiography of the depletion zone. *Plant Soil*, **38**:161–175.

Bhat, K. K. S. and P. H. Nye (1974). Diffusion of phosphate to plant roots in soil. III. Depletion around onion roots without root hairs. *Plant Soil*, **41**:383–394.

Bjorkman, O., M. R. Badger, and P. A. Armond (1980). Response and adaptation of photosynthesis to high temperatures. In *Adaptation of Plants to Water and High Temperature Stress* (N. C. Turner and P. J. Kramer, eds.). John Wiley, New York, pp. 233–249.

Bozarth, C. S., J. E. Mullet, and J. S. Boyer (1987). Cell wall proteins at low water potentials. *Plant Physiol.*, **85**:261–267.

Brown, M. S., E. S. B. Pereira, and B. J. Finkle (1974). Freezing of nonwoody plant tissues. 2. Cell damage and the fine structure of freezing curves. *Plant Physiol.*, **53**:709–711.

Buckley, R. (1982). Sand rhizosheath of an arid zone grass. *Plant Soil*, **66**:417–421.

Caldwell, M. M. and L. B. Camp (1974). Below ground productivity of two cool desert communities. *Oecologia*, **17**:123–130.

Cassab, G. I., J. J. Lin, L. S. Lin, and J. E. Varner (1988). Ethylene effect on extension and peroxidase distribution in the subapical region of pea epicotyls. *Plant Physiol.*, **88**:522–524.

Cheeseman, J. M. and J. B. Hanson (1979). Energy-linked potassium influx as related to cell potential in corn roots. *Plant Physiol.*, **64**:842–845.

Clarkson, D. T. (1974). *Ion Transport and Cell Structure in Plants*. McGraw Hill, London.

Clarkson, D. T. (1976). The influence of temperature on the exudation of xylem sap from detached root systems of rye (*Secale cereale*) and barley (*Hordeum vulgare*). *Planta*, **132**:297–304.

Clarkson, D. T. (1984). Calcium transport between tissues and its distribution in the plant. *Plant, Cell and Environ.*, **7**:449–456.

Clarkson, D. T., M. G. T. Shone, and A. V. Wood (1974). The effect of pretreatment temperature on the exudation of xylem sap by detached barley root systems. *Planta*, **121**:81–92.

Clarkson, D. T., K. C. Hall, and J. K. M. Roberts (1980). Phospholipid composition and fatty acid desaturation in the roots of rye during acclimation of low temperature. Positional analysis of fatty acids. *Planta*, **149**:464–471.

Clarkson, D. T., A. W. Robards, J. E. Stephens, and M. Stark (1987). Suberin lamellae in the hypodermis of maize (*Zea mays*) roots; development and factors affecting the permeability of hypodermal layers. *Plant, Cell and Environ*, **10**:83–93.

Cornish, K. and J. A. D. Zeevaart (1985). Abscisic acid accumulation by roots of *Xanthium strumarium* L. and *Lycopersicon esculentum* Mill. in relation to water stress. *Plant Physiol.*, **79**:653–658.

Currie, J. A. (1961). Gaseous diffusion in the aeration of aggregated soils. *Soil. Sci.*, **92**:40–45.

Dick, P. S. and T. ap Rees (1975). The pathway of sugar transport in roots of *Pisum sativum*. *J. Exp. Bot.*, **26**:305–314.

Drew, M. C. (1987). Function of root tissues in nutrient and water transport. In *Root Development and Function* (P. J. Gregory, J. V. Lake, and D. A. Rose, eds.). Cambridge University Press, Cambridge, pp. 71–101.

Drew, M. C. and J. M. Lynch (1980). Soil anaerobiosis, micro-organisms and root function. *Ann. Rev. Phytopathol.*, **18**:37–66.

Drew, M. C. and P. H. Nye (1969). The supply of nutrient ions by diffusion to plant roots in soil. II. The effect of root hairs on the uptake of potassium by roots of ryegrass (*Lolium multiflorum*). *Plant Soil*, **31**:407–424.

Drew, M. C. and P. H. Nye (1970). The supply of nutrient ions by diffusion to plant roots in soil. III. Uptake of phosphate by roots of onion, leek and ryegrass. *Plant Soil*, **33**:545–563.

Drew, M. C. and L. R. Saker (1975). Localized, compensatory increases in lateral root growth and rates of nitrate uptake when nitrate supply is restricted to only part of the root systems. *J. Exp. Bot.*, **26**:79–90.

Drew, M. C. and L. R. Saker (1978). Nutrient supply and the growth of the seminal root system in barley. III. Compensatory increases in growth of lateral roots and in rates of phosphate uptake in response to a localized supply of phosphate. *J. Exp. Bot.*, **29**:435–451.

Drew, M. C. and L. R. Saker (1984). Uptake and long distance transport of phosphate, potassium and chloride in relation to internal ion concentrations in barley: evidence of non-allosteric regulation. *Planta*, **160**:500–507.

Drew, M. C., P. H. Nye, and L. V. Vaidyanathan (1969). The supply of nutrient ions by diffusion to plant roots in soil. I. Absorption of potassium by cylindrical roots of onion and leek. *Plant Soil*, **30**:252–270.

Drew, M. C., L. R. Saker, and T. W. Ashley (1973). Nutrient supply and the growth of the seminal root system in barley. I. The effect of nitrate concentration on the growth of axes and laterals. *J. Exp. Bot.*, **24**:1189–1202.

Drew, M. C., M. B. Jackson, and S. Giffard (1979). Ethylene-promoted adventitious rooting and development of cortical air spaces (aerenchyma) in roots may be adaptive responses to flooding in *Zea mays* L. *Planta*, **147**:83–88.

Drew, M. C., L. R. Saker, S. A. Barber, and W. Jenkins (1984). Changes in the kinetics of phosphate and potassium absorption in nutrient-deficient barley roots measured by a depletion technique. *Planta*, **160**:490–499.

Drew, M. C., P. H. Saglio, and A. Pradet (1985). Higher adenylate energy charge and ATP/ADP ratios in aerenchymatous roots of *Zea mays* in anaerobic media as a consequence of improved internal oxygen transport. *Planta*, **165**:51–58.

Drew, M. C., C. J. He, and P. W. Morgan (1989). Decreased ethylene biosynthesis and induction of aerenchyma by nitrogen- or phosphate-starvation in adventitious roots of *Zea mays* L. *Plant Physiol.*, **91**:266–271.

Esau, K. (1965). *Plant Anatomy*, 2nd edn. John Wiley, New York.

Everard, J. D. and M. C. Drew (1987). Mechanisms of inhibition of water movement in anaerobically treated roots of *Zea mays* L. *J. Exp. Bot.*, **38**:1154–1165.

Everard, J. D. and M. C. Drew (1989). Mechanisms controlling changes in water movement through the roots of *Helianthus annuus* during continuous exposure to oxygen deficiency. *J. Exp. Bot.*, **40**:1–10.

Ferguson, I. B. and D. T. Clarkson (1976). Ion uptake in relation to the development of a root hypodermis. *New Phytol.*, **77**:11–14.

Foehse, D. and A. Jungk (1983). Influence of phosphate and nitrate supply on root hair formation of rape, spinach and tomato plants. *Plant Soil*, **74**:359–368.

Fushiya, S., K. Takahashi, S. Nakatsuyama, Y. Sato, S. Nozoe, and S. J. Takagi (1982). Co-occurrence of nictoinamine and avenic acids in *Avena sativa* and *Oryza sativa*. *Phytochem.*, **21**:1907–1908.

Gambrell, R. P. and W. H. Patrick (1978). Chemical and microbiological properties of anaerobic soils and sediments. In *Plant Life in Anaerobic Environments* (D. D. Hook and R. M. M. Crawford, eds.). Ann Arbor Science, Ann Arbor, Michigan, pp. 375–423.

Giaquinta, R. T., W. Lin, N. Sadler, and V. R. Franceschi (1983). Pathway of phloem unloading of sucrose in corn roots. *Plant Physiol.*, **72**:362–367.

Goodchild, D. J. and L. F. Myers (1987). Rhizosheaths—a neglected phenomenon in Australian agriculture. *Aust. J. Agric. Res.*, **38**:559–563.

Goss, M. J. (1977). Effects of mechanical impedance on root growth in barley (*Hordeum vulgare* L.). 1. Effects on elongation and branching of seminal roots. *J. Exp. Bot.*, **28**:96–111.

Goss, M. J. and R. S. Russell (1980). Effects of mechanical impedance on root growth in barley (*Hordeum vulgare* L.). *J. Exp. Bot.*, **31**:577–588.

Graecen, E. L., D. A. Farrell, and B. Cockroft (1968). Soil resistance to metal probes and plant roots. *Trans. 9th Int. Congr. Soil Sci.*, **1**:769–779.

Grinstead, M. J., M. J. Headley, R. E. White, and P. H. Nye (1982). Plant induced changes in the rhizosphere of rape (*Brassica napus* var. Emerald) seedlings. 1. pH change and the increase in P concentration in the soil solution. *New Phytol.*, **91**:19–29.

Guerrero, F. D. and J. E. Mullet (1988). Reduction of turgor induces rapid changes in leaf translatable RNA. *Plant Physiol.*, **88**:401–408.

Hackett, C. (1968). A study of the root system of barley. 1. Effects of nutrition on two varieties. *New Phytol.*, **67**:287–300.

Hamblin, A. P. (1985). The influence of soil structure on water movement, crop root growth, and water uptake. *Adv. Agron.*, **38**:95–158.

Head, G. C. (1973). Shedding of roots. In *Shedding of Plant Parts* (T. T. Kozlowski, ed.). Academic Press, New York, pp. 237–293.

Hedley, M. J., P. H. Nye, and R. E. White (1982). Plant-induced changes in the rhizosphere of rape (*Brassica napus* var. Emerald) seedlings. II. Origin of the pH change. *New Phytol.*, **91**:31–44.

Heikkila, J. J., J. E. T. Papp, G. A. Schultz, and J. D. Bewley (1984). Induction of heat shock protein messenger RNA in maize mesocotyls by water stress, abscisic acid, and wounding. *Plant Physiol.*, **76**:270–274.

Herold, A. and P. H. McNeil (1979). Restoration of photosynthesis in pot-bound tobacco plants. *J. Exp. Bot.*, **30**:1187–1194.

Hiatt, A. J. and R. H. Low (1967). Loss of organic acids, amino acids, K and Cl from barley roots treated anaerobically and with metabolic inhibitors. *Plant Physiol.*, **42**:1731–1736.

Hopkins, R. M. and W. H. Patrick (1969). Combined effect of oxygen content and soil compaction on root penetration. *Soil Sci.*, **108**:408–413.

Jackson, M. B., T. M. Fenning, M. C. Drew, and L. R. Saker (1985). Stimulation of ethylene production and gas space (aerenchyma) formation in adventitious roots of *Zea mays* L. by small partial pressures of oxygen. *Planta*, **165**:486–492.

Justin, S. H. F. and W. Armstrong (1987). The anatomical characteristics of roots and plant response to soil flooding. *New Phytol.*, **106**:465–495.

Kays, S. J., C. W. Nicklow, and D. H. Simons (1974). Ethylene in relation to the response of roots to physical impedance. *Plant Soil*, **40**:565–571.

Kinney, A. J., D. T. Clarkson, and B. C. Loughman (1987). Phospholipid metabolism and plasma membrane morphology of warm and cool rye roots. *Plant Physiol. Biochem.*, **25**:769–774.

Kolattukudy, P. E. (1980). Cutin, suberin and waxes. In *The Biochemistry of Plants: a Comprehensive Treatise, Vol. 4. Lipids: Structure and Function* (P. K. Stumpf, ed.). Academic Press, New York, pp. 571–645.

Kramer, P. J. (1983). *Water Relations of Plants*. Academic Press, Orlando, 489 pp.

Kramer, P. J. and H. C. Bullock (1966). Seasonal variations in the proportions of suberized and unsuberized roots of trees in relation to the absorption of water. *Am. J. Bot.*, **53**:200–204.

Lee, R. B. and M. C. Drew (1986). Nitrogen-13 studies of nitrate fluxes in barley roots: 2. Effect of plant N-status on the kinetic parameters of nitrate influx. *J. Exp. Bot.*, **37**:1768–1779.

Leshem, B. (1970). Resting roots of *Pinus halapensis*: structure, function, and reaction to water stress. *Bot. Gaz.*, **131**:99–104.

Levitt, J. (1980). *Responses of Plants to Environmental Stresses, Vol. 1*, 2nd edn. Academic Press, Orlando, 497 pp.

Lindberg, S. and S. Petterson (1985). Effects of mechanical stress on uptake and distribution of nutrients in barley. *Plant Soil*, **83**:295–309.

Lindow, S. E., D. C. Arny, and C. D. Upper (1982). Bacterial ice nucleation: a factor in frost injury to plants. *Plant Physiol.*, **70**:1084–1089.

Lyons, J. M. (1973). Chilling injury in plants. *Ann. Rev. Plant Physiol.*, **24**:445–466.

Marschner, H. (1986). *Mineral Nutrition of Higher Plants*. Academic Press, London, 674 pp.

Marschner, H. and V. Römheld (1983). In-vivo measurement of root-induced pH changes at the soil-root interface: effect of plant species and nitrogen source. *Z. Pflanzenphysiol*, **111**:241–251.

Martin, B. (1986). Arrhenius plots and the involvement of thermotropic phase transitions of the thylakoid membrane in chilling impairment of photosynthesis in thermophilic higher plants. *Plant, Cell and Environ.*, **9**:323–331.

Masle, J. and G. D. Farquar (1988). Effects of soil strength on the relation of water-use efficiency and growth to carbon isotope discrimination in wheat seedlings. *Plant Physiol.*, **86**:32–38.

Masle, J. and J. B. Passioura (1987). The effect of soil strength on the growth of young wheat plants. *Aust. J. Plant Physiol.*, **14**:643–656.

McCully, M. E. (1987). Selected aspects of the structure and development of field-grown roots with special reference to maize. In *Root Development and Function* (P. J. Gregory, J. V. Lake, and D. A. Rose, eds.). Cambridge University Press, Cambridge, pp. 53–70.

McKee, K. L., I. A. Mendelssohn, and M. W. Hester (1988). Reexamination of pore water sulfide concentrations and redox potentials near the aerial roots of *Rhizophora mangle* and *Avicennia germinalis*. *Amer. J. Bot.*, **75**:1352–1359.

Mirreh, H. F. and J. W. Ketcheson (1972). Influence of soil bulk density and matrix pressure on soil resistance to penetration. *Can. J. Soil Sci.*, **52**:477–483.

Mirreh, H. F. and J. W. Ketcheson (1973). Influence of soil water matric potential and resistance to penetration on corn root elongation. *Can. J. Soil Sci.*, **53**:383–388.

Moss, G. I., K. C. Hall, and M. B. Jackson (1988). Ethylene and the responses of roots of maize (*Zea mays* L.) to physical impedance. *J. Exp. Bot.*, **39**:303–311.

Munns, R. and R. W. King (1988). Abscisic acid is not the only stomatal inhibitor in the transpiration stream of wheat plants. *Plant Physiol.*, **88**:703–708.

Nichols, S. E. and G. G. Laties (1984). Ethylene-regulated gene transcription in carrot roots. *Plant Mol. Biol.*, **3**:393–401.

Nir, I., S. Klein, and A. Poljakoff-Mayber (1969). Effect of moisture stress on submicroscopic structure of maize roots. *Austr. J. Biol. Sci.*, **22**:17–33.

Nye, P. H. (1981). Changes in pH across the rhizosphere induced by roots. *Plant Soil*, **61**:7–26.

Nye, P. H. and P. B. Tinker (1977). *Solute Movement in the Soil–Root System*. Blackwell, Oxford, 342 pp.

Osmond, D. L., R. F. Wilson, and C. D. Raper (1982). Fatty acid composition and nitrate uptake of soybean roots during acclimation to low temperature. *Plant Physiol.*, **70**:1689–1693.

Pilet, P. E. and P. W. Barlow (1987). The role of abscisic acid in root growth and gravireaction: a critical review. *Plant Growth Reg.*, **6**:217–265.

Raison, J. K., J. A. Berry, P. A. Armond, and C. S. Pike (1980). Membrane properties in relation to the adaptation of plants to temperature stress. In *Adaptation of Plants to Water and High Temperature Stress* (N. C. Turner and P. J. Kramer, eds.). John Wiley, New York, pp. 261–273.

Renger, M. and O. Strebel (1979). Water and nutrient transport to plant roots as a function of depth and time under field conditions. In *Soils in Mediterranean Type Climates and their Yield Potential*. Proc. 14th Colloquium, Int. Potash Inst. Berne, Switzerland.

Richards, J. H. and M. M. Caldwell (1987). Hydraulic lift: substantial nocturnal water transport between soil layers by *Artemisia tridentata* roots. *Oecologia* (Berlin), **73**:486–489.

Riley, D. and S. A. Barber (1971). Effect of ammonium and nitrate fertilization on phosphorus uptake as related to root-induced pH changes at the root–soil interface. *Soil Sci. Soc. Am. Proc.*, **35**:301–306.

Roberts, J. K. M., J. Callis, O. Jardetsky, V. Walbot, and M. Freeling (1984). Cytoplasmic acidosis as a determinant of flooding intolerance in plants. *Proc. Natl. Acad. Sci.*, **81**:6029–6033.

Robertson, J. M., R. P. Pharis, Y. Y. Huang, D. M. Reid, and E. C. Yeung (1985). Drought-induced increases in abscisic acid levels in the root apex of sunflower. *Plant Physiol.*, **79**:1086–1089.

Roby, D., A. Toppan, and M. T. Esquerr-Tugay (1985). Cell surfaces in plant–microorganism interactions. 5. Elicitors of fungal and plant origin trigger the synthesis of ethylene and of cell wall hydroxyproline-rich glycoprotein in plants. *Plant Physiol.*, **77**:700–704.

Römheld, V. and H. Marschner (1983). Mechanism of iron uptake by peanut plants 1. Fe III reduction, chelate splitting, and release of phenolics. *Plant Physiol.*, **71**:949–954.

Russell, E. W. (1973). *Soil Conditions and Plant Growth*. 10th edn. Longman, London.

Sachs, M. M. and D. T. H. Ho (1986). Alternation of gene expression during environmental stress in plants. *Ann. Rev. Plant Physiol.*, **37**:363–376.

Sachs, M. M., M. Freeling, and R. Okimoto (1980). The anaerobic proteins of maize. *Cell*, **20**:761–767.

Saglio, P. H., P. Raymond, and A. Pradet (1980). Metabolic activity and energy charge of excised maize root tips under anoxia control by soluble sugars. *Plant Physiol.*, **66**:1053–1057.

Saglio, P. H., P. Raymond, and A. Pradet (1983). Oxygen transport and root respiration of maize seedlings. A quantitative approach using the correlation between ATP/ADP and the respiration rate controlled by oxygen tension. *Plant Physiol.*, **72**:1035–1039.

Saglio, P. H., F. Rancillac, F. Bruzan, and A. Pradet (1984). Critical oxygen for growth and respiration of excised and intact roots. *Plant Physiol.*, **76**:151–154.

Saglio, P. H., M. C. Drew, and A. Pradet (1988). Metabolic acclimation to anoxia induced by low (2–4 kPa partial pressure) oxygen pretreatment (hypoxia) in root tips of *Zea mays*. *Plant Physiol.*, **86**:61–66.

Seneratna, T., B. D. McKersie, and A. Borochov (1987). Desiccation and free radical mediated changes in plant membranes. *J. Exp. Bot.*, **38**:2005–2014.

Sharp, R. E. and W. J. Davies (1979). Solute regulation and growth by roots and shoots of water stressed maize plants. *Planta*, **147**:43–49.

Shone, M. G. T. and A. V. Flood (1980). Studies on uptake and loss of water by barley roots in relation to changes in root resistance. *J. Exp. Bot.*, **31**:1147–1159.

Shone, M. G. T. and A. V. Flood (1983). Effects of periods of localized water stress on subsequent nutrient uptake by barley roots and their adaptation by osmotic adjustment. *New Phytol.*, **94**:561–572.

Sinha, B. K. and N. T. Singh (1974). Effect of transpiration rate on salt accumulation around corn roots in a saline soil. *Agron. J.*, **66**:557–560.

Soon, Y. K. and M. H. Miller (1977). Changes in the rhizosphere due to NH_4^+ and NO_3^- fertilization and phosphorus uptake by corn seedlings (*Zea mays* L.). *Soil Sci. Soc. Am. Proc.*, **41**:77–80.

Strebel, O., W. H. M. Duynisveld, H. Grimme, M. Renger, and H. Fleige (1983). Wasserentzug durch Wurzeln und nitratenlieferung (Massenfluss, Diffusion) als funktion van Bodentiefe und zeit bein einem zukerrbenbestand. *Mitt. Dtsch. Bodenkd. Ges.*, **38**:153–158.

Taylor, H. M. and L. F. Ratliff (1969). Root elongation rates of cotton and peanuts as a function of soil strength and soil water content. *Soil Sci.*, **108**:113–119.

Trought, M. C. T. and M. C. Drew (1980). The development of waterlogging damage in young wheat plants in anaerobic solution culture. *J. Exp. Bot.*, **31**:1573–1585.

Tucker, M. L. and G. G. Laties (1984). Comparative effects of ethylene and cyanide on respiration, polysome prevalence, and gene expression in carrot roots. *Plant Physiol.*, **75**:342–348.

Van Bavel, C. H. M. and J. M. Baker (1985). Water transfer by plant roots from wet to dry soil. *Naturwissenschaften*, **72**:606–607.

Vartanian, N., C. Damerval, and D. de Vienne (1987). Drought-induced changes in protein patterns of *Brassica napus* var. *Oleifera* roots. *Plant Physiol.*, **84**:989–992.

Vigh, L., I. Horvath, J. Woltjes, T. Farkas, P. van Hassett, and P. J. C. Kuiper (1987). Combined electron spin resonance, X-ray diffraction studies on phospholipid vesicles obtained from cold-hardened wheats. 1. An attempt to correlate electron-spin-resonance spectral characteristics with frost resistance. *Planta*, **170**:14–19.

Welbank, P. J., M. J. Gibbs, P. J. Taylor, and E. D. Williams (1974). Root growth of cereal crops. *Rep. Roth. Exp. Sta. for 1973*, Part 2, pp. 26–66.

Westgate, M. E. and J. S. Boyer (1984). Transpiration and growth induced water potentials in maize. *Plant Physiol.*, **74**:882–889.

Westgate, M. E. and J. S. Boyer (1985). Osmotic adjustment and the inhibition of leaf, root, stem and silk growth at low water potentials in maize. *Planta*, **164**:540–549.

Whalen, M. C. and L. J. Feldmann (1988). The effect of ethylene on root growth of *Zea mays* seedlings. *Can. J. Bot.*, **66**:719–723.

White, P. J., D. T. Clarkson, and M. J. Earnshaw (1987). Acclimation of potassium influx in rye (*Secale cereale*) at low root temperatures. *Planta*, **171**:377–385.

Williams, J. P., M. V. Khan, K. Mitchell, and G. Johnson (1988). The effect of temperature on the level and biosynthesis of unsaturated fatty acids in diacylglycerols of *Brassica napus* leaves. *Plant Physiol.*, **87**:904–910.

Willigen, P. de and M. van Nordwick (1987). Roots, Plant Production and Nutrient Use Efficiency. PhD. Thesis, Agricultural University of Wageningen, Netherlands.

Wilson, A. J. and A. W. Robards (1978). The ultrastructural development of mechanically impeded barley roots. Effects on the endodermis and pericycle. *Protoplasma*, **95**:255–265.

Wilson, A. J. and A. W. Robards (1979). Some observations of the effects of mechanical impedance upon the ultrastructure of the root caps of barley. *Protoplasma*, **101**:61–72.

Wullstein, L. H. and S. A. Pratt (1981). Scanning electron microscopy of rhizosheaths of *Oryzopsis humenoides*. *Amer. J. Bot.*, **68**:408–419.

Yoshida, S. (1984). Chemical and biophysical changes in the plasma membrane during cold acclimation of mulberry bark cells (*Morus bombycis* Koidz. cv Goroji). *Plant Physiol.*, **76**:257–265.

Zhang, J. and W. J. Davies (1988). Increased synthesis of ABA in partially dehydrated root tips and ABA transport from roots to leaves. *J. Exp. Bot.*, **38**:2015–2023.

Zhang, J., V. Schurr, and W. J. Davies (1987). Control of stomatal behaviour by abscisic acid which apparently originates in the roots. *J. Exp. Bot.*, **38**:1174–1181.

4

CARBON ECONOMY

J. M. WHIPPS
AFRC Institute of Horticultural Research, Littlehampton, West Sussex BN17 6LP, UK

1. INTRODUCTION

Most microorganisms in the soil, with the exception of oligotrophs, remain dormant until an input of organic material is made (Wainwright, 1988). The loss of organic material from the roots as they grow through soil (rhizodeposition) stimulates microbial growth and provides the driving force for rhizosphere development (Whipps and Lynch, 1985, 1986). Over the last decade considerable interest has focused on the quantitative and qualitative aspects of the carbon economy of the rhizosphere and its relationship to the carbon balance of the whole plant (Fig. 1). Thus with time, rhizodeposition is measured as root loss and root respiration together with the accompanying microbial biomass utilizing the root-derived material, the CO_2 generated by these microorganisms and the portion entering the soil organic matter pool.

In general, the study of the carbon economy of the rhizosphere has been approached by scientists of two backgrounds: firstly, microbiologists and soil scientists interested in the ecology, physiology and population biology of microorganisms in the rhizosphere and soil; and secondly plant physiologists and agronomists concerned mainly with the growth and physiology of the higher plant. The latter group have often ignored the effects of rhizosphere microorganisms, although by necessity studies of mycorrhizal and *Rhizobium* symbioses have been more comprehensive. The two groups have consequently used quite different approaches for measuring carbon flows from the plant into the rhizosphere and this has resulted in very large variations in the quantities and types of estimations of carbon losses from the plant. In this chapter I hope to highlight the problems involved in quantifying rhizodeposition and to put

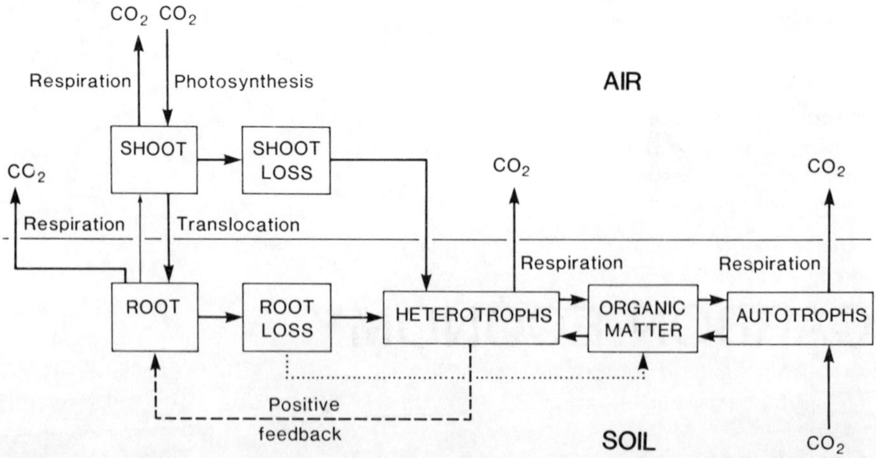

Fig. 1. Carbon flow in the plant and soil associated with rhizodeposition. In general, carbon flow through rhizodeposition occurs from the root into heterotrophs and it then is either released as respiratory CO_2 or enters the organic matter pool of the soil although some root material may enter the organic matter pool directly without the action of heterotrophy. Note that heterotrophs can directly affect the root, altering rhizodeposition.

some of the variation in the estimations of rhizodeposition into perspective. From this it may be possible to recognize firstly the limitations in the methods for estimating carbon budgets of plants and rhizodeposition, and secondly the significance of rhizodeposition to the plant and rhizosphere microorganisms under ecologically realistic conditions.

2. TYPES OF ORGANIC MATTER LOST FROM ROOTS—RHIZODEPOSITION

Almost all chemical components of the plant have been found to be lost from roots. They can be classified into four different groups depending on their mode of arrival:

1. Water-soluble *exudates* such as sugars, amino acids, organic acids, hormones and vitamins which leak from the root without the involvement of metabolic energy.
2. *Secretions* such as polymeric carbohydrates and enzymes which depend upon metabolic processes for their release.
3. *Lysates* released when cells autolyse, including cell walls, sloughed whole cells and, with time, whole roots.
4. *Gases* such as ethylene and CO_2.

Strictly these four classes represent all types of material lost from the root, but in reality, as microorganisms utilize many of these carbon sources immediately, quantification of losses must also include the consequent microbial biomass and accompanying respiratory CO_2. Thus the term "rhizodeposition" is used to describe all these losses arising from the plant root (Whipps and Lynch, 1985).

The whole gamut of compounds may be required to induce the general rhizosphere effect but often one type of material is used by, or stimulates, specific groups of microorganisms in the rhizosphere (see also Chapter 7). For instance, bacteria are capable of degrading the complete root surface gel (mainly polysaccharide material) from sand dune grasses (Old and Nicholson, 1975), and other bacteria, possibly pseudomonads, can penetrate intercellularly into the endorhizosphere of wheat utilizing cell wall components without causing disease (Greaves et al., 1986). Soluble exudates can attract zoospores of pathogens such as *Pythium aphanidermatum* to pea (*Pisum sativum*) roots (Royle and Hickman, 1964a, b) and stimulate hyphal elongation of vesicular-arbuscular mycorrhiza fungi such as *Glomus fasciculatus* (Elias and Safir, 1987). Adhesion of zoospores of *Phytophthora cinnamomi* may be mediated by carbohydrate components of the root surface gel of maize (*Zea mays*) (Hinch and Clarke, 1980) and adhesion of zoospores of *Pythium graminicola* and *Pythium arrhenomanes* is greater on host roots compared with non-host roots, implying host-specific recognition (Mitchell and Deacon, 1986, 1987), a phenomenon well established in bacterial root interactions (Edelman, 1983; see also Chapter 6).

Quantities and types of rhizodeposition vary with age of plant and position on the root. The initial loss must be derived from germinating seeds but this area has been reviewed (Hale and Moore, 1979) and it is not considered in detail here. Briefly, as a seed imbibes it immediately loses carbon, often in the form of water-soluble exudates and gases, and this loss continues as the root and shoot emerge. Carbon loss has been shown to be related to seed size (Vancura and Hanzlikova, 1972; Naim et al., 1976; Vancura and Stotzky, 1976; Nelson et al., 1986). Subsequently, as the root grows through the soil considerable numbers of cells are sloughed off from the root cap. Approximately 5000 cells per day are lost from maize (Moore and McClelen, 1983) accompanied by losses of exudates and polysaccharide gel at a rate of 35 μg per root per day in sterile cultures (Chaboud, 1983) representing a maximum of 60 mg material lost per gram of root (Newman, 1985). From ^{14}C-labelling experiments, the zone just behind the root tips of both primary and lateral roots of wheat appears to be the site of maximum exudation with the next greatest quantity being lost from the zone of elongation (McDougall and Rovira, 1970). However, in minirhizotron experiments with oats (*Avena sativa*) grown in axenic culture, rhizodeposition of carbon compounds only occurred at root tips (Trofymow et al., 1987). Other workers have found that bacteria are at the highest incidence on the root at cell junctions and this has been taken to indicate localized regions of greater exudation (Bowen and Rovira, 1973). Interestingly, the unusual amino acid analogue, beta-(isoxazolin-5-on-2yl) alanine is lost only from root tips of pea and not from other areas of the root (van Egeraat, 1978), and bean (*Phaseolus vulgaris*) plants release pipecolic acid from the roots of seedling plants but not from older plants (Vancura and Hanzlikova, 1972), illustrating the complexities of exudate patterns occurring spatially and temporally on the roots.

In cereals, cortical cells begin to autolyse in the region behind the root hair zone. As the cells die in the absence of microorganisms, such root cortical cell death is under the genetic control of the plant and is directly related to the age of the root (Holden, 1975; Henry and Deacon, 1981). It is not induced by microorganisms (Foster and Bowen, 1982). Hence this source of rhizodeposition is freely available for microorganisms and often cortical cells behind the root hair zone are packed with bacteria utilizing this carbon source. But, nevertheless, all roots gradually age and die and whole parts of the root system enter the soil and are utilized by the rhizosphere microflora. Each year, root loss must represent the greatest rhizodeposition input. Importantly, all types of rhizodeposition are probably continuing along the whole root system concomitantly and it is the proportion of the types of rhizodeposition occurring at any position on the root that controls the structure of the rhizosphere at that position. Because of this complexity, meaningful, detailed measurement of the quality of rhizodeposition from all parts of the roots for the whole of the growing season is virtually impossible.

Rhizodeposition is also affected by many external factors and these have been listed previously (Hale and Moore, 1979; Whipps and Lynch, 1986). Biological effects such as the presence of saprotrophs and pathogenic and mutualistic symbionts, environmental changes, and cultural and chemical practices of man can affect both positively and negatively the quantity and quality of organic compounds lost from the roots. Different cultivars of the same plant can also produce different amounts of rhizodeposition. As many of these factors act interdependently the majority of effects have been determined in nutrient culture, where the variables can be controlled and the factors under investigation manipulated more readily than in soil.

3. MEASUREMENT OF RHIZODEPOSITION IN NUTRIENT SOLUTION CULTURE

By necessity all experiments using nutrient solution culture to determine rhizodeposition are carried out in the laboratory under controlled environment conditions or within the glasshouse. Seeds are germinated and then supported above the nutrient solution; as the roots grow into the solution, exudates, sloughed cells, mucilage and accompanying microflora are collected, generally by replacing the nutrient solution at regular intervals. Sometimes seeds are surface sterilized and all operations are then carried out under sterile conditions, giving the opportunity to add microorganisms back to the system. This gnotobiotic approach has been widely used to study the effect of microorganisms on exudation. In general, adding microorganisms to the nutrient solution increases exudation and secretion (Prikryl and Vancura, 1980; Lee and Gaskins, 1982; Heulin et al., 1987). These forms of rhizodeposition have been examined quantitatively and qualitatively by many workers, some using extremely detailed, painstaking, chemical analyses (e.g. Schönwitz and Ziegler, 1982; Kraffczyk et al., 1984). Others have exposed plant shoots to $^{14}CO_2$ and measured the radioactivity in the culture medium to quantify the rhizodeposition. For instance, Beck and Gilmour (1983) grew wheat (*Triticum aestivum*) plants in nutrient solution in sand or agar in an atmosphere containing constant specific activity $^{14}CO_2$ for 14 days and found that approximately 120 mg of exudates were lost per gram dry weight of root and that this increased (but not significantly) to 130 mg per gram dry weight of root in the presence of a mixture of five N_2-fixing bacterial isolates. Similarly Lee and Gaskins (1982) fed $^{14}CO_2$ for 2 h to sorghum (*Sorghum bicolor*) seedlings and found that up to 2.4% of the total ^{14}C recovered (from root, shoot and nutrient solution 24 h after exposure) was present in root exudates in the absence of microorganisms but in the presence of *Azospirillum brasilense* this increased to 3.5%. Such techniques have been refined and extended to examine other factors affecting rhizodeposition. For example, the addition of glass ballotini to mimic the textural characteristics of soil was found to increase exudation of vitamins and sugars by maize from 94 to 280 mg g^{-1} dry weight of root over a 5-day period, and of sugar and amino acids by barley from 5 to 9% of the weight of the roots over a 3-week period (Barber and Gunn, 1974; Schönwitz and Ziegler, 1982). Artificial soils have even been developed for such purposes (Martin and Foster, 1985; Trofymow et al., 1987). The effects of anoxia, water stress, cultivar, nutrient sources and environmental parameters on rhizodeposition have all been similarly examined in nutrient solution-based systems (e.g. Hale and Moore, 1979; Wiedenroth and Poskuta, 1981; Lee and Gaskins, 1982; Shone et al., 1983; Kraffczyk et al., 1984; Kipe-Nolt et al., 1985; Smucker and Erickson, 1987).

An examination of the recent literature on these types of experiments together with previous reviews (Newman, 1985) has shown that values for rhizodeposition measured in this way do not exceed 600 mg g^{-1} root dry weight. In general the losses are considerably smaller than this, with soluble rhizodeposition always greater than insoluble amounts. However, these values must represent considerable underestimations of total rhizodeposition occurring naturally in the soil for several reasons. For example, the majority of experiments take no account of carbon losses through respiration. Clearly a large increase in rhizodeposition will be found if the CO_2 lost from root respiration and the respiration of accompanying microorganisms utilizing root-derived materials is included. Carbon loss by respiration is of central importance if carbon budgets of the plant and rhizosphere are being considered (Fig. 1). A few workers have measured the CO_2 component of rhizodeposition in these systems (Table 1). For instance, 30% of the total tracer transported to the roots of wheat seedlings was found as $^{11}CO_2$ in the nutrient solution 200 minutes after a pulse of $^{11}CO_2$ to the shoot (Minchin and McNaughton, 1984). By the use of a range of

TABLE 1. Examples of experiments measuring rhizodeposition in nutrient culture with or without sand where respiration has been measured

Plant	*System and comments*	*Reference*
Wheat	$^{14}CO_2$ fed to shoot; monitored $^{14}CO_2$ lost from roots ± microorganisms over 24 h period	Warembourg and Billes, 1979
Wheat	$^{11}CO_2$ fed to single leaf; monitored $^{11}CO_2$ lost from non-sterile roots over 3 h period ± various inhibitors	Minchin and McNaughton, 1984
Many	Continuous feeding $^{14}CO_2$ to plants for periods of time up to maturity	Sauerbeck et al., 1981
Maize	Sand culture; measured microbial respiration as a function of N_2O production using exudates as sole source of C	Haller and Stolp, 1985

inhibitors it was shown that at least 30% of $^{11}CO_2$ found in the nutrient solution was produced by microorganisms utilizing carbon sources coming from the roots. Warembourg and Billes (1979), using similar $^{14}CO_2$ pulse-chase experiments, also showed that microorganisms associated with the roots of wheat seedlings utilized root-derived carbon and subsequently respired a portion of it. In studies with 4-week-old maize seedlings grown in sand nutrient culture, the microbial respiration (calculated from the measurement of N_2O production—denitrification—by bacteria using exudates as sole carbon source) was separated from root respiration and on this basis it was suggested that 25% of the organic matter entering the root from the shoot entered the rhizosphere, which represents 7% of net photosynthates (Haller and Stolp, 1985). This value did not include CO_2 from root respiration. In longer term experiments with a range of plants, the ratio (carbon recovered as CO_2 from root respiration and root decomposition) : (carbon remaining in the roots and nutrient solution at harvest), a root turnover factor, ranged from 0.8 in spring wheat to 2.3 in ryegrass, with the majority of species having a value greater than 1.0 (Sauerbeck et al., 1981; Dormaar and Sauerbeck, 1983). This means that for most of the plants, considerably more carbon had been lost from the plant roots as CO_2 than there was carbon remaining as root material at harvest. Interestingly, as the respiration rate of wheat roots decreases with time (Sauerbeck and Johnen, 1977; Lambers et al., 1982), the relative proportion of respiration derived from wheat roots (as opposed to that of the accompanying microflora) dropped from 46% at tillering to only 5% at maturity. This implies that the majority of CO_2 respired from below ground during the growth of wheat was derived from microorganisms degrading root-derived material.

Even when respiration is considered, systems for measuring rhizodeposition based on nutrient solution culture still have many drawbacks. It is difficult to grow some plants to maturity satisfactorily. Nutrient film and rockwool techniques for culture of tomatoes, cucumbers and lettuces are established in the glasshouse, but to complete the life cycle of many plants such as perennial herbs and trees under environmentally controlled conditions is impossible. Notwithstanding the problems of the life cycle of the plant, many plants often look "sick" when grown under such conditions. Similarly it is very difficult to keep axenic plants sterile for long periods of time in these types of experiments and this explains the abundance of data on rhizodeposition (generally only exudates or secretions) from seedling plants. Nutrient solutions also lack the textural qualities of soil. As already mentioned, sand, glass ballotini or artificial soils have been used to obviate this problem, with corresponding increases in rhizodeposition. Under the electron microscope, losses of mucilage and root cells are clearly seen to be greater in soil compared with nutrient culture (Foster, 1986). Similarly, the addition of microorganisms to nutrient solutions to mimic the effect of the soil microflora can bear little resemblance to the complex mixture of microorganisms associated with the root and can have very different effects on rhizodeposition and root function, depending on the type of microorganisms added (Barber et al., 1976; Heulin et al., 1987). In conclusion, the only satisfactory method to measure rhizodeposition occurring in nature is to work in soil. This has its own set of problems.

4. MEASUREMENT OF RHIZODEPOSITION IN SOIL

4.1. Growth Cabinet Experiments

Because of the complex nature of soil and the microbial transformations of organic materials occurring in the rhizosphere, standard chemical techniques such as those used in nutrient solution studies cannot be realistically employed for analysing materials leaving the root under natural conditions. Those studies that have extracted chemicals from soil derived from the rhizosphere and non-rhizosphere zones just show differences in amounts of groups of compounds and give little impression of the true quality, quantity or dynamics involved in rhizodeposition (Bokhari et al., 1979). Consequently the only satisfactory system for measuring such carbon losses is to use ^{14}C tracer techniques. These can take two forms. Firstly, there is the classic pulse-chase system where the shoot of the plant is exposed to high specific activity $^{14}CO_2$ for a relatively short time (minutes–hours) and the distribution of radioactivity is followed with time as it spreads through the plant. Eventually, some fraction of the photosynthetically fixed carbon enters the rhizosphere and can be quantified (e.g. Warembourg and Paul, 1973; Kucey and Paul, 1982; Snellgrove et al., 1982; Warembourg et al., 1982; Bildusas et al., 1986). Many of these experiments have involved studies of mycorrhizal and *Rhizobium* symbiosis and will be discussed in detail later (see Section 5). This pulse-chase system has the advantage of ease and rapidity of use but produces a non-uniform distribution of label throughout the plant. Consequently, it can only give information on the movement and losses of recently fixed carbon, which may be useful for studying fluxes through different pools in the plant and losses of exudates and newly formed secretions but gives no information on the degradation of whole cells or dying parts of older roots. Even repeated pulse feeding cannot evenly label all pools of carbon in the plant. This lack of even labelling has resulted in the development of the second system. Here plants are grown from seedlings up to maturity, in the case of some annual species, in controlled environment growth chambers in an atmosphere containing constant specific activity $^{14}CO_2$ (Table 2). If the shoots are exposed to the $^{14}CO_2$ and the roots and soil separated from the $^{14}CO_2$ by a seal, any ^{14}C found in the soil or as $^{14}CO_2$ in the soil must have originated from the plant. As mentioned before, this soil CO_2 can be

TABLE 2. Examples of experiments where plants have been grown in soil under controlled environment conditions with shoots continually exposed to constant specific activity $^{14}CO_2$ to measure rhizodeposition

Plant	*Variables considered*	*Reference*
Wheat Barley	Sterility; 12 h or 16 h daylength; 21 day growth	Barber and Martin, 1976
Wheat	Temperature; sterility	Martin, 1977a
Wheat	Soil moisture	Martin, 1977b
Wheat Mustard	Effect of age; plants grown to maturity	Johnen and Sauerbeck, 1977; Sauerbeck and Johnen, 1977
Wheat	Cultivar	Martin and Kemp, 1980
Many	Range of environmental parameters	Sauerbeck et al., 1981
Maize	Distance of rhizodeposition from roots; 25 day growth	Helal and Sauerbeck, 1983, 1986
Blue grama	Temperature and daylength varied; 107 day growth	Dormaar and Sauerbeck, 1983
Wheat Barley	Temperature; daylength; 21 day growth	Whipps and Lynch, 1983; Whipps, 1984
Maize	CO_2 concentration; 28 day growth	Whipps, 1985
Maize	Soil type; 42 day growth	Merckx et al., 1987
Wheat	Soil type; 42 day growth	Merckx et al., 1985; Merckx et al., 1986
Tomato Pea	28 day growth	Whipps, 1987

derived from root respiration or from microbial respiration utilizing the root-derived material (Fig. 1) and in these systems cannot be separated directly.

Some studies have examined the distribution pattern of ^{14}C derived from the root in the soil using these constant feeding systems. For instance, ^{14}C derived from maize roots was found 20 mm away from the roots after 25-day growth, indicating that rhizodeposition can spread very rapidly from the root and must have considerable influence around the root (Helal and Sauerbeck, 1983). The ratio of ^{14}C in a water-soluble fraction : ^{14}C in an insoluble fraction decreases with distance from the root for several plant species, and, in the non-rhizosphere soil, the quantity of ^{14}C is always greater in the insoluble fraction compared with the soluble fraction (Martin and Kemp, 1980; Whipps, 1984, 1985, 1987). This suggests that relatively more soluble organic material is available close to the root. Merckx et al. (1987) studying rhizodeposition in soils of different nutrient availability showed that in soils with less nutrient availability, less organic rhizodeposition was incorporated into microbial biomass compared with soils with high nutrient availability, and it was suggested that a lack of nutrients may have limited microorganism degradation of root-derived organic material. Helal and Sauerbeck (1986) showed that rhizodeposition from maize acted as a primer for microbial degradation of existing soil carbon, illustrating the importance of interactions between rhizodeposition and soil processes in general.

A chemical analysis of the ^{14}C material in various soil fractions derived from wheat roots maintained in sterile or non-sterile soil for 21 days also showed that autolysis of roots was important in rhizodeposition (Martin, 1977c). In these experiments, as little difference in the chemical nature of the material in the fractions from sterile or non-sterile soils was found, much of the organic material recovered in the soil was apparently derived directly from root tissue without intervention of the soil microflora or microfauna. Fifteen per cent of the ^{14}C recovered in the soil was already present in the humic fraction in both soils indicating that considerable degradation of parts of the root-derived organic matter had already taken place. Using a fumigation technique, Merckx et al. (1985) showed that ^{14}C derived from wheat roots entered the microbial biomass and the total quantity of ^{14}C in the biomass increased throughout the 6-week period of the experiment. However, the proportion of ^{14}C in the biomass compared with the total ^{14}C in the soil, increased in a sandy soil but decreased in a clay loam soil. It was suggested that in the latter soil type, a shift in rhizodeposition with time from soluble exudates to less soluble materials and a binding of organic moieties to the clay, both decreasing the movement of organic material through the soil, would account for the differences in ^{14}C content in the biomass of the two soils.

The majority of these types of studies have been concerned with quantifying the amounts and proportions of ^{14}C both entering the roots and forming the rhizodeposition. Even here, different workers have expressed similar basic data in different ways making absolute comparisons difficult. Some useful proportional distribution derivations of data are given in Table 3. It should be noted that the technique cannot allow for carbon respired from the shoot and so distribution data can only be related to net fixed carbon rather than total fixed carbon. Overall, between 30 and 60% of net fixed carbon is transferred to roots in the annual plants studied although a lower range (16–46% net fixed carbon) was found for a group of nine different annual plants by Sauerbeck et al. (1981). These values agree with data for a large number of different plants grown in nutrient solutions under a range of conditions (Lambers, 1987). However, the respiration data of Barber and Martin (1976) must be questioned on both theoretical and experimental grounds (Whipps and Lynch, 1985; Lambers, 1987). When the results are compared with values obtained for root respiration, expressed as a proportion of the carbon transferred to the roots, for plants grown in nutrient solution culture with or without sand (Table 4) they are approximately 50% too small in the case of non-sterile conditions and ten times too small in the case of sterile plants. These percentage respiration losses become even smaller if the rhizodeposition which is known to be increased in soil is included. On theoretical grounds and assuming that 36 mol of ATP are produced in the oxidation of 1 mol of glucose (see Lambers, 1987), the absolute minimum amount of carbon lost in respiration is 100 mg for every gram of carbon incorporated into root dry matter. Taking maintenance respiration and the energy required for nutrient uptake into account, actual values must be higher. Values between 340 and 1400 mg C g^{-1} C in root dry matter have been found experimentally in nutrient culture conditions for a range of plants (Hansen and Jensen, 1977; Lambers and Steingroever, 1978b; Farrar, 1981; Lambers et al., 1983; Lambers, 1987). Barber and Martin (1976) showed that sterile wheat plants and non-sterile wheat plants had values of 42 and 240 mg C respired g^{-1} C in the root respectively whereas other data for soil-grown wheat, barley, maize, tomato and pea all fell within the limits or were greater than the values measured in nutrient culture conditions (Whipps and Lynch, 1983, 1985; Whipps, 1984, 1985, 1987). Consequently, as in most other cases, the proportion of carbon transferred to the roots and the respiration values are in line with values found in nutrient solution culture. This ^{14}C technique seems appropriate for measurements of rhizodeposition in soil under controlled conditions.

In experiments with wheat, maize, tomato (*Lycopersicon esculentum*) and pea, the percentage of carbon lost from the roots that is recovered in the soil decreases with time (Table 3; Whipps, 1985, 1987; Merckx et al., 1985). In the tomato and pea this is accompanied by an increase in the proportion of carbon recovered as CO_2, presumably where microorganisms are utilizing the root-derived organic matter. But in other cases, no clear pattern of respiration change is found and here less carbon may leave the root. In experiments with wheat and barley (*Hordeum vulgare*) the percentage of carbon lost from the roots that is recovered in the soil ranged from 17 to 31% in seedling plants (Barber and Martin, 1976; Whipps and Lynch, 1983; Whipps, 1984) but never exceeded 11% in the other experiments. These differences may simply reflect different experimental conditions as Martin and Kemp (1980) demonstrated that six cultivars of wheat grown for 28 days under the same conditions did not differ in their rhizodeposition. In the long-term experiments with wheat

TABLE 3. Percentage of net fixed carbon moved to roots and the percentage of net fixed carbon moved to the roots which is lost when shoots of plants are exposed continually to $^{14}CO_2$ (plants grown in soil)

	Conditions						*% of net fixed C transferred to root*[a]	*% of C transferred to root lost as:*				
Plant	*Age (days)*	*Temp. (°C)*	*Day-length (h)*	*Soil*[f]	*CO_2 conc. (%)*	*Sterility*[g]		*respiration (ignoring rhizo-deposition in soil)*[b]	*respiration (including rhizo-deposition in soil)*[c]	*rhizo-deposition in soil*[d]	*respiration and rhizo-deposition in soil*[e]	*Reference*
Wheat	21	18	16	1	0.03	S	32	4	3	20	23	Barber and Martin, 1976
	21	18	16	1	0.03	NS	48	19	16	17	33	Barber and Martin, 1976
	21	18	12	1	0.03	S	27	14	11	20	31	Barber and Martin, 1976
	21	18	12	1	0.03	NS	38	30	25	17	42	Barber and Martin, 1976
	21	18	16	2	0.03–0.10	NS	47	48	36	25	61	Whipps and Lynch, 1983
	21	15	16	2	0.03–0.10	NS	52	53	41	23	64	Whipps, 1984
	21	18–14	16	2	0.03–0.10	NS	59	55	39	29	68	Whipps, 1984
	21	18–14	12	2	0.03–0.10	NS	51	33	25	23	48	Whipps, 1984
	36	–	–	3	0.03	NS	–	44	43	4	47	Johnen and Sauerbeck, 1977
	98	–	–	3	0.03	NS	–	67	63	7	70	Johnen and Sauerbeck, 1977
	119	–	–	3	0.03	NS	–	72	69	5	73	Johnen and Sauerbeck, 1977
	153	–	–	3	0.03	NS	–	79	76	4	80	Johnen and Sauerbeck, 1977

Continued

TABLE 3. (*continued*)

	Conditions						% of net fixed C transferred to root[a]	% of C transferred to root lost as:				
Plant	Age (days)	Temp. (°C)	Day-length (h)	Soil[f]	CO_2 Conc. (%)	Sterility[g]		respiration (ignoring rhizo-deposition in soil)[b]	respiration (including rhizo-deposition in soil)[c]	rhizo-deposition in soil[d]	respiration and rhizo-deposition in soil[e]	Reference
Wheat	28	21–17	16	1	0.03	NS	51	41	37	9	46	Merckx et al., 1985
	35	21–17	16	1	0.03	NS	49	39	38	3	41	Merckx et al., 1985
	42	21–17	16	1	0.03	NS	50	40	39	2	41	Merckx et al., 1985
	28	21–17	16	4	0.03	NS	44	44	39	11	50	Merckx et al., 1985
	35	21–17	16	4	0.03	NS	48	36	33	10	43	Merckx et al., 1985
	42	21–17	16	4	0.03	NS	40	32	29	9	38	Merckx et al., 1985
Barley	21	18	16	1	0.03	S	32	6	4	31	35	Barber and Martin, 1976
	21	18	16	1	0.03	NS	43	27	21	23	44	Barber and Martin, 1976
	21	15	16	2	0.03–0.10	NS	54	60	48	21	69	Whipps, 1984
	21	18–14	16	2	0.03–0.10	NS	51	52	37	30	67	Whipps, 1984
	21	18–14	12	2	0.03–0.10	NS	27	51	35	31	66	Whipps, 1984
Maize	14	18–14	16	2	0.03–0.05	NS	28	32	16	52	67	Whipps, 1985
	28	18–14	16	2	0.03–0.05	NS	29	46	33	28	62	Whipps, 1985
	14	18–14	16	2	0.04–0.07	NS	26	50	33	34	67	Whipps, 1985
	28	18–14	16	2	0.04–0.07	NS	36	33	26	20	47	Whipps, 1985
	14	18–14	16	2	0.06–0.10	NS	36	38	31	19	50	Whipps, 1985
	28	18–14	16	2	0.06–0.10	NS	35	49	40	18	58	Whipps, 1985

Mustard	73	–	–	1	0.03	NS	–	75	66	13	78	Johnen and Sauerbeck, 1977
	73	–	–	1	0.03	NS	–	69	56	16	74	Johnen and Sauerbeck, 1977
	73	–	–	2	0.03	NS	–	60	52	12	65	Johnen and Sauerbeck, 1977
	73	–	–	3	0.03	NS	–	65	53	18	71	Johnen and Sauerbeck, 1977
Tomato	14	18–14	16	2	0.03–0.05	NS	43	67	20	70	90	Whipps, 1987
	14	18–14	16	2	0.03–0.05	NS	40	62	47	25	72	Whipps, 1987
Pea	14	18–14	16	2	0.03–0.05	NS	65	55	31	43	75	Whipps, 1987
	28	18–14	16	2	0.03–0.05	NS	44	74	53	29	82	Whipps, 1987

$$a = \frac{\mu\text{Ci in root} + \text{rhizodeposition} + CO_2 \text{ from root and soil microorganisms}}{\text{net } \mu\text{Ci fixed}} \times 100; \ \mu\text{Ci} \simeq \text{amount of C}$$

$$b = \frac{\mu\text{Ci in } CO_2 \text{ from root and soil microorganisms}}{\mu\text{Ci in root} + CO_2 \text{ from root and soil microorganisms}} \times 100$$

$$c = \frac{\mu\text{Ci in } CO_2 \text{ from root and soil microorganisms}}{\mu\text{Ci in root} + \text{rhizodeposition in soil} + CO_2 \text{ from root and soil microorganisms}} \times 100$$

$$d = \frac{\mu\text{Ci in rhizodeposition in soil} \times 100}{\mu\text{Ci in root} + \text{rhizodeposition in soil} + CO_2 \text{ from root and soil microorganisms}}$$

$$e = \frac{\mu\text{Ci in } CO_2 \text{ from root and soil microorganisms} + \text{rhizodeposition in soil}}{\mu\text{Ci in root} + \text{rhizodeposition in soil} + CO_2 \text{ from root and soil microorganisms}} \times 100 \equiv \frac{(\text{Total } ^{14}\text{C lost from the root})}{^{14}\text{C moved to the root}} \equiv c + d$$

$e = c + d$

[f] 1 = sand; 2 = sandy loam; 3 = loamy sand; 4 = silty clay loam.
[g] S = sterile; NS = non-sterile

TABLE 4. Percentage of net fixed carbon moved to the roots which is lost as CO_2 (plants grown in nutrient solution systems). After Lambers (1987)

Plant	*% of C transferred to root lost as respiration (ignoring rhizodeposition in solution)*[a]	*Reference*
Wheat	67	Barneix et al., 1984
	52	Lambers et al., 1982
Sunflower	67	Hatrick and Bowling, 1973
	35–44	Szaniawski and Kielkiewicz, 1982
Maize	49	Veen, 1980
Senecio spp.	50–70	Lambers and Steingröever, 1978a

$$^{a} = \frac{\text{g C in CO}_2 \text{ from root (+nutrient solution microorganisms)} \times 100}{\text{g C in root} + \text{g C in CO}_2 \text{ from root (+nutrient solution microorganisms)}}$$

(Johnen and Sauerbeck, 1977) the percentage of carbon lost from roots recovered as respiration increased throughout the growing season, and as root respiration decreases with time (Sauerbeck and Johnen, 1977), microorganisms must have utilized the organic material for their growth. In comparison with similar experiments carried out in nutrient solution (Sauerbeck et al., 1976), 40% more organic material was lost from the root and turned over in soil. At harvest, the total quantity of carbon recovered mechanically in root material was only 20% of the quantity of carbon which had entered the roots in wheat and 20–35% of that in mustard (*Sinapsis alba*). With the assumption that 30% of $^{14}CO_2$ was derived from root respiration, it has been calculated that organic rhizodeposition (^{14}C in soil + ^{14}C from microbial respiration of root-derived material) is between 0.5 and 2.8 t C ha^{-1} in one year for mustard and wheat (Johnen and Sauerbeck, 1977).

All the experiments considered so far have been concerned with annual plants. The only perennial studied under these types of experimental conditions is blue grama (*Bouteloua gracilis*), a major constituent of the prairie ecosystem in Canada (Dormaar and Sauerbeck, 1983). Here plants were grown from seed in soil and, once established, were subjected to three climatic periods in the growth chamber to simulate summer, autumn-winter and spring (Table 5). During the summer period 1, 78% of net fixed carbon was transferred below ground and 22% of net fixed carbon was lost as CO_2 from root respiration and microbial respiration of the organic material derived from the root. During the winter period 2, the soil in the pots eventually froze and respiration decreased with concomitant build-up of ^{14}C in the soil as microorganisms were inhibited from degrading the organic matter present and more root material entered the soil pool. As the conditions ameliorated in period 3, carbon was again moved to the root but CO_2 production was slow in recovering. Root losses of 30–60% of young, non-suberized roots have been observed after cessation of early growth in blue grama using other techniques (Ares, 1976). These results illustrate the massive input of photosynthetic products from the shoot to below ground during the year that must be required by this plant to develop and maintain a root

TABLE 5. Percentage distribution of net fixed carbon in plant, soil and CO_2 after growth of blue grama (*Bouteloua gracilis*) with shoots exposed to constant specific activity $^{14}CO_2$ during three environmental periods. After Dormaar and Sauerbeck (1983)

	Environmental period		
	1:	*2:*	*3:*
		days 31–52; 10 h daylength; 14–3 °C;	
Compartment	*days 1–31; 14 h daylength; 25–15 °C*	*days 52–81; 10 h daylength; −5 °C constant*	*days 81–107; 14 h daylength; 25–15 °C*
Shoots	22	29	18
Roots	33	16	25
Soil (organic rhizodeposition)	23	53	54
CO_2	22	2	3

system in semi-arid climatic conditions. This is in marked contrast to annual plants growing under temperate conditions where only approximately 50% of net fixed carbon was transferred below ground (see above).

Even though these techniques are valuable in giving estimations of rhizodeposition in soil they still suffer from a range of problems. For instance, the plants are grown in pots or containers with inherent difficulties of water and gas exchange, nutrient limitation and root growth constraints. The pots may not be leak-proof and respiratory CO_2 may escape from the pot or radioactive CO_2 may enter. The seals around the plants may affect plant growth and some plants may still appear "sick" due to growth in unnatural conditions. Harvests have to be sequential, which often limits the number of plants available for each harvest, and some damage to plant roots may occur during the recovery of roots from the soil. The seedlings often have considerable reserves of carbon in the seed so that it takes time for all the plant parts to become evenly labelled, especially as some workers leave plants for several days before sealing them into the growth cabinets. Similarly, because the shoot respiration cannot be measured *in situ*, distribution data always refer to net fixed carbon. Perhaps the most frustrating problem, and one that holds for nutrient solution as well, is the inability to study rhizodeposition in developing or mature stands of crops or trees as would occur in nature. Consequently, methods have been developed to study rhizodeposition and carbon flow in the field.

4.2. Field Experiments

There are two techniques for estimating the rhizodeposition occurring in the field and both have been used to study a range of species in various ecosystems (Table 6).

TABLE 6. Examples of experiments where rhizodeposition has been measured in the field

Plant	*System and comments*	*Reference*
Annual systems		
Wheat	Crop study	Gregory et al., 1978
	Crop study. CO_2 flux measured from seedling to flowering at 10-day intervals	Martin and Puckridge, 1982
	$^{14}CO_2$ pulse-chase to plots at tillering/flowering (no flowers allowed to develop). Rhizodeposition measured from pulse to harvest	Buyanavsky and Wagner, 1987
	Single plants pulse-fed $^{14}CO_2$ at 7 or 10 weeks. Rhizodeposition measured from pulse to flowering ± soil sterilization	Martin and Kemp, 1986
	Single plants pulse fed $^{14}CO_2$ throughout growing season	Keith et al., 1986
Soybean	Crop study	Sivakumar et al., 1977
Maize	Crop study	Barber, 1971
Perennial systems		
Prairie		
Bouteloua gracilis	$^{14}CO_2$ pulse-fed. Measured only root and soil ^{14}C	Singh and Coleman, 1974
Short prairie grass mixture	Crop study	Sims and Singh, 1971

Agropyron/Koeleria-dominated mixture	$^{14}CO_2$ pulse-fed at 6 sampling dates during growing season. Measured rhizodeposition at intervals	Warembourg and Paul, 1977
Tall prairie grass (large mixture of species)	Crop study	Kucera et al., 1967
Forest		
Coniferous		
Pinus cembra	Crop study. CO_2 flux and photosynthesis measured	Tranquillini, 1964
Scots pine	Crop study. CO_2 flux and photosynthesis measured	Agren et al., 1980
Douglas fir	Crop study	Fogel and Hunt, 1983
Deciduous		
Oak	Crop study	Ovington et al., 1963
Yellow poplar	Crop study	Edwards and Harris, 1977

The first and oldest, termed crop studies, involves the measurement of components of crop growth. These can include carbon gained by photosynthesis and carbon lost from roots, shoots and soil respirations, enabling calculation of gross and net productivity. Other procedures measure changes in shoot and root masses and carbon contents of soils at various harvest intervals. All these procedures are fraught with problems as far as measurement of rhizodeposition is concerned, particularly in forest ecosystems. For instance, measurement of respiration losses can only be made infrequently as it is impossible to obtain respiration data from the roots of a single tree. Similarly, soil coring procedures to recover roots often miss fine roots and associated mycorrhizas leading to considerable underestimate of below ground rhizodeposition (Fogel, 1983). Nevertheless they are the only realistic methods available for established trees. These crop studies are also unable to quantify movement and distribution of carbon within the plant and into the rhizosphere with time. Consequently the second technique involving $^{14}CO_2$ pulse-chase experiments has been developed and used in both crop and grassland ecosystems. A single $^{14}CO_2$ pulse-chase experiment has been carried out on a mature lodgepole pine (*Pinus contorta*) tree (Read et al., 1985).

4.2.1. Annual Crop Plants

Rhizodeposition measurement in the field has concentrated on wheat. Some general derivations of distribution of carbon movements and losses for this crop by a range of workers using both techniques are given in Table 7. The percentage of net fixed carbon moved to the root decreases with time and may explain some of the variations in distribution found by different workers using plants of different ages. The proportion of net fixed C moved to the root at 42 days, the youngest age sampled in these field experiments (Keith et al., 1986), is very similar to that found for seedlings in controlled environment experiments (Whipps and Lynch, 1983; Whipps, 1984) and can be related to the values found in nutrient culture. Interestingly, even though the respiration of the roots decreases with time (Johnen and Sauerbeck, 1977), and the percentage of net fixed C which is lost from the root decreases with age (Keith et al., 1986), between 45 and 65% of the carbon transferred to the root is lost as CO_2 in these experiments, which is in close agreement for most of the values determined in a range of growth cabinet experiments (Table 3). The quantity of rhizodeposition calculated from these field experiment data, which include losses by root and associated soil respiration and root residues, ranges between 1.2 and 2.9 t C ha^{-1} $year^{-1}$; this agrees well with values of 2.3 t C ha^{-1} $year^{-1}$ for wheat and 0.5–2.9 t C ha^{-1} $year^{-1}$ for mustard in $^{14}CO_2$ growth cabinet experiments (Johnen and Sauerbeck, 1977). This represents approximately 15% of the total fixed carbon (Keith et al., 1986). Assuming the organic constituents of roots contain an average of 40% carbon, annual production of roots of soyabean (*Glycine max*), wheat and maize measured by crop growth systems

TABLE 7. Carbon movement to and rhizodeposition from the roots of wheat measured in field experiments

Measurement period	*% net fixed C transferred to root*		*% of C transferred to root that is lost*		*Rhizodeposition (root C + soil C + CO_2 C) (t C ha^{-1} $year^{-1}$)*	*Reference*
			Soil	*CO_2*		
Seedling–flowering (10–30 days)	30		–	–	1.5	Martin and Puckridge, 1982
Late tillering–harvest	–		–	53	2.9	Buyanavsky and Wagner, 1987
Early tillering–flowering (49 days)	40		–	65	–	Martin and Kemp, 1986
Late tillering–flowering (63 days)	9		–	45	–	Martin and Kemp, 1986
42 days	50					Keith et al., 1986
56 days	20	mean	mean	mean		
84 days	15	15%	25	50	1.3	
119 days	3					
154 days	2					
Whole season	–		–	–	1.2	Jenkinson and Rayner, 1977

without allowance for respiration losses are in the range 0.24–1.6 t C ha^{-1} (Barber, 1971; Sivakumar et al., 1977; Gregory et al., 1978).

Several other features related to rhizodeposition can be discovered from these field results. For instance, Martin and Puckridge (1982) obtained a value of 1.92 : 1 for the ratio of quantity of C in CO_2 lost : quantity of C in roots + soil for wheat, in close agreement with the values obtained by Sauerbeck et al. (1981). But the value for this ratio in $^{14}CO_2$ growth cabinet experiments (Martin, 1977a; Barber and Martin, 1976) was only 0.29. As discussed earlier, this may be due to low respiration values in these growth cabinet experiments but Martin and Puckridge (1982) suggest that a lack of soil animals in the growth cabinet experiments resulted in little breakdown in insoluble material. In the $^{14}CO_2$ pulse-chase field experiments for wheat (Martin and Kemp, 1986), total soil respiration rates (including those of roots), averaged over the whole crop, were approximately 90 mg CO_2 m^{-2} h^{-1}, which is within the range reported for several annual crops in a crop study (Monteith et al., 1964). In soils partially sterilized with chloropicrin, total soil respiration (including roots) was decreased by almost half, and this decrease was equated with a decrease in carbon in the roots and soil fraction. Here, chloropicrin could have affected both microbial populations and root growth (Martin and Kemp, 1986).

4.2.2. Grassland Perennial Ecosystems

Grassland prairies have been studied less than annual crop production systems. From a range of studies, the input of carbon to the soil has been found to be cyclical, corresponding to growth of roots in spring and summer and a gradual cessation of root growth and decomposition over winter and early spring. Using $^{14}CO_2$ pulse-feeding experiments in late spring, ^{14}C was initially found in the roots of blue grama in the region of actively growing roots in the top 20 cm of soil (Singh and Coleman, 1974). This early growth was dependent upon carbon stored in the overwintered roots. With time, ^{14}C from later pulses of $^{14}CO_2$ moved to lower regions of the soil as the roots grew deeper. Similar patterns of carbon transfer were also found in the *Agropyron/Koeleria*-dominated grassland. Only limited data on the distribution of carbon into the rhizosphere are available (Table 8). In the *Agropyron/Koeleria*-dominated grassland 35–50% of net fixed C was transferred below ground (Warembourg and Paul, 1977), which is less than in growth cabinet experiments using blue grama (*Bouteloua gracilis*) (Dormaar and Sauerbeck, 1983) and other field experiments (Kucera et al., 1967; Sims and Singh, 1971). Only between 20 and 28% of the carbon entering the roots of the *Agropyron/Koeleria*-dominated grassland was respired; this value is low compared with annual plants grown in other systems and may reflect the lack of growth and decomposition during the winter period. Consequently the calculated annual rhizodeposition in this system (roots + soil + respiration) of 1.3 t C ha^{-1}, which agrees well with annual systems, should be considered a minimum. The value for rhizodeposition

TABLE 8. Carbon movement to and rhizodeposition from the roots of prairie plants in field experiments

Species	*% net fixed C transferred below ground*	*% of C transferred to roots that is lost*		*Rhizodeposition (root C + soil C + CO_2 C) (t C ha^{-1} $year^{-1}$)*	*Reference*
		Soil	*CO_2*		
Agropyron/Koeleria-dominated mixture	35–50	–[a]	20–28	1.3	Warembourg and Paul, 1977
Tall prairie grass (large mixture of species)	50	–	–	2.1[b]	Kucera et al., 1967
Short prairie grass *(Bouteloua gracilis/Buchloe dactyloides*-dominated mixture)	80	–	–	–	Sims and Singh, 1971

[a] No value available.
[b] This is an underestimate as the value does not include CO_2 respired. It is calculated assuming roots are 40% carbon.

of tall prairie grass mixture based on a crop study not including respiratory losses was approximately 2.0 t C ha^{-1} $annum^{-1}$ (Dahlman, 1968).

4.2.3. Forest Perennial Systems

More information on carbon flow is available for tree species growing in the field than for perennial grasslands, and some estimates of carbon distribution are given in Table 9. Between 40 and 73% of net fixed carbon is passed below ground and this results in a net increase of 1.4–4.4 t C ha^{-1} in root biomass per annum. Annual losses in the order of 30 to 90% occur in fine white roots of a number of species (Fogel, 1983, 1985) and this loss, together with the loss of associated mycorrhizal fungi, has been calculated to range between 5.8 and 7.5 t C ha^{-1} $annum^{-1}$ (Fogel and Hunt, 1983). This was two to five times more important in returning organic matter to the soil than the aerial biomass.

Overall, perennial ecosystems contain plants which pass more of their fixed carbon below ground compared with annual systems. In part, this may be due to perennials having to invest more of their productivity to survive all year under stress. Root sloughing with subsequent microbial immobilization could help retain nutrients in the upper horizon when the plants are stressed and could thus be an adaptive strategy (Harris et al., 1980).

5. RHIZODEPOSITION ASSOCIATED WITH SYMBIOSES

Mycorrhizas and *Rhizobium* are classic examples of mutualistic biotrophs (Lewis, 1973). Plants benefit from mycorrhizal associations by improved mineral nutrition (particularly of P) and drought resistance, and legumes benefit from *Rhizobium* infection by improved nitrogen nutrition derived from nitrogen fixation. In return the heterotrophs gain organic material from the endorhizosphere of their hosts. Most of the plant species growing in field soil have mycorrhizal associations, although the crucifers and several families in the subclass Caryophyllidae are exceptions. Indeed many plant species are ecologically dependent on the mycorrhizas for their survival. Similarly, legumes grown in the field will normally become infected with appropriate *Rhizobium* species and in some instances may become infected simultaneously with mycorrhizal fungi. Other mutualistic symbiotic root associations, such as those between *Frankia* and non-leguminous angiosperms (e.g. *Myrica* and *Alnus* spp.), and between cyanobacteria and cycads, are also known. The carbon cost in the synthesis and maintenance of *Rhizobium* and mycorrhizal associations is well researched.

5.1. Rhizobium

Hiltner (1904) was studying the loss of nitrogen compounds from nodules of legumes when he first coined the term rhizosphere, and since then there have been

TABLE 9. Carbon movement to and rhizodeposition from the roots of trees in the field

Species	*% net fixed C transferred below ground (to roots + mycorrhizas)*	*% of C transferred to roots that is lost (white root decomposition*	*Rhizodeposition (root C) (t C ha^{-1} $year^{-1}$)*	*Reference*
Deciduous forest				
Oak	–[a]	52	–	Ovington et al., 1963
Yellow poplar	40	42	–	Edwards and Harris, 1977; Harris et al., 1980
Coniferous forest				
Douglas fir	73	40–47	5.8–7.5	Sanantonio, 1979; Fogel and Hunt, 1983
Scots pine	60	66	–	Persson, 1978; Agren et al., 1980
Pacific silver fir	60–71	–	–	Grier et al., 1981
Pinus cembra	70	–	–	Tranquillini, 1964

[a]No value available.

many reports of measurements of exudates and secretions from legume roots. Measurements of total rhizodeposition from legumes in soil are, however, scant. Losses of up to 82% of the carbon transferred to the root have been reported for pea seedlings, with up to 43% appearing in soil (Whipps, 1987). But the majority of experiments involving legumes have been concerned with efficiency and cost of nitrogen fixation rather than losses to the rhizosphere soil. The carbon costs of *Rhizobium*-legume symbioses are due to synthesis and maintenance of the nodules and reduction of gaseous nitrogen to ammonia. On theoretical grounds, the overall cost of fixation of 1 mol N_2 to 2 mol ammonium is at least 25 ATP (Minchin et al., 1981). Stoichiometrically:

$$N_2 + 4NADH + 6H + 16ATP \rightarrow 2NH_4 + H_2 + 4NAD + 16ADP$$

This rises to 32 ATP equivalent assuming that the 2 mol NH_4 are assimilated into 1 mol asparagine—the main nitrogenous product of many legumes (Minchin et al., 1981). If the legume also has an ATP-dependent evolution of H_2 catalysed by nitrogenase (which may account for 50% of total electron flux) and does not have an H_2 recycling system, the cost rises to 49 ATP equivalent per mol asparagine produced from N_2 fixation, ignoring the cost of subsequent transport.

Some values for distribution and losses of photosynthetically fixed carbon in nodulated legumes are given in Table 10. In young plants, when N_2 fixation rates were increasing, between 41 and 65% of net photosynthate was moved to the roots and between 13 and 28% of net fixed carbon entered the nodules. Averaged throughout the life cycle, between 31 and 50% of net fixed C entered the roots. In young plants 19–40% of net fixed C was subsequently lost as respiration (5–10% derived from nodules), representing 46–76% of the total carbon entering the roots. In older plants, respiration losses from roots increased to 25–40% of net fixed C, representing 63–79% of carbon entering the root. These respiration losses also include that derived from microbial respiration in the rhizosphere, which must generally have been small because of the experimental conditions used. Some carbon was re-exported to the shoot in the form of nitrogen-containing compounds. Overall the roots consumed 8–20 mg C mg^{-1} N_2 fixed and respired 4–11 mg C mg^{-1} N_2 fixed, although the nodules were more efficient.

From the evidence obtained during the growth of a range of non-nodulated legumes supplied with NO_3 as N source, the cost of N_2 fixation (based on CO_2 lost g^{-1} root or CO_2 lost g^{-1} N assimilated) was up to twice that when growing on NO_3 (Ryle et al., 1979; Pate et al., 1979; Ryle et al., 1983). Also, without nodules plants with adequate NO_3-N produce plants with larger tops and smaller roots (Mahon and Child, 1979; Minchin et al., 1981). This suggests that nitrogen fixation incurs an additional carbon usage on underground parts which for cowpea (*Vigna unguiculata*) and soybean may be equivalent to 10–15% of photosynthate available for growth in comparison with nitrate-fed plants (Schubert and Ryle, 1980).

Many of the estimates for efficiency and cost of N_2 fixation are based upon gross root respiration measurements and these give no information on the

TABLE 10. Distribution of photosynthates (C) in nodulated legumes and ratios of C respired or consumed:N fixed (mg mg^{-1}). After Minchin et al. (1981)

% C fixed		
To roots	Young[a]	41–65
	Total[b]	31–50
To nodules		13–28
Lost as root respiration	Young	19–40
	Total	25–40
Lost as nodule respiration		5–10
% root C lost		
Exported to shoot	Young	8–20
	Total	11–14
Respired	Young	46–76
	Total	63–79
C respired:N_2 fixed (mg mg^{-1})		
Roots	Young	4–11
	Total	6–15
Nodules		1–4
C consumed:N_2 fixed (mg mg^{-1})		
Roots	Young	8–17
	Total	8–20
Nodules		3–7

[a]Measurements taken during the period of increasing nitrogen fixation rates; pea, white lupin and cowpea.
[b]Measurements throughout the life cycle; white lupin and cowpea.

relative rates of respiration of nodules in comparison with the rest of the root system. Mean values of 13.4, 7.1 and 9.2 mg CO_2 respired g^{-1} root h^{-1} have been found for whole nodulated roots, roots (separated from nodules) and roots from non-nodulated nitrate-grown plants respectively for a range of legumes (Ryle et al., 1983). It has been argued that this increased respiration rate in non-nodulated roots in comparison wih roots from nodulated plants may be a response to different patterns of carbon movement. This results in an increased quantity of carbohydrate in the root which stimulates electron flow via a cyanide-resistant pathway in the non-nodulated roots (Lambers, 1980) or encourages excretion of greater quantities of carbohydrates from nodulated roots (see also Chapter 7). In roots of non-nodulated plants growing with NO_3-N, more energy would be available in comparison with nodulated roots and an "overflow metabolism" may act to remove excess carbohydrate not required for growth from the root. The alternative, cyanide-resistant pathway is much more active in non-nodulated plants and this would ensure continued carbohydrate metabolism and synthesis of organic acid skeletons from the TCA cycle. It would also remove excess NADH even though the normal cytochrome system is repressed by high amounts of ATP (Lambers, 1980, 1985). Similar increases in activity of the alternative respiratory pathway have been observed in plants exposed to environmental conditions that induce sugar build-up in the root. This has significance to the relationship between rhizodeposition and plant growth, which is discussed later (Section 6).

Several other systems for studying simultaneous $^{15}N_2$ fixation and $^{14}CO_2$ fixation and distribution by legumes growing in soil have been devised (e.g. Warembourg et al., 1982, 1984) and although technically interesting the procedure has not been extended to consider rhizodeposition in soil.

5.2. Mycorrhizas

The relationship between exudation and secretion from roots and vesicular-arbuscular (V-A) mycorrhizal infection has received considerable attention recently. Work by Menge's group and others (Table 11) has suggested that V-A mycorrhizal infection is directly related to the amount of exudation of reducing sugars, and in some cases amino acids, that occurs prior to infection. Thus conditions which increase membrane permeability or increase exudation *per se* are likely to stimulate V-A mycorrhizal infection. For instance, wheat plants with low exudation rates had lower levels of V-A mycorrhizal infection than wheat plants with higher exudation rates (Azcon and Ocampo, 1981) and plant species that did not form associations with V-A mycorrhizas had lower exudation rates than plants that did (Schwab et al., 1984). However, in other studies no relationship was found between the degree of V-A mycorrhizal infection and total sugar content of root exudates from several plant species with differing degrees of mycorrhizal susceptibility (Azcon and Ocampo, 1984). In none of these experiments was a relationship found between the amount of reducing sugars and amino acids in the plant roots and V-A mycorrhizal infection. However, a close correlation was found between concentration of reducing sugars within the root of *Trifolium subterraneum* and the percentage root length infected by *Glomus fasciculatum* (Same et al., 1983) and no clear explanation for these differences is apparent.

When infection by *G. fasciculatum* occurred in *Sorghum vulgare* growing under low P availability and in tomato growing under ozone stress, exudation from the root was decreased in comparison with non-mycorrhizal plants (Graham et al., 1981; McCool and Menge, 1983). In the first instance, it was suggested that P nutrition was improved following V-A mycorrhizal infection and that the membranes of the roots had become less leaky; in the latter case it was proposed that the mycorrhizal fungus competed for the available carbohydrate so less leaked out. It should be noted that in all the experiments of Menge's group where exudation was measured, plants were removed from the soil, washed and placed in a solution containing $CaCl_2$ and antibiotics to eliminate microorganisms before measuring exudation in fresh $CaCl_2$ solution for 22 h (Graham et al., 1981). In view of the trauma involved with the removal of the roots and the effect that microorganisms are known to have on exudation, these results must be treated with caution.

TABLE 11. Examples of experiments examining rhizodeposition associated with mycorrhizal symbioses

Experimental details	*Reference*
Vesicular-arbuscular mycorrhizas	
a) Exudates and secretion measured chemically	
1) *Sorghum vulgare–Glomus fasciculatum* Effect of P content	Graham et al., 1981
2) *S. vulgare–G. fasciculatum* Effect of light intensity and temperature	Graham et al., 1982
3) *Chrysanthemum morifolium–G. fasciculatum* Effect of plant age	Johnson et al., 1982a
4) *Citrus sinensis–G. fasciculatum* Effect of photoperiod	Johnson et al., 1982b
5) *Lycopersicon esculentum–G. fasciculatum* Ozone stress	McCool and Menge, 1983
6) Host and non-host plants–*G. fasciculatum*	Schwab et al., 1984
7) Wheat cultivar effect	Azcon and Ocampo, 1981
b) $^{14}CO_2$ pulse-chase experiments	
1) *Allium porrum–Glomus mosseae*	Snellgrove et al., 1982
2) *Vicia faba–G. mosseae*	Paul and Kucey, 1981; Pang and Paul, 1980; Kucey and Paul, 1982
3) *Citrus aurantium–G. intraradices* *Poncirus trifoliata* × *C. sinensis–G. intraradices*	Koch and Johnson, 1984
4) *Glycine max–G. fasciculatum*	Harris et al., 1985
Ectomycorrhizas	
a) $^{14}CO_2$ pulse-chase experiments	
1) *Pinus taeda* *Pisolithus tinctorius* + *Pinus contorta* *Suillus granulatus*	Reid et al., 1983
b) Crop studies	
Douglas fir	Fogel and Hunt, 1983

Quantitative studies of carbon distribution and rhizodeposition in soil have been carried out with V-A mycorrhizal and non-mycorrhizal leek (*Allium porrum*) and nodulated faba bean (*Vicia faba*) plants using $^{14}CO_2$ pulse-chase techniques (Table 12). In both species more carbon moved to the roots and respiration from the roots and associated microorganisms was greater in mycorrhizal plants compared with non-mycorrhizal plants. In the leek experiment, greater amounts of ^{14}C were found in the soil of the mycorrhizal treatment compared with the non-mycorrhizal treatment and it was argued that this was due to an increase in exudation from fungal material. This is in contrast with the exudation experiments mentioned earlier. The cost of mycorrhizal

TABLE 12. Percentage distribution of net fixed C in mycorrhizal and non-mycorrhizal leek and nodulated faba bean

Species	*Period of chase with* $^{12}CO_2$	*% fixed* ^{14}C *moved to root*		*% distribution of fixed* ^{14}C *below ground*						*Reference*
				CO_2		*Soil*		*Roots*		
		m[a]	*nm*[a]	*m*	*nm*	*m*	*nm*	*m*	*nm*	
Leek	214 h	44.0	37.1	23.1	18.4	5.3	2.5	15.7	16.2	Snellgrove et al., 1982
Faba bean	48 h	47.0	36.0	30.0	18.0	–[b]	–	17.0	18.0	Pang and Paul, 1980

[a]m = mycorrhizal, nm = non-mycorrhizal.
[b]Soil not analysed.

infection in terms of the extra movement of carbohydrate to the roots (approx. 7–11% of C fixed) could have serious effects on the growth of mycorrhizal plants and there are reports of decreased growth of mycorrhizal plants at some growth stages under some environmental conditions, such as high soil P availability or low light (Bethlenfalvay et al., 1982; Bethlenfalvay and Pacovsky, 1983; Bildusas et al., 1986). But in these leek experiments the loss of photosynthate in the shoot was partially compensated for by a lower percentage dry matter in the leaves, which produced a greater net assimilation rate per unit dry matter (Snellgrove et al., 1982).

$^{14}CO_2$ pulse-chase experiments with the sour orange (*Citrus aurantium*) and the carrizo citrange (*Poncirus trifoliata* × *C. sinensis*) grown with split-root systems, half infected with *Glomus intraradices* and half non-mycorrhizal, have also been carried out (Koch and Johnson, 1984). Results suggested that the cost of sustaining completely mycorrhizal roots may be between 6 and 10% of net fixed carbon, in close agreement with the other studies.

These types of $^{14}CO_2$ pulse-chase studies have been extended to consider all combinations of V-A mycorrhiza–*Rhizobium*–legume interactions in *Vicia faba* using additional P to produce control plants of similar size to mycorrhizal plants (Paul and Kucey, 1981; Kucey and Paul, 1982). Mycorrhizal fungi of both nodulated and non-nodulated roots consumed approximately 4% of carbon fixed; 1% entered fungal biomass and 3% was respired. Nodules consumed approximately 7% of carbon fixed in non-mycorrhizal plants but 12% in mycorrhizal plants. Here, 5% was lost as respiration in nodules of non-mycorrhizal plants and 9% in nodules of mycorrhizal plants. These respiration values agree well with observations of respiration costs in other legumes obtained by other systems (see Section 5.1). The nodulated root systems of mycorrhizal plants fixed more N_2 than nodulated root systems of non-mycorrhizal plants. This was related to an increase in nodule biomass for doubly infected plants. Here, it was suggested that the legume partly compensated for the cost of the symbionts by increased rates of photosynthesis. Alternatively it may only reflect morphological adaptations that increase specific leaf areas, similar to the leek and soybean (Snellgrove et al., 1982; Harris et al., 1985). Nevertheless, compensatory stimulation of photosynthetic efficiency (μmol CO_2 fixed $dm^{-2}\ h^{-1}$) has also been observed following defoliation of soybean (Bayne et al., 1984), mycorrhizal infection of *Bouteloua gracilis* (Allen et al., 1981) and mycorrhizal infection of *Bromus inermis* (Bildusas et al., 1986), implying a complex relationship between carbon availability in roots and mycorrhizal synthesis.

Rhizodeposition in ectomycorrhizas has been less frequently studied, in part because of the technical difficulties involved. However, Reid et al. (1983) studied a range of factors concerned with the associations between pines and their mycorrhizas in gnotobiotic culture for up to 10 months, including some $^{14}CO_2$ pulse-chase experiments (Table 13). Considerably greater amounts of carbon were transferred to the roots in mycorrhizal compared with non-mycorrhizal plants. After 2 months, mycorrhizal roots contained a greater proportion of

TABLE 13. Percentage distribution of net fixed C in *Pisolithus tinctorius*-infected and non-mycorrhizal *Pinus taeda*. From Reid et al. (1983)

Age of plants (months)	*% fixed ^{14}C moved to root*		*% distribution of fixed ^{14}C below ground*					
			CO_2		*Exudates*[b]		*Roots*	
	m[a]	*nm*[a]	*m*	*nm*	*m*	*nm*	*m*	*nm*
2	38.7	16.3	6.02	2.84	0.12	0.10	32.53	13.4
10	38.4	26.4	23.5	11.66	0.4	0.3	14.48	14.46

[a]m = mycorrhizal; nm = non-mycorrhizal.
[b]Only exudates measured as the roots were transferred from vermiculite to nutrient solution for $^{14}CO_2$ feeding (5 h) and $^{12}CO_2$ chase (48 h).

^{14}C than non-mycorrhizal roots but this difference had disappeared by 10 months. However, the percentage of ^{14}C from respiration of the mycorrhizal roots was approximately twice that in non-mycorrhizal roots and this was maintained up to 10 months. This heavy drain, between 12 and 22% of photosynthate, was partially compensated for by an increase in photosynthetic rate. No differences in exudation were detected in this nutrient solution system.

The only other techniques for estimating carbon distribution and rhizodeposition in ectomycorrhizal trees involve crop studies. For Douglas fir, it has been estimated that the total standing crop of active and inactive ectomycorrhizas comprises 6% of the total tree biomass (Fogel and Hunt, 1983) and that standing crops of ectomycorrhizas from a range of host trees have biomasses in the range 0.4–10.0 t C ha^{-1} (Fogel, 1985).

6. THE SIGNIFICANCE OF RHIZODEPOSITION TO PLANTS

At first sight the large quantities of rhizodeposition appear to be very wasteful to the plant, but these may be the necessary cost for roots to grow and function in the soil. However, the availability of carbon to the root for rhizodeposition is dependent on the relative strengths of sources and sinks acting on assimilate partitioning at any one time (Gifford and Evans, 1981). There is evidence that under some circumstances the plant may have excess source potential, particularly if suitable sinks are lacking (Herold, 1980). Under these conditions several phenomena can occur. For instance, when root growth is limited by environmental conditions, an excess of carbohydrate and reducing power can build up in the roots and switch on "overflow metabolism" via an alternative, cyanide-resistant respiratory pathway (Lambers and van de Dijk, 1979; Lambers, 1980, 1985; Lambers et al., 1981; Barneix et al., 1984). An alternative strategy open to the plant could be to increase rhizodeposition, but this idea appears not to have been tested. Similarly, it has been suggested (Harris et al., 1985) that in P-deficient, non-mycorrhizal plants, growth is limited and photosynthetic end-products accumulate. By a series of feedback controls, phosphate levels would fall in the shoot and would restrict photosynthetic rates by reducing exchange of phosphate and triose phosphates across the chloroplast membrane (Giaquinta, 1980) and by directly affecting the carbon reduction cycle via the ATP/ADP ratio (Walker and Robinson, 1978). Understandably, any change in the plant status which relieves the limitation on growth would stimulate photosynthesis and, dependent on the type and degree of relief, would increase overall growth. This would therefore account for plant growth stimulation in the field in response to fertilizer applications (Cooke, 1983). It would also account for the stimulation of photosynthetic rate which occurs following mycorrhizal and *Rhizobium* infection of host plants (Allen et al., 1981; Reid et al., 1983; Bildusas et al., 1986) following defoliation (Bayne et al., 1984) and in uninfected leaves of plants infected with rust fungi (Paul and Ayres, 1984). Hence rhizodeposition from the root may not have such serious consequences for the plant as first thought. However, many of these effects depend on the severity of the carbon drain and the environmental conditions prevailing. For instance, a stimulation of photosynthesis was found in mycorrhizal-infected blue grama in comparison with non-mycorrhizal plants, but an increase in dry matter production was not evident (Allen et al., 1981). Also V-A mycorrhizal, nodulated soybean showed stimulated photosynthetic rate following defoliation but this was not sufficient to make up for the loss of shoot material (Bayne et al., 1984). Similarly, assimilation partitioning differs with plant age. For example, when plants flower and fruit, considerably more of the newly fixed carbon passes to the developing reproductive structures compared with the roots. Carbon balances within the plant and the rhizosphere are thus not simply described.

Irrespective of these source-sink considerations, the rhizodeposition may still fulfil a role for the plant. For example, by improving phosphate nutrition in soils of low phosphate availability and by enhancing growth under conditions of water stress, the carbon passed to the mycorrhizas must give these plants an ecological advantage. Similarly, by the improved nitrogen economy derived by the *Rhizobium* symbiosis in N-deficient soils, the carbon "cost" must be seen as a good investment for the plant. Continual sloughing of root cap cells and secretion of mucilage aids the growth of roots through the soil, easing the passage between soil particles. The accompanying loss of exudates may be a process that the plant has to endure to grow in soil. The carbon loss can be beneficial by attracting mutualistic organisms to the root, but correspondingly it can attract both pathogens and neutral rhizosphere microorganisms. Such phenomena have evolved over the millennia and it is difficult to talk meaningfully in terms of the cost of rhizodeposition to the plant growing in the natural environment. Perhaps only in the case of root pathogens can the term "loss" realistically be used. Eventually, all roots die and that is the final rhizodeposition of any plant.

7. THE FUTURE

Away from the natural environment, we have been conditioned in the last few years to think in terms of agricultural efficiency and how to improve the growth of crop plants. Consequently, the question arises as to whether it is possible to decrease rhizodeposition or utilize it to obtain enhanced plant growth. One possibility is to use plant breeding. For instance, in some experiments concerned with the breeding of resistance to the pathogen *Cochliobolus sativus* in wheat, a substitution of a chromosome from a resistant to a susceptible variety changed the quantity and quality of the rhizosphere population (Neal et al., 1970). It was suggested that this was related to changes in exudation. Similar changes in rhizosphere population were observed in cultures of tomato differing in resistance to tracheomycosis (Bourbos and Skoudridakis, 1987). Breeding for varieties of crop plants with decreased rhizodeposition could lead to an increase of carbon within the plant available for food production. Modern legumes, bred from wild species which originally were adapted to colonize and survive in nitrogen-deficient soils and put relatively little of their photosynthetic production into seeds, now yield protein-rich seeds in abundance and fix nitrogen more efficiently than before (Minchin et al., 1981). Here, breeding for altered carbon balances and perhaps rhizodeposition may have occurred inadvertently.

Another possibility is to treat the foliage of the plant with chemicals to either decrease rhizodeposition or to alter its composition to suit beneficial micoorganisms. Such treatment with fertilizers, hormones and fungicides has been shown to alter the rhizodeposition of several plants (Balasubramanian and Rangaswami, 1969, 1973; Jalili, 1976; Wyse et al., 1976; Hale et al., 1977).

There is also the possibility of selecting specific strains or engineering new strains of useful microorganisms to utilize the existing rhizodeposition more efficiently. Enhanced plant growth has been obtained by inoculation of bacteria and fungi onto seeds or plant roots, which act either directly on the root or by controlling soil-borne pathogens (Burr and Caesar, 1984; Kerr and Tate, 1984; Chang et al., 1986; Whipps, 1986). It is even feasible to breed plants with rhizodeposition to fit new microorganisms. Many of these steps have already been taken with *Rhizobium* species and legumes, and it may only be a question of time before suitable genetically engineered microorganisms are released into the rhizosphere.

REFERENCES

Agren, G. I., B. Ayelsson, J. G. Flower-Ellis, S. Linder, H. Persson, H. Staaf, and E. Troeng (1980). Annual carbon budget for a young Scots Pine. *Ecol. Bull. (Stockholm)*, **32**:307–313.

Allen, M. F., W. K. Smith, T. S. Moore, and M. Christensen (1981). Comparative water relations and photosynthesis of mycorrhizal and non-mycorrhizal *Bouteloua gracilis* H. B. K. Lag Ex Steud. *New Phytol.*, **88**:683–693.

Ares, J. (1976). Dynamics of the root system of blue grama (*Bouteloua gracilis* (H. S. K.) Lag.). *J. Range Management*, **29**:208–213.

Azcon, R. and J. A. Ocampo (1981). Factors affecting the vesicular-arbuscular infection and mycorrhizal dependency of thirteen wheat cultivars. *New Phytol.*, **87**:677–685.

Azcon, R. and J. A. Ocampo (1984). Effect of root exudation on VA mycorrhizal infection at early stages of plant growth. *Plant & Soil*, **82**:133–138.

Balasubramanian, A. and G. Rangaswami (1969). Studies on the influence of foliar nutrient sprays on the root exudation pattern in four crop plants. *Plant & Soil*, **30**:210–220.

Balasubramanian, A. and G. Rangaswami (1973). Influence of foliar application of chemicals on the root exudation and rhizosphere microflora of *Sorghum vulgare* and *Crotalaria juncea*. *Folia Microbiol.,* **18**:492–498.

Barber, D. A. and K. B. Gunn (1974). The effect of mechanical forces on the exudation of organic substances by roots of cereal plants grown under sterile conditions. *New Phytol.*, **73**:39–45.

Barber, D. A. and J. K. Martin (1976). The release of organic substances by cereal roots into soil. *New Phytol.*, **76**:69–80.

Barber, D. A., G. D. Bowen, and A. D. Rovira (1976). Effects of microorganisms on absorption and distribution of phosphate in barley. *Austr. J. Plant Physiol.*, **3**:801–808.

Barber, S. A. (1971). Effect of tillage practice on corn (*Zea mays* L.) root distribution and morphology. *Agronomy Journal*, **63**:724–726.

Barneix, A. J., H. Breteler, and S. C. van de Geijn (1984). Gas and ion exchanges in wheat roots after nitrogen supply. *Plant Physiol.*, **61**:357–362.

Bayne, H. G., M. S. Brown, and G. J. Bethlenfalvay (1984). Defoliation effects on mycorrhizal colonization, nitrogen fixation and photosynthesis in the *Glycine–Glomus–Rhizobium* symbiosis. *Physiol. Plant.*, **62**:576–580.

Beck, S. M. and C. M. Gilmour (1983). Role of wheat root exudates in associative nitrogen fixation. *Soil. Biol. Biochem.*, **15**:33–38.

Bethlenfalvay, G. J. and R. S. Pacovsky (1983). Light effects in mycorrhizal soybeans. *Pl. Physiol.*, **73**:969–972.

Bethlenfalvay, G. J., R. S. Pacovsky, M. S. Brown, and G. Fuller (1982). Mycotrophic growth and mutualistic development of host plant and fungal endophyte in an endomycorrhizal symbiosis. *Plant & Soil*, **68**:43–54.

Bildusas, I. J., R. K. Dixon, F. L. Pfleger, and E. L. Stewart (1986). Growth, nutrition and gas exchange of *Bromus inermis* inoculated with *Glomus fasciculatum*. *New Phytol.*, **102**:303–311.

Bokhari, U. G., D. C. Coleman, and A. Rubink (1979). Chemistry of root exudates and rhizosphere soils of prairie plants. *Can. J. Bot.*, **57**:1473–1477.

Bourbos, V. A. and M. T. Skoudridakis (1987). Das Verhalten einiger pilzlicher Antagonisten in der Rhizophäre resistenter und anfälliger Gewächshaustomaten. *J. Phytopathology*, **120**:193–198.

Bowen, G. D. and A. D. Rovira (1973). Are modelling approaches useful in rhizosphere biology? In *Modern Methods in the Study of Microbial Ecology* (T. Rosswall, ed.). *Bull. Ecol. Res. Com. (Stockholm)*, **17**:443–450.

Burr, T. J. and A. Caesar (1984). Beneficial plant bacteria. *CRC Critical Reviews in Plant Science Vol. 2* (B. V. Conger, ed.). CRC Press Inc., Boca Raton, Florida, pp. 1–20.

Buyanavsky, G. A. and G. H. Wagner (1987). Carbon transfer in a winter wheat (*Triticum aestivum*) ecosystem. *Biol. Fertil. Soils*, **5**:76–82.

Chaboud, A. (1983). Isolation, purification and chemical composition of maize root cap slime. *Plant & Soil*, **73**:395–402.

Chang, Y.-C., R. Baker, O. Kleifield, and I. Chet (1986). Increased growth of plants in the presence of the biological control agent *Trichoderma harzianum*. *Plant Disease*, **70**:145–148.

Cooke, G. W. (1983). Constraints imposed by soil on crop production. *Agricultural Research Council Letcombe Laboratory Annual Report for 1982*, pp. 84–102.

Dahlman, R. C. (1968). Root production and turnover of carbon in the root–soil matrix of a grassland ecosystem. In *Methods of Productivity Studies in Root Systems and Rhizosphere Organisms* (M. S. Ghilarov, V. A. Kovda, L. N. Novichkova-Ivanovo, L. E. Rodin, and V. M. Sveshnikova, eds.). USSR Academy of Sciences, Leningrad, pp. 11–21.

Dormaar, J. F. and D. R. Sauerbeck (1983). Seasonal effects of photoassimilated carbon-14 in the root system of blue grama and associated soil organic matter. *Soil Biol. Biochemistry*, **15**:475–479.

Edelman, G. M. (1983). Cell adhesion molecules. *Science*, **219**:450–457.

Edwards, N. T. and W. F. Harris (1977). Carbon cycling in a mixed deciduous forest floor. *Ecology*, **58**:431–437.

Elias, K. S. and G. R. Safir (1987). Hyphal elongation of *Glomus fasciculatus* in response to root exudates. *Appl. Environ. Microbiol.*, **53**:1928–1933.

Farrar, J. F. (1981). Respiration rate of barley roots: its relation to growth, substrate supply and the illumination of the shoot. *Ann. Bot.*, **48**:53–63.

Fogel, R. (1983). Root turnover and productivity of coniferous forests. *Plant & Soil*, **71**:75–85.

Fogel, R. (1985). Roots as primary producers in below-ground ecosystems. In *Ecological Interactions in Soil. Plants, Microbes and Animals* (A. H. Fitter, D. Atkinson, D. J. Read, and M. B. Usher, eds.). Blackwell Scientific Publications, Oxford, pp. 23–26.

Fogel, R. and G. Hunt (1983). Contribution of mycorrhizae and soil fungi to nutrient cycling in a Douglas-fir ecosystem. *Can. J. For. Res.*, **12**:219–232.

Foster, R. C. (1986). The ultrastructure of the rhizoplane and rhizosphere. *Ann. Rev. Phytopathol.*, **24**:211–234.

Foster, R. C. and G. D. Bowen (1982). Plant surfaces and bacterial growth: the rhizosphere and rhizoplane. In *Phytopathogenic Prokaryotes* (M. S. Mount and G. H. Lacy, eds.). Academic Press, New York, pp. 159–185.

Giaquinta, R. T. (1980). Translocation of sucrose and oligosaccharides. In *The Biochemistry of Plants Vol. 3*. Academic Press, New York, pp. 280–336.

Gifford, R. M. and L. T. Evans (1981). Photosynthesis, carbon partitioning and yield. *Ann. Rev. Plant Physiol.*, **32**:485–509.

Graham, J. T., R. T. Leonard, and J. A. Menge (1981). Membrane-mediated decrease in root exudation responsible for phosphorus inhibition of vesicular-arbuscular mycorrhiza formation. *Plant Phys.*, **68**:548–552.

Graham, J. T., R. T. Leonard, and J. A. Menge (1982). Interaction of light intensity and soil temperature with phosphorus inhibition of vesicular-arbuscular mycorrhiza formation. *New Phytol.*, **91**:683–690.

Greaves, M. P., J. A. Sargent, and J. M. Whipps (1986). Herbicide-induced interactions between cereal roots and fluorescent *Pseudomonas* spp. In *Iron, Siderophore and Plant Disease*. Proceedings of NATO Advanced Research Workshop, Wye, July 1985 (T. R. Swinburne, ed.). Plenum Press, New York, pp. 189–202.

Gregory, P. J., M. McGowan, P. V. Biscoe, and B. Hunt (1978). Water relations of winter wheat. 1. Growth of the root system. *J. Agric. Sci.*, **91**:91–102.

Grier, C. G., K. A. Vogt, M. R. Keyes, and R. L. Edmonds (1981). Biomass distribution and above and below ground production in young and mature *Abies amabilis* zone ecosystems of the Washington cascades. *Can. J. For. Res.*, **11**:155–167.

Hale, M. G. and L. D. Moore (1979). Factors affecting root exudation. II: 1970–1978. *Adv. Agron.*, **31**:93–124.

Hale, M. G., D. M. Orcutt, and L. D. Moore (1977). GA_3 and 2,3,-D effects on free sterol and fatty acid content of peanut. *Plant Physiol. Suppl.*, **59**:30.

Haller, T. and M. Stolp (1985). Quantitative estimation of root exudation of maize plants. *Plant & Soil*, **86**:207–216.

Hansen, G. K. and C. R. Jensen (1977). Growth and maintenance respiration in whole plants, tops and roots of *Lolium multiflorum*. *Plant Physiol.*, **39**:155–164.

Harris, D., R. S. Pacovsky, and E. A. Paul (1985). Carbon economy of soybean–*Rhizobium*–*Glomus* association. *New Phytol.*, **101**:427–440.

Harris, W. F., D. Santantonio and D. McGinty (1980). The dynamic below-ground ecosystem. In *Forests: Fresh Perspectives from Ecosystem Analysis* (R. H. Waring, ed.). Oregon State University Press, Corvallis, pp. 119–129.

Hatrick, A. A. and D. J. F. Bowling (1973). A study of the relationship between root and shoot metabolism. *J. Exp. Bot.*, **24**:607–613.

Helal, H. M. and D. R. Sauerbeck (1983). Method to study turnover processes in soil layers of different proximity to roots. *Soil Biol. Biochem.*, **15**:223–225.

Helal, H. M. and D. R. Sauerbeck (1986). Effect of plant roots on carbon metabolism of soil microbial biomass. *Z. Pflanzenernaehr. Bodenk.*, **149**:181–188.

Henry, C. M. and J. W. Deacon (1981). Natural (non-pathogenic) death of the cortex of wheat and barley seminal roots, as evidenced by nuclear staining with acridine orange. *Plant & Soil*, **60**:255–274.

Herold, A. (1980). Regulation of photosynthesis by sink activity—the missing link. *New Phytol.*, **86**:131–144.

Heulin, T., A. Guckert, and J. Balandreau (1987). Stimulation of root exudation of rice seedlings by *Azospirillum* strains: carbon budget under gnotobiotic conditions. *Biol. Fertil. Soils*, **4**:9–14.

Hiltner, L. (1904). Über neuere Erfahrungen und Probleme auf dem Gebiet der Bodenbakteriologie und unter besonderer Berücksichtigung der Gründüngung und Brache. *Arbeiten der Deutschen Landwirtschaftsgesellschaft Berlin*, **98**:59–78.

Hinch, J. M. and A. E. Clarke (1980). Adhesion of fungal zoospores to root surfaces is mediated by carbohydrate determinants of the root slime. *Physiol. Pl. Pathol.*, **16**:303–307.

Holden, J. (1975). Use of nuclear staining to assess rates of cell death in cortices of cereal roots. *Soil Biol. Biochem.*, **7**:333–334.

Jalili, B. L. (1976). Biochemical nature of root exudates in relation to root rot of wheat. III. Carbohydrate shifts in response to foliar treatment. *Soil Biol. Biochem.*, **8**:127–129.

Jenkinson, D. S. and J. M. Rayner (1977). The turnover of soil organic matter in some of the Rothamsted Classical Experiments. *Soil Science*, **123**:298–305.

Johnen, B. G. and D. R. Sauerbeck (1977). A tracer technique for measuring growth, mass and microbial breakdown of plant roots during vegetation. In *Soil Organisms as Components of Ecosystems* (V. Lohm and T. Persson, eds.). *Proc. VI Int. Soil Zoo. Coll. Ecol. Bull. Stockholm*, **25**:366–373.

Johnson, C. R., J. H. Graham, R. T. Leonard, and J. A. Menge (1982a). Effect of flower bud development in chrysanthemum on vesicular-arbuscular mycorrhiza formation. *New Phytol.*, **90**:671–675.

Johnson, C. R., J. A. Menge, S. Schwab, and I. P. Ting (1982b). Interaction of photoperiod and vesicular-arbuscular mycorrhizae on growth and metabolism of sweet orange. *New Phytol.*, **90**:665–669.

Keith, H., J. M. Oades, and J. K. Martin (1986). Input of carbon to soil from wheat plants. *Soil Biol. Biochem.*, **18**:445–449.

Kerr, A. and M. E. Tate (1984). Agrocins and the biological control of crown gall. *Microbiol. Sci.*, **1**:1–4.

Kipe-Nolt, J. A., U.K. Avalakki, and P. J. Dart (1985). Root exudation of sorghum and utilisation of exudates by nitrogen-fixing bacteria. *Soil Biol. Biochem.*, **17**:859–863.

Koch, K. E. and C. R. Johnson (1984). Photosynthate partitioning in split-root citrus seedlings with mycorrhizal and non-mycorrhizal root systems. *Pl. Phys.*, **75**:26–30.

Kraffczyk, I., G. Trolldenier, and H. Beringer (1984). Soluble root exudates of maize: influence of potassium supply and rhizosphere micro-organisms. *Soil Biol. Biochem.*, **16**:315–322.

Kucera, C. L., R. C. Dahlman, and M. R. Koelling (1967). Total net productivity and turnover on an energy basis for tallgrass prairie. *Ecology*, **48**:536–541.

Kucey, R. M. N. and E. A. Paul (1982). Carbon flow, photosynthesis and N_2 fixation in mycorrhizal and nodulated faba beans (*Vicia faba* L). *Soil Biol. Biochem.*, **14**:407–412.

Lambers, H. (1980). The physiological significance of cyanide-resistant respiration in higher plants. *Plant Cell Environ.*, **3**:293–302.

Lambers, H. (1985). Respiration in intact plants and tissues: its regulation and dependence on environmental factors, metabolism and invaded organisms. In *Encylopedia of Plant Physiology, Vol. 18* (R. Douce and D. A. Day, eds.). Springer Verlag, Berlin, pp. 418–473.

Lambers, H. (1987). Growth, respiration, exudation and symbiotic associations: the fate of carbon translocated to the roots. In *Root Development and Function* (P. J. Gregory, J. V. Lake and D. A. Rose, eds.). Cambridge University Press, Cambridge, pp. 125–145.

Lambers, H. and E. Steingröever (1978a). Efficiency of root respiration of a flood-tolerant and a flood-intolerant *Senecio* species as affected by low oxygen tension. *Physiol. Plant.*, **42**:179–184.

Lambers, H. and E. Steingröever (1978b). Growth respiration of a flood-tolerant and a flood-intolerant *Senecio* species: correlation between calculated and experimental values. *Physiol. Plant.*, **43**:219–224.

Lambers, H. and S. J. van de Dijk (1979). Cyanide resistant root respiration and tap root formation in two subspecies of *Hypochaeris radicata* L. *Physiol. Plant.*, **45**:235–239.

Lambers, H., F. Posthumus, I. Stulen, L. Lanting, N. J. van de Dijk, and R. Hofstra (1981). Energy metabolism of *Plantago major* ssp. *major* as dependent on the supply of mineral nutrients. *Physiol. Plant.*, **51**:245–252.

Lambers, H., R. J. Simpson, V. C. Beilharz, and M. J. Dalling (1982). Translocation and utilization of carbon in wheat (*Triticum aestivum*). *Physiol. Plant.*, **56**:18–22.

Lambers, H., R. K. Szaniawski, and R. deVisser (1983). Respiration for growth, maintenance and ion uptake. An evaluation of concepts, methods, values and their significance. *Physiol. Plant.*, **58**:556–563.

Lee, K. J. and M. H. Gaskins (1982). Increased root exudation of ^{14}C-compounds by sorghum seedlings inoculated with nitrogen-fixing bacteria. *Plant & Soil*, **69**:391–399.

Lewis, D. H. (1973). Concepts in fungal nutrition and the origin of biotrophy. *Biol. Rev.*, **48**:261–278.

Mahon, J. D. and J. J. Child (1979). Growth response of inoculated peas (*Pisum sativum*) to combined nitrogen. *Can. J. Bot.*, **57**:1687–1693.

Martin, J. K. (1977a). Factors influencing the loss of organic carbon from wheat roots. *Soil Biol. Biochem.*, **9**:1–7.

Martin, J. K. (1977b). Effects of soil moisture on the release of organic carbon from wheat roots. *Soil Biol. Biochem.*, **9**:303–304.

Martin, J. K. (1977c). The chemical nature of the carbon-14-labelled organic matter released into soil from growing wheat roots. In *Soil Organic Matter Studies Vol. 1.* International Atomic Energy Agency, Vienna, pp. 197–203.

Martin, J. K. and R. C. Foster (1985). A model system for studying the biochemistry and biology of the root/soil interface. *Soil Biol. Biochem.*, **17**:261–269.

Martin, J. K. and J. R. Kemp (1980). Carbon loss from roots of wheat cultivars. *Soil Biol. Biochem.*, **12**:551–554.

Martin, J. K. and J. R. Kemp (1986). The measurement of C transfers within the rhizosphere of wheat grown in field plots. *Soil Biol. Biochem.*, **18**:103–107.

Martin, J. K. and D. W. Puckridge (1982). Carbon flow through the rhizosphere of wheat crops in South Australia. In *The Cycling of Carbon, Nitrogen, Sulfur and Phosphorus in Terrestrial and Aquatic Ecosystems* (I. E. Galbelly and J. R. Freney, eds.). Australian Academy of Science, Canberra, pp. 77–82.

McCool, P. M. and J. A. Menge (1983). Influence of ozone on carbon partitioning in tomato: potential role of carbon flow in regulation of the mycorrhizal symbiosis under conditions of stress. *New Phytol.*, **94**:241–247.

McDougall, B. M. and A. D. Rovira (1970). Sites of exudation of ^{14}C-labelled compounds from wheat roots. *New Phytol.*, **69**:999–1003.

Merckx, R., A. den Hartog, and J. A. van Veen (1985). Turnover of root-derived material and related microbial biomass formation in soils of different texture. *Soil Biol. Biochem.*, **17**:565–569.

Merckx, R., J. H. van Ginkel, J. Sinnaeve, and A. Cremers (1986). Plant-induced changes in the rhizosphere of maize and wheat. I. Production and turnover of root-derived material in the rhizosphere of maize and wheat. *Plant & Soil.*, **96**:85–93.

Merckx, R., A. Dijkstra, A. den Hartog, and J. A. van Veen (1987). Production of root-derived material and associated microbial growth in soil at different nutrient levels. *Biol. Fertil. Soils*, **5**:126–132.

Minchin, F. R., R. J. Summerfield, P. Hadley, E. H. Roberts, and S. Rawsthorne (1981). Carbon and nitrogen nutrition of nodulated roots of grain legumes. *Plant Cell. Environ.*, **4**:5–26.

Minchin, P. E. H. and G. S. McNaughton (1984). Exudation of recently fixed carbon by non-sterile roots. *J. Exp. Bot.*, **35**:74–82.

Mitchell, R. T. and J. W. Deacon (1986). Differential (host-specific) accumulation of zoospores of *Pythium* on roots of graminaceous and non-graminaceous plants. *New Phytol.*, **102:** 113–122.

Mitchell, R. T. and J. W. Deacon (1987). Differential adhesion of zoospore cysts of *Pythium* on roots of graminaceous and non-graminaceous plants. *Trans. Br. Mycol. Soc.*, **88:**401–403.

Monteith, J. L., G. Szeicz, and K. Yabuki (1964). Crop photosynthesis and the flux of carbon dioxide below the canopy. *J. Appl. Ecol.*, **1:**321–337.

Moore, R. and C. E. McClelen (1983). A morphometric analysis of cellular differentiation in the root cap of *Zea Mays. Am. J. Bot.*, **70:**611–617.

Naim, M. S., A. F. Afifi, and A. A. El-Grindy (1976). Effect of seed and root exudates on some crops as well as their constituents singly, on the spore germination of some soil micro-mycetes. *Ind. Phytopath.*, **29:**412–417.

Neal, J. L., T. G. Atkinson, and R. I. Larson (1970). Changes in the rhizosphere microflora of spring wheat induced by disomic substitution of a chromosome. *Can. J. Microbiol.*, **16:**153–158.

Nelson, E. B., W-L. Chao, J. M. Norton, G. T. Nash, and G. E. Harman (1986). Attachment of *Enterobacter cloacae* to hyphae of *Pythium ultimum*: possible role in the biological control of *Pythium* preemergence damping-off. *Phytopathology*, **76:**327–335.

Newman, E. I. (1985). The rhizosphere: carbon sources and microbial populations. In *Ecological Interactions in Soil: Plants, Microbes and Animals* (A. H. Fitter, D. Atkinson, D. J. Read, and M. B. Usher, eds.). Blackwell Scientific Publications, Oxford, pp. 107–121.

Old, K. M. and T. H. Nicholson (1975). Electron microscopical studies of the microflora of roots of sand dune grasses. *New Phytol.*, **74:**51–58.

Ovington, J. D., D. Heitkamp, and D. B. Lawrence (1963). Plant biomass of prairie, savanna, oak wood and maize field ecosystems in central Minnesota. *Ecology*, **44:**52–63.

Pang, P. C. and E. A. Paul (1980). Effects of vesicular-arbuscular mycorrhiza on ^{14}C and ^{15}N distribution in nodulated faba beans. *Can. J. Soil Sci.*, **60:**241–250.

Pate, J. S., D. B. Layzell, and C. A. Atkins (1979). Economy of carbon and nitrogen in a nodulated and non-nodulated (NO_3-grown) legume. *Plant Physiol.*, **64:**1083–1088.

Paul, E. A. and R. M. N. Kucey (1981). Carbon flow in plant microbial association. *Science*, **213:**473–474.

Paul, N. D. and P. G. Ayres (1984). Effects of rust and post-infection drought on photosynthesis, growth and water relations in groundsel. *Plant Pathol.*, **33:**561–569.

Persson, H. (1978). Root dynamics in young Scots pine stand in central Sweden. *Oikos*, **30:** 508–519.

Prikryl, Z. and V. Vancura (1980). Root exudates of plants. VI. Wheat root exudation as dependent on growth, concentration gradient of exudates and the presence of bacteria. *Plant & Soil*, **57:**69–83.

Read, D. J., R. Francis, and R. D. Finlay (1985). Mycorrhizal mycelia and nutrient cycling in plant communities. In *Ecological Interactions in Soil. Plants, Microbes and Animals* (A. H. Fitter, D. Atkinson, D. J. Read, and M. B. Usher, eds.). Blackwell Scientific Publications, Oxford, pp. 193–217.

Reid, O. P. P., F. A. Kidd, and S. A. Ekwebelam (1983). Nitrogen nutrition, photosynthesis and carbon allocation in ectomycorrhizal pine. *Plant & Soil*, **71:**415–431.

Royle, D. J. and C. J. Hickman (1964a). Analysis of factors governing *in vitro* accumulation of zoospores of *Pythium aphanidermatum* on roots. I. Behaviour of zoospores. *Can. J. Microbiol.*, **10:**151–162.

Royle, D. J. and C. J. Hickman (1964b). Analysis of factors governing *in vitro* accumulation of zoospores of *Pythium aphanidermatum* on roots. II. Substances causing response. *Can. J. Microbiol.*, **10:**201–219.

Ryle, G. J. A., C. E. Powell, and A. J. Gordon (1979). The respiratory costs of nitrogen fixation in soyabean, cowpea and white clover. II. Comparisons of the cost of nitrogen fixation and the utilization of combined nitrogen. *J. Exp. Bot.* **30:**145–153.

Ryle, G. J. A., R. A. Arnott, C. E. Powell, and A. J. Gordon (1983). Comparisons of the respiratory effluxes of nodules and roots in six temperate legumes. *Ann. Botany*, **52:**469–477.

Same, B. I., A. D. Robson, and L. K. Abbott (1983). Phosphorus, soluble carbohydrates and endomycorrhizal infection. *Soil Biol. Biochem.*, **15**:593–597.

Santantonio, D. (1979). Seasonal dynamics of fine roots in mature stands of Douglas-fir of different water regimes: a preliminary report. In *Root Physiology and Symbiosis* (A. Reidacker and J. Gagnaie-Michard, eds.). Proceedings IUFRO Symposium on Root Physiology and Symbiosis, Nancy, France, 1978, Vol. 6. CNRF Champenoux, France, pp. 90–203.

Sauerbeck, D. R. and B. G. Johnen (1977). Root formation and decomposition during plant growth. In *Soil Organic Matter Studies, 1.* IAEA Vienna, pp. 141–148.

Sauerbeck, D. R., B. G. Johnen, and R. Six (1976). Atmung, Abbau und Ausscheidungen von Weizenwurzeln im Laufe ihrer Entwicklung. *Lanwirtsch. Forsch. Sonderh.*, **32**:49–58.

Sauerbeck, D. R., S. Nonnen, and J.-L. Allard (1981). Assimilateverbrauch und Umsatz im Warzelraum in Abhängigkeit von Pflanzenart und Anzucht. *Landwirtsch. Forsch. Sonderh.*, **37**:207–216.

Schönwitz, R. and H. Ziegler (1982). Exudation of water-soluble vitamins and of some carbohydrates by intact roots of maize seedlings (*Zea mays* L.) into a mineral nutrient solution. *Z. Pflanzenphysiol.*, **107**:7–14.

Schubert, K. R. and G. J. A. Ryle (1980). The energy requirements for nitrogen fixation in nodulated legumes. In *Advances in Legume Science* (R. J. Summerfield and A. H. Bunting, eds.). HMSO, London, pp. 85–96.

Schwab, S. M., R. T. Leonard, and J. A. Menge (1984). Quantitative and qualitative comparison of root exudates of mycorrhizal and non-mycorrhizal plant species. *Can. J. Bot.*, **62**:1227–1231.

Shone, M. G. T., J. M. Whipps, and A. V. Flood (1983). Effects of localized and overall water stress on assimilate partitioning in barley between shoots, roots and root exudates. *New Phytol.*, **95**:625–634.

Sims, P. L. and J. S. Singh (1971). Herbage dynamics and net primary production in certain ungrazed and grazed grasslands in North America. In *Preliminary Analysis of Structure and Function in Grasslands* (N. R. French, ed.). Range Science Department Science Series No. 10. Colorado State University, Fort Collins, Colorado, pp. 59–124.

Singh, J. S. and D. C. Coleman (1974). Distribution of photo-assimilated ^{14}carbon in the root system of a shortgrass prairie. *J. Ecol.*, **62**:359–365.

Sivakumar, M. V. K., H. M. Taylor, and R. H. Shaw (1977). Top and root relations of field-grown soybeans. *Agronomy Journal*, **69**:470–473.

Smucker, A. J. M. and A. E. Erickson (1987). Anaerobic stimulation of root exudates and disease of peas. *Plant & Soil*, **99**:423–433.

Snellgrove, R. C., W. E. Splittstoesser, D. P. Stribley, and P. B. Tinker (1982). The distribution of carbon and the demand of the fungal symbiont in leek plants with vesicular-arbuscular mycorrhizas. *New Phytol.*, **92**:75–87.

Szaniawski, R. K. and M. Kielkiewicz (1982). Maintenance and growth respiration in shoots and roots of sunflower plants grown at different root temperatures. *Physiol. Plant.*, **53**:500–504.

Tranquillini, W. (1964). Photosynthesis and dry matter production of trees at high altitudes. In *Formation of Wood in Forest Trees* (M. H. Zimmermann, ed.). Academic Press, New York, pp. 505–518.

Trofymow, J. A., D. C. Coleman, and C. Cambardella (1987). Rates of rhizodeposition and ammonium depletion in the rhizosphere of axenic oat roots. *Plant & Soil*, **97**:333–344.

Vancura, V. and A. Hanzlikova (1972). Root exudates of plants. IV. Differences in chemical composition of seed and seedlings exudates. *Plant & Soil*, **36**:271–282.

Vancura, V. and G. Stotzky (1976). Gaseous and volatile exudates from germinating seeds and seedlings. *Can. J. Bot.*, **54**:518–532.

van Egeraat, A. W. S. M. (1978). Root exudates of pea seedlings and their effect upon *Rhizobium leguminosarum*. In *Microbial Ecology* (M. W. Loutit and J. A. R. Miles, eds.). Springer-Verlag, Berlin, pp. 384–389.

Veen, B. W. (1980). Energy cost of ion transport. In *Genetic Engineering of Osmoregulation. Impact on Plant Productivity for Food, Chemicals and Energy* (D. W. Rains, R. C. Valentine, and A. Hollaender, eds.). Plenum Press, New York, pp. 187–195.

Wainwright, M. (1988). Metabolic diversity of fungi in relation to growth and mineral cycling in soil—a review. *Trans. Br. Mycol. Soc.*, **90:**159–170.

Walker, D. A. and S. P. Robinson (1978). Regulation of photosynthetic carbon assimilation. In *Photosynthetic Carbon Assimilation* (H. W. Siegelman and G. Hinds, eds.). Plenum Press, New York, pp. 43–59.

Warembourg, F. R. and G. Billes (1979). Estimating carbon transfers in the plant rhizosphere. In *The Soil-Root Interface* (J. L. Harley and R. S. Russel, eds.). Academic Press, London, pp. 183–196.

Warembourg, F. R. and E. A. Paul (1973). The use of $^{14}CO_2$ canopy techniques for measuring carbon transfer through the plant-soil system. *Plant & Soil*, **38:**331–345.

Warembourg, F. R. and E. A. Paul (1977). Seasonal transfers of assimilated ^{14}C in grassland: plant production and turnover, soil and plant respiration. *Soil Biol. Biochem.*, **9:**295–301.

Warembourg, F. R., D. Montange, and R. Bardin (1982). The simultaneous use of $^{14}CO_2$ and $^{15}N_2$ labelling techniques to study the carbon and nitrogen economy of legumes grown under natural conditions. *Physiol. Plant.*, **56:**46–55.

Warembourg, F. R., B. Haegel, M. Fernandez, and D. Montange (1984). Distribution et utilisation des assimilats corbones en relation avec la fixation symbiotique d'azote chez le soya (*Glycine max* L. Merrill). *Plant & Soil*, **82:**163–178.

Whipps, J. M. (1984). Environmental factors affecting the loss of carbon from the roots of wheat and barley seedlings. *J. Exp. Bot.*, **35:**767–773.

Whipps, J. M. (1985). Effect of CO_2 concentration on growth, carbon distribution and loss of carbon from the roots of maize. *J. Exp. Bot.*, **36:**644–651.

Whipps, J. M. (1986). Use of microorganisms for biological control of vegetable diseases. *Aspects Appl. Biol.* **12:**75–94.

Whipps, J. M. (1987). Carbon loss from the roots of tomato and pea seedlings grown in soil. *Plant & Soil*, **103:**95–100.

Whipps, J. M. and J. M. Lynch (1983). Substrate flow and utilization in the rhizosphere of cereals. *New Phytol.*, **95:**605–623.

Whipps, J. M. and J. M. Lynch (1985). Energy losses by the plant in rhizodeposition. *Ann. Proc. Phytochem. Soc. Eur.*, **26:**59–71.

Whipps, J. M. and J. M. Lynch (1986). The influence of the rhizosphere on crop productivity. *Adv. Microb. Ecol.*, **9:**187–244.

Wiedenroth, E. and J. Poskuta (1981). The influence of oxygen deficiency in roots on CO_2 exchange rates of shoots and distribution of ^{14}C-photoassimilates of wheat seedlings. *Z. Pflanzenphysiol.*, **103:**459–467.

Wyse, D. L., W. F. Meggitt, and D. Penner (1976). Factors affecting EPTC injury to navy bean. *Weed Sci.*, **24:**1–4.

5

POPULATION DYNAMICS AND RHIZOSPHERE INTERACTIONS

M. J. BAZIN, P. MARKHAM, E. M. SCOTT

Microbial Physiology Research Group, King's College (University of London), Campden Hill Road, London W8 7AH, UK

J. M. LYNCH

AFRC Institute of Horticultural Research, Littlehampton, West Sussex BN17 6LP, UK

1. INTRODUCTION

Within the rhizosphere there is an exchange of materials between the plant root and the surrounding microbial population. These materials may inhibit or promote growth of the plant or the microorganisms. In addition, within the microbial population itself there is likely to be a multiplicity of interactions similar to those found in other ecosystems. The question addressed in this chapter is what might be the likely effect of the rhizosphere environment on the dynamics of these interactions. We will discuss this topic in relation to what might be called an "ideal" rhizosphere, i.e. a gross simplification of the field situation. Such an approach is adopted in order to determine the most significant characteristics that control the system and to develop hypotheses that can be tested experimentally, either in the laboratory or in the field. Initially, the major physico-chemical properties that distinguish the rhizosphere from other microbial ecosystems are discussed, and then the types of biological interactions that occur in the root region are summarized. We then give an account of rhizosphere dynamics and, finally, speculate on how these might modify microbial behaviour and interactions in the rhizosphere.

2. PHYSICO-CHEMICAL PROPERTIES OF THE RHIZOSPHERE

Microorganisms may interact with the physical component of the root's environment and the result may alter the physical nature of the soil. For example, soil bacteria such as *Enterobacter cloacae* produce extracellular polysaccharides which can serve to bind soil particles together to improve soil aggregation properties and therefore soil stability (Lynch, 1987). However, the most significant physical property of the rhizosphere is that, in common with most other ecosystems, it is thermodynamically open. That is to say it has an input and output of mass and energy (Richter, 1987). This results from the downward movement of material by leaching and radial movement to and from the root by diffusion. This directional flow of solute and solvent gives rise to gradients in metabolites and in microbial biomass.

Data of Rovira and Davey (1974) show microbial population density as a function of distance from the root and are reproduced diagrammatically in Fig. 1, allowing an arbitrary definition of the soil volume occupied by the rhizosphere. The microbial population density, x, decreases with respect to radial distance, r from the root. Thus dx/dr is decreasingly negative and tends towards zero as r increases. We take the point at which this derivative is indistinguishable from zero to define the limit of the rhizosphere. Many factors may affect this point and thus the rhizosphere volume. The static situation illustrated in Fig. 1 will pertain when there is either a constant flow of material passing down the soil column and through the rhizosphere or when there is no flow at all. For variable flow we would expect the point at which dx/dr equates to zero to vary as a function of flow rate. Thus the volume of the rhizosphere as we have defined it will vary. The nature of the microbial metabolites diffusing from

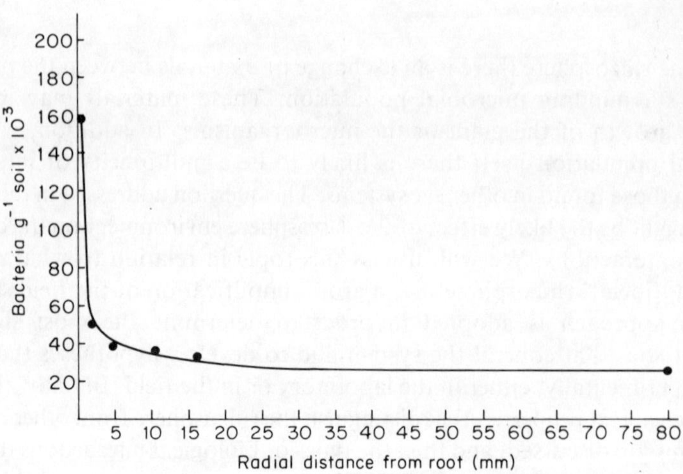

Fig. 1. Idealized rhizosphere region defined in terms of microbial density. Constructed using the data of Rovira and Davey (1974).

the root would also be expected to change the rhizosphere volume, and so the volume will be dependent on plant species and position down the root. The rate at which a solute diffuses and therefore the potential distance it can diffuse away from the root and through the microbial population is controlled by its diffusion coefficient. This is a function of the size and charge of the ion or molecule that is diffusing, the type of soil through which it is diffusing, the moisture content of the soil and the temperature (Nye and Tinker, 1977). Therefore, the effective rhizosphere volume will be a function of these parameters. Particular species of microorganisms will have effectively different rhizosphere volumes, depending on their ability to utilize or resist the effects of the materials diffusing from the root. This idea is expanded in Fig. 2, which attempts to classify microbial populations on the basis of their position in the rhizosphere. Many other factors will affect rhizosphere volume. The rhizosphere is therefore a dynamic system resulting at least in part from the physical properties of the ecosystem. On the other hand, the attachment of microorganisms to soil particles is likely to ameliorate changes in population densities. Because of this adhesion, such microorganisms are more likely to remain within the ecosystem and although their metabolic activity might vary as a function of nutrient availability, fluctuations of their population densities will be less due to the factors we have discussed.

In addition to the factors imposed by the physical structure of the rhizosphere environment, there are other interactions of a physico-chemical nature that affect the population dynamics of rhizosphere microorganisms. These may be considered to be of two types: those that are of a general nature and modify the behaviour of microorganisms in a wide variety of ecosystems, and those that are more specific to the root or rhizosphere environment. In the former category may be placed nutrient availability, oxygen tension, inhibitor concentration, temperature and pH. Although values of these parameters may be modified in the rhizosphere, the basic mechanisms by which they control or modify microbial growth and metabolism are similar in all environments. Laboratory experiments with pure cultures are often eschewed by soil micro-biologists (Dommergues, 1978), but such studies have increased considerably our knowledge of microbial physiology, particularly with respect to these environmental factors. The Monod relationship between growth-limiting nutrient concentration and the specific growth rate of a microbial population (Monod, 1942) is a classic example of such a study and has been modified, expanded and replaced with alternatives by many workers (Pirt, 1975; Howell, 1983; Roels, 1983; Bailey and Ollis, 1986). It is unlikely that the basic kinetics suggested by these and other workers are fundamentally different for microorganisms growing in the rhizosphere or in a lake or in the alimentary canal of a mammal. Similarly, the relationships suggested for the way in which temperature and specific growth rate are related (Topiwala and Sinclair, 1971; Ratowsky et al., 1982, 1983; McMeekin et al., 1988) are likely to be generally applicable. The effect of pH is not quite so clear cut. The fundamental effect of hydrogen ion concentration on microbial metabolism is species

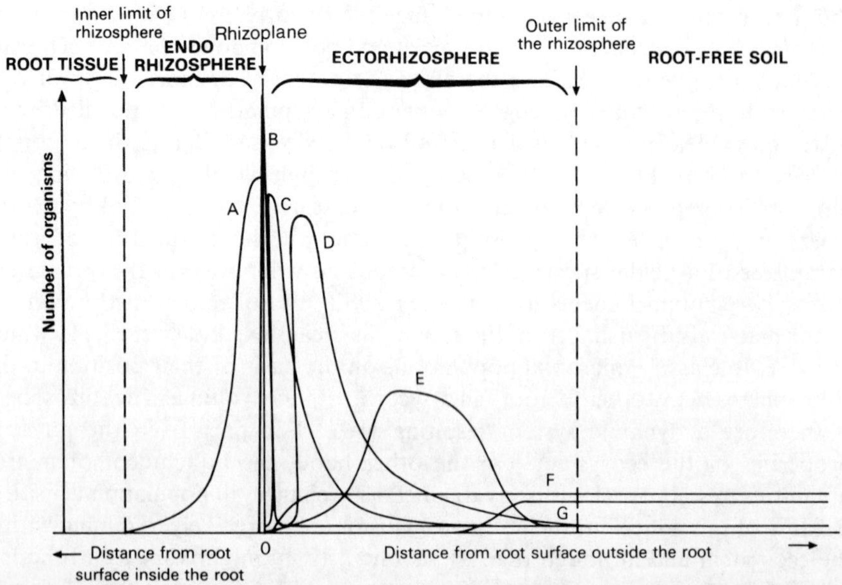

Fig. 2. Distribution of different types of microorganism around the plant root. The *rhizoplane* = the interface between the soil and the root surface. The *ectorhizosphere* = the region external to the root, within which the root directly or indirectly perturbs the concentration of one or more microbial species compared to the root-free environment. The *endorhizosphere* = the region internal to the root, within which microorganisms other than pathogens or symbionts occur. The *outer limit of the rhizosphere* is defined as the point at which concentrations of *all* microorganisms in the environment surrounding the root have returned to the level found in an identical but root-free environment. The *inner limit of the rhizosphere* is defined as the point within the root at which concentrations of *all* microorganisms other than pathogens and symbionts fall to zero.

Types of rhizosphere microorganisms:

Root Penetrating:

Type A = microorganisms which grow in the intracellular spaces within the root and whose concentration falls to that found in root-free environments immediately outside the root.

Type B = microorganisms which grow on the rhizoplane and whose concentrations fall to basal levels at very short distances from the root surface and to zero at very short distances into the root.

Not Root Penetrating

Type C = microorganisms physically excluded from the rhizoplane by type B organisms, but which utilize organic material originating from the root or products of type B organisms, or both.

Type D = microorganisms physically or biochemically excluded by type B and C organisms but which utilize products of type C organisms, or complex or intractable organic materials originating from the root and not utilized by type B and C organisms due to their inability to metabolize them or due to catabolite repression in those organisms by the simpler organic molecules they do utilize from the plant.

Type E = organisms which utilize secondary products originating from rhizosphere organisms, but which are at a competitive disadvantage with respect to types B, C and D and therefore are unable to compete for molecules originating directly from the root.

Type F = general soil microorganisms physically excluded from niches on or close to the rhizoplane, but not sensitive to any inhibitory products or activities of the types B, C, D and E organisms (e.g. antibiotics, iron chelation, etc.).

Type G = General soil microorganisms excluded from most of the rhizosphere by the activities and inhibitory products of type B, C, D, E and F organisms (i.e. sensitive to antibiotics and unable to compete for carbon source and essential elements, e.g. iron).

None of these classes is meant to represent the likely distribution of any given microbial species, but each represents the possible distribution of a species with the type of behaviour described. It is clear that any given species will have its own specific distribution pattern which will only approximate to one of those shown. Similarly it is quite conceivable that some species may show a combination of the behaviours described and therefore a combination of two or more types of distribution pattern.

specific. However, Nye (1986) and Hedley et al., (1982) present convincing evidence that imbalance between nutrient cation and anion uptake across the root–soil interface results in a decrease in rhizosphere pH by as much as two units. As a result, phosphorus, iron and manganese may become more available to the plant (and to the microbial population) and consequently have a beneficial effect, while the resulting increase in aluminium might be detrimental. Some pathogens, such as *Gauemannomyces graminis*, are sensitive to pH (Chapters 9 and 15) while some species of *Pseudomonas* are more tolerant and outcompete the pathogen under such conditions.

Physico-chemical factors of the rhizosphere include water availability, the cation exchange properties of soil and growth of microbial populations on the surface of soil particles. An excellent introduction to the subject of soil water content is given by Flegmann and George (1975). Several forces are responsible for retaining water in soil and these must be overcome before a plant root or a microbial population can absorb water. Similarly, the cation exchange properties of soil might decrease the availability of these ions to plant and microorganisms. This is particularly significant when ammonium rather than nitrate is a nitrogen source. The effect of microbial attachment to surfaces is discussed by Fletcher (1984) and Savage (1984). Dazzo (1984) emphasizes the importance of attachment to root surfaces as the first step in infection by symbionts and pathogens and the influence of microbes on plant morphogenesis and nutrition.

3. BIOLOGICAL INTERACTIONS IN THE RHIZOSPHERE

The great majority of organisms in the rhizosphere are microorganisms, and these include representatives of all the major groups: bacteria, actinomycetes, fungi, protozoa and microalgae. In addition, microfauna, such as nematodes and mites, are found at significantly elevated concentrations in the rhizosphere compared to root-free soil. A convenient way of expressing the relative concentrations of organisms in the rhizosphere as compared to their basal concentration in root-free soil is by determining their rhizosphere : soil (R : S) value. This is simply the ratio of the number of any given species present in one unit mass of rhizosphere soil to the number present in equivalent amount of root-free soil. Typical R : S values for unicellular bacteria are 10–20, although the figure can be greater than 100. For other groups of organisms the R : S values decrease in the general order: unicellular bacteria > actinomycetes and fungi > protozoa and algae > microfauna (Katznelson, 1965). This order of relative densities of the major groups of organisms in the rhizosphere does not differ from that found in the soil as a whole, nor does it apparently change significantly with different plant types, but absolute numbers of organisms do change. For example, legumes appear to support considerably larger rhizosphere populations than non-leguminous plants (Katznelson, 1965).

Whilst the biomass concentration per gram of soil in the rhizosphere is greater than that in root-free soil (Lynch and Panting, 1980), species diversity shows the opposite trend. For example, the number of fungal species detectable is least on the rhizoplane, greater within the rhizosphere, and greatest in the root-free soil (Srivastava, 1973).

Within the rhizosphere there is interaction between the plant root and the microbial community, and between the microbial populations that compose the community. Both root and microorganisms will compete for nutrients when these are in short supply. Similar kinetic expressions for nutrient uptake by a bacterial population and by the plant root have been suggested. In both cases the specific rate of nutrient uptake is expressed as a function of nutrient concentration (S) and saturates at a maximum value. For microbial specific growth rate (μ) the Monod function (Monod, 1942) is employed:

$$\mu = \frac{\mu_m S}{K_s + S} \tag{1}$$

Here μ_m is the maximum specific growth rate and K_s is a saturation constant. The analogous expression for nutrient uptake by a root (Nye and Tinker, 1977) is:

$$V = -\frac{V_m S}{K + S} \tag{2}$$

where K is the saturation constant. Alternative ways of expressing this relationship for a root used by Nye and Tinker (1977) are:

$$I = 2\pi\alpha a S \tag{3}$$

and

$$C = \frac{2\alpha S}{a} \tag{4}$$

Here I is uptake rate per unit length of root, C is the unit absorption rate, α is the root absorbing power and a is the radius of the root axis. A range of alternatives have been suggested for the kinetics of nutrient uptake by microorganisms (Pirt, 1975; Howell, 1983, Roels, 1983; Bailey and Ollis, 1986).

Considerable amounts of the carbon fixed by plants are lost through the roots. Estimates of this loss being as high as 40% have been made (see Chapter 4). Rovira and Davey (1974) classify this material as diffusible volatile, diffusible water soluble and non-diffusible. The nature of the material will affect its availability as nutrient for microorganisms. Environmental factors that affect the release of this carbonaceous material include temperature (Rovira, 1959; Brown, 1972; Whipps, 1984), light intensity (Rovira, 1959), water stress (Reid, 1974), pH (Rovira and Ridge, 1973), root impedance and injury (Barber and

Gunn, 1974; D'Arcy, 1982; Schonwitz and Ziegler, 1982) and the stage of development of the plant (Rovira, 1956).

Although the flow of carbon and other materials from the root must be of considerable significance in the rhizosphere, little information about the kinetics of loss seems to be available. By contrast, product formation by microorganisms has been studied extensively (Pirt, 1975). This may be growth-linked so that

$$\frac{dP}{dt} = q_p X \qquad (5)$$

where P is the concentration of product, X is the microbial biomass concentration and q_p is the specific rate of product formation. Non-growth-linked product formation results either when q_p is independent of growth rate or when it varies with specific growth rate in what Pirt (1975) describes as a complex way.

The types of interaction that may occur within the rhizosphere can be classified in two ways, either by using the standard categories for interactions between organisms—mutualism, commensalism, symbiosis, parasitism, etc.—or by considering the nature of the interacting organisms and grouping them accordingly as plant : microbe, microbe : microbe, microfauna : microbe, etc. This latter scheme seems more useful in the context of the rhizosphere as the precise nature of the interaction is often difficult to identify and can more conveniently be considered in terms of the organisms involved. We begin this section with such a scheme and then describe microbial interactions in rather more general terms. Some of the classic symbioses and parasitisms are excluded from our discussion as these are documented in detail elsewhere in this book.

Interactions between organisms that involve physical contact, such as in predation and parasitism, are said to be direct (Fredrickson and Stephanopoulos, 1981). Indirect interactions are those in which no physical contact occurs but some non-living part of the environment is a necessary intermediary, such as in antagonism—the inhibition of one organism by the growth product of another. If the presence of one organism promotes the growth of another, the interaction is said to be positive; if it is detrimental, the interaction is negative. In the rhizosphere examples of all these types of interaction have been observed.

3.1. Plant–Microbe Interactions

Indirect interactions occur when the microbial population produces biologically active substances which affect plant growth. These may be positive when these substances promote growth, as seems to be the case when biotin and pantothenic acid are produced (Rovira, 1959; Brown, 1972) or when iron uptake by the root is enhanced by agrobactin production by *Agrobacterium tumefaciens* (see Chapter 7). Some species of *Pseudomonas* produce cyanide, resulting in an indirect negative interaction with respect to the root (Schippers et al., 1987). Reduction of pH by the root can result in both indirect positive and negative interactions within the microbial population as indicated above. The most widely studied interactions between the root and microorganisms are direct. This is because of their economic importance. These are negative when parasitism by bacteria or fungi causes disease, or positive when the result is beneficial, such as with legume–*Rhizobium* associations.

3.2. Microbe–Microfauna Interactions

Probably the most widespread interaction between microbes and microfauna results from predation by the latter on the former. This is a direct interaction, negative with respect to the prey and positive with respect to the predator. It is possible that this interaction results in the higher concentration of such organisms as nematodes, which prey on bacteria, and mites, which eat fungi, in the rhizosphere than in root-free soil. There seems to be little evidence that this microfauna utilizes low molecular weight exudates from roots. It probably feeds opportunistically on root debris; nematodes can certainly be devastatingly invasive of root systems. Such high-level invasion of the rhizosphere and the root tissue is, however, the exception, but it is not clear to what extent the root's own defences normally prevent such invasion or to what extent microbe : microfauna interactions in the rhizosphere provide protection for the root due to microbial antagonism of the microfauna.

Ingham et al. (1985) undertook an extensive investigation of predation of fungi and bacteria by nematodes in microcosms with and without plants. They interpreted their results on the basis of what they called a "conceptual" model of the systems under study. They found that bacteria, fungi and nematodes were more abundant in the rhizosphere than in root-free soil; the bacterial density was higher in the presence of nematodes; and that plants in the presence of bacteria plus nematodes grew faster and initially took up more nitrogen than they did in the presence of bacteria alone.

3.3. Microbe–Microbe Interactions

Between microorganisms the most widely studied interactions are those which secondarily either benefit or are detrimental to the plant. In the former category, the direct negative interactions between *Enterobacter cloacae* and *Pythium* and between *Trichoderma* and many pathogenic fungi result in a decrease in root disease (Lynch, 1987). Protozoa are usually more numerous in planted soils than in unplanted soils, and the R : S ratio for protozoa associated with spring wheat has been reported to be 2 : 1 (Darbyshire and Greaves, 1973). Heznen et al. (1988) reported that protozoa reduced the numbers of *Rhizobium leguminosarium* in a loamy sand but introduction of bentonite clay seemed to protect the bacteria. Indirect interactions benefiting the plant occur when a harmful microbial species is inhibited by antibiotic production by another species. Such appears to be the case with *Pseudomonas fluorescens* and *Bacillus mycoides*, which produce antibiotics inhibitory to *Gauemannomyces graminis*, the causative organism of take-all of cereals and grasses. A slightly more complex interaction occurs between *Agrobacterium rhizogenes* and *A. radiobacter*, which causes crown gall of fruit trees. The pathogen induces the plant to produce opines, which are catabolized after being transported into the cells by a specific permease. The antagonist produces an antibiotic, agrocin, which is transported by the same permease and to which the pathogen is sensitive (Chapters 6 and 7).

It appears superficially then that the most important type of microbe : microbe interaction in the rhizosphere is antagonism, although this may simply reflect the fact that in general such activity is easier to detect and investigate than the other major types of interaction, such as mutualism and competition. As indicated above, one of the most significant factors involved in antagonistic interactions between rhizosphere microbes is the extent to which they produce and are sensitive to antibiotics. Another factor appears to be competition for metal ions, especially Fe^{3+}.

Whilst at least some fungi do produce iron-scavenging molecules, the predominant competition for Fe^{3+} in the rhizosphere seems to be between different species or strains of bacteria (Chapter 7). This may either take the form of direct competition between a species which produces an iron-scavenging siderophore and species which do not have such a capacity, or probably more likely, between two different types of siderophore molecules produced by two competing species. The competition will then depend on the relative affinities of the two types of siderophore molecules for the Fe^{3+} ions, but also on the capacity of the bacterial cell surface and membrane complex to release the bound iron from the siderophore molecule. Thus the successful species in such a competition will be either the species which produces the siderophore molecule with the highest affinity for Fe^{3+} ions, or the species which is able to release the iron not only from its own siderophore molecules but from those produced by one or more of its competitors. In this way some species of bacteria may be able to pirate Fe^{3+} ions chelated by siderophores in the soil which originated from competitor species (Schippers et al., 1987).

In a similar way the classic antagonistic interaction between bacteria mediated by antibiotic molecules will depend not only on the ability of a given species to produce such molecules, but also on the relative sensitivity of that species to antibiotic molecules produced by other organisms in the rhizosphere. Thus a bacterium with a high level of resistance to a range of antibiotics is more likely to be a successful competitor in such an environment than a species which is capable of producing large quantities of one or more highly active antibiotics, but which has a marked sensitivity to even a single antibiotic molecule it is likely to encounter. Antibiotic resistance is thus likely to be the most important attribute for a successful bacterium in the rhizosphere, especially in view of the fact that not only will other bacteria be competitors producing antibiotics, but far more significantly, filamentous fungi will also be competitors in this respect. Such fungi have the advantage of being sensitive to fewer antibiotics than bacteria and because they consist of ramifying networks of hyphae rather than single cells, they have greater resources in biomass and stored nutrients to employ both for offensive and defensive strategies.

3.4. The Dynamics of Microbial Interactions

The ways in which microorganisms interact with each other have been reviewed several times (Slater and Bull, 1978; Bazin, 1981; Slater, 1988). The types of interaction between two species can be described by considering growth to be logistic and writing equations for the two populations at concentrations X_1 and X_2 as:

$$\frac{dX_1}{dt} = \alpha_1 X_1 - \beta_{12} X_1 X_2 - \beta_{11} X_1^2 \quad (6)$$

$$\frac{dX_2}{dt} = \alpha_2 X_2 - \beta_{21} X_1 X_2 - \beta_{22} X_2^2 \quad (7)$$

Here the α_i terms are specific growth rates and the β_{ii} terms are related to the carrying capacity of the environment. The β_{ij} terms, which need not be constants, define the interaction between the two populations. In particular, the sign of the β_{ij} terms defines the simple two-species interactions as indicated in Table 1. Of these interactions, commensalism, mutualism and amensalism are of considerable ecological importance and are reviewed by Slater and Bull (1978).

Competition has been admirably reviewed by Fredrickson and Stephanopoulos (1981) in an article which neatly classifies the way this interaction can be expressed. From a theoretical point of view a long-recognized dilemma concerning competition has existed. This is that in a well-mixed open ecosystem, competition results in the survival of only one of the species. A simple proof of this conjecture is given by Burghes and Borrie (1981). This idea is extended in Hutchinson's paradox of the plankton. Essentially, this poses the question

TABLE 1. Two-species interactions defined by the signs of their coefficients of interaction, that is the β_{ij} terms in equations (6) and (7)

Coefficient of interaction		*Type of interaction*
β_{12}	β_{21}	
0	0	Neutralism
0	>0	Commensalism
>0	0	
0	<0	Amensalism (antagonism)
<0	0	
<0	>0	Predator–prey; host–parasite
>0	<0	
>0	>0	Mutualism
<0	<0	Competition

of why so many different species of metabolically similar algae are capable of coexisting in lakes. There seems to be considerable confusion in the microbiological literature about competition. In a closed environment, under the assumptions we have made here, both species will, of course, survive. The relative concentrations of each species at stationary phase will depend on their inoculum sizes and their competitive ability. The situation is not that described by Slater (1988) who in his equations for competition in batch culture does not include an interaction term. By most definitions, competition does not take place between species if they do not interact.

Theoretical aspects of predator–prey activity have been investigated exhaustively, but relatively few experimental observations have been made on such systems. For microorganisms, a Monod function has been suggested for the predator specific growth rate (Canale, 1970). That is:

$$\gamma = \frac{\gamma_m X}{K_2 + X} \tag{8}$$

Here X represents the population density of the prey, γ_m is the maximum specific growth rate and K_2 is the saturation constant. When it is assumed that prey growth is also a Monod function, equations representing change in prey substrate, change in prey density and change in predator density can be written. Analysis by numerical integration (Curds, 1971) or linearization near equilibrium (Canale, 1969, 1970) shows that the behaviour of such a system depends on the rate of flow through it, or more precisely its residence time. For increasing residence times the following pattern of qualitatively different types of behaviour is predicted: both prey and predator wash out of the system; the prey only survives; the two populations coexist and their densities change smoothly and monotonically before steady state is reached; steady state occurs after the populations exhibit damped oscillations; no steady state is achieved but the two populations oscillate continuously. Tsuchiya et al. (1972) grew slime mould amoebae as the predator organisms and *Escherichia coli* as the prey in chemostat cultures. They interpreted their results in terms of the last mode of behaviour—sustained limit-cycle oscillations. Dent et al. (1976) using a similar system observed damped oscillations. Closer analysis of these data showed that the predator specific growth rate was a function of both the prey and the predator population densities rather than just the prey density (Bazin and Saunders, 1978). This conclusion was in agreement with earlier work of Curds and Cockburn (1968) using batch cultures and was later reinforced by results from two-stage chemostat experiments (Bazin et al., 1983). Thus, rather than a Monod-type specific growth rate function, it appears that an expression introduced by Contois (1959) for bacterial growth might be a better approximation for predation kinetics. This takes the form:

$$\gamma = \frac{\gamma_m X/P}{K_2 + X/P} = \frac{\gamma_m X}{K_2 P + X} \tag{9}$$

where P is the predator density. As indicated by this relationship it appears that the predator specific growth rate depends on the amount of prey per predator, something akin to the ration for each predator.

Host–parasite interactions in well-mixed systems can be treated dynamically in a manner similar to that for prey and predator populations. Although there may be differences in the kinetic functions employed, model formulation and analysis are analogous. Reviews of this approach are given by Anderson and May (1979) and May and Anderson (1979).

Although we have been considering relationships between species as though each association can have only one interaction, this is not the case. Under different environmental conditions the type of interaction between two populations can change. This has been shown experimentally by Megee et al. (1972) who grew *Saccharomyces cerevisiae* and *Lactobacillus casei* together in continuous culture.

3.5. Multispecies Systems

Interactions between more than two species under defined laboratory conditions have received little attention except for theoretical investigations. Probably this is due to the technical difficulties associated with problems of experimentation, in particular, sampling methods.

Equations (6) and (7) may be abbreviated to:

$$\dot{X}_i = \alpha_i X_i + \sum_{j=1}^{n} \beta_{ij} X_i X_j \tag{10}$$

when n, the number of species present, is equal to 2 and the dot over a symbol represents its time derivative. To represent more than two species interacting all that is required is to increase the value of n appropriately. These equations are nonlinear and have no known solution. However, their behaviour near equilibrium may be determined by approximating them with linear equations. Equilibrium is defined by setting the differential equations equal to zero, i.e.:

$$\dot{X}_i = \alpha_i \tilde{X}_i + \sum_{j=1}^{n} \beta_{ij} \tilde{X}_i \tilde{X}_j \tag{11}$$

where the tilde over a variable indicates its equilibrium value. Linearization near equilibrium requires new variables to be defined such that the new equilibrium values are all zero. The transformation

$$x_i = X_i - \tilde{X}_i \tag{12}$$

achieves this. Because we are considering the situation near equilibrium, all values of x will be small. Second and higher order values of x will therefore be very small and linearization depends on equating these to zero. For example, the first member of a three-species system changes according to:

$$\dot{X}_1 = \alpha_1 X_1 + \beta_{11} X_1{}^2 + \beta_{12} X_1 X_2 + \beta_{13} X_1 X_3 \tag{13}$$

Applying the transformation given in equation (12) gives:

$$\dot{x}_1 = \alpha_1(x_1 + \tilde{X}_1) + \beta_{11}(x_1 + \tilde{X}_1)^2 + \beta_{12}(x_1 + \tilde{X}_1)(x_2 + \tilde{X}_2) + \beta_{13}(x_1 + \tilde{X}_1)(x_3 + \tilde{X}_3) \quad (14)$$

and multiplying out the terms in the brackets results in:

$$\begin{aligned} \dot{x}_1 = {} & \alpha_1 x_1 + \alpha_1 \tilde{X}_1 + \beta_{11} x_1{}^2 + 2\beta_{11}\tilde{X}_1 x_1 + \beta_{11}\tilde{X}_1{}^2 \\ & + \beta_{12}x_1x_2 + \beta_{12}\tilde{X}_2 x_1 + \beta_{12}\tilde{X}_1 x_2 + \beta_{12}\tilde{X}_1\tilde{X}_2 \\ & + \beta_{13}x_1x_3 + \beta_{13}\tilde{X}_3 x_1 + \beta_{13}\tilde{X}_1 x_3 + \beta_{13}\tilde{X}_1\tilde{X}_3 \end{aligned} \quad (15)$$

At equilibrium, by definition,

$$0 = \alpha_1\tilde{X}_1 + \beta_{11}\tilde{X}_1{}^2 + \beta_{12}\tilde{X}_1\tilde{X}_2 + \beta_{13}\tilde{X}_1\tilde{X}_3 \quad (16)$$

and therefore,

$$0 = \alpha_1 x_1 + \beta_{11}\tilde{X}_1 x_1 + \beta_{12}\tilde{X}_2 x_1 + \beta_{13}\tilde{X}_3 x_1 \quad (17)$$

Applying these equalities to equation (15) and eliminating second-order terms gives:

$$\dot{x}_1 = \beta_{11}\tilde{X}_1 x_1 + \beta_{12}\tilde{X}_1 x_2 + \beta_{13}\tilde{X}_1 x_3 \quad (18)$$

Similar treatment of the equations for the other two species results in the following set of equations written, for convenience, in matrix notation:

$$\begin{bmatrix} \dot{x}_1 \\ \dot{x}_2 \\ \dot{x}_3 \end{bmatrix} = \begin{bmatrix} \beta_{11}\tilde{X}_1 & \beta_{12}\tilde{X}_1 & \beta_{13}\tilde{X}_1 \\ \beta_{21}\tilde{X}_2 & \beta_{22}\tilde{X}_2 & \beta_{23}\tilde{X}_2 \\ \beta_{31}\tilde{X}_3 & \beta_{32}\tilde{X}_3 & \beta_{33}\tilde{X}_3 \end{bmatrix} \begin{bmatrix} x_1 \\ x_2 \\ x_3 \end{bmatrix} \quad (19)$$

The solution of such equations is well known and takes the form

$$x_i = \mathrm{A}_i e^{\lambda_1 t} + \mathrm{B}_i e^{\lambda_2 t} + \mathrm{C}_i e^{\lambda_3 t} \quad (20)$$

Here A, B, and C are constants dependent on the initial conditions of the system and are not important for our purposes here. What is important are the values of λ_1, λ_2 and λ_3. If all of these eigenvalues (as they are called) are negative then the x_i terms will tend to zero as time increases. Thus X_1, X_2, and X_3 move towards their equilibrium values. In this case the system is said to be stable. If just one of the eigenvalues is positive, however, x_i will increase without limit. The system moves away from equilibrium and is said to be unstable. In order to judge whether or not a system is stable it is necessary to know how the eigenvalues are determined. This is done by writing the central matrix in equation (19), often called the stability matrix, as a determinant, subtracting λ from each of the diagonal elements and equating the result to zero, i.e.,

$$\begin{vmatrix} \beta_{11}\tilde{X}_1-\lambda & \beta_{12}\tilde{X}_1 & \beta_{13}\tilde{X}_1 \\ \beta_{21}\tilde{X}_2 & \beta_{22}\tilde{X}_2-\lambda & \beta_{23}\tilde{X}_2 \\ \beta_{31}\tilde{X}_3 & \beta_{32}\tilde{X}_3 & \beta_{33}\tilde{X}_3-\lambda \end{vmatrix} = 0 \qquad (21)$$

The diagonal elements of the stability matrix correspond to the intrinsic stabilities of the parts. The off-diagonal elements represent the interaction terms between the species; whether they are positive or negative depends on the sorts of interactions taking place. If they are zero a neutralistic association is implied, that is to say no interaction is taking place. In fact, as the actual numerical values of these need not be calculated in order to determine whether or not a system is stable, Routh–Hurwitz criteria may be applied to the stability matrix. These are commonly part of computer libraries and can be simply applied to test for stability.

It has long been held in ecology that the stability of a system is dependent on its complexity; the more complex the system the more stable it is supposed to be. Gardner and Ashby (1970) showed that at least for linearly connected systems this is not the case. They performed calculations in which they estimated the probability of stability in such a system and plotted this probability against the connectance (C) of the system and the number of independent variables (n). In terms of the current discussion, these are equivalent to the number of interactions between species and the number of species present. Mathematically, this is the number of non-zero, non-diagonal elements and the magnitude of the stability matrix. They did this by giving randomly chosen negative values to the diagonal elements of an $n \times n$ matrix between -1 and -0.1. This ensured that no eigenvalues would be *a priori* positive. They then randomly assigned numbers between -1 and $+1$ to C% of the off-diagonal elements. Then they determined whether or not the resulting matrix was stable. This procedure was repeated a sufficient number of times to determine the probability of stability for systems with given values for C and n. By changing these two values and repeating the procedure Gardner and Ashby (1970), and at a later date, Daniels and Mackay (1974) clearly showed that by increasing either C or n, the number of interactions or the number of interacting species, the probability of stability was reduced. Of course, this does not imply that ecologically this is the case. It might be that by discarding nonlinear terms an essential feature of community structure is eliminated and these may be of fundamental importance for ecosystem stability. However, the analysis does indicate that stability is not the result of simple interactions between quasi-random assemblages of organisms.

A less theoretical study of multi-species microbial interactions was undertaken by Rashit and Bazin (1987). The purpose of this study was to determine the effects of fluctuating environments on the structure of microbial communities. They subjected mixed populations of protozoa and bacteria to fluctuations in the supply of nutrient to chemostat cultures. They varied both the magnitude of the fluctuations and the concentration of nutrient supplied. The response of the system was measured in terms of species diversity after several weeks of culture at which time the cultures were judged to have equilibrated. The species diversity of an ecosystem reflects both the number of species present and the number of individuals per species. A system with the same number of species but with a more even distribution of individuals is more diverse than one with unevenly distributed individuals. Several indices of species diversity have been suggested (Pielou, 1969). The one adopted by the authors was the Shannon–Wierner index, H, given by:

$$H = -\sum p_i \log p_i \tag{22}$$

where p_i is the proportion of the community belonging to the ith species. Rashit and Bazin (1987) found that the effect of the perturbations depended on the concentration of the input medium to the system or, in ecological terms, the productivity of the environment. Under low productivity conditions, fluctuations decreased the diversity, whereas under high productivity conditions, fluctuations increased species diversity.

4. RHIZOSPHERE DYNAMICS

What we have discussed briefly above is essentially a qualitative description of some aspects of microbial activity in the rhizosphere. However, as we indicated in the discussion of our working definition of the rhizosphere, the system is dynamic; it changes with respect to time. It also changes with respect to position, that is distance from the root and depth in the soil. The consequences of these changes can realistically only be described and investigated by resorting to mathematical models of the system. As Moorby (1987) has pointed out in a wider context, we now have sufficient quantitative information about crop growth to produce and validate mathematical models which can increase our understanding of how the process works and how it can be controlled. Fortunately, some excellent work of this type has been carried out with soil–root systems (Nye and Tinker, 1977; Newman and Watson, 1977).

Mathematical models of the rhizosphere are based usually on the two major factors contributing to the movement of solutes in the soil water. These are downward flow by leaching and diffusion from and to the root. All such models are idealizations of the rhizosphere system but attempts have been made to incorporate data from the field and from laboratory observations (Nye and Tinker, 1977; Newman and Watson, 1977).

A review of leaching models has been undertaken by Addiscott and Wagenet (1985). For a homogeneous, root-free soil through which there is a downward flow of water at a constant rate, the one-dimensional continuity equation is:

$$\frac{\partial S}{\partial t} = D^* \frac{\partial^2 S}{\partial z^2} - v\frac{\partial S}{\partial z} \pm \text{sink terms} \tag{23}$$

where S is the solute concentration, z is depth and v is the pore water velocity obtained by dividing the volumetric water flux by the volume of water present, both quantities assumed to be constant. The apparent diffusion coefficient, D^*, is supposed to represent the effects both of mechanical dispersion and chemical diffusion, and is also taken to be constant. When there is a sink or a source for the solute, an appropriate extra term on the right-hand side of equation (23) is included. Relatively simple equations of this form have been solved analytically or at least partially analytically for herbicide movement and leaching under field conditions (Smith et al., 1984). A nitrification system with a constant concentration of nitrifiers has been analysed by Cho (1971), and nitrification with microbial growth but no dispersion or diffusion was investigated by Saunders and Bazin (1973). Warrick et al. (1971) derived two partially analytic methods for solving one-dimensional continuity equations which describe the displacement of solutes during infiltration of a soil column allowing for variable velocity. However, equations of this complexity usually require the application of numerical methods.

Newman and Watson (1977) developed a model for microbial growth in the rhizosphere in which growth was assumed to be nutrient limited and nutrient was supplied to the microbial population by diffusion from the root. Change in substrate at concentration S was described by:

$$\frac{\partial S}{\partial t} = F_D + F_G + F_M + F_I \tag{24}$$

where F_D is the diffusion term, F_G and F_M represent loss due to microbial growth and maintenance respectively, and F_I is a term representing an indigenous supply of nutrient from the soil. The diffusion term used is as given in equation (24), growth is assumed to follow Monod kinetics and substrate utilization for maintenance is proportional to microbial biomass. A constant supply of nutrient from the soil capable of supporting a quantity of microbial biomass equivalent to that present in root-free soil is also included. The equations describing change in growth-limiting nutrient and biomass at density X are:

$$\frac{\partial S}{\partial t} = \frac{D}{r}\frac{\partial}{\partial r}\left(r\frac{\partial S}{\partial r}\right) - \frac{\mu_m XS}{Y(K_s\theta + S)} + CX_0 \tag{25}$$

$$\frac{\partial X}{\partial t} = \frac{\mu_m XS}{K_s\theta + S} - CYX \tag{26}$$

where C is the maintenance coefficient, X_0 is the endogenous biomass density of microorganisms supported in root-free soil, Y is the growth yield and θ is the soil water content (the rest of the terms have already been defined). The authors made considerable efforts to estimate realistic values for the coefficients in their equations from both field and experimental data. The boundary conditions for the equations are:

$$-D\frac{\partial S}{\partial r} = Q(t)$$

at the root surface, and

$$-D\frac{\partial S}{\partial r} = 0$$

at the outer boundary of substrate diffusion. Here $Q(t)$ is a function describing the rate of exudation of substrate at the root surface. This was taken to be either constant or a linear function of time.

The equations were solved numerically using the Crank–Nicolson finite difference method. The accuracy of the procedure was checked by analytic solution with $C=X=0$ and $Q=$constant. The authors present graphical representations of change in substrate and biomass concentrations as functions of time and distance from the root for a variety of different conditions.

The Newman–Watson model is well founded in terms of both its mathematical and biological bases. It is a great pity that it has neither been applied more widely to field data in order to develop predictive equations nor tested more rigorously by laboratory experimentation.

A simpler approach forms the basis of so-called "layer" models. In these, the soil column is regarded as being subdivided horizontally into an arbitrary number of theoretical compartments. Essentially each compartment is considered to be well mixed and to transfer leachate to the compartment below it. Such models have the advantage that for many biologists they are conceptually more tractable than analytic models, and certainly the mathematics involved is easier to understand. Such an approach has been used by Burns (1974) and Addiscott (1977) for studying the redistribution of salts applied to the surface of soils; by Theodorou et al. (1981) to study cellulolysis; and by Prosser and Gray (1977) to analyse their laboratory experiments on nitrification in column reactors.

If material moves by diffusion from the root alone, the rate of change in concentration is given by Fick's second law. For the rhizosphere this may be extended to give:

$$\frac{\partial S}{\partial t} = \frac{D}{r}\frac{\partial}{\partial r}\left(r\frac{\partial S}{\partial r}\right) \pm \text{sink terms} \tag{27}$$

where r is the radial distance from the root and D is the coefficient of diffusion. Newman and Watson (1977) and Nye and Tinker (1977) utilized this model in their extensive studies on the root–soil system. With respect to microbial activity, the significant features in equations (23) and (27) are what we have called "sink terms". For a full description of the fate of a metabolic substrate, even in an idealized rhizosphere, what is required is a combination of equations (23) and (27) to represent flow down the soil column and diffusion from and to the root, their representation in three dimensions and a "sink" term which reflects the way in which the microbial population and the root interact with the substrate. Included in such a term will be a representation of the way in which the microorganisms interact with each other.

5. MICROBIAL INTERACTIONS IN THE RHIZOSPHERE

As indicated above, the "sink" terms of equations (23) and (27) will be functions of microbial activity, including the interactions between microbial populations. Our purpose in this section is to use the theoretical and experimental results we described earlier to form the basis of a description of how microorganisms might grow and interact in the rhizosphere. In other words, to describe qualitatively these "sink" terms. As we have already said, the rhizosphere is not well mixed but has a characteristic structure. Although the kinetic functions we have described might be suitable for modelling both homogeneous and heterogeneous environments, the question that arises is how does the particular structure of the rhizosphere alter the dynamics of the system as a whole? There is as yet no definitive answer to this question and we will mostly add to the speculations of other workers.

In Fig. 3 we have represented diagrammatically a layer model of the rhizosphere. Here two cylinders are depicted, the inner one representing a root and the outer one the rhizosphere. This is a gross simplication of the real situation of course, but our aim is to consider an idealized system. After such basic features have been described modifications may be made to reflect biological and physical activities and structures in greater detail. The rhizosphere region is subdivided by radial, vertical and horizontal lines to delineate compartments of arbitrarily small volume. Each compartment is considered to be well mixed. The movement of solute into a compartment is by flow from the compartment above and by diffusion from the compartment nearest the root. Diffusion from other compartments is considered to be negligible and, at least in part, balanced by outward diffusion. Movement of solute out of the compartment is by flow to the compartment below and by diffusion into the compartment distal to the root. We emphasize here that we do not consider it appropriate to set up a difference model based on the above considerations in order to make quantitative predictions. For such purposes the appropriate partial differential equations should be used. Rather we use these ideas to infer what the qualitative behaviour of the system might be on the basis of what we know about the kinetics of single-species growth and microbial interactions. In this light it is important to notice

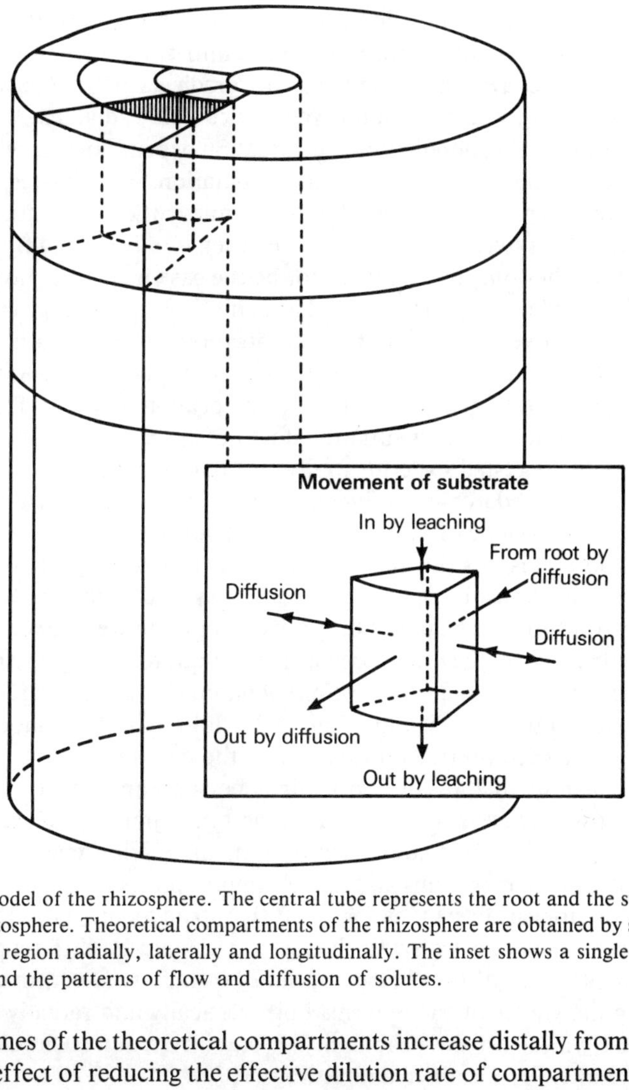

Fig. 3. Layer model of the rhizosphere. The central tube represents the root and the surrounding cylinder the rhizosphere. Theoretical compartments of the rhizosphere are obtained by subdividing the rhizosphere region radially, laterally and longitudinally. The inset shows a single theoretical compartment and the patterns of flow and diffusion of solutes.

that the volumes of the theoretical compartments increase distally from the root. This has the effect of reducing the effective dilution rate of compartments further away from the root.

On the basis of this rather heuristic model, we now speculate on what the effects of this idealized rhizosphere environment might have been on the population dynamics of the microorganisms present. We will consider only a single growth-limiting substrate being secreted by the root and contained in the solution flowing down the soil column. Although a simple modification only is needed to represent utilization of substrate by the root, this will not be taken into account here.

For single-species population growth with some washout of the microorganisms, the situation is relatively clear. A supply of nutrient will be received by each compartment from above and this will decrease in concentration with distance down the column due to utilization by organisms higher up. Radial diffusion will supply a relatively high concentration of nutrient close to the root and this will decrease further away from the root. Thus gradients in the population densities of the organisms should occur along the vertical and radial axes of the system.

With two competing populations, the situation depends on the relative specific growth rates of the populations. If one population grows faster, and thereby uses nutrient at a greater rate for all limiting nutrient concentrations, only one population will survive. This will be the case even if only a small fraction of each population washes out, but it might not be the case if this fraction is different for the two populations. On the other hand, if one population grows faster than the other at relatively low nutrient concentrations while the other population grows faster at relatively high concentrations then the two populations would not be competing as they would be in different locations: ecologically, they would occupy different niches. This situation does not seem to be an unlikely one and the idea could be tested experimentally comparatively easily.

In the case of predator–prey dynamics, the important feature of the system is the increase in volume of the theoretical compartments with respect to distance from the root. This has the analogous effect of increasing the residence time of a thermodynamically open system such as a chemostat. Based on our discussion of Monod-type predator–prey dynamics in well-mixed systems, this might result in both prey and predator populations changing smoothly and monotonically close to the root, exhibiting damped oscillations further out and oscillating continuously distally. This behaviour would be modified by the decreasing concentration of nutrient down the column so that in some cases vertical behaviour of the same sort might be superimposed on the system.

For multi-species systems, the situation becomes more complex. The stability matrix in equation (19) is composed only of interaction coefficients, steady-state population density terms and the self-limitation coefficients. Growth terms (α_i) do not appear except as part of the steady-state population values. With nutrient-dependent growth the α_i terms would be different in different compartments, but all the β terms should be similar. Determination of how decreasing the values of the α_i terms both vertically and radially in the model rhizosphere will affect the probability of stability requires simulations of the type undertaken by Gardner and Ashby (1970).

The effect of fluctuating the rate of nutrient supply to a rhizosphere-containing soil column might be guessed at using the experimental results of Rashit and Bazin (1987). Low productivity conditions will occur distally to the root, while close to it relatively high productivity will obtain. Similarly, near the top of the column more nutrient will be available than further down. Therefore, on the basis of the supposition on p. 113, one might suppose that fluctuations in the rate of nutrient supply—a situation analogous to that which occurs naturally—would increase the species diversity near the root and decrease it distally; moreover, nearer the top of the column diversity would be higher relative to further down. However, as we indicate in Section 3, this prediction does not accord with what has been observed.

6. GENE TRANSFER IN THE RHIZOSPHERE

Components of rhizosphere communities could potentially interact by three gene transfer mechanisms—conjugation, transformation and transduction. Conjugation is generally considered to be the most likely and universal process of horizontal gene flow in bacterial (but not fungal) communities. The comparable process in fungal communities would be anastomosis where hyphal fusing would lead to heterokaryon formation. Evidence for anastomosis and conjugation requires the identification of suitable genetic markers, and there are many more markers available in bacterial than fungal systems. Transduction is less likely because it requires a bacteriophage to multiply, lyse a host cell and then carry genes into the next recipient. Host DNA packaging by this mechanism has low likelihood and the host ranges of phage are narrow. In transformation, DNA released by one cell is taken up and incorporated into a recipient cell genome, the process probably being confined to a few bacterial species. Exceptional circumstances, such as the release of protoplasts, may lead to this process in fungi, but its frequency is likely to be very low.

In rhizosphere communities gene transfer is most likely to be mediated by plasmids in the process of bacterial conjugation. There is thus considerable concern that genetically engineered microorganisms (GEMs) derived by recombinant DNA (rDNA) technologies will exchange rDNA within the rhizosphere and the environment at large. In reviewing this possibility Stotzky and Babich (1986) provided few definitive examples of plasmid (gene) transfer. However, a recent example has been the transfer of the plasmid encoding the production of the antibiotic agrocin 84 from the antagonist *Agrobacterium rhizogenes* strain 84 to *Agrobacterium radiobacter*, the pathogen which causes crown gall of fruit trees (Chapters 6 and 7).

In Holland, van Elsas and co-workers have studied experimentally the dynamics of gene transfer between strains of *Pseudomonas* and strains of *Bacillus*. Much of their work is summarized by van Elsas et al. (1988a). Donor strains employed were plasmid-bearing and the mechanism of gene transfer in their studies appears to have been by conjugation. In experiments in which soil was not pre-sterilized, transfer frequencies were low unless nutrients were supplied to the system. Plasmid exchange was found to be higher in soil chambers (Dijkstra et al., 1987) than in culture tubes, indicating dependence on the physical nature of the environment, and in both experimental systems genetic exchange was higher in the rhizosphere. Other factors which affected the rate of gene transfer included the presence of bentonite (which increased it) and reducing the temperature and moisture content to suboptimal levels (which reduced it) (van Elsas et al., 1987, 1988b).

A simple model for gene exchange in the rhizosphere (and also the phyllosphere) has been proposed by Knudsen et al. (1988). It takes the form of a set of difference equations describing change in the population densities of a plasmid donor strain (D), a recipient (R) and the transconjugant (T). These populations are assumed to die at constant exponential rates β_1 and β_2 and the plasmid conjugational transfer rate is supposed to be a constant, τ. The model employed by the authors, therefore, takes the form:

$$D_t = 10^{\beta_1} D_{t-1} \quad (28)$$

$$R_t = 10^{\beta_2} R_{t-1} - \tau R_{t-1} (D_{t-1} + T_{t-1}) \quad (29)$$

$$T_t = 10^{\beta_1} T_{t-1} + \tau R_{t-1} (D_{t-1} + T_{t-1}) \quad (30)$$

This closed model for gene exchange, in which neither flow nor diffusion are taken into account, has the advantage of mathematical tractability and heuristic simplicity. The authors solved the equations on a digital computer using experimentally determined values for the coefficients β_1 and β_2. They then performed experiments with a plasmid-bearing strain of *Pseudomonas cepacia* and compared the results to those predicted by their model. On the basis of this comparison, the authors accepted the validity of the basic assumptions of the model. Wellington (personal communication) has used a similar model to that of Knudsen et al. (1988) to investigate plasmid transfer dynamics of streptomycetes in root-free soil. Again, the model has the advantage of simplicity, though it is not quite so tractable as it is posed in terms of differential equations. The experimental results match those predicted quite well. However, there is no term in the model that makes it distinguishable from one that could be applied to a well-mixed closed ecosystem as opposed to the heterogeneous situation in the soil. A similar criticism can be applied to the work of Knudsen et al. (1988) and additionally in this case, no aspect of their model identified it as being specific for the rhizosphere.

7. EXPERIMENTAL APPROACHES

Our emphasis in this chapter has been on investigating microbial interactions in the rhizosphere by representing them with idealized hypothetical systems which take the form of mathematical models. These models may incorporate either specific or general hypotheses about the system under study and, because they are posed in mathematical terms, analytical methods are usually available to allow the generation of well-defined predictions which can be tested by experiment. Such experiments may be undertaken in the field or in the laboratory; the location is not relevant but it is important to ensure as far as is possible that the prediction is tested unambiguously. The problem of data collection from environments such as the rhizosphere now arises. Non-destructive sampling of the rhizosphere for identification of microbial species, the major metabolites present and the concentrations of each of these is a goal yet to be achieved. Currently an investigator is required to design an experiment which will test a particular hypothesis or hypotheses as well as is practicable with the technology available to him. We emphasize that the term "experiment" is used in a fairly restrictive sense, i.e. an operation designed to generate data that can be compared to the predictions generated from a hypothesis. Few experiments of this kind have been carried out in the study of the rhizosphere. However, the following provides some examples of relevant experimental approaches in the absence of such hypotheses. They highlight some of the key physical and biological aspects of the rhizosphere which affect design and interpretation of experiments and the need for labelling of individual microorganisms and some methods for their subsequent detection.

7.1. Physical Aspects of the Rhizosphere

7.1.1. Soil Complexity

The physical, chemical and biological complexity of soil can make it a difficult medium in which to understand the basic principles of the microbe/root interaction and the microbe/microbe interaction in the rhizosphere. Consequently as a preliminary approach to understanding such population biology, simpler substitutes for soil are often used. For example, in an experiment reported by Bennett and Lynch (1981a) plants were grown in tubes of sterile sand. Three bacteria (*Curtobacterium* sp., *Mycoplana* sp., *Pseudomonas* sp.), either singly or in combination, were introduced into the rhizosphere of gnotobiotic wheat, barley and maize plants. The colonization by the bacteria, which could be easily distinguished by plate counting, was recorded as $\log_{10}$ [no. of viable bacteria (mg^{-1} dry weight of root)]. In the wheat rhizosphere, for example, each bacterium reached a similar population density on the root, but when all three were introduced together, *Pseudomonas* sp. dominated the *Mycoplana* sp., while the *Curtobacterium* sp. failed to grow. The final population of a particular bacterium might simply reflect the inoculum size. This was investigated but found not to be the situation for the single bacterium/plant interaction, the final population being independent of inoculum size.

7.1.2. Water Content

Water availability is also critical in determining the population size and *in vitro* osmotolerance of seven *Pseudomonas* strains correlated with their abilities to establish stable populations in the rhizosphere of potato. Soil water status will affect bacterial movement down a growing root system. Parke et al. (1986) investigated the development of *Pseudomonas fluorescens* strain 2-79 RN_{10} (resistant to rifampicin and naladixic acid) on the roots of wheat seedlings and found that increased soil water content, and consequently soil matrix potential, promoted colonization.

7.2. Biology of Rhizosphere Organisms

7.2.1. Binding of Organisms to Root Surfaces

A critical factor determining population development in the rhizosphere will be the capacity of the microorganism to bind to the root surface, in both the short and long term. Adherence has been determined by dipping the seedling roots in a range of bacterial suspensions and dilution-plating from homogenates of washed roots (James et al., 1985). Binding was rapid (within minutes), irreversible and dependent on concentration. Irrespective of source and habitat, the different bacteria showed similar concentration-dependent binding. *Pseudomonas fluorescens* E6-22 binding was promoted by divalent cations but monovalent cations (Na^+ and K^+) had little effect. This indicates that, at least in the short term, electrostatic phenomena are likely to be involved in the binding.

7.2.2. Colonization Potential

The saturation level of population has been termed the "colonization potential" of the root and is likely to be characteristic of both the plant and microbial species (Bennett and Lynch, 1981b). This concept of *in vitro* colonization potential provided a stimulation for further *in vivo* studies in complex soil systems with a multitude of strains as potential colonizers of the rhizosphere. This can only be achieved satisfactorily by using single- or multiple-antibiotic resistance markers in the bacteria in order that the inoculants can be satisfactorily recovered on agar media containing antibiotics. It must always be remembered, however, that such strains may carry multiple genetic modifications and can, therefore, become less successful ecologically than wild-type strains. For example, Loper et al. (1985) investigated the population dynamics of rifampicin-resistant soil pseudomonads in the rhizosphere of potato by inoculating seed pieces. Unlike the colonization potential concept applied to gnotobiotic systems, the rhizosphere populations of two strains were quantitatively related to seed piece inoculum levels. Both strains grew more rapidly in the rhizosphere at 24 °C than at 12 °C but the *in vivo* colonization potential was greatest at 12 °C and 18 °C.

7.2.3. Antibiotic Production

Clearly the capacity to produce antibiotics *in vivo* would potentially be useful to establish an inoculant population within the rhizosphere community. *Erwinia carotovora* subsp. *betavesculorum*, a wound pathogen which causes a vascular necrosis and root rot of sugarbeet, produces a broad-spectrum antibiotic *in vitro*. It has also been shown that the antibiotic can be produced in the infection count of potato, enabling this subspecies to gain ascendancy over competing antibiotic-sensitive *Erwinia carotovora* subsp. *carotovora* strains (Axelrood et al., 1988).

This has some analogy to the competition between *Agrobacterium* strains mediated by the production of agrocin 84 promoting the biological control of the pathogen (Chapters 6 and 7), but in the *Erwinia* situation the antibiotic-producer is also virulent. Clearly the population biology of the rhizosphere affords great potential in studying infection and biological disease control provided it is linked to the population biology of pathogenesis (Chapter 8).

7.3. Labelling and Detection of Rhizosphere Organisms

A great advance in studying the population biology of introduced organisms is to use transposon mutants and to trace their fate by the most-probable-number-DNA hybridization procedure and antibiotic resistance (Fredrickson et al. 1988). Isolates of *Rhizobium* spp. and *Pseudomonas putida* carrying the transposon Tn5 were added to sterile soil and enumerated by colony hybridization, most-probable-number-DNA hybridization procedure, plate counts, plant infectivity most probable number, and fluorescent antibody counts (Chapter 1). The population values compared well for all methods after 5 and 30 days, although the fluorescent antibody technique tended to overestimate the population size.

Clearly these modern approaches are extremely relevant to tracking the release of GEMs into the environment, as opposed to merely tracking their fate in laboratory microcosms. In one of the major field trials in the USA the recombinant soil and rhizosphere bacterium *Pseudomonas aureofaciens* 3732RNL11 marked with *Escherichia coli* lac ZY tracking system was used (Drahos et al., 1988, Chapter 1). This also provided a "dummy run" for the introduction of a GEM into the environment under carefully controlled conditions. Both parent strains and GEM had the same colonization potentials of the root, but although the organism survived for at least 10 weeks there was little movement of the organism from the root.

Clearly detection systems for assessing bacterial population dynamics in the rhizosphere are advancing rapidly, but detection systems for fungi have lagged far behind. Recently it has been shown that the biomass of *Trichoderma* spp. associated with roots can be tracked non-specifically by measuring chitin or ATP levels; these levels are inversely related to the colony-forming units of the fungus (Lumsden et al., 1989). The fungus can be detected more specifically using polyclonal antibodies raised to cell proteins with detection by a dot blotting procedure (Carter and Lynch, unpublished). Another promising approach is to develop DNA probes for fungi.

8. CONCLUSIONS

Now that suitable experimental procedures are being developed to assess rhizosphere colonization, the mathematical theories governing microbial interactions can be evaluated in terms of their applicability within this environment. Mathematical models will provide hypotheses that can be tested experimentally and will be used in two ways. Firstly, they will provide a structured framework within the complexities of the rhizosphere system and so constitute a basis for the design of experiments which are sufficiently circumscribed to answer specific questions. Secondly, they will inevitably suggest areas for investigation which would not be apparent from an exclusively empirical or subjective approach. In particular, with subsequent modification and development, it should prove possible to establish predictions of the fate of natural populations and the perturbations which might be generated by the introduction of GEMs or wild-type organisms at elevated levels. In addition, consequences of agrochemical usage or agronomic practices on rhizosphere stability might also be evaluated.

REFERENCES

Addiscott, T. M. (1977). A simple computer model for leaching in structured soils. *J. Soil Sci.*, **28**:554–563.

Addiscott, T. M. and R. J. Wagenet (1985). Concepts of solute leaching in soils: a review of modelling approaches. *J. Soil Sci.*, **36**:411–424.

Anderson, R. M. and R. M. May (1979). Population biology of infectious disease, Part 1. *Nature*, **280**:361–367.

Axelrood, P. E., M. Rella, and M. N. Schroth (1988). Role of antibiosis in competition of *Erwinia* strains in potato infection counts. *Appl. Environ. Microbiol.*, **54**:122–129.

Bailey, J. and D. F. Ollis (1986). *Biochemical Engineering Fundamentals* (2nd edn.). McGraw-Hill, New York.

Barber, D. A. and K. B. Gunn (1974). The effect of mechanical forces on the exudation of organic substances by the roots of cereal plants grown under sterile conditions. *New Phytol.*, **73**:39–45.

Bazin, M. J. (1981). Mixed culture kinetics. In *Mixed Culture Fermentations* (M. E. Bushell and J. H. Slater, eds.). Academic Press, London, pp. 25–51.

Bazin, M. J. and P. T. Saunders (1978). Determination of critical variables in a microbial predator–prey system by catastrophe theory. *Nature*, **275**:52–54.

Bazin, M. J., C. Curds, A. Dauppe, B. A. Owen, and P. T. Saunders (1983). Microbial predation dynamics. In *Foundations of Biochemical Engineering Kinetics and Thermodynamics in Biological Systems* (H. W. Blanch, E. T. Papoutsakis and G. Stephanopoulos, eds.). American Chemical Society, Washington, D.C., pp. 253–264.

Bennett, R. A. and J. M. Lynch (1981a). Bacterial growth and development in the rhizosphere of gnotobiotic cereal plants. *J. Gen. Microbiol.*, **125**: 95–102.

Bennett, R. A. and J. M. Lynch (1981b). Colonization potential of bacteria in the rhizosphere. *Curr. Microbiol.*, **6**:137–138.

Brown, M. E. (1972). Plant growth substances produced by microorganisms of soil and rhizosphere *J. Appl. Bact.*, **35**:443–451.

Burghes, D. N. and M. Borrie (1981). *Modelling with Differential Equations*. Ellis Horwood, Chichester.

Burns, I. G. (1974). A model for predicting the redistribution of salts applied to fallow soils after excess rainfall or evaporation *J. Soil Sci.*, **25**:165–178.

Canale, R. P. (1969). Predator–prey relationships in a model for the activated process. *Biotech. Bioeng.*, **11**:887–907.

Canale, R. P. (1970). An analysis of models describing predator–prey interaction. *Biotech. Bioeng.*, **12:** 353–378.

Cho, C. M. (1971). Convective transport of ammonium with nitrification in soil. *Can. J. Soil Sci.*, **51**:339–350.

Contois, D. E. (1959). Kinetics of bacterial growth; relationship between population density and specific growth rate of continuous cultures. *J. Gen. Microbiol.*, **21**:40–50.

Curds, C. R. (1971). A computer-simulation study of predator–prey relationships in a single-stage continuous-culture system. *Water Research*, **5**:793–812.

Curds, C. R. and A. Cockburn (1968). Studies on the growth and feeding of *Tetrahymena pyriformis* in axenic and monoxemic culture. *J. Gen. Microbiol.*, **54**:343–358.

Daniels, J. and A. L. Mackay (1974). The stability of connected linear systems. *Nature*, **251**:49–50.

Darbyshire, J. F. and M. P. Greaves (1973). Bacteria and protozoa in the rhizosphere. *Pesticide Science*, **4**:349–360.

D'Arcy, A. L. (1982). Study of Soja and Lens exudates 1. Kinetics of exudation of phenolic compounds, amino acids, and sugars in the first days of plant growth. *Plant and Soil*, **68**:399–403.

Dazzo, F. B. (1984). Bacterial adhesion to plant root surfaces. In *Microbial Adhesion and Aggregation* (K. C. Marshall, ed.). Springer-Verlag, Berlin, pp. 85–93.

Dent, V. E., M. J. Bazin, and P. T. Saunders (1976). Behaviour of *Dictyostelium discoideum* amoebae and *Escherichia coli* grown together in chemostat culture. *Arch. Microbiol.*, **109**:187–194.

Dijkstra, A. F., J. M. Govaert, G. H. N. Scholten, and J. D. van Elsas (1987). A soil chamber for studying the bacterial distribution in the vicinity of roots. *Soil Biol. Biochem.*, **19**:351–352.

Dommergues, Y. R. (1978). The plant–microorganism system. In *Interaction Between Non-Pathogenic Soil Microorganisms and Plants* (Y. R. Dommergues and S. V. Krupa, eds.). Elsevier, Amsterdam, pp. 1–37.

Drahos, D. J., G. F. Barry, B. C. Hemming, E. J. Brandt, H. D. Skipper, E. L. Kline, D. A. Kluepfel, T. A. Hughes, and T. D. Gooden (1988). Pre-release testing procedures: US field test of a lacZY-engineered soil bacterium. In *The Release of Genetically Engineered Micro-organisms* (M. Sussman, C. H. Collins, F. A. Skinner, and D. E. Stewart-Tull, eds.). Academic Press, London, pp. 181–192.

Flegmann, A. W. and R. A. T. George (1975). *Soils and Other Growth Media.* Macmillan, London.

Fletcher, M. (1984). Comparative physiology of attached and free-living bacteria. In *Microbial Adhesion and Aggregation* (K. C. Marshall, ed.). Springer-Verlag, Berlin, pp. 223–232.

Fredrickson, A. G. and G. Stephanopoulos (1981). Microbial competition. *Science*, **213**:972–979.

Fredrickson, J. K., D. F. Bezdicek, F. J. Brockman, and S. W. Li (1988). Enumeration of Tn5 mutant bacteria in soil using a most-probable-number-DNA hybridization procedure and antibiotic resistance. *Appl. Environ. Microbiol.*, **54**:446–453.

Gardner, M. R. and W. R. Ashby (1970). Connectance of large dynamic (cybernetic) systems: critical values for stability. *Nature*, **228**:784–785.

Hedley, M. J., P. H. Nye, and R. E. White (1982). Plant-induced changes in the rhizosphere of rape (*Brassica napus* var. emerald) seedlings II. Origin of the pH change. *New Phytol.*, **91**:31–44.

Heznen, C. E., J. D. van Elsas, P. J. Kuikman, and J. A. van Veen (1988). Dynamics of *Rhizobium leguminosiarum* biovar *trifolii* introduced into soil; the effect of bentonite clay on predation by protozoa. *Soil Biol. Biochem.*, **20**:483–488.

Howell, J. A. (1983) Mathematical models in microbiology: mathematical tool-kit. In *Mathematics in Microbiology* (M. Bazin, ed.). Academic Press, London, pp. 37–76.

Ingham, R. E., J. A. Trofymow, and E. R. Ingham (1985). Interactions of bacteria, fungi and their nematode grazers: effects on nutrient cycling and plant growth. *Ecological Monographs*, **55**:119–140.

James, D. W., T. V. Suslow, and K. E. Steinback (1985). Relationship between rapid, firm adhesion and long-term colonization of roots by bacteria. *Appl. Environ. Microbiol.*, **50**:392–397.

Katznelson, H. (1965). Nature and importance of the rhizosphere. In *Ecology of Soil Borne Plant Pathogens—Prelude to Biological Control* (K. F. Baker and W. C. Snyder eds.). University of California Press, Berkeley, pp. 187–209.

Knudsen, G. R., M. V. Walter, L. A. Porteous, V. J. Prince, J. L. Armstrong, and R. J. Seidler (1988). Predictive model of conjugative plasmid transfer in the rhizosphere and phyllosphere. *Appl. Environ. Microbiol.*, **54**:343–347.

Loper, J. E., C. Haack, and M. N. Schroth (1985). Population dynamics of soil pseudomonads in the rhizosphere of potato (*Solanum tuberosum* L.). *Appl. Environ. Microbiol.*, **49**: 416–422.

Lumsden, R. D., J. P. Carter, J. M. Whipps, and J. M. Lynch (1989). Comparison of biomass and viable propagule measurements in the antagonism of *Trichoderma harzianum* against *Pythium ultimum*. *Soil Biol. Biochem.*, in press.

Lynch, J. M. (1987). Microbial interactions in the rhizosphere. *Soil Microorg. (Japan)*, **30**:33–41.

Lynch, J. M. and E. Bragg (1985). Microorganisms and soil aggregate stability. In *Advances in Soil Science, Vol. 2* (B. A. Stewart, ed.). Springer-Verlag, New York, pp. 133–172.

Lynch, J. M. and L. M. Panting (1980). Cultivation and the soil biomass. *Soil. Biol. Biochem.*, **12**:29–33.

May, R. M. and R. M. Anderson (1979). Population biology of infectious diseases. Part II. *Nature*, **280**:455–461.

McMeekin, T. A., J. Olley, and D. A. Ratowsky (1988). Temperature effects on bacterial growth rates. In *Physiological Models in Microbiology, Vol. 1* (M. Bazin and J. Prosser, eds.). CRC Press, Boca Raton, Florida, pp. 75–90.

Megee, R., J. Drake, A. G. Fredrickson, and H. M. Tsuchiya (1972). Studies in intermicrobial symbiosis. *Saccharomyces cerevisiae* and *Lactobacillus casei*. *Can. J. Microbiol.*, **18**:1733–1742.

Monod, J. (1942). *Recherches sur la Croissance Bacteriennes*. Herman et Cie, Paris.

Moorby, J. (1987). Can models hope to guide change? *Annals of Botany*, **60**:175–188.

Newman, E. I. and A. Watson (1977). Microbial abundance in the rhizosphere: A computer model. *Plant and Soil*, **48**:17–56.

Nye, P. H. (1986). Acid–base changes in the rhizosphere. *Adv. Plant Nutr.*, **2**:129–153.

Nye, P. H. and P. B. Tinker (1977). *Solute Movement in the Soil Root System*. Blackwell Scientific Publications, Oxford.

Parke, J. L., R. Moen, A. D. Rovira, and G. D. Bowen (1986). Soil water flow affects the rhizosphere distribution of a seed-borne biological control agent *Pseudomonas fluorescens*. *Soil Biol. Biochem.*, **18**:583–588.

Pielou, E. C. (1969). *An Introduction to Mathematical Ecology*. Wiley Interscience, New York.

Pirt, S. J. (1975). *Principles of Microbe and Cell Cultivation*. Blackwell Scientific Publications, Oxford.

Prosser, J. I. and T. R. G. Gray (1977). Use of finite difference method to study a model system of nitrification at low substrate concentrations. *J. Gen. Microbiol.*, **102**:119–128.

Rashit, E. and M. Bazin (1987). Environmental fluctuations, productivity, and species diversity: An experimental study. *Microb. Ecol.*, **14**:101–112.

Ratowsky, D. A., J. Olley, T. A. McMeeken, and A. Ball (1982). Relationship between temperature and growth rate of bacterial cultures. *J. Bact.*, **149**:1–5.

Ratowsky, D. A., R. K. Lowry, T. A. McMeekin, A. N. Stokes, and R. E. Chandler (1983). Models for bacterial culture growth rate throughout the entire biokinetic temperature range. *J. Bact.*, **154**:1222–1226.

Reid, C. P. P. (1974). Assimilation, distribution, and root exudation of ^{14}C by ponderosa pine seedlings under induced water stress. *Plant Physiol.*, **54**:44–49.

Richter, J. (1987). *The Soil as a Reactor: Modelling Processes in the Soil*. Catena Verlag, Crealingen, West Germany.

Roels, J. (1983). *Energetics and Kinetics in Biotechnology*. Elsevier, Amsterdam.

Rovira, A. D. (1956). Plant root excretions in relation to the rhizosphere effect. I. The nature of root exudates from oats and peas. *Plant and Soil*, **7**:178–194.

Rovira, A. D. (1959). Root excretions in relation to the rhizosphere effect. IV. Influence of plant species, age of plant, light, temperature and calcium nutrition on exudation. *Plant and Soil*, **11**:53–64.

Rovira, A. D. and C. B. Davey (1974). Biology of the rhizosphere. In *The Plant Root and its Environment* (E. W. Cursova, ed.). University of Virginia Press, Charlottesville, pp. 153–204.

Rovira, A. D. and E. H. Ridge (1973). Exudation of C-labeled compounds from wheat roots: Influence of nutrients, micro-organisms and added organic compounds. *New Phytol.*, **72**:1081–1087.

Saunders, P. T. and M. J. Bazin (1973). Nonsteady state studies of nitrification in soil: Theoretical considerations. *Soil Biol. Biochem.*, **5**:545–557.

Savage, D. C. (1984). Activities of microorganisms attached to living surfaces. In *Microbial Adhesion and Aggregation* (K. C. Marshall, ed.). Springer-Verlag, Berlin, pp. 233–249.

Schippers, B., A. W. Bakker, and P. A. H. M. Bakker (1987). Interactions of deleterious and beneficial rhizosphere microorganisms and the effect of cropping practices. *Ann. Rev. Phytopathol.*, **25**:339–358.

Schonwitz, R. and H. Ziegler (1982). Exudation of water soluble vitamins and of some carbohydrates by intact roots of maize seedlings (*Zea mays* L.) into a mineral nutrient solution. *Zeitschrift für Pflanzenphysiologie*, **107**:7–14.

Slater, J. H. (1988). Microbial population and community dynamics. In *Microorganisms in Action: Concepts and Applications in Microbial Ecology* (J. M. Lynch and J. B. Hobbie, eds.). Blackwell Scientific Publications, Oxford, pp. 51–74.

Slater, J. H. and A. T. Bull (1978). Interactions between microbial populations. In *Companion to Microbiology* (A. T. Bull and P. M. Meadows, eds.). Longman, London, pp. 181–206.

Smith, S. J., L. R. Ahuja, and J. D. Ross (1984). Leaching of a soluble chemical under field crop conditions. *Soil Sci. Soc. Amer. J.*, **48**:252–258.

Srivastava, V. B. (1973). Investigations into rhizosphere microflora IV. Fungal association in different root regions of some rainy-season crops. *Acta Socieatis Botanicarum Poloniae*, **42**:409–422.

Stotzky, G. and H. Babich (1986). Survival of and genetic transfer by, genetically engineered bacteria in natural environments. *Adv. Appl. Microbiol.*, **31**:93–138.

Theodorou, M. K., M. J. Bazin, and A. P. J. Trinci (1981). Growth of *Trichoderma reesei* and the production of cellulolytic enzymes in a continuous flow column. *Arch. Microbiol.*, **130**:372–380.

Topiwala, M. and C. Sinclair (1971). Temperature relationship in continuous culture. *Biotech. Bioeng.*, **13**:795–813.

Tsuchiya, H. M., J. F. Drake, J. L. Jost, and A. G. Fredrickson (1972). Predator–prey interactions of *Dictyostelium discoideum* and *Escherichia coli* in continuous culture. *J. Bact.*, **110**:1147–1153.

van Elsas, J. D., J. M. Govaert, and J. A. van Veen (1987). Transfer of plasmid pFT30 between bacilli in soil as influenced by bacterial population dynamics and soil conditions. *Soil Biol. Biochem.*, **19**:639–647.

van Elsas, J. D., J. T. Trevors, and M. E. Starodub (1988a). Plasmid transfer in soil and rhizosphere. In *Risk Assessment for Deliberate Release* (W. Klingmuller, ed.). Springer-Verlag, Berlin, pp. 89–99.

van Elsas, J. D., J. T. Trevors, and M. E. Starodub (1988b). Bacterial conjugation between pseudomonads in the rhizosphere of wheat. *FEMS Microb. Ecol.*, **53**:299–306.

Warrick, A. W., J. W. Biggar, and D. R. Nielsen (1971). Simultaneous solute and water transfer for an unsaturated soil. *Water Resource Res.*, **7**:1216–1225.

Whipps, J. M. (1984). Environmental factors affecting the loss of carbon from the roots of wheat and barley seedlings. *J. Exp. Bot.*, **35**:767–773.

Whipps, J. M. and J. M. Lynch (1985). Energy losses by the plant in rhizodeposition. *Ann. Proc. Phytochem. Soc. Europe*, **26**:59–71.

6

GENETIC ASPECTS OF RHIZOSPHERE INTERACTIONS

R. W. HEDGES* and E. MESSENS

Rijksuniversiteit Gent, Laboratorium Genetika, Ledeganckstraat 35, B-9000 Gent, Belgium

1. INTRODUCTION

The structure of any ecosystem is based on complex exchanges of signals between constituent organisms. The systems of signals and receptors constitute the transfer of information which integrates the ecosystem just as the transfer of information within and between cells ensures integrity of the individual. Although the rhizosphere seems a relatively simple niche and as such may prove a useful model for more complex systems, we have only just begun the analysis of the interactions. Most of our current understanding concerns the signals exchanged between bacteria, and especially between bacteria and the cells of plant roots, and this chapter must perforce concentrate on these exchanges.

1.1. Chemotaxis

In general, the bacteria involved in rhizosphere ecology seem to be relatively ordinary organisms, often closely related to strains found in very different locales, though often provided with a few extra potentialities; the genes for these are often clustered as a cointegrated set and may be located on an accessory genetic structure of extrachromosomal DNA, or plasmid. Quite frequently, these

*Deceased, February 1988

ecologically specialized potentialities are simply extensions of pre-existing functions. For example, many motile bacteria can migrate towards a range of chemical attractants (chemotaxis) (Gaworzewska and Carlile, 1982). When a bacterium is attracted by a particular product of plant metabolism we may expect that this bacterium has a typical motility and chemotactic gene set *plus* an extra determinant whose product integrates into the normal cellular machinery and adds a new specificity to the set of recognized attractants. The production of such an attractant may be the first signal from a plant to a potential rhizosphere colonist.

The attractant may contain specific information concerning the quality of the host. For example, the attractant recognized by agrobacteria is produced by damaged tissue (Stachel et al., 1985). Atsatt (1977) postulated that parasitic seed plants used host defensive compounds as recognition cues and this may be rather generally true. Plants are likely to concentrate chemical defences around vulnerable sites and a parasite may profitably recognize such a signal as an indication of an exploitable site. Thus, phenolics related to the *Agrobacterium* attractant are well-known antimicrobials (Rice, 1979) and their production at sites of damage is presumably a defence mechanism. Similarly, flavonoids (the nodulation inducers of rhizobia) have antibacterial activity (Fottrell et al., 1964; Mori et al., 1987; D'Arcy-Lameta and Jay, 1987).

Rhizobia have been reported to undergo taxis toward increasing concentrations of a glycoprotein exuded by roots of birdsfoot trefoil (Currier and Strobel, 1977). This is a surprisingly large molecule to act as an attractant (the authors suggest it may form a steep and long-lasting gradient around roots) and it is tempting to postulate a small molecule, perhaps a flavonoid, which could diffuse further from the root and act as the primary stimulus inducing rhizobia to enter the zone within which the glycoprotein was active. Bacteria may recognize and be attracted non-specifically towards compounds (e.g. sugars) secreted by roots.

Mandimba et al. (1986) have suggested some degree of specificity of such chemotaxis in that although bacteria (enterics and azospirillae) isolated from maize or rice roots all migrated towards mucilage from maize roots, those from rice roots were less strongly attracted than those which originated on maize.

Within the genus *Azospirillum* there is some indication of host specificity (Döbereiner and Boddey, 1981) in that strains classified as *A. lipoferum* were most commonly found on roots of plants with the C4 photosynthetic pathway (e.g. maize and kallar grass) while C3 plants (e.g. wheat and barley) more frequently carried strains of *A. brasilense*. This preferential association suggests that different plant species attract different bacteria. Reinhold et al. (1985) have shown that even different strains belonging to a single species can respond differentially. Thus of two strains of *A. brasilense* one, isolated from maize roots, was strongly attracted to malate (produced by maize roots) but not to oxalate (the acid most abundantly excreted by wheat) whereas an isolate from wheat showed the reverse specificity. Once within the rhizosphere, positive chemotaxis should lead bacteria towards the highest concentrations of soluble nutrients and it seems likely that many become anchored to plant structures, forming microcolonies at the borders between cells at the root surface where nutrients may leak out of the root (Lynch, 1982).

1.2. Specific Binding

Current opinion favours the view that specific binding of bacteria to other biological structures (such as plant cells) usually involves interaction between carbohydrate groups and proteins (lectins) which recognize very specific saccharide configurations. It is important to note that although such interactions are chemically specific their effects may be rather non-specific biologically. Thus, the combining site of type 1 fimbriae (see below) recognizes specifically carbohydrates with alpha-mannose groupings; yet such pili can effect binding to yeasts, mammalian or plant cells.

Fimbriae are hair-like projections of bacterial cells with lectin activity (Sharon, 1987). They have essential roles in pathogenic interactions of bacteria in humans and other animals, and the binding of fimbriae (including type 1) to root hairs has been reported as a major factor in the association of plant cells with Enterobacteriaceae (Haahtela et al., 1985, 1986) and fluorescent pseudomonads (Vesper, 1987). Apparently, the fimbriae recognize carbohydrate arrays common to many plant species so little host specificity is involved, but, since the binding is rather specific for root hairs it seems that the receptors are tissue-specific glycoproteins (Diaz et al., 1986). At all events, the bacteria are associated with cells which are highly morphogenetically and metabolically active and microscopic investigation shows that their binding can influence the growth pattern of the root hairs.

Since a major part of the bacterial outer membrane is composed of lipopolysaccharide and glycoproteins it is not surprising that plant produced lectins have been proposed as significant agents for the specific recognition of bacteria. Nor, given the importance of the legume–*Rhizobium* interaction, is it suprising that much of the attention has focused upon this system. There can be little doubt that lectin–carbohydrate interactions play an important role in the binding (and determination of specificity) of these systems and several studies (e.g. Dazzo et al., 1976; Stacey et al., 1980; Diaz et al., 1986) have presented apparently unanswerable evidence for the involvement of such interactions. Yet, other evidence (reviewed by Bauer, 1981) does not fit easily with this interpretation. The most likely conclusion seems to be that interaction between bacterial carbohydrate groups and the lectin of the plant cell walls facilitates interaction between appropriate partners, but that such binding may not be essential for the initiation of interaction and is almost certainly only one of several stages at which discrimination against an unsuitable partner may be exerted.

1.3. Competition

1.3.1. Antibiotic

The rhizosphere is a relatively rich ecological niche and we may expect that organisms exploiting this resource will have evolved mechanisms to defeat potential competitors. Beyond competition by metabolic efficiency, we may expect other forms of competition. The ecological significance of production of antibiotics has long been debated (e.g. Williams and Vickers, 1986). Antibiotics are produced primarily by organisms growing in rich media, and many are adsorbed by soil particles or inactivated in other ways. We have noted the formation of microcolonies at sites of nutrient supply, and these may permit accumulation of significant levels of antibiotics. Such a "structured habitat" offers an ideal site for the activity of an "anticompetitor toxin" (Chao and Levin, 1981).

Plants, too, may secrete antibiotic compounds into the rhizosphere and thereby influence the microbial community (and other plants). For example, Wilson and Bell (1978a) demonstrated the release from germinating *Glycine wightii* seeds of the arginine analogue canavanine.

1.3.2. Nutritional

The physiological state of the shoot is a major determinant of the amount and composition of root exudation (e.g. Hale and Moore, 1979). For example, leaf inoculation with *Beijerinckia* spp. increased root exudation (Bhat et al., 1971). These bacteria may be found on seeds, in the rhizosphere and on other plant-associated sites. If, as the plant develops, the seed-borne bacteria come to be distributed over the surface of the plant, those bacteria which reach the leaf surface may be able to augment the supply of nutrients to those clone members which become associated with the roots. The clonal organization of bacterial populations implies that the progeny of one cell is as closely related to a "clone sister" as to its own progeny (apart from mutation, etc.) and the "altruistic" provision of benefits may be a rather general phenomenon. Once the cells are bound to the root cells they may follow any one of a number of lifestyles. For example, they may adopt a passive strategy nourished by materials spontaneously liberated into the rhizosphere by plant cells (pure saprophytism). But others have evolved mechanisms to induce liberation of such nutrients. Thus, plants grown under sterile conditions liberate far less organic nutrients into the soil than do plants grown in non-sterile soil (Schönwitz and Ziegler, 1986). Pure cultures of *Pseudomonas putida* and *Azospirillum* spp. are capable of inducing nutrient supplies (Přikryl and Vančura, 1980; Heulin et al., 1987) and there is a hint of specificity since azospirillae isolated from rice roots were more active on rice than on maize. Within the rhizosphere ecosystem, the root is the primary supplier of nutrients and most of the roles which can be played by bacteria in that niche exploit this source.

1.4. Coexistence

In the following sections we shall discuss a few of the systems of exploitation which have evolved among rhizosphere microorganisms. These vary from the direct enzymatic attack upon plant tissues typified by the root rotting erwiniae, to the rhizobia with their remarkable symbiosis whereby they and their plant hosts profit from the association.

These represent, however, no more than the extremes of a continuum and since both extremes have been the subject of numerous recent reviews (Long, 1984; Pugsley and Schwartz, 1985; Gresshoff and Delves, 1986; Kondorosi and Kondorosi, 1986; Collmer, 1987), it may be advantageous to emphasize those bacteria which coexist with plant roots in less easily classified fashion, showing neither the very specialized morphogenetic and physiological interactions of the latter nor the unambiguous parasitism of the former.

2. CROWN GALL INDUCTION

Agrobacterium tumefaciens is capable of interactions with host plants (which include most dicots, several gymnosperms and some monocots). It has the ability to introduce DNA sequences into the chromosomes of plant cells. This DNA, the T-DNA, forms a part of a plasmid, the Ti plasmid, in the bacterium. Once inserted into the plant chromosome, genes of the T-DNA are transcribed and induce cell multiplication (introducing a new pathway of auxin synthesis and a new enzyme catalysing a critical step in cytokinin synthesis) leading to formation of a gall (a plant cancer). The cells carrying the integrated T-DNA (termed transformed cells) produce and secrete one or more "opines", compounds not made by normal plant cells, which act as nutrients for *A. tumefaciens* cells which thrive around the gall tissues. The details and especially the evidence for these assertions have been presented in numerous reviews (Kahl and Schell, 1982; Depicker et al., 1983; Nester et al., 1984; Gheysen et al., 1985; Gheysen et al., 1989).

2.1. Tumorigenesis

For simplicity we may consider a single *A. tumefaciens* strain C58, as a representative crown gall former. This strain carries a 200 kb Ti plasmid (pTiC58) required for tumorigenesis and a "cryptic" plasmid of 410 kb which is not required for this function (Rosenberg and Huguet, 1984).

As well as the T-DNA sequence, pTi C58 carries a number of virulence (*vir*) genes whose activities are either essential for tumorigenesis or increase the efficiency of transformation of plant cells. Other genes involved in this process are usually regarded as having chromosomal locations and will, for convenience, be described as chromosomal, but it should be emphasized that in most cases a location on the cryptic plasmid has not been excluded.

2.1.1. Recognition

Damaged plant tissues secrete a range of phenolic compounds believed to be precursors or breakdown products of lignin. At least one of these, acetosyringone (AS), is a chemotactic attractant for strain C58 (Ashby et al., 1987) with an optimal concentration of 10^{-7} M, which implies the constitutive expression of a very efficient specific recognition system in the bacterial cell. Since cells lacking the Ti plasmid do not respond to AS (though fully capable of other chemotactic responses) the protein which responds to this compound is determined by the plasmid. This specific receptor must be so efficiently integrated into the (chromosomally determined) bacterial chemotaxis and motility functioning that the whole system operates efficiently and provides a good example of the principle that the chromosomes of these bacteria (belonging to the Rhizobiaceae) determine mechanisms to permit interactions with plant cells, whilst the plasmids determine the specificity both of signalling and the recognition of plant signals. The ability of AS to act as an attractant has been disputed (Parke et al., 1987), but by selection of motile populations of agrobacteria prior to chemotaxis experiments, the response has now undeniably been shown to be Ti plasmid encoded (Ashby et al., 1988).

Recent studies on strains of *A. tumefaciens* with a narrow host range (typical of grape vines) suggest that these are unable to respond to AS. The *vir*A genes of these strains (which show marked differences from the homologues in typical wide host range strains) do not respond to AS (or to exudates from a plant which acts as host only for wide host range strains), which implies that the *vir*A gene product is a major host specificity determinant (Leroux et al., 1987). Nothing is yet known of the product recognized by the narrow host range strains or of their chemotactic specificity.

2.1.2. Virulence

At the wound site, the concentration of AS should be sufficiently high (optimum concentration about 10^{-5} M) to induce the *vir* genes. These are the genes of the Ti plasmid whose functions are involved in the introduction of the T-DNA sequence into the plant chromosome. The *vir* genes belong in five transcription units. Of these, *vir*A is expressed constitutively; *vir*E is also expressed constitutively, but the level of expression can be raised by induction with AS; the others (*vir*B, C and D) are normally repressed but transcription can be induced by AS. Stachel and Zambryski (1986) showed that expression of *vir*B, C and D requires the presence of *vir*A and G, but that these latter can be transcribed regardless of the state of the former genes. A plausible interpretation of these results is that *vir*A determines a membrane protein (permease) which draws AS into the cell, and *vir*E encodes an autoregulated positive regulatory function necessary for transcription of *vir*B, C and D, enhancing the basal level of transcription of *vir*A and also permitting expression of yet one more operon, *pin*F, whose function is unknown. The genes of operons *vir*B, C and D function to insert T-DNA into sites on the plant chromosome. Mechanisms are only gradually being understood.

2.2. Transformation

Transformed cells differ from typical root cortical cells in two major respects: they undergo growth and cytokinesis leading to the formation of a plant tumour (crown gall); and they produce and secrete compounds, termed opines, not produced by non-transformed plant cells. Genes of the T-DNA determine both characters. Genes 1 and 2 determine enzymes which catalyse the conversion of tryptophan to indole acetic acid (via indole acetamide), whilst gene 4 is responsible for an enzyme in the biosynthetic pathway of a cytokinin (which presumably catalyses the rate-limiting step in their production). The result of this overproduction is the somewhat unorganized growth that leads to gall formation. Here, the T-DNA, itself a "signal" from the agrobacterium, causes the plant to produce secondary signals which intervene in the morphogenic pathways of the root itself (Weiler and Schröder, 1987).

So far as is known, all Ti plasmids carry these three genes responsible for hyperplasia but there is much more diversity in respect of the genes for opine metabolism. pTiC58 has two genes for opine production. One, *nos,* determines the synthesis of nopaline; the other, *acs,* encodes the enzyme responsible for agrocinopine synthesis. Both enzymes catalyse the conversion of rather ordinary cellular constituents—sugars and amino acids present in reasonable abundance in plant cells—into unusual compounds not catabolized within the plant cell. Ryder et al. (1984) concluded that all opines "comprise two common plant cell constituents linked in an uncommon manner by the minimum number of enzymes". Opines are secreted from transformed cells (Messens et al., 1985).

2.3. Response of the Bacterium

Thus, as a result of a system of message exchanges between plant and bacterium, genetic information originating in the Ti plasmid is added to the genome of the plant cell and in that new site functions (using the information transfer systems of the host) to send signals to the plant (to modulate the morphogenetic programme) and the bacterium (specifically to Ti plasmid-encoded functions in the bacterium).

The response of the bacterium to the supply of opines is entirely determined by the Ti plasmid. Ti plasmid-free segregants are not affected by these compounds.

The opines have three effects on the bacteria:

1. They are positive chemotropic agents stimulating the bacteria to migrate towards the source of the opines (the galls).
2. They induce the production of enzymes involved in metabolism of the opines.
3. They induce synthesis of conjugation-pili (and probably of other components of the plasmid transfer apparatus) (Ellis et al., 1982a).

In the case of strain C58, nopaline can induce the synthesis of the catabolic (*noc*) enzymes. Agrocinopine A is the inducer of two groups of genes, agrocinopine catabolic (*agc*) and plasmid transfer (*tra*) genes. Mutants constitutive for transfer functions have been identified and these are also constitutive for agrocinopine A catabolic functions (Ellis et al., 1982b). Conjugation functions are important in the ecology of any plasmid and the finding that opines regulate the expresion of these genes in many (perhaps all) types of Ti plasmid must have a major effect on the biology of both the bacterium and its plasmid. Transfer of Ti plasmids in nature is limited to sites of opine concentration, i.e. to the surfaces of galls. This is precisely the environment in which the plasmid can confer a selective advantage (ability to metabolize the opine) upon recipient bacteria and offers a particularly efficient way for the Ti plasmid to colonize new hosts. At first sight, it should be disadvantageous to the pTi$^+$ bacteria, which must share the supply of nutrients with the (unrelated) new recipients of the plasmids, and there may be a conflict of selective advantage between the bacteria (more precisely, the chromosome and megaplasmid) and that of the pTi plasmid. But the situation has not been subjected to much analysis and the ecological balance may be more subtly weighted.

2.4. Ecology

It must be energetically expensive for the bacterium to maintain the rather large Ti plasmid and (as discussed earlier) there is need for energy economy during growth of the cell in the soil. In fact, quite a large proportion of agrobacterial isolates lack Ti plasmids. It may be that in soil the Ti plasmids are discarded rather easily and only when the survival of the bacteria depends upon exploitation of an opine as supply for nitrogen or carbon does transfer of the plasmid occur.

Some other bacteria, particularly fluorescent pseudomonads, proved able to metabolize opines and, in a few cases, it has been shown that the determinants were located upon transmissible plasmids. These organisms, exploiting the nutritional supply provided by the crown galls, may be regarded as hyper-parasites. Whether these opine metabolizing determinants are derived from Ti plasmids or are separately evolved is not known.

2.4.1. Competition and Co-operation

A particular subtle mechanism by which a strain may intervene in an ecological interaction is demonstrated by an avirulent agrobacterium (strain 84). This carries three plasmids. One, pAt84a, is 186 kb and confers the ability to metabolize nopaline. Although there are no reports in the literature, it seems likely that, like pTiC58, it also confers the ability to use agrocinopine (Ellis et al., 1982a).

A second plasmid, 45 kb pAgK84, produces a bacteriocin, agrocin 84 (Chapter 7), which is a disubstituted fraudulent analogue of adenosine. It also encodes a mechanism of immunity to this bacteriocin. The bacteriocin is lethal to agrobacteria carrying a nopaline/agrocinopine A Ti plasmid. A permease, whose normal function is the uptake of agrocinopine A, is required for sensitivity and presumably acts as primary receptor for agrocin 84 (Ellis et al., 1979, 1982b).

It is interesting that these two plasmids are separate since their biological roles are so intricately connected. Near a tumour producing nopaline or agrocinopine A, the chemotactic functions of pAt84a should ensure that the strain approaches the tumour. In the tumorosphere (the zone surrounding the tumour, in which its nutritional influence is powerful) the genes for opine uptake and catabolism will be induced so as to make use of the nutrients supplied. Plasmid pAgK84 produces the agrocin which should kill non-immune agrobacteria carrying an agrocinopine A permease thus eliminating competitors for the nutrient. Agrocin 84 resistant mutants lack the permease and hence cannot compete with strain 84 for the agrocinopine A (though they can still compete for the supply of nopaline). It is interesting that pAt84a is not able to transform plant cells on its own account so that the bacteria must rely upon other agrobacteria to carry out tumorigenesis. Such close co-operation between two separate plasmids is unusual—though agrobacteria in which two plasmids are required to catabolize the opines formed by transformed cells have been reported (Petit et al., 1983)—and it would be interesting to study a number of isolates producing comparable agrocins to see whether such co-operation is the rule or whether "self-sufficient" plasmids have evolved, each with a complete set of tumour-exploitation genes.

3. THE *RHIZOBIUM* SYMBIOSIS WITH LEGUMINOUS PLANTS

The rhizobia are rather closely related to agrobacteria but have adopted a completely distinct pattern of interaction with plants. Where the agrobacteria interact with a wide range of host species, the rhizobia can establish symbiosis only with plants of the family Fabaceae and a few species of the *Parasponia* (Ulmaceae). Rather little is known of the ecology of rhizobia as free-living rhizosphere organisms though there may be rhizosphere organisms perhaps able to fix nitrogen without nodule formation (Hess et al., 1983; Ramaswamy and Bal, 1987).

3.1. Host Range

For any particular strain, the host range is precisely delimited, sometimes to a single genus or small group of genera. The best studied rhizobia are *R. leguminosarum,* which interacts with the common pea; *R. meliloti,* which nodulates lucerne (alfalfa); and *R. trifolii* which nodulates clover. These are often regarded as biotypes of a single species, the differences being determined by symbiosis (Sym) plasmids (Jarvis et al.; 1986). These plasmids determine the specificity and most of the functions needed for nodulation and the genes determining proteins required for nitrogen fixation (Young and Matthews, 1982). There is now evidence from laboratory studies that plasmid exchange between rhizobial strains occurs in nature (Young, 1985; Young et al., 1987; Broughton et al., 1987; Schofield et al., 1987).

3.2. Analogies With Other Bacteria

The normal pathway of nodulation has been outlined as attachment to root hairs, root hair curling, infection thread formation within root hairs, growth of the thread toward the inner root cortex, initiation of a nodule meristem and nodule organogenesis (Debellé et al., 1986).

Attachment of *Rhizobium meliloti* to host roots is a very different process from that of *A. tumefaciens* in that the former binds specifically nascent root hairs (Law and Strijdom, 1984; Diaz et al., 1986) whilst infection by the latter requires a wound site in the process of regeneration. Yet both processes involve homologous genes, termed *ndv*A and *ndv*B in rhizobia and *chv*A and *chv*B in agrobacteria. They are completely interchangeable between the genera (Dylan et al., 1986). Since DNA homologous with the *chv* genes has been detected in *Azospirillum brasilense* (Waelkens et al., 1987) it seems likely that these genes are highly conserved in evolution and may prove useful markers of bacteria whose ecology involves association with plant tissues. There is no indication that the *ndv*-mediated interaction is in any way involved in determination of host specificity. Such specificity seems to be controlled by Sym plasmid genes.

3.3. Specificity

3.3.1. Binding

The specific recognition of an appropriate plant root by a *Rhizobium* strain has been the subject of intense research. Details are still unclear (and it remains possible that different pairs of symbionts interact by entirely different mechanisms) but it is hard to avoid the conclusion that specific interactions between proteins (lectins) on the plant root cell surfaces and carbohydrate sequences in the bacterial capsules are involved. The evidence is particularly strong in the interaction of the clover lectin with polysaccharides of the *R. trifolii* capsule (Dazzo and Brill, 1979).

It is attractive to suppose that for such an essential feature of the rhizobial life cycle as specific binding to host cells, more than one mechanism might operate. For example, there may be enzymes capable of building particular molecular configurations into more than one polysaccharide of the cell wall, and all of these structures may play appropriate roles in lectin-mediated binding.

3.3.2. Nodulation

The genes directly responsible for induction of nodulation (*nod* genes) located on the Sym plasmids fall into two categories. Certain genes (the common *nod* genes) are functionally interchangeable in the sense that a strain of *R. meliloti* which has lost one or more such genes, regains its ability to nodulate lucerne if the homologous *nod* genes of *R. leguminosarum* are introduced on a compatible plasmid. Other *nod* genes are specific in that these functions can only be supplied from an *R. meliloti* Sym plasmid.

The best characterized common *nod* genes are *nod*A, B and C, which constitute a transcription unit positively regulated by the product of *nod*D. The common *nod* genes are unusual in that mutation in these genes determines absolute deficiency in nodulation ability whereas most of the defects in other genes result in reduced efficiency, delay or temperature sensitivity of nodulation. Mutations in these genes abolish completely all aspects of nodulation ability (Debellé et al., 1986). Normally *nod*A, B and C are repressed whereas *nod*D is expressed constitutively, but if rhizobia are exposed to plant exudates the ABC operon is transcribed. This induction requires the *nod*D gene, which is autoregulated (Rossen et al., 1985). Fractionation of the plant exudate showed that the most active inducer was luteolin (3′,4′,5,7-tetrahydroxyflavone) but that other active components, probably also flavonoids, were also present (Peters et al., 1986). It seems that *nod*D may be a permease or a positive control element (or both). That luteolin should be the inducer was rather surprising, in that luteolin is produced by plants of several families, most of which we never see nodulated by rhizobia. However, subsequent studies showed that flavones (4′,7-dihydroxy derivatives) produced only by leguminous plants, were active inducers (Redmond et al., 1986) and that different leguminous plants produced different sets of flavonoid compounds (D'Arcy-Lameta, 1986).

It used to be considered that, like the operon it regulates, the *nod*D gene was essentially similar in the different rhizobia; indeed, the genes do show extensive homologies. However, Spaink et al. (1987) showed that the *nod*D gene products of different rhizobia show different specificities when faced with different flavonoids or plant exudates such that although all *nod*D products respond to luteolin (so as to permit transcription of the ABC operon) the *nod*D product of *R. meliloti* did not respond to 5,7,4′-trihydroxyflavanone though the genes from *R. leguminosarum* and *R. trifolii* did. Most strikingly, an exudate from *Trifolium pratense* was recognized by the *nod*D gene product of *R. trifolii* but not by those of the other two species, and a strain of *R. trifolii* lacking its own *nod*D function but carrying the corresponding gene from either of the other two species was unable to nodulate *T. pratense* (though such strains could nodulate another host). Thus, part of the specificity of rhizobia may lie in the range of inducers recognized by its *nod*D product. A single Sym plasmid may encode several separate *nod*D genes each, perhaps, with a distinct specificity so that the rhizobium can exploit a range of hosts with distinct flavonoid secretion patterns (Horvath et al., 1987).

3.4. Subsequent Signals

The products of the *nod*ABC operon are not yet characterized, but it is clear that induction leads to the secretion from the bacteria of products with an influence on the growth of root hairs. These outgrowths develop characteristic curled forms ("shepherd's crooks"). The nature of the bacterial activity is unknown. Root hair curling by *R. meliloti* (on lucerne) is also controlled by another *nod* gene (*nod*H) and this is believed to regulate the hair curling effect, promoting the activity upon an appropriate host and reducing it on an unsuitable plant (Debellé et al., 1986). This seems to represent a separate host specificity recognition system. The *nod*H gene is unique to *R. meliloti* (it cannot be replaced by genes from Sym plasmids of *R. trifolii* or *R. leguminosarum*) and is therefore one of the host-specific nodulation genes of that species. It seems likely to be a second control system for the *nod*ABC operon, but there is no evidence for its action.

3.5. Organogenesis Signals

3.5.1. From the Bacterium

The initiation and morphogenesis of the nodule must be regarded as a process of organogenesis comparable in complexity with any other which the plant can perform. It seems certain that the primary stimulus is provided by the bacterium and plausible that the process involves several steps, each characterized by signal exchanges. But rather little is known of the nature of the signals.

3.5.2. From the Host

As might be expected in such a programme there is a synthesis of protein species not found in typical root cortex cells. Many of these are, no doubt, found in other meristems but some, the "nodulins", are specific to this organ. Host plant genes are undoubtedly required for effective nodulation. Studies of genetic variation within natural populations and of induced mutations affecting this property have been presented recently (Gresshoff and Delves, 1986; Verma and Brisson, 1987). Verma et al. (1986) have reviewed the soybean nodulins and the genes encoding these proteins. The best known (and most abundant) nodulins are the leghaemoglobins, whose function is believed to regulate the supply of oxygen to the bacteroids. Genes for several of the nodulin proteins have been cloned and some (but not all) have been shown to have 5′ sequences (Mauro et al., 1985), which are plausibly interpreted as regulatory regions.

The development programme for a nodule has been the subject of much interest recently (for review see Gresshoff and Delves, 1986). Much of the specificity of this programme is determined by the plant, a particularly clear example being provided by strains of *Bradyrhizobium* able to form totally different sorts of nodules on legumes and *Parasponia* spp. (Lancelle and Torrey, 1984). Such divergent morphogenic patterns suggest that the bacteria provide an initial stimulus causing the plant to initiate a developmental programme which could, in principle, proceed without further active participation by the bacteria. At the other extreme, each step in the developmental programme might involve a signal to the bacterium, evoking a response which would stimulate the plant cells to perform the following step. Evidence in favour of the latter model comes from the findings that strains of bacteria can fail at a number of distinct steps in the nodulations (e.g. Moerman et al., 1987).

3.6. An Intriguing Difference with Cyanobacteria

As previously discussed, many of the early stages in nodulation seem to be direct responses of the plant to the diffusible products of the *nod*ABC operon of the bacterium. But it seems well established that this initial stimulus is merely the first stage in a complex exchange of signals between bacterium and host plant (e.g. Morrison and Verma, 1987). This contrasts intriguingly with an apparently analogous system in which nitrogen-fixing cyanobacteria (*Nostoc* spp.) (Haselkorn, 1978; Haselkorn et al., 1987; Golden et al., 1987) are associated with modified (coralloid) roots of cycads. Lamont and Ryan (1977) showed that apparently typical (though presumably non nitrogen-fixing) coralloid roots could be produced without infection. Here it seems that organogenesis can be initiated by some relatively simple signal (such as a fluctuation in phytohormone balance) and once initiated can go to completion without further stimulus.

3.7. Plant Factors

The description of the initiation of nodulation treats the plant as an essentially passive partner, but this is clearly not the whole story. In fact, susceptibility to infection and the subsequent development of effective nodules are tightly regulated. Kapulnik et al. (1987) showed that addition of luteolin to the rhizosphere of a strain of lucerne increased the efficiency of nodulation (and the amount of nitrogen fixed). They also showed that the short-term selection gave plants with increased luteolin production and higher nodulation efficiency. Thus, it is clear that nodulation efficiency in this strain is limited by the rate of flavone production and that the plant population contains significant amounts of genetic variation in respect of the production of these compounds.

Induction of the *nod*ABC may be antagonized by flavonoids produced by plants (including leguminous plants susceptible to nodulation) (Firmin et al., 1986). The ratio between inducers and inhibitory analogues produced by a plant may have a role in modulating nodule initiation. Pierce and Bauer (1983) showed that one of the consequences of effective invasion of soybean root hairs by *R. japonicum* was a marked reduction in subsequent infection. Kosslak and Bohool (1984) showed that the whole root system, not merely the region near the first invasion site, was affected.

3.8. Soil Factors

The exogenous supply of fixed nitrogen (especially the nitrate concentration) can also reduce susceptibility to nodulation. The mechanism whereby the presence of nitrate in the rhizosphere reduces nodulation efficiency is not known but Ligero et al. (1987) presented evidence that excess nitrate may stimulate the synthesis of ethylene and that this compound reduces nodulation efficiency. Since the development and maintenance of a nodule is energetically expensive, it must be advantageous for the plant to be able to regulate the number of effective rhizobium infection sites. A few species of non-nodulating leguminous plants are known. Presumably these normally grow in soil sufficiently rich in nitrogen. Thus rhizobial infection would be valueless. A number of these were shown to produce diffusible inhibitors of rhizobia, which may be responsible for preventing nodulation (Rao et al., 1973). Fottrell et al. (1964) and Van Egeraat (1975) showed that roots of legumes capable of nodulation also produce diffusible inhibitors of rhizobium (*R. leguminosarum*). Whether these have roles in the regulation of nodulation is not known. From some leguminous species, mutants have been isolated with increased nodulation susceptibility. Certain mutants of the pea produce more nodules than wild type, but nodule production remains susceptible to nitrate inhibition and a mutant has been isolated which was "largely resistant to nitrate" (Jacobsen and Feenstra, 1984). Thus, it appears that there may be two separate mechanisms for regulation of nodulation (by compatible rhizobia). Host genes responsible for failure of effective nodulation have been described in several plant species and these are reviewed by Vance (1983) and Verma and Long (1983). It is clear that the nodulation process may fail due to incompatibility at any one of several steps.

3.9. Plant Defence Reactions

Several of these failures to nodulate may be regarded as host specificity steps by which plants may reject potential symbionts. For example, Pankhurst and Jones (1979) report a correlation between ability to nodulate roots of *Lotus* spp. containing a high concentration of flavolans (proanthocyanidins) and resistance of otherwise suitable rhizobia to these compounds. Ineffective nodules (containing few if any bacteroids) on *Lotus* contained ten times more flavolan than effective ones, suggesting that nodulation may be aborted by induction of phytoalexins. Why some invasions avoid this fate is unclear but the separation of bacteria from plant cytoplasm by the peribacteroid membrane may reduce the probability of defence mechanisms being initiated.

3.10. Coexistence

Since there is little reason to believe that bacteroids can revert to bacillary forms capable of independent existence, the advantage gained by the bacterium (or its plasmid) by nodulation ability has long been obscure. Yet it is clear that rhizobia thrive in soils planted with legumes and their populations fall if these plants are replaced by other plants (Parker et al., 1977). And there is an excellent geographical correlation between growth of varieties of legumes and the distribution of rhizobia capable of nodulating these, which suggests that the bacteria are favoured by the presence of suitable host plants (Lie et al., 1987).

It has recently been reported that in some cases nodules synthesize "opine-like compounds" which can feed free-living rhizobia (Murphy et al., 1987; Scott et al., 1987). These examples suggest a role for nodule formation in the ecology of rhizobia, but it remains true that many rhizobia induce nodules which seem unable to form such compounds. L-3-O-methyl-*scyllo*-inosamine, the "opine-like compound" in induced nodules of some *R. meliloti* strains on lucerne (Murphy et al., 1987), is remarkable. It does not meet the criterion (Ryder et al., 1984) that an opine should be formed by an unusual type of bond between two ordinary small molecule metabolites. It is, in fact, a close relative of O-methyl-*scyllo*-inositol, which is accumulated by the leaves deprived of water (Ford, 1982). It is interesting to suppose that this may reflect an early stage in the evolution of opine synthesizing systems where the bacterium caused the plant to oversynthesize compounds normally produced in response to physiological stress. These "stress compounds" (with or without small enzymatic modifications) should be exploited by the bacteria.

4. RELATIONSHIPS BETWEEN AGROBACTERIA AND RHIZOBIA

The Rhizobiaceae, notable for their ability to colonize plant tissues, seem to constitute a natural group but with a notable degree of diversity (De Smedt and De Ley, 1977; Hooykaas et al., 1982; Jarvis et al., 1986). The classical subdivisions of this group were differentiated by their characteristic interactions with host plant species and these characters are often determined by Ti, Ri or Sym plasmids. The question must arise whether we should think of rhizobia and agrobacteria as being essentially similar chromosomally and differentiated only by the carriage of one or another plasmid which "programmes" the bacterium so as to be able to play a specified ecological role.

The rhizobia can be separated into the fast-growing (*Rhizobium sensu stricto*) and slow-growing (*Bradyrhizobium*) strains. It seems that *Bradyrhizobium* strains carry nodulation and nitrogen-fixing genes as a group on the chromosome (Haugland and Verma, 1981). This sort of organization suggests that these organisms are more totally committed to the nodule-forming lifestyle, and indeed such bacteria do seem to have diverged from the other members of the family (as judged by ribosomal RNA sequences). The genus *Agrobacterium* also seems heterogeneous, with most (but not all) isolates falling into one of two subgroups (Kersters et al., 1973). The significance of this pattern is unclear, but the outcome of these studies seems to be that there is a close similarity between the fast-growing rhizobia and one of the two subdivisions of the agrobacteria (De Smedt and De Ley, 1977).

4.1. Plasmid Transfer Experiments

Transfer of a Ti plasmid into various rhizobia showed that the recipients were able to induce entirely normal crown galls (Hooykaas et al., 1977). This fits well with the idea that the Ti plasmid can programme a range of bacteria (remember the diversity of agrobacteria) to function as tumour inducers.

On the other hand, transfer of Sym plasmids (or cloned fragments of such plasmids) into agrobacteria produced diverse and complex results. Most of the combinations produced bacteria able to initiate nodulation but unable to complete the process. Failure occurs at various stages and these defects have been used to characterize critical steps in the interactions involved in nodule formation, especially the induction of nodulin proteins (e.g. Truchet et al., 1984; Moerman et al., 1987). Martinez et al. (1987) have reported the transfer to *A. tumefaciens* of a Sym plasmid from an *R. phaseoli* isolate permitting the transcipient to form nitrogen-fixing nodules on appropriate hosts. Even in this case the efficiency of nodulation, and the acetylene reducing activity of the nodules, was notably less than that of the original strain. It is difficult to draw precise conclusions but it does seem that Sym plasmids differ and that some either lack genetic functions required for production of nitrogen-fixing nodules (the required genes being provided by the chromosome or by "cryptic" plasmids), or that they can only express these genes effectively in a limited range of host species.

4.2. New Ecological Roles?

It is rather unfortunate that the agrobacterial strains which have been the main subjects of genetic study all belong to the subgroup clearly separated from *Rhizobium* and most if not all the studies on expression of Sym plasmids have been done using such strains. It would be interesting to know how the strains which seem more closely related to rhizobia might function if provided with Sym plasmids. Thus, although it seems clear that many bacterial clones are committed towards the nodulation or the tumorigenic lifestyle there may also be uncommitted forms capable of existence as free-living saprophytes but susceptible to being programmed as rhizobia or agrobacteria. Indeed, other ecological roles may be open to such bacteria, e.g. as associates of leaves, calyces and shoot tips of certain plant associates (Miller et al., 1984).

The report that *Azospirillum brasilense* carries genetic sequences homologous with the agrobacterial *chv* genes (Waelkens et al., 1987) suggests that ecologically significant genetic information may be conserved over substantial taxonomic distances. It has been noted that several members of the "rRNA superfamily IV" (to which both genera belong) are ecologically involved with plants (Jarvis et al., 1986) and it may be that *chv* and other genes involved in interactions with plant structures have been conserved within this group.

5. UPTAKE OF IRON

It is likely that competition for limited minerals plays a significant role in the rhizosphere (see also Chapter 7). Competition for iron has been the most studied case. Most of the root will be in an oxidizing environment where iron will exist in the extremely insoluble Fe^{3+} form. To ensure its own growth, each organism must convert part of the iron in the root region into a form to which it has access and (if that form is inaccessible to competing organisms) the iron acquisition mechanism can also function to deny this essential material to a competitor. In this way, a single mechanism may serve two quite distinct functions. It is, however, possible that the competitor may evolve a receptor system capable of binding the ferrated form of the siderophore and thus benefit from this system. Such "parasitism" of an iron uptake mechanism is shown by *E. coli* and relatives, which can take up iron from ferrated ferrichromes. Ferrichromes are fungal siderophores which have not as yet been demonstrated to be made by *E. coli* (Neilands and Leong, 1986). Although enteric bacteria possess their own enterobactin-dependent iron transport system, they are also able to take up iron from ferrichromes. Under low iron conditions the tonA (fhuA) protein (78 kD) in the outer membrane (which also binds phages T1, T5 and ϕ 80 as well as colicin M and albomycin), functions as a ferrichrome receptor. The *fhu*A mutants are unable to take up iron via ferrichrome (Hantke and Braun, 1975; Wayne and Neilands, 1975).

5.1. Mechanism

Typically, an organism produces a repressor which regulates its iron uptake mechanism (O'Brien et al., 1971; Page and Huyer, 1984). When there is no shortage of the metal, the repressor will be active (presumably, bound iron acting as corepressor). If the cell is starved of iron, the derepressed genes are expressed, producing enzymes required for the synthesis of a siderophore (a small soluble molecule capable of chelating Fe^{3+}). The siderophore is actively secreted and diffuses into the environment. It can bind ferric iron so powerfully as to remove this from most inorganic complexes likely to be found in soil. The ferric–siderophore complex is recognized very specifically by a receptor in the cell membrane (itself the product of a gene or genes regulated by iron) of the siderophore producer.

5.2. Ecology and Uptake Modulation

The siderophores of different organisms are a remarkably diverse set of molecules, presumably implying that iron uptake mechanisms have evolved independently on many different occasions. It is clear that certain types of siderophores are particularly well adapted to particular ecological situations. *Azotobacter vinelandii* is highly adapted to exploiting the iron resources of soils (Page and Huyer, 1984). Unstressed it forms a cell wall able to complex iron (which can be used by the cell) and produces a weakly binding chelator (with reductive properties). When relatively short of iron it produces azotochelin (a more complex chelator) and under extreme iron shortage a third (and apparently most efficient) chelator, azotobactin (plus membrane proteins probably required for the uptake of ferriazotobactin), is produced. Thus, there is a logical progression of activities from iron solubilization by reduction up to a very efficient siderophore system.

5.3. Competition

The most studied rhizosphere siderophores are those encoded by fluorescent pseudomonads. These are pyoverdines (Meyer and Abdallah, 1978). Genetic studies on the determinants involved in the synthesis of the pyoverdines show that they constitute at least four groups (?operons) all apparently chromosomal

(Loper et al., 1984). Genes determining the (protein) receptors for the ferrated siderophores have also been identified (Cody and Gross, 1987).

Although fluorescent pseudomonads all produce pyoverdines under certain growth conditions (Leisinger and Margraff, 1979) (and presumably all producers have receptors which recognize the ferrated forms of their own pyoverdines), the products of different strains are not functionally equivalent (Buyer and Leong, 1986; Hohnandel and Meyer, 1986). In low iron medium, the siderophore produced by one strain may inhibit growth of a second pseudomonad by withholding iron (the effect may be reciprocal or unidirectional). The ability of one strain to inhibit another by denying iron correlated well with the ability to suppress growth in the rhizosphere.

Just as possession of a permease for agrocinopine A can confer sensitivity to agrocine 84, so carriage of a receptor for ferrated siderophores can permit adsorption of bacteriophages and bacteriocines. So far, this has only been studied in *E. coli* and close relatives (Braun, 1985; Govan, 1986), but it seems likely that any such surface receptor (especially if produced in large numbers) may be the receptor for a toxic agent. And as we have discussed, the rhizosphere is just the sort of location in which we might expect biochemical and genetical warfare by DNA transfer mechanisms, such as phage transduction and plasmid conjugation, to be profitable.

5.4. A Surprising Control Failure

It seems that competition for iron plays a vital role in the competition between bacteria and fungi in the rhizosphere (Schroth and Hancock, 1982). Although siderophore-mediated inhibition of fungi by bacteria has been assumed to be the result of a denial of iron, the reverse mechanism may also be exploited. Ahl et al. (1986) showed that the inhibition of *Thielaviopsis basicola* by a fluorescent pseudomonad was due to the bacterial pyoverdine loading the fungus with iron until a toxic concentration was reached. Such overloading by the siderophore of another organism may be a risk for any organism which exploits the iron-uptake processes of another species (since the recipient has no way of regulating the level of production of the chelator). But it does seem surprising that control at the level of synthesis or functioning of the uptake mechanism does not operate to protect the recipient from iron-poisoning.

5.5. Iron and Siderophores as Signals

The influence of siderophores on germinating fungal spores also shows how organisms can make use of fluctuations in iron concentration. Baker et al. (1986) showed that pyoverdines can inhibit germination of chlamydospores of *Fusarium oxysporum* sufficiently to reduce the probability of infection of growing root tips. They suggest that increase in spore size and hence in iron content may provide a defence against such inhibition. On the other hand, germination of spores of several phytopathogenic fungi may be inhibited by Fe^{3+}. Because of the withdrawal of the Fe^{3+} by bacterial (over)production of siderophores, such as pyoverdines, these spores are stimulated to germinate (Swinburne, 1986). In this way, the spores can recognize regions of high biological activity and thus, indirectly, the growing plant whose tissues nourish the bacteria which produce the siderophores.

6. NECROTROPHIC BACTERIA

Whereas the exploitation of plants by rhizobia and agrobacteria requires the long-term survival of host cells, several other parasites adopt a more aggressive approach involving the destruction of tissue and, often, the death of the entire plant (necrotrophism). Typical examples are members of genera such as *Erwinia* and *Xanthomonas,* which induce ''rotting'' diseases that kill seedlings or cause major lesions on established plants. Such bacteria produce extracellular enzymes capable of hydrolysing plant cell wall and membrane components, including cellulases and proteases. Most attention has been paid to the enzymes involved in degradation of pectic polymers (chains of 1,4-linked α-D-galacturonic acid and methoxylated derivatives).

It is believed that degradation of pectins weakens cell walls leaving the cells osmotically fragile. Endopectate lyases, enzymes involved in the hydrolysis of pectin polymer to water-soluble dimers, are the primary agents responsible for maceration (rotting) of tissues by *Erwinia* spp. (Chatterjee and Starr, 1977; Collmer et al., 1982).

Strains of *E. chrysanthemi* secrete multiple endopectate lyases distinguishable by isoelectric focusing, and from this species five distinct *pel* genes encoding distinguishable endopectate lyases have been cloned into plasmids. Carriage of such plasmids by *Escherichia coli* confers at least limited ability to macerate plant tissues. The five *pel* genes are organized into two clusters (Kotoujansky et al., 1985; Van Gijsegem et al., 1985). One cluster consists of two genes, *pel*B and C, which are highly homologous but unrelated to the other three *pel* genes A, D and E (Schoedel and Collmer, 1986), which constitute the second cluster.

6.1. Primary Plant Signals

Production of all these lyases is normally repressed but their synthesis can be induced by plant cells, preparations of plant cell wall purified pectic polymer, or various soluble derivatives produced by the breakdown of this polymer (Collmer and Bateman, 1981). Molecular details of the regulation are not yet well defined.

6.2. Bacterial Regulation

Since the *pel* genes are expressed constitutively if transferred to *E. coli* (Keen et al., 1984), it seems likely that they are subject to negative control (repression) by a gene present in *E. chrysanthemi* but absent from *E. coli*. Those genes so far characterized appear to be transcribed separately to form monocistronic mRNAs (Keen and Tamaki, 1986). Discovery of constitutive mutants causing constitutive expression of all five *pel* genes (Diolez et al., 1986) implies that there is at least one common element in their regulation and suggests that they constitute a regulon. Whether any of the other enzymes apparently involved in the destruction of plant components (e.g. cellulases, cutinases) show similar regulation is not known. Andro et al. (1984) report that cellulase synthesis is induced in a medium containing galacturonates but lacking cellulose (or anything which seems a plausible candidate as a specific cellulase inducer).

6.3. Enzyme Expulsion

Secretion of proteins from Gram-negative bacteria is a relatively rare phenomenon and though several examples have been reported these all involved special (not very well defined) processes; hence the secretion of a whole battery of enzymes is a remarkable occurrence. *E. coli* cells carrying cloned *pel* genes produce their appropriate lyase but these are retained in the periplasm (perhaps explaining the rather poor macerating ability of these bacteria), suggesting the requirement for an active system for liberating the enzymes into the medium. This was confirmed by the isolation of mutants of *E. chrysanthemi* which produced but failed to secrete either pectate lyases or cellulases. Such mutants are non-pathogenic (Andro et al., 1984).

The other enzymes involved in the breakdown of plant tissues could be controlled by the same repressor as pectate lyase; they could have comparable but distinct regulatory pathways, or could even be produced constitutively. The mechanism which ensures that the whole battery of enzymes is assembled will surely be of great interest. These enzymes must be secreted into the medium and it is clear that those examined have N-terminal signal peptide sequences appropriate for entering the periplasm (Keen and Tamaki, 1986). The ability to pass through the outer membrane into the extracellular medium is probably effected by a common mechanism and *E. coli* is unable to effect such expulsion.

6.4. Host–Pathogen Signals

The hydrolysis of pectate may be a signal to the plant since some of the oligosaccharide breakdown products induce the formation of antimicrobial agents (Davis et al., 1984) and *E. carotovora* subsp. *atroseptica* is susceptible to rishitin, a phytoalexin from potato (Lyon and Bayliss, 1975). This signal exchange may represent a defence reaction of the host.

One should recall that although the secreted enzymes are the most immediately active agents in the interaction with plant tissues, the intracellular organization of the bacteria is critical since it is this which ensures that the supply of enzymes will be maintained (and modulated as the environment changes). The bacteria will make maximal advantage of the environmental opportunities (e.g. the area of colonization expanded and new sites invaded) and multiplication will ensure that an increasing number of cells can be recruited both to exploit the current site and to increase the chances of new sites being found.

The very large numbers of enzymes involved in the conversion of pectic polymers to soluble products (which can be taken up efficiently by the bacteria) and the variation in the sets of enzymes between different erwiniae has been the subject of much comment (e.g. Collmer, 1986). One possibility is that these are adaptations to minimize the level of production of inducers of phytoalexins. If different hosts recognize different inducers the different sets of pectin solubilizing enzymes may be adaptations to different host plants.

A few of the genes involved in the disruption of the structural molecules and the metabolism of the products (direct, such as pectin derivatives, and indirect, e.g. the materials released by osmotic lysis of the protoplasts of weakened cells) have been cloned and mutants in several others are now known. Since genetic maps of several erwiniae are being prepared (e.g. Leary and Fulbright, 1982) it seems likely that information on the genetic organization of the genes involved in host–pathogen interactions may soon be forthcoming.

7. CONTROL OF GROWTH PATTERNS OF PLANT ROOT SYSTEMS

7.1. Plant Genotype

The only well-established case where genetic control of root organization has a clear effect on rhizosphere population concerns a gene (or genes) on chromosome 5 of spring wheat. Plants carrying one form of this chromosome supported larger rhizosphere bacterial populations than those with the homologue (Neal et al., 1973). This chromosome also had a dramatic influence upon the ability of roots to support fixation of nitrogen by bacterial associates (although other chromosomes were also involved in this effect) (Larson and Neal, 1978; Rennie and Larson, 1979). The plants with high rhizosphere populations were susceptible to "common root rot" caused by weakly pathogenic fungi (Neal et al., 1973) and Deacon and Lewis (1982) showed that in these plants cortical cells underwent natural (non-induced) senescence to a very much greater degree than in other plants. They proposed that in the plants showing cortical senescence, saprophytic and weakly pathogenic fungi establish large populations and that these may protect their hosts against infection with major pathogens such as take-all fungus (Kirk and Deacon, 1987).

7.2. Bacterial Control

Since it is clear that the growth pattern of a plant root may be modified so as to increase the supply of nutrient without significant disadvantage (and perhaps in certain environments even positive advantage) to the host, we need not be surprised to find that some root organisms are able to modulate the growth schemes of plant roots. Doubtless, there are many diverse mechanisms by which bacteria are able to influence the growth of roots, but it may be best to concentrate upon a single easily observed phenomenon. The formation of lateral roots involves the destruction (enzymatic and mechanical) of large numbers of cells of the cortex of the primary root (Bonnett, 1969) supplying nutrients and access into the root interior for the rhizosphere bacteria (Allen and Allen, 1940; Mosse, 1975; Jacobs et al., 1985). Thus it seems plausible to postulate that rhizosphere bacteria might be able to increase their nutrient supply by inducing the formation of secondary roots.

Many root-associated bacteria can influence growth of plant roots (e.g. Brown, 1975) (see also Chapter 7). Experiments using *Azospirillum lipoferum* (Martin and Glatzle, 1982; Marschner et al., 1986) have shown that root length, the root/shoot ratio and the number of lateral roots formed could all be increased by bacteria that produce auxins and other plant growth regulators. The production of a plant hormone by a free-living rhizosphere organism may influence the development of the host in ways likely to increase the food supply to the bacterium. It has been shown that plant growth regulators (especially gibberellic acid) permitted germination of barley seeds in the presence of inhibitory concentrations of sodium chloride (Kabar, 1987). Thus, spermosphere bacteria may be able to promote germination under unfavourable conditions (and such germination will be accompanied by exudation).

A particularly impressive but little investigated phenomenon is the formation of dense clusters of roots by certain plants (especially members of the Proteaceae) (Lamont, 1982). These clusters may be very efficient in the solubilization and uptake of insoluble minerals (Marschner et al., 1986) and their formation is inhibited in soils rich in nutrients. Yet, their initiation requires microorganisms, so far uncharacterized (Malajczuk and Bowen, 1974; Lamont and McComb, 1974).

7.3. Vesicular-arbuscular Mycorrhiza Infection

The requirement for both a biotic and a physiological stimulus in the formation of proteoid roots is reminiscent of the establishment of vesicular-arbuscular mycorrhiza (VAM) symbiosis. Graham et al. (1981) observed that in phosphorus-poor soils sorghum roots excrete increased amounts of sugars and that excretion was less in soils enriched with exogenous phosphate. VAM infection was much more pronounced in the former plants. Apparently, the excreted sugars promoted germination of spores of VAM fungi and may have aided growth of germlings into the plant cells. Such a phenomenon can be considered in two ways. Either one assumes that phosphorus-deficient plant cells are naturally leaky (membranes low in phospholipids?) or one could imagine that the ability to exploit VAM symbiosis at time of nutritional stress (but not when mineral nutrients are abundant) has been the outcome of evolutionary pressure.

7.4. Generation of Attachment Sites

Root hairs are sites of excretion of amino acids (Pearson and Parkinson, 1961) and are thus attractive to motile bacteria. They may contain superficial proteins distinct from those of other parts of the root epidermal surface (Law and Strijdom, 1984) which provide efficient binding sites for bacteria (Yao and Vincent, 1969; Haahtela et al., 1986). As active secretory structures (Werker and Kislev, 1978) they provide obvious targets for bacterial exploitation and klebsiellae (Haahtela et al., 1986), azospirillae (Jain and Patriquin, 1985) and rhizobia (Yao and Vincent, 1969) are able to induce deformations of growth of these structures, probably through production of plant hormones. Azospirillae are also able to induce superabundant production of root hairs (Martin and Glatzle, 1982).

7.5. Co-operation for Symbiosis

An interaction between bacterial root-associates was reported by Yahalom et al. (1987) who found that prior infection of roots of leguminosae with azospirillae increased the efficiency of nodulation by rhizobia. They suggest that newly forming root hairs, induced by the azospirillae, provide suitable infection sites for the rhizobia. There are other reports of a free-living rhizosphere organism playing a role in establishing symbiosis. For example, *Endogone* spores failed to infect plant roots under aseptic conditions but VAM infection was consistently observed if pure cultures of soil pseudomonads were added to the soil (Mosse, 1962). Similarly, cultures of *Frankia* sp. were almost completely incapable of nodulating *Alnus* roots under gnotobiotic conditions but infection proceeded efficiently in the presence of isolates of *Pseudomonas cepacia* (Knowlton et al., 1980). Presumably all such cases reflect modifications of young roots, inducing regions to enter physiological states in which they are susceptible to infection by potential symbionts. One may wonder whether sensitivity to pathogenic fungi may be similarly inducible.

7.6. Iron

We may reasonably expect that rhizosphere bacteria manipulate the activities of roots in many diverse ways. It is known, for example, that graminaceous plants secrete phytosiderophores (e.g. mugineic acid) when iron-deficient (Sugiura and Nomoto, 1984) and it could be that bacteria may be able to withhold iron from these plants with their own siderophores, thus inducing production of phytosiderophores, and subsequently catabolize to their own benefit the phytosiderophores secreted by the plants (see Chapter 7).

8. THE FACTORS REGULATING EXUDATION

In part, the loss of organic compounds from plant roots is a passive process due to leakage of materials through cell walls, accidental abrasions, and so on (see Chapter 4). But in part, it is subject to discernible control influenced by genotype and physiology.

The simplest cases are provided by the genetic variations within species. For example, Azcón and Ocampo (1981) showed that wheat varieties varied markedly in their level of sugar exudation and that the concentration in exudates was not a reflection of the sugar concentration in the root tissue. Strains with low sugar exudation showed little susceptibility to mycorrhizal infection, even when the concentration within the root was high. Thus, it seems that the interstrain differences reflect genetic polymorphisms with potential ecological significance. Plants with low sugar exudation are less "wasteful" of sugars but fail to attract mycorrhizal symbionts, whilst plants more prodigal of organic materials are more susceptible to such infection. Genetic and ecological studies have not been carried out but an obvious suggestion is that the former strategy might be adaptive to growth on good soils (e.g. abundant phosphorus compounds) under conditions not favourable to photosynthesis (e.g. inadequate sunlight), whilst reversal of the environmental characteristics might favour the latter genotypes. Although this argument is surely an oversimplification, it is hard to believe that so striking a difference between the two patterns of rhizosphere interactions does not have ecological significance.

The work on wheat varieties shows that the concentration of sugar within the root is not simply correlated with the amount exuded; this was confirmed by Vietor (1982) who showed that *Paspalum notatum* translocated 7% of the photosynthate to its roots and exuded 1.1%, whereas *Panicum coloratum* and *Pennisetum americanum* translocated 12% but exuded only 0.4%. It may be more than coincidental that *P. notatum* is characteristic of poor soils in subtropical South America (Döbereiner, 1970) and so falls within the class of plants most likely to benefit from mycorrhizal infection.

Wilson and Bell (1978b) studied two species of *Lathyrus, L. aphaca,* which produces homoarginine, and *L. odoratus,* which produces beta-amino-propionitrile; each was inhibited by the other's product but relatively (or totally) resistant to its own product. Similarly, both were sensitive to canavanine (produced by *Vicia benhalensis* and itself almost totally unable to inhibit its own producing species). Thus, the plants can produce the compounds without seriously inhibiting their own growth. Most attention has focused upon the ability of the chemical analogues to inhibit germination and growth of competing plants, but the effects on associated microorganisms have been little studied. However, it is at least clear that the plants have evolved so as to benefit from the production of these chemical analogues without suffering inhibition.

9. ASSOCIATIVE NITROGEN FIXATION

That legumes can benefit from the association with rhizobia seems beyond doubt and though less intensively studied, the associations of cycads with cyanobacteria and of various flowering plants with *Frankia* spp. are probably beneficial to the host plant (see Chapter 12). In all of the cases, the bacteria are housed in special structures produced by modification of plant morphogenetic processes. As well as these clearly defined symbiotic relationships, many of the bacteria to be found in association with plant roots are capable of nitrogen fixation, and the question of whether the plant can benefit from such association is one of the most hotly debated points in rhizosphere microbiology (see Brown, 1982).

Kraffczyk et al. (1984) and others have noted the very low N:C ratio in compounds of plant exudates. Growth on such a food source might be expected to favour organisms capable of nitrogen fixation and to provide an environment where such ability would be expressed. There may even be a positive induction of nitrogenase by products of plant root metabolism (e.g. Hess, 1982). An obvious problem is the route for transfer of fixed nitrogen from bacteria to host plant. In the rhizobial symbioses the peribacteroid membrane seems to be adapted for this transfer, but is anything comparable available in associations between free-living bacteria and plants? The association between azospirillae and plant roots takes many forms. The earliest seems to be a covering of the surface of young roots (Whallon et al., 1985) but in well-established infections the bacteria are commonly observed in intracellular spaces within the cortex (Patriquin et al., 1983).

It seems likely that this area will remain a subject of debate.

10. SOIL NUTRIENT STATUS

10.1. Chemical and Physical State

The soil is the major source of most of the nutrients required for plant growth and the microbial soil flora plays a crucial part in determining the physical and chemical state of the soil. So it seems plausible that plants might have evolved to maximize the beneficial aspects of microbial activity and to reduce activities detrimental to their own functioning. A great deal of attention has been paid to solubilization of phosphates (phosphate solubilizing strains of bacteria are common in the rhizosphere). Mycorrhizal associations are probably important agents in transferring phosphates (and perhaps other relatively immobile minerals), and the hyphae of such fungi may maintain the physical structure of soils. The fixation of nitrogen by symbionts and associates is another aspect of the role of microorganisms in the biochemistry of soils.

10.2. Inhibitory Substances

Rice and Pancholy (1972) proposed that climax vegetation (e.g. oak and pine woodland) was characterized by soils inhibitory to nitrification (the bacterial oxidation of ammonium compounds to nitrites and nitrates). This inhibition is due to production (partly as exudates, partly as litter) of a variety of inhibitory substances (Rice and Pancholy, 1974). The use of ammonium compounds is much less demanding energetically than is that of nitrate, so there are real advantages for the plants in preventing nitrification. It is clear that forest soils do contain compounds strongly inhibitory to *Nitrosomonas* (Rice and Pancholy, 1974). These compounds have other biological effects (e.g. inhibition of germination) and the significance of their role in modulating the oxidation level of nitrogenous compounds in soil is an open question.

10.3. Detoxification

A somewhat analogous proposal was that of Pitts et al. (1972) who suggested that rice plants growing in paddy fields might be in need of protection from sulphide ions generated by bacterial sulphate reduction. This may be provided by *Beggiatoa* (or related bacteria) which oxidize the sulphide. Growth of these bacteria in the rhizosphere is promoted by exudation of catalase, which prevents the accumulation of autotoxic peroxides.

This interaction illustrates what may prove a rather general effect, in that under favourable conditions the plant is independent of the symbiont and it is only under physiological stress (sulphide toxicity) that the interaction benefits the plant. We have already seen that similar arguments apply to mycorrhizal and rhizobial symbioses and it would be interesting to learn whether catalase exudation is increased by sulphides in the rhizosphere.

A few other cases have been reported where rhizosphere microorganisms are believed to influence the availability of minerals such as manganese (Timonin, 1947) and molybdenum (Loutit and Brooks, 1970) but their significance remains obscure.

11. SPECIFICITIES OF RHIZOSPHERE INTERACTIONS

11.1. Host Preferences of Rhizobia

The recognition of appropriate hosts by rhizobia is much the best example of specific interaction between rhizosphere organisms. This process seems to require a series of interactions between plant and rhizobium (e.g. flavone induction of morphogenetically active products, binding of sugar components by lectins and the ability to induce nodulin synthesis), where each interaction is a necessary step in a highly coordinated pathway.

11.2. Other Bacteria With Host Preferences

Other interactions seem to be simpler, though this may be an illusion based on lack of information; several hints of precise specificity have been found, for example the association of *Azotobacter paspali* with an ecotype of *Paspalum notatum* (Döbereiner, 1970). Döbereiner and Boddey (1981) report that the different species of *Azospirillum* colonize the rhizospheres of different host species. Dramatic evidence of extremely specific recognition was provided by Tabary et al. (1984) who isolated from rice embryos a lectin capable of binding a strain of *Beijerinckia* sp. and one *A. lipoferum* strain (both isolated from the rhizosphere of a rice plant) but unable to bind isolates from the maize rhizosphere. If further research confirms these preliminary observations and the lectin proves to be present at the root surface, these results may indicate a higher degree of specificity of interaction between plant roots and rhizosphere associates in the absence of gross morphological alterations. A further level of specificity of interaction between plant roots and azospirillae has already been mentioned. Heulin et al. (1987) report that some but not all azospirillae can stimulate exudation of catabolizable organic material from rice seedlings. It is hard not to doubt that such an ability is of great value to a rhizosphere organism. Whether these bacteria are specifically adapted to a single species (or group of species) or whether they can stimulate exudation from a wide range of hosts is, as yet, unknown. Studying the colonization of two different rice varieties by a single *A. lipoferum* strain, Murty and Ladha (1987) showed different patterns of spread over the root surface. One variety produced abundant mucigel and in the earliest phases of infection most of the bacteria were associated with this material, whilst the other variety produced much less mucigel and the root hairs were colonized much more rapidly than those of the former strain. These results hint at the possibility of several different patterns of association between azospirillae and the roots of even a single plant species. If cells can bind to mucigel or root hairs (and perhaps other sites) we are almost forced to postulate different molecular processes, each of which may show its own pattern of specificity; it may prove very difficult to analyse one type of binding when others are also operating.

11.3. Host Preferences by Chemotaxis

The work of Reinhold et al. (1985) demonstrating differences between strains of *Azospirillum* in their chemotactic responses to root exudation of carboxylic acids has already been mentioned. But this is not the only way in which these bacteria discriminate between host species. Martin and Glatzle (1982) found an excellent correlation between the rate of liberation of biotin from roots of grasses of different species and the ability of these hosts to support the growth of (and nitrogen fixation by) their strain of *A. lipoferum*. If different species (or strains within the species level) have different requirements this could explain the host preferences described above.

11.4. Genetics of the Plant

The ability of different host plants to interact with azospirillae must have a genetic basis, and this was clear in the case of spring wheat (Rennie and Larson, 1979). Variety Cadet was strongly stimulated by a strain of *A. brasilense* to which the var. Rescue showed no response. Substitution of either chromosome 5B or 5C from var. Cadet into the genome of Rescue permitted the plants to benefit from the association. The replacement of chromosome 5B (though not replacement of 5D) by the homologue from Rescue abolished the ability of Cadet to be stimulated. A *Bacillus* sp. isolated from the wheat rhizosphere showed a similar response, and the 2D chromosome of Cadet was also able to stimulate Rescue plants to respond.

12. BIOCIDES IN THE RHIZOSPHERE

It is difficult to assess the ecological role of antibiotic molecules transferred into the rhizosphere by plants. One situation in which such assessment is possible is where biologically active compound producers and non-producing plant lines exist so that comparison of the ecological effect is possible. An optimal situation for comparison would be that both plant lines exist where one isomer (or analogue) is produced in the former, the other isomer (or analogue) in the latter.

12.1. Fraudulent Analogues

A few natural isomers of possibly biologically active compounds are known (e.g. production of thymol and isothymol in *Coridothymus capitatus*) (Fleisher et al., 1984) but the significance of these for interaction with rhizosphere organisms has received little study.

The production of analogues of important substances (usually amino acids) by plants is not uncommon and offers opportunity for experiment because the effects of the analogue can be (at least partially) reversed by addition of the normal compound. These play diverse roles in plant–microbe interactions.

Hildebrand (1973) observed that of 15 species of pseudomonads only *P. syringae* pv. *pisi* could metabolize homoserine, which was toxic to the other strains. This amino acid is present in high concentrations in peas (especially in seedlings) and may be responsible for the fact that of the 15 pseudomonads, only the resistant strain was pathogenic for that host. Such a resistance mechanism can probably be overcome by relatively simple genetic change (e.g. overproduction of the normal amino acid or alteration of specificity of the enzyme which recognizes the analogue), and it is probable that specialized pathogens will so evolve. But it may be a powerful inhibitor of an organism with wide host range which does not often encounter the analogue producer.

Several phytopathogenic bacteria produce amino acid analogues. The best studied case is the production of phaseolotoxin (an analogue of ornithine) by *P. syringae* pv. *phaseolicola.*

12.2. Methionine

Several observations have suggested that methionine may have a specially important role in the rhizosphere. Wallace and Lochhead (1950) found that methionine requirement was particularly common amongst rhizosphere bacteria. This observation was at least partly explained by the observation that *Pseudomonas maltophilia* was especially abundant in some rhizospheres, such as those of cruciferous plants, which produce particularly high levels of sulphur-containing compounds; *P. maltophilia* requires methionine (Debette and Blondeau, 1980).

Fredrickson et al. (1987) isolated from wheat roots pseudomonads that inhibited the growth of such roots; they showed that this inhibition was much reduced by exogenous methionine, which indicates that the inhibitor may be a methionine analogue. Production of such analogues by pseudomonads (Sahm et al., 1973) and by enterobacteria (where the character is plasmid determined)

(Aguilar et al., 1982) is already well known. Some strains of *Rhizobium japonicum* produce the wide-range biocide, rhizobiotoxin, which inhibits the same pathway (Owens, 1969). It may be significant that none of the rhizosphere *Pseudomonas putida* strains tested by Vančura (1980) could metabolize methionine, though they could use a wide range of other amino acids. This suggests that there may be advantages in not taking up this compound or its analogues. Further, methionine is a critical substrate in the formation of the growth regulator, ethylene (Chapter 7).

12.3. Inhibitory Substances

We may be sure that no toxins, sets of toxins or any other products can prevent the growth of a microbial community around a source of nutrients such as a plant root. Nonetheless, the plant can influence the composition, spatial organization and physiological activity of its microbial associates so as to favour beneficial organisms and to gain the greatest benefit from their activities.

A somewhat analogous case concerns the production of coumarin (an effective inhibitor of plant growth) by *Anthoxanthum odoratum*. Rivière and Chaussat (1966) showed that the rhizosphere of this grass was strongly enriched for bacteria of several genera that were capable of metabolizing the compound. The genetic basis of this ability was not tested but it may be notable that the ability of pseudomonads to metabolize such bioactive compounds as xylene, camphor and salicylate is plasmid determined. The transmissibility of such genetic elements means that in a rhizosphere where such a compound is produced, a single metabolically competent bacterium can donate the plasmid to a range of recipients capable of exploiting the compound in ecologically diverse fashions. It would be interesting to know whether (like agrocinopine) these compounds can induce the conjugal transfer genes of the various plasmids.

Conversely, Kaminsky (1981) has suggested that phytotoxic compounds found in soil under the canopy of *Adenostoma fasciculatum* may be produced by microbial conversion of non-toxic compounds washed off the leaves into soil. These toxins are believed to be important in maintaining the dominance of these shrubs in the Californian chaparral. Hence it seems that an ecosystem may be maintained by a symbiosis in which the plant provides nutrients for the bacteria whilst they inhibit germination and growth of competing plants.

12.4. Fungal Inhibition

Hopkins et al. (1987) studied the susceptibility to *Fusarium* wilt of some varieties of water melon and concluded that one particularly resistant line owed its insensitivity to the activities of some microorganism whose growth was favoured in the vicinity of these plants.

12.4.1. Enzymatic Origin of Fungal Inhibitor

Rice (1979) has summarized a research programme of C.Y. Li and colleagues upon red maple (which is relatively resistant to root infection by the fungus *Poria weirii*) and Douglas fir (which is very sensitive). The difference lies in the production by the former, but not the latter, of an enzyme which catalyses the conversion of *p*-coumaric acid to caffeic acid and of *p*-hydroxyphenylacetic acid to 3,4-dihydroxylphenylacetic acid (two effective inhibitors of the fungus). All of these phenolics pass into the soil around the roots. It is interesting to reflect that a root-associated fungus or bacterium capable of producing the enzyme might be able to protect the roots of the inherently sensitive plant.

13. PRODUCTION OF ANTIFUNGALS

The production of antifungal agents by bacteria, especially by bacteria associated with plant tissues, has been the subject of intense study in recent years (Cook and Baker, 1983; Cook, 1986), particularly for biocontrol purposes. Recent success on the subject with *Trichoderma* as a biocontrol agent makes a bright future (Gear, 1988) (see Chapters 7 and 10). Nevertheless here we will focus mainly on bacterial agents.

It is easy to find bacteria which have large inhibitory zones in respect of different fungi (Becker, 1984). Judging by the zone size these must be secreting antagonists, and from the different patterns of species affected there must be many different compounds (Becker et al., 1985). The difficulty in finding some products could be due to the fact that they are very unstable or volatile (e.g. Dennis and Webster, 1971; Claydon et al., 1987) or tightly bound. We can imagine that in the case where antifungal activity is of enzymatic origin, e.g. chitinase (Roberts and Selitrennikoff, 1988) the observation of biological effects demands isolation procedures with defined non-denaturing conditions with or without suitable cofactor(s), coenzymes and substrates (e.g. Section 12.4.1). Nevertheless many antifungals have been isolated and identified.

13.1. Small Molecules

Particular attention has been paid to pseudomonads, e.g. *P. cepacia*. In some cases, the active agents have been characterized as small molecules, e.g. fluopsin (= antibiotic YC73) (Egawa et al., 1970), antibiotic BD–2073 (Kanda et al., 1975), pyrrolnitrin (Howell and Stipanovic, 1979), pyoluteorin (Howell and Stipanovic, 1980), tropolone (Lindberg, 1981) and phenazines (Gurusiddaiah et al., 1986; Turner and Messenger, 1986) (Chapters 7 and 15), and many unidentified compounds come from these species.

13.2. Peptides

Some bacteria produce peptides (or compounds with a major peptide component) (Umezawa et al., 1978) with antifungal activity. Several *Bacillus* isolates produce cyclic peptides (Wakayama et al., 1984, and references therein; Babasaki et al., 1985; Loeffler et al., 1986). These typically contain lipid moieties (Besson et al., 1977; Vanittanakom et al., 1986). Amongst the Gram-negative bacteria a few pseudomonads produce antifungal peptides. *Pseudomonas antimycetica* produces comirin (Forsyth, 1955). *P. syringae* pv. *syringae* produces two phytotoxins (syringomycin and syringotoxin), both containing major peptide components, which also inhibit fungal growth (Gross et al., 1977; Gross and Cody, 1985). All of these peptides contain modified (or unidentified) amino acids. The siderophores produced by fluorescent pseudomonads (which inhibit fungi by withholding iron) are also modified peptides (Teintze et al., 1981; Wendenbaum et al., 1983; Buyer et al., 1986).

13.3. Enterobacteriaceae

Pseudomonads are, however, not the only plant-associated bacteria; Enterobacteriaceae are an important component of the phytosphere flora (Patriquin, 1981). Antifungal products of these bacteria have received little attention. There are several examples of the production of the same antibiotics by members of different genera of microorganisms (Lechevalier, 1975). Independent evolution could be due to "horizontal" transfer of the character by plasmid transfer (Hedges, 1972) or by independent evolution by relatively few modification steps of precursors present in every living cell (Floss et al., 1978).

Several *Serratia* (Wasserman et al., 1962; Gerber, 1975) and *Erwinia* (Winkelmann et al., 1980) isolates have proved to be antibacterial and sometimes good antifungal agents, and recently we found in a screening programme for antifungal agents a *Serratia* strain producing pyrrolnitrin (R. W. Hedges, unpublished results). As enteric bacteria they may prove especially convenient to handle since knowledge of the genetics of these bacteria and techniques for genetic manipulation are especially well advanced in this group.

Erwinia herbicola may prove a particularly suitable organism. It associates intimately with plant surfaces but shows no pathogenic abilities (Starr, 1981). Recent evidence suggests that, at least in temperate regions, *E. herbicola* is one of the most important nitrogen-fixing organisms in the rhizosphere (Rennie, 1980; Patriquin, 1981) and this suggests that it may be a suitable agent for seed treatment. Herbicolin A and B, two closely related depsinonapeptides recently isolated from *E. herbicola* strain A 111 (Aydin et al., 1985), are non-toxic for bacteria (Winkelmann et al., 1980), but are potent wide-spectrum antifungals. These depsipeptides are not general biocides (as, e.g., is syringomycin), but are relatively specific antifungal agents, promising for biocontrol purposes.

Relatively little is known concerning the production and secretion of peptides by enteric bacteria, but recent studies on microcins (Martinez and Pérez-Diáz, 1986) are providing information. Production of peptides with antifungal activity may be widespread among enterobacteria of soil. Whether the fungi that are beneficial to plant growth, notably mycorrhizas and other fungal biocontrol agents, e.g. *Trichoderma* (Ricard and Highley, 1988), are subject to inhibition by these peptides has to be a primary consideration before engineering these antifungal characteristics into efficient phytosphere colonizers. However, mutants of the most active *Trichoderma* strains, resistant to fungicides but active against mildew and grey mould of grape, could be selected (Gullino and Garibaldi, 1986).

Noteworthy is a little group of the Enterobacteriaceae that are on the borderlines of being classified with the family (Ramia et al., 1982). These are the *Xenorhabdus* spp. (Boemare and Akhurst, 1988), which are associated with insect pathogenic nematodes (Thomas and Poinar, 1979). Their highly specialized symbiosis may provide particular hints for biocontrol. Such bacteria have been shown to produce different combinations of several antibiotics active against insects, bacteria and yeasts: hydroxyl and acetoxyl-bearing indole derivatives have come from a *X. nematophilus* strain and two trans-stilbene compounds from a *X. luminescens* (Paul et al., 1981), while further 1,2,-dithiolo(4,3-b)pyrrole derivatives (designated xenorhabdins) have been isolated from both *X. luminescens* and *X. nematophilus* (Rhodes et al., 1987). Screening of these *Xenorhabdus* species for antifungal activity showed strains which were highly active against filamentous fungi, including phytophathogens. Interestingly, as far as we have gone (R. W. Hedges, unpublished results) the antifungal compounds were found to be in all cases identical to the antibiotics identified previously.

14. CONCLUSIONS

Considering the main groups of plant-associated bacteria we have to conclude that these are, on the whole, not too different from their non-plant-associated relatives. Thus the Erwiniae show many resemblances to the other Enterobacteriaceae and are sometimes reported as animal pathogens. Moreover, it would be hard to separate the plant pathogenic fluorescent pseudomonads from the rhizosphere saprophytes or the forms adapted to other environments (without direct pathogenicity tests). It seems rather that when we consider such strains we find bacteria with the potentials much like those of related bacteria from other niches but with a few extra features (rather like a standard car frame to which a variety of "optional extras" may be added to exploit the different sets of customers). Probably this is necessarily so. Few bacteria can be so certain of transport between ideal sites that they can adapt exclusively to life in such niches. A rhizosphere organism must be able to survive in soil (as in water as wherever) and may have to survive saprophytically on decaying vegetable matter, in the digestive tracts of insects (Harrison et al., 1980; Venette, 1982) or nematodes (Chantanao and Jensen, 1969), or other sites. In such bacteria, the genes required solely for the rhizosphere-based lifestyle may be switched on and off as a group and may be most efficiently organized as a cluster of genes on a plasmid or on the chromosome.

At the genetic level, this is especially clear amongst the agrobacteria. Lacking Ti plasmids, they are totally non-pathogenic and presumably exist as saprophytic rhizosphere associates. The plasmid genes confer an entirely different way of exploiting such a host (though the cost of this option, i.e. maintenance of the plasmid, must be significant and a large proportion of agrobacteria lack Ti plasmids) (Schroth et al., 1971).

At the physiological level also, *A. tumefaciens* provides attractive model systems. Like most flagellated bacteria it has a (chromosomally determined) chemotactic apparatus ensuring that environmental stimuli (nutrients and antagonists) can be monitored and the information transmitted to the structures regulating flagellar activity. The Ti plasmid modifies this system by inserting into the pre-existing information transfer system the ability to recognize a wound associated factor, acetosyringone. Thus, the generalized potentialities of the cell are redirected so as to be available for the new function.

Analogously, the ability to produce pectic lyase is not unique to phytopathogenic erwiniae (being required for the breakdown of plant tissue by saprophytes) but the ability to produce and secrete a whole battery of such enzymes (and others attacking other plant structural polymers) is not observed in non-phytopathogenic bacteria (Chatterjee et al., 1979) and rhizobia (Fåhraeus and Ljunggren, 1959; Lopez and Singer, 1987). Enzymes are secreted into the periplasm by the typical signal-peptide recognition system (found in other enterobacteriaceae) but the system for moving them into the extracellular environment, in large quantities, may be a critical adaptation.

The ability to adjust metabolically to a set of sharply distinct environments may be required of many microorganisms. A different physiological balance may be required in each situation and evidence is accumulating for the existence of "global regulatory networks" (Gottesman, 1984) which permit major switches of physiological state in response to environmental stress. As occupants of transient niches, plant-associated bacteria (especially those in climatic zones with marked seasonal changes) may prove to be extreme examples of this tendency. This may involve cyclic genetic shifts, as in the agrobacteria where, as discussed earlier, loss and reacquisition of a Ti plasmid may be a regular element of the life cycle. More typically, however, the adjustments may be purely physiological.

Our understanding of the ecology of phytopathogens is in its infancy but consider the strains of *Erwinia carotovora* described by Venette (1982). These infect onions and have mutualistic interactions with several insects, including the onion maggot (*Hylemya antiqua*). The insect larvae cannot develop fully without bacteria. Bacteria overwinter in insect puparia and the adult insects lay upon host plants eggs externally contaminated with bacteria. After hatching, the larvae invade the plant causing lesions suitable for bacterial infection.

Thus, the bacterium must be able to switch between niches as diverse as the exploitation of tissues of an infected plant, the digestive tract of an actively feeding insect larva (where its physiological activities seem to be essential for the well-being of the host), an undefined location within the nonfeeding pupal stage of the insect, the surface of the insect egg, and the invasion of a new host plant (plus all the necessary intermediate phases). It seems almost indisputable that such a life cycle must be reflected in the genetic organization of a bacterium. There must be a set of switch mechanisms permitting major readjustment of the physiological organization in response to environmental stress.

The production by plant-associated microorganisms of plant growth regulators is striking. In a few cases, notably *Agrobacterium* spp. and *Pseudomonas savastanoi,* the physiological (ecological) consequences are understood (at least partially), but in most we may only guess that the bacterium is acting to modulate the growth of the host plant.

It may prove that the ability to respond to plant stimuli by the production of proteins (especially extracellular or outer membrane components) may be the most general characteristic of rhizosphere organisms. Hence it could be that screening strains by PAGE after growth in media with and without extracts of plant roots may prove a useful technique for distinguishing rhizosphere bacteria from other soil organisms (and for recognizing proteins involved in rhizosphere functions in the former organisms). One must, of course, recall that plant extracts will include substances (e.g. sugars) which may induce the synthesis of permeases and catabolic enzymes in bacteria quite unrelated to a rhizosphere niche. It may be possible to circumvent these problems by the use of "standard" strains of bacteria from diverse groups.

The presence of particular genes (identifiable by DNA homology) may prove a useful technique for identifying bacterial species able to interact with plants. The very wide taxonomic distribution of genes encoding nitrogenases is well established and the presence of *chv*-related sequences in *Agrobacterium, Rhizobium* and *Azospirillum* suggests that the products of these sequences may be applicable to diverse modes of plant-associated ecologies. The involvement of bacterial fimbriae in the binding to plant tissues may also imply the presence of homologous genes in diverse organisms. Although fimbriae play diverse roles in the ecologies of bacteria, so that similar structures may have dissimilar functions, it may be possible to detect ecologically significant functions from genetic homologies.

Studies indicating that roots lost more organic matter if grown in the presence of rhizosphere bacteria suggest that this ability could be a useful criterion to identify bacteria able to exploit the root as a source of nutrition. It is likely that different sorts of bacteria use different mechanisms to stimulate the plant to transfer catabolizable material from the root. Some stimulate activities of the root hairs, others hasten the breakdown of cortical cells; yet others produce enzymes capable of degrading cell wall polymers, and so on. Such abilities must be correlated with the ability to thrive in this niche.

Another possible clue is the observation that so many plant-asssociated bacteria produce plant hormones. Before one can evaluate the significance of this observation it will be necessary to characterize an adequate number of non-plant-associated isolates. So far as we know, no such test has been done, but, if production of such hormones (or of particular classes, e.g. gibberellins) should be rare among the latter organisms, such production might prove a useful indicator of the former group.

More directly, it could be interesting to assay the ability to modify the growth pattern of roots. Bacteria able to inhibit or stimulate root growth, to promote secondary root initiation or to modify root hair configuration are candidates.

Analysis of the clonal organization of bacterial species can hardly fail to throw light upon the specific adaptation of particular bacteria to particular hosts (not only to host species but to particular genotypes within those species). As the diversity within a plant species becomes more apparent, especially the race-specific hypersensitivity reactions, so we must seek to understand a corresponding diversity in phytopathogenic microorganisms.

But, considering the rhizobacteria as a whole, we have to accept that they have evolved (and very likely are even now evolving) from very diverse bacterial groups, that they exploit the rhizosphere environment in many different manners, and that their environment is not only very diverse but (as the direct and indirect product of interrelated biological activities) it must be in a constant flux. We should feel completely confident that all our generalizations will encounter abundant exceptions since nature is more diverse and opportunistic than we can imagine.

REFERENCES

Aguilar, A., J. C. Perez-Diaz, F. Baquero, and C. Asensio (1982). Microcin 15m from *Escherichia coli*: mechanism of antibiotic action. *Antimicrob. Agents Chemother.*, **21**:381–386.

Ahl, P., C. Voisard, and G. Defago (1986). Iron bound siderophores, cyanic acid and antibiotics involved in suppression of *Thielaviopsis basicola* by a *Pseudomonas fluorescens* strain. *J. Phytopathol.*, **116**:121–134.

Allen, O. N. and E. K. Allen (1940). Response of the peanut plant to inoculation with rhizobia, with special reference to morphological development of the nodules. *Bot. Gaz.*, **102**:121–142.

Andro, T., J.-P. Chambost, A. Kotoujansky, J. Cattaneo, Y. Bertheau, F. Barras, F. Van Gijsegem, and A. Coleno (1984). Mutants of *Erwinia chrysanthemi* defective in secretion of pectinase and cellulase. *J. Bacteriol.*, **160**:1199–1203.

Ashby, A. M., M. D. Watson, and C. H. Shaw (1987). A Ti-plasmid determined function is responsible for chemotaxis of *Agrobacterium tumefaciens* towards the plant wound product acetosyringone. *FEMS Microbiol. Lett.*, **41**:189–192.

Ashby, A. M., M. D. Watson, G. J. Loake, and C. H. Shaw (1988). Ti plasmid-specified chemotaxis of *Agrobacterium tumefaciens* C58C[1] toward *vir*-inducing phenolic compounds and soluble factors from monocotyledonous and dicotyledonous plants. *J. Bacteriol.*, **170**:4181–4187.

Atsatt, P. R. (1977). The insect herbivore as a predictive model in parasitic seed plant biology. *Am. Nat.*, **111**:579–586.

Aydin, M., N. Lucht, W. A. König, R. Lupp, G. Jung, and G. Winkelmann (1985). Structure elucidation of the peptide antibiotic Herbicolin A and B. *Liebigs Ann. Chem.*, **1985**:2285–2300.

Azcón, R. and J. A. Ocampo (1981). Factors affecting the VA infection and mycorrhizal dependency of thirteen wheat cultivars. *New Phytol.*, **87**:677–685.

Babasaki, K., T. Takao, Y. Shimonishi, and K. Kurahashi (1985). Subtilosin A, a new antibiotic peptide produced by *Bacillus subtilis* 168: isolation, structural analysis, and biogenesis. *J. Biochem.*, **98**:585–603.

Baker, R., Y. Elad, and B. Sneh (1986). Physical, biological and host factors in iron competition in soils. In *Iron, Siderophores and Plant Disease* (T. R. Swinburne, ed.). Plenum Press, New York, pp. 77–83.

Bauer, W. D. (1981). Infection of legumes by rhizobia. *Ann. Rev. Plant Physiol.*, **32**:407–449.

Becker, J. O. (1984). Isolation and characterization of antimycotic bacteria from rhizosphere soil. In *Pests and Diseases, Vol. 1.* British Crop Protection Council Publications, Croydon, pp. 365–370.

Becker, J. O., R. W. Hedges, L. Van Royen, E. Messens, M. F. Van Outryve, B. Lambert, F. Leyns, P. Tenning, V. Gosselé, M. Zabeau, and F. Gosselé (1985). The production of antifungal compounds by phytosphere bacteria. *Med. Fac. Landbouww. Rijksuniv. Gent*, **50**:1135–1139.

Besson, F., F. Peypoux, G. Michel, and L. Delcambe (1977). Structure de la bacillomycine L, antibiotique de *Bacillus subtilis. Eur. J. Biochem.*, **77**:61–67.

Bhat, J. V., K. S. Limaye, and V. N. Vasantharajan (1971). The effect of the leaf surface microflora on the growth and root exudation of plants. In *Ecology of Leaf Surface Microorganisms* (T. F. Preece and C. H. Dickinson, eds.). Academic Press, New York, pp. 581–595.

Boemare, N. E. and R. J. Akhurst (1988). Biochemical and physiological characterization of colony form variants in *Xenorhabdus* spp. (Enterobacteriaceae). *J. Gen. Microbiol.*, **134**:751–761.

Bonnett, H. T. Jr. (1969). Cortical cell death during lateral root formation. *J. Cell Biol.*, **40**:144–159.

Braun, V. (1985). The unusual features of the iron transport systems of *Escherichia coli. Trends Biochem. Sci.*, **10**:75–78.

Broughton, W. J., V. Samrey, and J. Stanley (1987). Ecological genetics of *Rhizobium meliloti*: symbiotic plasmid transfer in the *Medicago sativa* rhizosphere. *FEMS Microbiol. Lett.*, **40**:251–255.

Brown, M. E. (1975). Rhizosphere micro-organisms—opportunists, bandits or benefactors. In *Soil Microbiology* (N. Walker, ed.). Butterworths, London, pp. 21–38.

Brown, M. E. (1982). Nitrogen fixation of free-living bacteria associated with plants—fact or fiction? In *Bacteria and Plants* (The Society for Applied Bacteriology Symposium Series, Vol. 10) (M. E. Rhodes-Roberts and F. A. Skinner, eds.). Academic Press, London, pp. 25–41.

Buyer, J. S. and J. Leong (1986). Iron transport-mediated antagonism between plant growth-promoting and plant-deleterious *Pseudomonas* strains. *J. Biol. Chem.*, **261**:791–794.

Buyer, J. S., J. M. Wright, and J. Leong (1986). Structure of pseudobactin A214, a siderophore from a bean-deleterious *Pseudomonas. Biochemistry*, **25**:5492–5499.

Chantanao, A. and H. J. Jensen (1969). Saprozoic nematodes as carriers and disseminators of plant pathogenic bacteria. *J. Nematol.*, **1**:216–218.

Chao, L. and B. R. Levin (1981). Structured habitats and the evolution of anticompetitor toxins in bacteria. *Proc. Natl. Acad. Sci. USA*, **78**:6324–6328.

Chatterjee, A. K. and M. P. Starr (1977). Donor strains of the soft rot bacterium *Erwinia chrysanthemi* and conjugational transfer of the pectolytic capacity. *J. Bacteriol.*, **132**:862-869.

Chatterjee, A. K., G. E. Buchanan, M. K. Behrens, and M. P. Starr (1979). Synthesis and excretion of polygalacturonic acid *trans*-eliminase in *Erwinia, Yersinia* and *Klebsiella* species. *Can. J. Microbiol.*, **25**:94–102.

Claydon, N., M. Allan, J. R. Hanson, and A. G. Avent (1987). Antifungal alkyl pyrones of *Trichoderma harzianum. Trans. Brit. Mycol. Soc.*, **88**:503-513.

Cody, Y. S. and D. C. Gross (1987). Outer membrane protein mediating iron uptake via pyoverdin$_{pss}$, the fluorescent siderophore produced by *Pseudomonas syringae* pv. *syringae. J. Bacteriol.*, **169**:2207–2214.

Collmer, A. (1986). The molecular biology of pectic enzyme production and bacterial soft rot pathogenesis. In *Biology and Molecular Biology of Plant-Pathogen Interactions* (J. A. Bailey, ed.). Springer-Verlag, Berlin, pp. 277–289.

Collmer, A. (1987). Pectic enzymes and bacterial invasion of plants. In *Plant-Microbe Interactions: Molecular and Genetic Perspectives, Vol. 2* (T. Kosuge and E. W. Nester, eds.). Macmillan, New York, pp. 253–284.

Collmer, A. and D. F. Bateman (1981). Impaired induction and self-catabolite repression of extracellular pectate lyase in *Erwinia chrysanthemi* mutants deficient in oligogalacturonide lyase. *Proc. Natl. Acad. Sci. USA*, **78**:3920–3924.

Collmer, A., P. Berman, and M. S. Mount (1982). Pectate lyase regulation and bacterial soft rot pathogenesis. In *Phytopathogenic Procaryotes* (M. S. Mount and G. H. Lacy, eds.). Academic Press, New York, pp. 395–422.

Cook, R. J. (1986). Plant health and the sustainability of agriculture with special reference to disease control by beneficial microorganisms. *Biol. Agric. & Horticult.*, **3**:211–232.

Cook, R. J. and K. F. Baker (1983). *The Nature and Practice of Biological Control of Plant Pathogens.* American Phytopathological Society, St. Paul MN, 439 pp.

Currier, W. W. and G. A. Strobel (1977). Chemotaxis of *Rhizobium spp.* to a glycoprotein produced by birdsfoot trefoil roots. *Science*, **196**:434–436.

D'Arcy-Lameta, A. (1986). Study of soybean and lentil root exudates. II. Identification of some polyphenolic compounds, relation with plantlet physiology. *Plant & Soil*, **92**:113–123.

D'Arcy-Lameta, A. and M. Jay (1987). Study of soybean and lentil root exudates. III. Influence of soybean isoflavonoids on the growth of rhizobia and some rhizospheric microorganisms. *Plant & Soil*, **101**:267–272.

Davis, K. R., G. D. Lyon, A. G. Darvill, and P. Albersheim (1984). Host–pathogen interactions. XXV. Endopolygalacturonic acid lyase from *Erwinia carotovora* elicits phytoalexin accumulation by releasing plant cell wall fragments. *Plant Physiol.*, **74**:52–60.

Dazzo, F. B. and W. J. Brill (1979). Bacterial polysaccharide which binds *Rhizobium trifolii* to clover root hairs. *J. Bacteriol.*, **137**:1362–1373.

Dazzo, F. B., C. A. Napoli, and D. H. Hubbell (1976). Adsorption of bacteria to roots as related to host specificity in the *Rhizobium*-clover symbiosis. *Appl. Environm. Microbiol.*, **32**:168-171.

De Smedt, J. and J. De Ley (1977). Intra- and intergeneric similarities of *Agrobacterium* ribosomal ribonucleic acid cistrons. *Int. J. Syst. Bacteriol.*, **27**:222–240.

Deacon, J. W. and S. J. Lewis (1982). Natural senescence of the root cortex of spring wheat in relation to susceptibility to common root rot (*Cochliobolus sativus*) and growth of a free-living nitrogen-fixing bacterium. *Plant & Soil*, **66**:13–20.

Debellé, F., C. Rosenberg, J. Vasse, F. Maillet, E. Martinez, J. Dénarié, and G. Truchet (1986). Assignment of symbiotic developmental phenotypes to common and specific nodulation (*nod*) genetic loci of *Rhizobium meliloti. J. Bacteriol.*, **168**:1075–1086.

Debette, J. and R. Blondeau (1980). Présence de *Pseudomonas maltophilia* dans la rhizosphère de quelques plantes cultivées. *Can. J. Microbiol.*, **26**:460–463.

Dennis, C. and J. Webster (1971). Antagonistic properties of species groups of *Trichoderma.* II. Production of volatile antibiotics. *Trans. Brit. Mycol. Soc.*, **57**:41–48.

Depicker, A., M. Van Montagu, and J. Schell (1983). Plant cell transformation by *Agrobacterium* plasmids. In *Genetic Engineering of Plants—An Agricultural Perspective* (T. Kosuge, C. P. Meredith, and A. Hollaender, eds.). Plenum Press, New York, pp. 143–176.

Diaz, C. L., P. C. van Spronsen, R. Bakhuizen, G. J. J. Logman, E. J. J. Lugtenberg, and J. W. Kijne (1986). Correlation between infection by *Rhizobium leguminosarum* and lectin on the surface of *Pisum sativum* L. roots. *Planta*, **168**:350–359.

Diolez, A., F. Richaud, and A. Coleno (1986). Pectate lyase gene regulatory mutants of *Erwinia chrysanthemi. J. Bacteriol.*, **167**:400–403.

Döbereiner, J. (1970). Further research on *Azotobacter paspali* and its variety specific occurrence in the rhizosphere of *Paspalum notatum. Zentbl. Bakt. Parasitkde Abt. II*, **124**:224–230.

Döbereiner, J. and R. M. Boddey (1981). Nitrogen fixation in association with gramineae. In *Current Perspectives in Nitrogen Fixation* (A. H. Gibson and W. E. Newton, eds.). Elsevier/North Holland, Amsterdam, pp. 305–312.

Dylan, T., L. Ielpi, S. Stanfield, L. Kashyap, C. Douglas, M. Yanofsky, E. Nester, D. R. Helinski, and G. Ditta (1986). *Rhizobium meliloti* genes required for nodule development are related to chromosomal virulence genes in *Agrobacterium tumefaciens. Proc. Natl. Acad. Sci. USA*, **83**:4403–4407.

Egawa, Y., K. Umino, S. Awataguchi, Y. Kawano, and T. Okuda (1970). Antibiotic YC73 of pseudomonas origin. I. Production, isolation, and properties. *J. Antibiot.,* **23**:267–270.

Ellis, J. G., A. Kerr, M. Van Montagu, and J. Schell (1979). *Agrobacterium*: genetic studies on agrocine 84 production and the biological control of crown gall. *Physiol. Plant Pathol.*, **15**:311–319.

Ellis, J. G., A. Kerr, A. Petit, and J. Tempé (1982a). Conjugal transfer of nopaline and agropine Ti-plasmids—the role of agrocinopines. *Mol. Gen. Genet.*, **186**:269–274.

Ellis, J. G., P. J. Murphy, and A. Kerr (1982b). Isolation and properties of transfer regulatory mutants of the nopaline Ti-plasmid pTi-C58. *Mol. Gen. Genet.*, **186**:275–281.

Fåhraeus, G. and H. Ljunggren (1959). The possible significance of pectic enzymes in root hair infection by nodule bacteria. *Physiol. Plant.*, **12**:145–154.

Firmin, J. L., K. E. Wilson, L. Rossen, and A. W. B. Johnston (1986). Flavonoid activation of nodulation genes in *Rhizobium* reversed by other compounds present in plants. *Nature* (London), **324**:90–92.

Fleisher, A., Z. Fleisher, and S. Abu-Rukun (1984). Chemovarieties of *Coridothymus capitatus* L. Rchb. growing in Israel. *J. Sci. Food Agric.*, **35**:495–499.

Floss, H. G., C.-J. Chang, O. Mascaretti, and K. Shimida (1978). Studies on the biosynthesis of antibiotics. *J. Med. Plant Res.*, **34**:349–380.

Ford, C. W. (1982). Accumulation of O-methyl-inositols in water-stressed *Vigna* species. *Phytochemistry,* **21**:1149–1151.

Forsyth, W. G. C. (1955). Comirin: isolation and properties. *Biochem. J.*, **59**:500–506.

Fottrell, P. F., S. O'Connor, and C. L. Masterson (1964). Identification of the flavonol myricetin in legume seeds and its toxicity to nodule bacteria. *Irish J. Agr. Res.*, **3**:246–249.

Fredrickson, J. K., L. F. Elliot, and J. C. Engibous (1987). Crop residues as substrate for host-specific inhibitory pseudomonads. *Soil Biol. Biochem.*, **19**:127–134.

Gaworzewska, E. T. and M. J. Carlile (1982). Positive chemotaxis of *Rhizobium leguminosarum* and other bacteria towards root exudates from legumes and other plants. *J. Gen. Microbiol.*, **128**:1179–1188.

Gear, A. (1988). Introduction. *Trichoderma Newsletter*, **4**:2.

Gerber, N. N. (1975). Prodigiosin-like pigments. *CRC Crit. Rev. Microbiol.*, **1**:469–485.

Gheysen, G., P. Dhaese, M. Van Montagu, and J. Schell (1985). DNA flux across genetic barriers: the crown gall phenomenon. In *Genetic Flux in Plants (Advances in Plant Gene Research, Vol. 2)* (B. Hohn and E. S. Dennis, eds.). Springer Verlag, Wien, pp. 11–47.

Gheysen, G., L. Herman, P. Breyne, M. Van Montagu, and A. Depicker (1989). *Agrobacterium tumefasciens* as a tool for the genetic transformation of plants. In *Genetic Transformation and Expression* (L. O. Butler, ed.). Intercept, London, pp. 151–174.

Golden, S. S., J. Brusslan, and R. Haselkorn (1987). Genetic engineering of the cyanobacterial chromosome. In *Recombinant DNA, part D (Methods in Enzymology, Vol. 153)* (R. Wu and L. Grossman, eds.). Academic Press, San Diego, pp. 215–231.

Gottesman, S. (1984). Bacterial regulation: global regulatory networks. *Ann. Rev. Genet.*, **18**:415–441.

Govan, J. R. W. (1986). *In vivo* significance of bacteriocins and bacteriocin receptors. *Scand. J. Infect. Dis.*, Suppl. **49**:31–37.

Graham, J. H., R. T. Leonard, and J. A. Menge (1981). Membrane-mediated decrease in root exudation responsible for phosphorus inhibition of VAM formation. *Plant Physiol.*, **68**:548–552.

Gresshoff, P. M. and A. C. Delves (1986). Plant genetic approaches to symbiotic nodulation and nitrogen fixation in legumes. In *A Genetic Approach to Plant Biochemistry* (A. D. Blonstein and P. J. King, eds.). Springer-Verlag, Berlin, pp. 158–206.

Gross, D. C. and Y. S. Cody (1985). Mechanisms of plant pathogenesis by *Pseudomonas* species. *Can. J. Microbiol.*, **31**:403–410.

Gross, D. C., J. E. De Vay, and F. H. Stadtman (1977). Chemical properties of syringomycin and syringotoxin: toxigenic peptides produced by *Pseudomonas syringae. J. Appl. Bacteriol.*, **43**:453–463.

Gullino, M. L. and A. Garibaldi (1986). Biological and integrated control of gray mould of grape in Italy. *Phytopathology,* **76**:1065.

Gurusiddaiah, S., D. M. Weller, A. Sarkar, and R. J. Cook (1986). Characterization of an antibiotic produced by a strain of *Pseudomonas fluorescens* inhibitory to *Gaeumannomyces graminis* var. *tritici* and *Phythium* spp. *Antimicrob. Agents Chemother.*, **29**:488–495.

Haahtela, K., E. Tarkka, and T. K. Korhonen (1985). Type 1 fimbria-mediated adhesion of enteric bacteria to grass roots. *Appl. Environm. Microbiol.*, **49**:1182–1185.

Haahtela, K., T. Laakso, and T. K. Korhonen (1986). Associative nitrogen fixation by *Klebsiella* spp.: adhesion sites and inoculation effects on plant roots. *Appl. Environm. Microbiol.*, **52**:1074–1079.

Hale, M. G. and L. D. Moore (1979). Factors affecting root exudation II: 1970–1978. *Adv. Agron.*, **31**:93–123.

Hantke, K. and V. Braun (1975). Membrane receptor dependent iron transport in *Escherichia coli. FEBS Lett.*, **49**:301–305.

Harrison, M. D., J. W. Brewer, and L. D. Merrill (1980). Insect involvement in the transmission of bacterial pathogens. In *Vectors of Plant Pathogens* (K. F. Harris and K. Maramorosch, eds.). Academic Press, New York, pp. 201–292.

Haselkorn, R. (1978). Heterocysts. *Ann. Rev. Plant Physiol.*, **29**:319–344.

Haselkorn, R., J. W. Golden, P. J. Lammers, and M. E. Mulligan (1987). Rearrangement of *nif* genes during cyanobacterial heterocyst differentiation. *Phil. Trans. R. Soc. Lond.*, **B317**:137–181.

Haugland, R. and D. P. S. Verma (1981). Interspecific plasmid and genomic DNA sequence homologies and localization of *nif* genes in effective and ineffective strains of *Rhizobium japonicum. J. Mol. Appl. Genet.*, **1**:205–217.

Hedges, R. W. (1972). The pattern of evolutionary change in bacteria. *Heredity*, **28**:39–48.

Hess, D. (1982). Induction of nitrogenase activity in *Azospirillum* by wheat. In *Azospirillum: Genetics, Physiology, Ecology* (W. Klingmüller, ed.). Springer-Verlag, Berlin, pp. 69–74.

Hess, D., H.-A. Federolf, A. Würfel, and K. Dressler (1983). Induction of nitrogenase activity in *Rhizobium* 32H1 by conditioned media from association of wheat with *Rhizobium* 32H1 and *Rhizobium trifolii. Z. Pflanzenphysiol.*, **111**:203–211.

Heulin, T., A. Guckert, and J. Balandreau (1987). Stimulation of root exudation of rice seedlings by *Azospirillum* strains: carbon budget under gnotobiotic conditions. *Biol. Fertil. Soils*, **4**:9–14.

Hildebrand, D. C. (1973). Tolerance of homoserine by *Pseudomonas pisi* and implications of homoserine in plant resistance. *Phytopathol.*, **63**:301–302.

Hohnandel, D. and J. M. Meyer (1986). Pyoverdine facilitated iron uptake among fluorescent pseudomonads. In *Iron, Siderophores, and Plant Diseases* (T. R. Swinburne, ed.). Plenum Press, New York, pp. 119–129.

Hooykaas, P. J. J., P. M. Klapwijk, M. P. Nuti, R. A. Schilperoort, and A. Rörsch (1977). Transfer of the *Agrobacterium tumefaciens* Ti-plasmid to avirulent agrobacteria and to *Rhizobium* ex planta. *J. Gen. Microbiol.*, **98**:477–484.

Hooykaas, P. J. J., R. Peerbolte, A. J. G. Regensburg-Tuïnk, P. De Vries, and R. A. Schilperoort (1982). A chromosomal linkage map of *Agrobacterium tumefaciens* and a comparison with the maps of *Rhizobium* spp. *Mol. Gen. Genet.*, **188**:12–17.

Hopkins, D. L., R. P. Larkin, and G. W. Elmstrom (1987). Cultivar-specific induction of soil suppressiveness to *Fusarium* wilt of water melon. *Phytopathol.*, **77**:607-611.

Horvath, B., C. W. B. Bachem, J. Schell, and A. Kondorosi (1987). Host-specific regulation of nodulation genes in *Rhizobium* is mediated by a plant-signal, interacting with the nodD gene product. *EMBO J.*, **6**:841–848.

Howell, C. R. and R. D. Stipanovic (1979). Control of *Rhizoctonia solani* on cotton seedlings with *Pseudomonas fluorescens* and with an antibiotic produced by the bacterium. *Phytopathol.*, **69**:480–482.

Howell, C. R. and R. D. Stipanovic (1980). Suppression of *Pythium ultimum*-induced damping-off of cotton seedlings by *Pseudomonas fluorescens* and its antibiotic, pyoluteorin. *Phytopathol.*, **70**:712–715.

Jacobs, M. J., W. M. Bugbee, and D. A. Gabrielson (1985). Enumeration, location and characterization of endophytic bacteria within sugar beet roots. *Can. J. Bot.*, **63**:1262–1265.

Jacobsen, E. and W. J. Feenstra (1984). A new pea mutant with efficient nodulation in the presence of nitrate. *Plant Sci. Lett.*, **33**:337–344.

Jain, D. K. and D. G. Patriquin (1985). Characterization of a substance produced by *Azospirillum* which causes branching of root hairs. *Can. J. Microbiol.*, **31**:206–210.

Jarvis, B. D. W., M. Gillis, and J. De Ley (1986). Intra- and intergeneric similarities between the ribosomal ribonucleic acid cistrons of *Rhizobium* and *Bradyrhizobium* species and some related bacteria. *Int. J. System. Bacteriol.*, **36**:129–138.

Kabar, K. (1987). Alleviation of salinity stress by plant growth regulators on seed germination. *J. Plant Physiol.*, **128**:179–182.

Kahl, G. and J. Schell (1982). *Molecular Biology of Plant Tumors.* Academic Press, New York, 615 pp.

Kaminsky, R. (1981). The microbial origin of the allelopathic potential of *Adenostoma fasciculatum* H. & A. *Ecology Monographs*, **51**:365–382.

Kanda, N., N. Ishizaki, N. Inouye, M. Oshima, A. Handa, and T. Kitahara (1975). DB-2073, a new alkylresorcinol antibiotic. I. Taxonomy, isolation, and characterization. *J. Antibiot.*, **28**:935–942.

Kapulnik, Y., C. M. Joseph, and D. A. Phillips (1987). Flavone limitations to root nodulation and symbiotic nitrogen fixation in alfalfa. *Plant Physiol.*, **84**:1193–1196.

Keen, N. T. and S. Tamaki (1986). Structure of two pectate lyase genes from *Erwinia chrysanthemi* EC16 and their high level expression in *Escherichia coli. J. Bacteriol.*, **168**:595–606.

Keen, N. T., D. Dahlbeck, B. Staskawicz, and W. Belser (1984). Molecular cloning of the pectate lyase genes from *Erwinia chrysanthemi* and their expression in *Escherichia coli. J. Bacteriol.*, **159**:825–831.

Kersters, K., J. De Ley, P. H. A. Sneath, and M. Sackin (1973). Numerical taxonomic analysis of *Agrobacterium. J. Gen. Microbiol.*, **78**:227–239.

Kirk, J. J. and J. W. Deacon (1987). Control of take-all fungus by *Microdochium bolleyi*, and interactions involving *M. bolleyi, Phialophora graminicola* and *Periconia macrospina* on cereal roots. *Plant & Soil*, **98**:231–237.

Knowlton, S., A. Berry, and J. G. Torrey (1980). Evidence that associated soil bacteria may influence root hair infection of actinorhizal plants by *Frankia. Can. J. Microbiol.*, **26**:971–977.

Kondorosi, E. and A. Kondorosi (1986). Nodule induction on plant roots by *Rhizobium. Trends Biochem. Sci.*, **11**:296–299.

Kosslak, R. M. and B. B. Bohool (1984). Suppression of nodule development of one side of split-root system of soybeans caused by prior inoculation of the other side. *Plant Physiol.*, **75**:125–130.

Kotoujansky, A., A. Diolez, M. Boccara, Y. Bertheau, T. Andro, and A. Coleno (1985). Molecular cloning of *Erwinia chrysanthemi* pectinase and cellulase structural genes. *EMBO J.*, **4**:781–785.

Kraffczyk, I., G. Trolldenier, and H. Beringer (1984). Soluble root exudates of maize: influence of potassium supply and rhizosphere microorganisms. *Soil Biol. Biochem.*, **16**:315-322.

Lamont, B. (1982). Mechanisms for enhancing nutrient uptake in plants, with special reference to Mediterranean, South Africa and Western Australia. *Bot. Rev.*, **48**:597–689.

Lamont, B. B. and A. J. McComb (1974). Soil microorganisms and the formation of proteoid roots. *Austral. J. Bot.*, **22**:681–688.

Lamont, B. B. and R. A. Ryan (1977). Formation of coralloid roots by cycads under sterile conditions. *Phytomorphol.*, **27**:426–429.

Lancelle, S. A. and J. G. Torrey (1984). Early development of *Rhizobium*-induced root nodules of *Parasponia rigida.* I. Infection and early nodule initiation. *Protoplasma*, **123**:26–37.

Larson, R. I. and J. L. Neal (1978). Selective colonization of the rhizosphere of wheat by nitrogen-fixing bacteria. *Ecol. Bull.*, **26**:331–342.

Law, I. J. and B. W. Strijdom (1984). Role of lectins in the specific recognition of *Rhizobium* by *Lotononis bainseii. Plant Physiol.*, **74**:779–785.

Leary, J. V. and D. W. Fulbright (1982). Chromosomal genetics of *Pseudomonas* spp. and *Erwinia* spp. In *Phytopathogenic Prokaryotes* (M. S. Mount and G. H. Lacy, eds.). Academic Press, New York, pp. 229–279.

Lechevalier, H. A. (1975). Production of the same antibiotics by members of different genera of microorganisms. *Adv. Appl. Microbiol.*, **19**:25–45.

Leisinger, T. and R. Margraff (1979). Secondary metabolites of fluorescent pseudomonads. *Microbiol. Rev.*, **43**:422–442.

Leroux, B., M. F. Yanofsky, S. C. Winans, J. E. Ward, S. F. Ziegler, and E. W. Nester (1987). Characterization of the *virA* locus of *Agrobacterium tumefaciens*: a transcriptional regulator and host range determinant. *EMBO J.*, **6**:849–856.

Lie, T. A., D. Göktan, M. Engin, J. Pijnenborg, and E. Anlarsal (1987). Co-evolution of the legume–*Rhizobium* association. *Plant & Soil*, **100**:171–181.

Ligero, F., C. Lluch, and J. Olivares (1987). Evolution of ethylene from roots and nodulation rate of alfalfa (*Medicago sativa* L.) plants inoculated with *Rhizobium meliloti* as affected by the presence of nitrate. *J. Plant Physiol.*, **129**:461–467.

Lindberg, G. D. (1981). An antibiotic lethal to fungi. *Plant Disease*, **65**:680–683.

Loeffler, W., J. S.-M. Tschen, N. Vanittanakom, M. Kugler, E. Knorpp, T.-F. Hsieh, and T.-G. Wu (1986). Antifungal effects of bacilysin and fengymycin from *Bacillus subtilis* F-29-3. A comparison with activities of other *Bacillus* antibiotics. *J. Phytopathol.*, **115**:204–213.

Long, S. R. (1984). Genetics of *Rhizobium* nodulation. In *Plant–Microbe Interactions: Molecular and Genetic Perspectives, Vol. 1* (T. Kosuge and E. W. Nester, eds.). Macmillan, New York, pp. 265–306.

Loper, J. E., C. S. Orser, N. J. Panopoulos, and M. N. Schroth (1984). Genetic analysis of fluorescent pigment production in *Pseudomonas syringae* pv. *syringae. J. Genet. Microbiol.*, **130**:1507–1515.

Lopez, M. F. and E. R. Singer (1987). Degradative enzymes in *Rhizobium meliloti.* In *Molecular Genetics of Plant–Microbe Interactions* (D. S. Verma and N. Brisson, eds.). Martinus Nijhoff, Dordrecht, pp. 185–187.

Loutit, M. W. and R. R. Brooks (1970). Rhizosphere organisms and molybdenum concentrations in plants. *Soil Biol. Biochem.*, **2**:131–135.

Lynch, J. M. (1982). Interactions between bacteria and plants. *Symp. Soc. Appl. Bacteriol.*, **10**:1–23.

Lyon, G. D. and C. E. Bayliss (1975). The effect of rishitin on *Erwinia carotovora* var. *atroseptica* and other bacteria. *Physiol. Plant Pathol.*, **6**:177–186.

Malajczuk, N. and G. D. Bowen (1974). Proteoid roots are microbially induced. *Nature* (London), **251**:316–317.

Mandimba, G., T. Heulin, R. Balley, A. Guckert, and J. Balandreau (1986). Chemotaxis of free-living nitrogen-fixing bacteria towards maize mucilage. *Plant & Soil,* **90**:129–139.

Marschner, H., V. Romheld, W. J. Horst, and P. Martin (1986). Root-induced changes in the rhizosphere: importance for the mineral nutrition of plants. *Z. Pflanzenernähr. Bodenk.*, **149**:441–456.

Martin, P. and A. Glatzle (1982). Mutual influences of *Azospirillum* spp. and grass seedlings. In *Azospirillum: Genetics, Physiology, Ecology* (W. Klingmüller, ed.). Springer-Verlag, Berlin, pp. 108–120.

Martinez, E., R. Palacios, and F. Sanchez (1987). Nitrogen-fixing nodules induced by *Agrobacterium tumefaciens* harboring *Rhizobium phaseoli* plasmids. *J. Bacteriol.*, **169**:2828–2834.

Martínez, J. L. and J. C. Pérez-Díaz (1986). Isolation, characterization and mode of action on *Escherichia coli* strains of microcin D93. *Antimicrobiol. Agents Chemother.*, **29**:456–460.

Mauro, V. P., T. Nguen, P. Katinakis, and D. P. S. Verma (1985). Primary structure of the soybean nodulin-23 gene and potential regulatory elements in the 5′-flanking regions of nodulin and leghemoglobin genes. *Nucl. Acids Res.*, **13**:239–249.

Messens, E., A. Lenaerts, M. Van Montagu, and R. W. Hedges (1985). Genetic basis for opine secretion from crown gall tumour cells. *Mol. Gen. Genet.*, **199**:344–348.

Meyer, J. M. and M. A. Abdallah (1978). The fluorescent pigment of *Pseudomonas fluorescens*: biosynthesis, purification and physicochemical properties. *J. Gen. Microbiol.*, **107**:319–328.

Miller, I. M., A. Scott, and I. C. Gardner (1984). The occurrence of calyx nodules in *Psychotria* spp. (*Rubiaceae*). *Protoplasma,* **121**:199–208.

Moerman, M., J.-P. Nap, F. Govers, R. Schilperoort, A. van Kammen, and T. Bisseling (1987). *Rhizobium nod* genes are involved in the induction of two early nodulin genes in *Vicia sativa* root nodules. *Plant Mol. Biol.*, **9**:171–179.

Mori, A., C. Nishino, N. Enoki, and S. Tawata (1987). Antibacterial activity and mode of action of plant flavonoids against *Proteus vulgaris* and *Staphylococcus aureus. Phytochem.*, **26**:2231–2234.

Morrison, J. and D. P. S. Verma (1987). A block in the endocytosis of *Rhizobium* allows cellular differentiation in nodules but affects the expression in some peribacterioid membrane nodulins. *Plant Mol. Biol.*, **9**:185–196.

Mosse, B. (1962). Establishment of VAM under aseptic conditions. *J. Gen. Microbiol.*, **27**:509–520.

Mosse, B. (1975). A microbiologist's view of root anatomy. In *Soil Microbiology* (N. Walker, ed.). Butterworth, London, pp. 39–66.

Murphy, P. J., N. Heyncke, Z. Banfalvi, M. E. Tate, F. de Bruijn, A. Kondorosi, J. Tempé and J. Schell (1987). Genes for the catabolism and synthesis of an opine-like compound in *Rhizobium meliloti* are closely linked and on the Sym plasmid. *Proc. Natl. Acad. Sci. USA*, **84**:493–497.

Murty, M. G. and J. K. Ladha (1987). Differential colonization of *Azospirillum lipoferum* on roots of two varieties of rice (*Oryza sativa* L.). *Biol. Fertil. Soils*, **4**:3–7.

Neal, J. L. Jr., R. I. Larson, and T. G. Atkinson (1973). Changes in rhizosphere populations of selected physiological groups of bacteria related to substitution of specific pairs of chromosomes in spring wheat. *Plant & Soil*, **39**:209–212.

Neilands, J. B. and S. A. Leong (1986). Siderophores in relation to plant growth and disease. *Ann. Rev. Plant Physiol.*, **37**:187–208.

Nester, E. W., M. P. Gordon, R. M. Amasino, and M. F. Yanofsky (1984). Crown gall: a molecular and physiological analysis. *Ann. Rev. Plant Physiol.*, **35**:387–413.

O'Brien, I. G., G. B. Cox, and F. Gibson (1971). Enterochelin hydrolysis and iron metabolism in *Escherichia coli. Biochem. Biophys. Acta*, **237**:537–549.

Owens, L. D. (1969). Toxins in plant disease: structure and mode of action. *Science*, **165**:18–25.

Page, W. J. and M. Huyer (1984). Depression of the *Azotobacter vinelandii* siderophore system, using iron-containing minerals to limit iron repletion. *J. Bacteriol.*, **158**:496–502.

Pankhurst, C. E. and W. T. Jones (1979). Effectiveness of *Lotus* root nodules. II. Relationship between root nodule effectiveness and *in vitro* sensitivity of fast-growing *Lotus* rhizobia to flavolans. *J. Exp. Bot.*, **30**:1095–1107.

Parke, D., L. N. Ornston, and E. W. Nester (1987). Chemotaxis to plant phenolic inducers of virulence genes is constitutively expressed in the absence of the Ti plasmid in *Agrobacterium tumefaciens. J. Bacteriol.*, **169**:5336–5338.

Parker, C. A., M. J. Trinick, and D. L. Chatel (1977). Rhizobia as soil and rhizosphere inhabitants. In *A Treatise on Dinitrogen Fixation: Section 4* (E. W. Hardy and A. H. Gibson, eds.). John Wiley & Sons, New York, pp. 311–352.

Patriquin, D. G. (1981). New developments in grass–bacteria associations. In *Advances in Agricultural Microbiology* (N.S. Subba Rao, ed.). Oxford and IBH Publishing Co., New Delhi, pp. 139–190.

Patriquin, D. G., J. Döbereiner, and D. K. Jain (1983). Sites and processes of association between diazotrophs and grasses. *Can. J. Microbiol.*, **29**:900–915.

Paul, V. J., S. Frautschy, W. Fenical, and K. H. Nealson (1981). Antibiotics in microbial ecology. *J. Chem. Ecol.*, **7**:589–598.

Pearson, R. and D. Parkinson (1961). The sites of excretion of ninhydrin-positive substances by broad bean seedlings. *Plant & Soil*, **13**:391–396.

Peters, N. K., J. W. Frost, and S. R. Long (1986). A plant flavone, luteolin, induces expression of *Rhizobium meliloti* nodulation genes. *Science*, **233**:977–980.

Petit, A., C. David, G. A. Dahl, J. G. Ellis, P. Guyon, F. Casse-Delbart, and J. Tempé (1983). Further extension of the opine concept: plasmids in *Agrobacterium rhizogenes* cooperate for opine degradation. *Mol. Gen. Genet.*, **190**:204–214.

Pierce, M. and W. D. Bauer (1983). A rapid regulatory response governing nodulation in soybean. *Plant Physiol.*, **73**:286–290.

Pitts, G., A. I. Allam, and J. P. Hollis (1972). *Beggiatoa*: occurrence in the rice rhizosphere. *Science*, **178**:990–991.

Přikryl, Z. and V. Vančura (1980). Root exudates of plants. VI. Wheat root exudation as dependent on growth, concentration gradient of exudates and the presence of bacteria. *Plant & Soil*, **57**:69–83.

Pugsley, A. P. and M. Schwartz (1985). Export and secretion of proteins by bacteria. *FEMS Microbiol. Reviews*, **2**:3–38.

Ramaswamy, P. and A. K. Bal (1987). Asymbiotic nitrogen fixation by *Rhizobium* sp. 127E15 in culture and in the lima bean rhizosphere. *Current Microbiol.*, **15**: 223–228.

Ramia, S., E. Neter, and D. J. Brenner (1982). Production of enterobacterial common antigen as an aid to classification of newly discovered species of the families Enterobacteriaceae and Vibrionaceae. *Int. J. System. Bacteriol.*, **33**:395–398.

Rao, V. R., N. S. S. Rao, and K. G. Mukerji (1973). Inhibition of *Rhizobium in vitro* by non-nodulating legume roots and root extracts. *Plant & Soil*, **39**:449–452.

Redmond, J. W., M. Batley, M. A. Djordjevic, R. W. Innes, P. L. Kuempel, and B. G. Rolfe (1986). Flavones induce expression of nodulation genes in *Rhizobium. Nature* (London), **323**:632–635.

Reinhold, B., T. Hurek, and I. Fendrik (1985). Strain-specific chemotaxis of *Azospirillum* spp. *J. Bacteriol.*, **162**:190–195.

Rennie, R. J. (1980). Dinitrogen-fixing bacteria; computer-assisted identification of soil isolates. *Can. J. Microbiol.*, **26**:1275–1283.

Rennie, R. J. and R. I. Larson (1979). Dinitrogen fixation associated with disomic chromosome substitution lines of spring wheat. *Can. J. Bot.*, **57**:2771–2775.

Ricard, J. and T. L. Highley (1988). Biocontrol of decay or pathogenic fungi in wood and trees. *Trichoderma Newsletter*, **4**:9–15.

Rice, E. L. (1979). Allelopathy—an update. *Bot. Rev.*, **45**:15–109.

Rice, E. L. and S. K. Pancholy (1972). Inhibition of nitrification by climax ecosystems. *Amer. J. Bot.*, **59**:1033–1040.

Rice, E. L. and S. K. Pancholy (1974). Inhibition of nitrification by climax ecosystems. III. Inhibitors other than tannins. *Amer. J. Bot.*, **61**:1095–1103.

Rhodes, S. H., G. R. Lyons, R. P. Gregson, R. J. Akhurst, and M. J. Lacey (1987). Process of making xenorhabdin antibiotics. U.S. patent 4672130 (June 9 1987) *Off. Gaz. US Pat. Trademark Off Pat.* 1079 : 1062 (*Chem. Abs.* **101**:189727k).

Rivière, J. and R. Chaussat (1966). Destruction de la coumarine dans la rhizosphère. *Ann. Inst. Pasteur*, Suppl. **111**:155–167.

Roberts, W. K. and C. P. Selitrennikoff (1988). Plant and bacterial chitinases differ in antifungal activity. *J. Gen. Microbiol.*, **134**:169–176.

Rosenberg, C. and T. Huguet (1984). The pAtC58 plasmid of *Agrobacterium tumefaciens* is not essential for tumour induction. *Mol. Gen. Genet.*, **196**:533–536.

Rossen, L., C. A. Shearman, A. W. B. Johnston, and J. A. Downie (1985). The *nodD* gene of *Rhizobium leguminosarum* is autoregulatory and in the presence of plant exudate induces the *nodA,B,C* genes. *EMBO J.*, **4**:3369–3373.

Ryder, M. H., M. E. Tate, and G. P. Jones (1984). Agrocinopine A, a tumor-inducing plasmid-coded enzyme product, is a phosphodiester of sucrose and L-arabinose. *J. Biol. Chem.*, **259**:9704–9710.

Sahm, U., G. Knobloch, and F. Wagner (1973). Isolation and characterization of the methionine antagonist l-2-amino-4-methoxy-trans-butenoic acid from *Pseudomonas aeruginosa* grown on n-paraffin. *J. Antibiot.*, **26**:389–390.

Schoedel, C. and A. Collmer (1986). Evidence of homology between pectate lyase-encoding *pel*B and *pel*C genes in *Erwinia chrysanthemi*. *J. Bacteriol.*, **167**:117–123.

Schofield, P. R., A. H. Gibson, W. F. Dudman, and J. M. Watson (1987). Evidence for genetic exchange and recombination of *Rhizobium* symbiotic plasmids in a soil population. *Appl. Environm. Microbiol.*, **53**:2942–2947.

Schönwitz, R. and H. Ziegler (1986). Quantitative and qualitative aspects of a developing rhizosphere microflora of hydroponically grown maize seedlings. *Z. Pflanzenernäh. Bodenk.*, **149**:623–634.

Schroth, M. N. and J. G. Hancock (1982). Disease-suppressive soil and root-colonizing bacteria. *Science*, **216**:1376–1381.

Schroth, M. N., A. R. Weinhold, A. H. McCain, D. C. Hildebrand, and N. Ross (1971). Biology and control of *Agrobacterium tumefaciens*. *Hilgardia*, **40**:537–552.

Scott, D. B., R. Wilson, G. J. Shaw, A. Petit, and J. Tempé (1987). Biosynthesis and degradation of nodule-specific *Rhizobium loti* compounds in *Lotus* nodules. *J. Bacteriol.*, **169**:278–282.

Sharon, N. (1987). Bacterial lectins, cell-cell recognition and infectious disease. *FEBS Lett.*, **217**:145–157.

Spaink, H. P., C. A. Wijffelman, E. Pees, R. J. H. Okker, and B. J. J. Lugtenberg (1987). *Rhizobium* nodulation gene *nodD* as a determinant of host specificity. *Nature* (London), **328**:337–340.

Stacey, G., A. S. Paau, and W. J. Brill (1980). Host recognition in the *Rhizobium*-soybean symbiosis. *Plant Physiol.*, **66**:609–614.

Stachel, S. E. and P. C. Zambryski (1986). *virA* and *virG* control the plant-induced activation of the T-DNA transfer process of *A. tumefaciens*. *Cell*, **46**:325–333.

Stachel, S. E., E. Messens, M. Van Montagu, and P. Zambryski (1985). Identification of the signal molecules produced by wounded plant cells that activate T-DNA transfer in *Agrobacterium tumefaciens*. *Nature* (London), **318**:624–629.

Starr, M. P. (1981). The genus *Erwinia*. In *The Prokaryotes* (M. P. Starr, H. Stolp, H. G. Trüper, A. Balows, and H. G. Schlegel, eds.). Springer-Verlag, Berlin, pp. 1260–1271.

Sugiura, Y. and K. Nomoto (1984). Phytosiderophores. *Struct. Bonding*, **58**:107–135.

Swinburne, T. R. (1986). Stimulation of disease development by siderophores and inhibition by chelated iron. In *Iron, Siderophores and Plant Disease* (T. R. Swinburne, ed.). Plenum Press, New York, pp. 217–225.

Tabary, F., J. Balandreau, and R. Bourrillon (1984). Purification of the rice embryolectin and its binding to nitrogen-fixing bacteria from the rhizosphere of rice. *Biochem. Biophys. Res. Commun.*, **119**:549–555.

Teintze, M., M. B. Hossain, C. L. Barnes, J. Leong, and D. van der Helm (1981). Structure of ferric pseudobactin, a siderophore from a plant growth promoting *Pseudomonas*. *Biochemistry*, **20**:6446–6457.

Thomas, G. M. and G. O. Poinar (1979). *Xenorhabdus* gen. nov., a genus of entomopathogenic nematophilic bacteria of the family Enterobacteriaceae. *Int. J. Syst. Bacteriol.*, **29**:352–360.

Timonin, M. I. (1947). Microflora of the rhizosphere in relation to the manganese-deficiency disease of oats. *Soil Sci. Soc. Amer. Proc.*, **11**:284–292.

Truchet, G., C. Rosenberg, J. Vasse, J.-S. Julliot, S. Camut, and J. Dénarié (1984). Transfer of *Rhizobium meliloti* Sym genes into *Agrobacterium tumefaciens*: host-specific nodulation by atypical infection. *J. Bacteriol.*, **157**:134–142.

Turner, J. M. and A. J. Messenger (1986). Occurrence, biochemistry and physiology of phenazine pigment production. *Adv. Microbial Physiol.*, **27**:211–275.

Umezawa, H., T. Shiba, and T. Takita (1978). *Bioactive Peptides Produced by Microorganisms*. John Wiley & Sons, Tokyo, 275 pp.

Van Egeraat, A. W. S. M. (1975). The growth of *Rhizobium leguminosarum* on the root surface and in the rhizosphere of pea seedlings in relation to root exudates. *Plant & Soil*, **42**:367–379.

Van Gijsegem, F., A. Toussaint, and J. Schoonejans (1985). *In vivo* cloning of the pectate lyase and cellulase genes of *Erwinia chrysanthemi*. *EMBO J.*, **4**:787–792.

Vance, C. P. (1983). *Rhizobium* infection and nodulation: a beneficial plant disease? *Ann. Rev. Microbiol.*, **37**:399–424.

Vančura, V. (1980). Fluorescent pseudomonads in the rhizosphere of plants and their relation to root exudates. *Folia Microbiol.*, **25**:168–173.

Vanittanakom, N., W. Loeffler, U. Koch, and G. Jung (1986). Fengycin—a novel antifungal lipopeptide antibiotic produced by *Bacillus subtilis* F-29-3 *J. Antibiot.*, **39**:888–901.

Venette, J. R. (1982). How bacteria find their hosts. In *Phytopathogenic Prokaryotes* (M. S. Mount and G. H. Lacy, eds.). Academic Press, New York, pp. 1–30.

Verma, D. P. S. and N. Brisson (1987). *Molecular Genetics of Plant-Microbe Interactions.* Martinus Nijhoff, Dordrecht, 340 pp.

Verma, D. P. S. and S. Long (1983). The molecular biology of *Rhizobium*-legume symbiosis. In *Intracellular Symbiosis* (Review of Cytology, Suppl. 14) (K. W. Jeon, ed.). Academic Press, New York, pp. 211–245.

Verma, P. D. S., M. G. Fortin, J. Stanley, V. P. Mauro, S. Purohit, and N. Morrison (1986). Nodulins and nodulin genes of *Glycine max*, a perspective. *Plant Mol. Biol.*, **7**:51–61.

Vesper, S. J. (1987). Production of pili (fimbriae) by *Pseudomonas fluorescens* and correlation with attachment to corn roots. *Appl. Environm. Microbiol.*, **53**:1397–1405.

Vietor, D. M. (1982). Variation of carbon 14-labelled photosynthate recovery from roots and rooting media of warm season grasses. *Crop Sci.*, **22**:362–366.

Waelkens, F., M. Maris, C. Verreth, J. Vanderleyden, and A. Van Gool (1987). *Azospirillum* DNA shows homology with *Agrobacterium* chromosomal virulence genes. *FEMS Microbiol. Lett.*, **43**:241–246.

Wakayama, S., F. Ishikawa, and K. Oishi (1984). Mycocerein, a novel antifungal peptide antibiotic produced by *Bacillus cereus. Antimicrob. Agents Chemother.*, **26**:939–940.

Wallace, R. H. and A. G. Lochhead (1950). Qualitative studies of soil microorganisms. IX. Amino acid requirements of rhizosphere bacteria. *Can. J. Res.*, **28C**:1–6.

Wasserman, H. H., J. J. Keggi, and J. E. McKeon (1962). The structure of serratamolide. *J. Amer. Chem. Soc.*, **84**:2978–2982.

Wayne, R. and J. B. Neilands (1975). Evidence from common binding sites for ferrichrome compounds and bacteriophage ϕ80 in the cell envelope of *Escherichia coli. J. Bacteriol.*, **121**:497–503.

Weiler, E. W. and J. Schröder (1987). Hormone genes and crown gall disease. *Trends Biochem. Sci.*, **12**:271–275.

Wendenbaum, S., P. Demange, A. Dell, J. M. Meyer, and M. A. Abdallah (1983). The structure of pyoverdine Pa, the siderophore of *Pseudomonas aeruginosa. Tetrahedron Lett.*, **24**:4877–4880.

Werker, E. and M. Kislev (1978). Mucilage on the root surface and root hairs of *Sorghum*: heterogeneity in structure, manner of production and site of accumulation. *Ann. Bot.*, **42**:809–816.

Whallon, J. H., G. Acker, and H. El-Khawas (1985). Electron microscopy of young wheat roots inoculated with *Azospirillum.* In *Azospirillum III: Genetics, Physiology, Ecology* (W. Klingmüller, ed.). Springer-Verlag, Berlin, pp. 222–229.

Williams, S. T. and Vickers, J. C. (1986). The ecology of antibiotic production. *Microbial Ecol.*, **12**:43–52.

Wilson, M. F. and E. A. Bell (1978a). Determination of the changes in the free amino acid content of the eluate from germinating seeds of *Glycine wightii* (L.) and its effect on the growth of lettuce fruits. *J. Exp. Bot.*, **29**:1243–1247.

Wilson, M. F. and E. A. Bell (1978b). Amino acids and β-aminoproprionitrile as inhibitors of seed germination and growth. *Phytochem.*, **17**:403–406.

Winkelmann, G., R. Lupp, and G. Jung (1980). Herbicolins—new peptide antibiotics from *Erwinia herbicola. J. Antibiot.*, **33**:353–358.

Yahalom, E., Y. Okun, and A. Dovrat (1987). *Azospirillum* effects on susceptibility to *Rhizobium* nodulation and on nitrogen fixation of several forage legumes. *Can. J. Microbiol.*, **33**:510–514.

Yao, P. Y. and J. M. Vincent (1969). Host specificity in the root hair curling factor of *Rhizobium* spp. *Austral. J. Biol. Sci.*, **22**:413–423.

Young, J. P. W. (1985). *Rhizobium* population genetics: enzyme polymorphism in isolates from peas, clover, beans and lucerne grown at the same site. *J. Gen. Microbiol.*, **131**:2399–2408.
Young, J. P. W. and P. Matthews (1982). A distinct class of peas (*Pisum sativum* L.) from Afghanistan that show strain specificity for symbiotic *Rhizobium*. *Heredity*, **48**:203–210.
Young, J. P. W., L. Demetriou, and R. G. Apte (1987). *Rhizobium* population genetics: enzyme polymorphism in *Rhizobium leguminosarum* from plants and soil in a pea crop. *Appl. Environm. Microbiol.*, **53**:397–402.

7

MICROBIAL METABOLITES

J. M. LYNCH
AFRC Institute of Horticultural Research, Littlehampton, West Sussex BN17 6LP, UK

1. INTRODUCTION

Previous reviews on microbial metabolites in soil (Lynch, 1976, 1985) have catalogued the range of organic compounds which can be formed by microorganisms; the lists included plant growth regulators, phytotoxins, antibiotics and soil stabilizers. These substances can affect the biology of the rhizosphere whether they are formed by microbial activity within the region or whether they are formed remotely, for example on plant residues, and diffuse towards roots. However, minimum boundary conditions must be satisfied for the substance so formed to be of significance to plants:

Biological activity must be observed in the form and in the concentration in which the substance is present in the rhizosphere.

In some studies extraction procedures have been used which can modify organic matter, generating monomers which would otherwise have been present in polymeric (and often inactive) forms. Concentration is important, especially with growth regulators, because commonly low concentrations can be stimulatory to plant growth whereas higher concentrations can be inhibitory. It is disappointing that very few studies have satisfied the boundary conditions indicated above.

Frequently it is assumed that the modification of root development by a microorganism known to produce material whose activity can be demonstrated by bioassay on a plant organ is evidence of growth regulatory action in the rhizosphere. However, to prove such a hypothesis it would be necessary to show (1) that the chemical is produced by the microorganisms in pure culture, and (2) that concentrations of the substance are present at physiologically active concentrations in the rhizosphere. Such studies can be easier in hydroponic culture of plants, where a large but relatively well-defined microbial population can develop. Large microbial biomasses can be associated with roots because of the substantial flow of carbon growth substrates to the root region (Chapter 4).

It is often assumed by gardeners that high microbial activity around roots is associated with rich brown soils and is the ideal growing medium for plants. Such observations have never been proven scientifically but the brown coloration is almost certainly due to humic substances. Generally there appears to be a beneficial effect of humic substrates on plant growth (Vaughan and Malcolm, 1985). They are derived from the microbial and free-enzymic decay of plant residues (especially the lignin and other polyphenolic components) in soil. However, such humification can be regarded as chemical transformation rather than true metabolite production. Humic acids are complex chemical structures. *Humic acid* can be extracted from soil with weak alkali. The non-extracted fraction is *humin*, whereas *fulvic acid* is soluble in both acid and alkali. The polymers consist of three main aryl aldehydes (*p*-hydroxybenzaldehyde, vanillin and syringaldehyde). Although microorganisms oxidatively couple phenols and amino acids to yield melanin (another brown substance), its contribution to the total humic fraction of soil is unclear.

2. PLANT GROWTH REGULATORS

The main groups of plant growth regulators are outlined in Fig. 1. Their formation by microorganisms has been reviewed in the past (Lynch, 1976, 1985). It was indicated then that few investigators had shown that the formation of these substances by microorganisms exogenous to the root had any significance to the plant itself. Since those reviews, many reports have again suggested that microorganisms stimulate plant growth by the production of plant growth regulators. But nobody has done the critical experiments of extracting and characterizing the metabolite and then reapplying it to the root at a natural dose range. Bacteria which penetrate the cortex and form products in the endorhizosphere could be more significant for plant growth than those formed in the ectorhizosphere. Whereas the observations of root morphology following bacterial inoculation would be consistent with growth regulator action, no definite evidence has been offered in support and, therefore, the idea must be regarded as hypothesis at this stage.

Ethylene is probably formed mainly at the sites of decomposing organic matter in soil, such as plant residues or straw. However, it can easily diffuse to the rhizosphere. It accumulates mainly in waterlogged soils where there is a sufficiently low redox potential but also where diffusion from the site of production is restricted because of the lack of continuity of air-filled porosity.

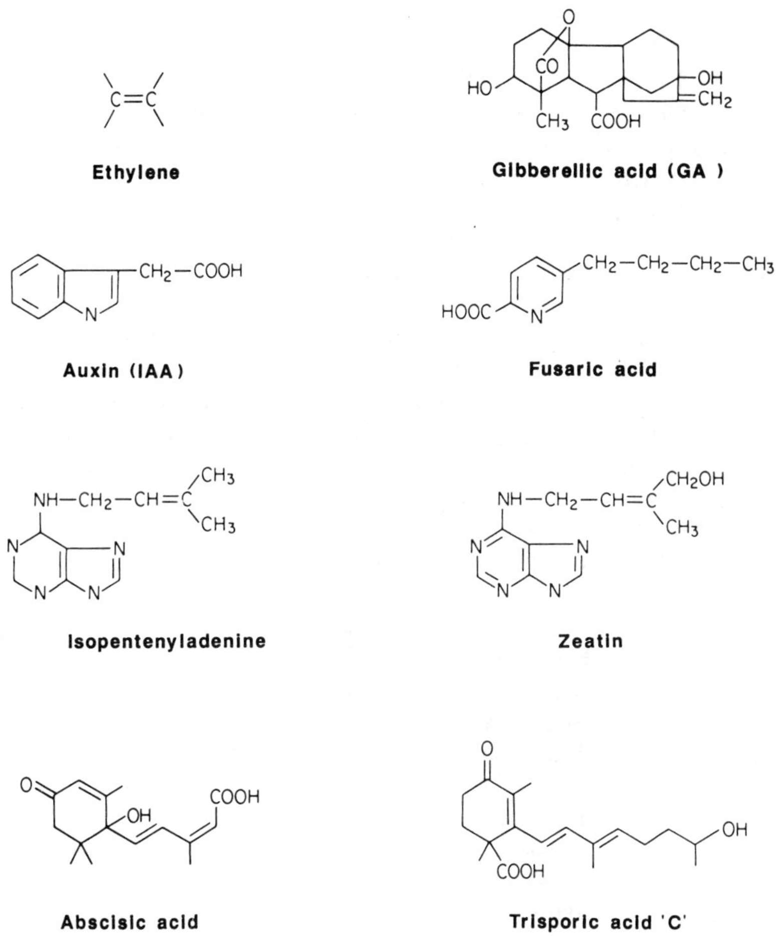

Fig. 1. Plant growth regulators.

However, under these conditions, the root itself becomes stressed and endogenous production of the gas can also become important (Drew and Lynch, 1980) (see also Chapter 3). However, the relative contribution of microorganisms and plants to the total rhizosphere ethylene pool is unclear. One response of plants, especially rice, to ethylene is to develop aerenchyma (air spaces in the cortex), although it is not clear if ethylene or anaerobiosis is the trigger for aerenchyma formation. The extra air spaces allow a more rapid movement of air from the shoots to the roots and thereby the amelioration of the anaerobic stress.

There has been considerably less study on the microbial formation of the non-volatile plant growth regulators (gibberellins, auxins, cytokinins and abscisic acid). This is probably largely due to the relative ease of measuring ethylene.

However, it should always be remembered that gibberellic acid (GA_3) was the first of the gibberellins to be characterized and that stemmed from the observation by Kurosawa (1926) in Taiwan that *Gibberella fujikuroi* (an imperfect form of *Fusarium moniliforme*) killed rice plants by causing foot rot after initially causing them to become abnormally tall. *G. fujikuroi* also produces fusaric acid, which is inhibitory to rice growth. The list of gibberellins known to occur in plants now exceeds 50, but few of them have been searched for in root-infecting or free-living rhizosphere organisms. By contrast auxin is a single structure and is formed readily by a wide range of microorganisms from tryptophan. Even bearing this in mind, however, the identities of auxins and gibberellins as microbial metabolites involved in the regulation of plant growth have been based on *in vitro* bioassays (Lynch, 1976, 1985).

Cytokinins promote cell division in plant tissues. As aminopurines they can link to phosphate esters of sugars to form nucleosides and nucleotides. Whereas there is much evidence that bacteria degrade tRNA, there have been few studies on cytokinin formation *per se* (van Andel and Fuchs, 1972). An interesting factor in the pathogenicity of *Agrobacterium tumefaciens*, however, is that oncogene 4 of the wide host-range plasmid pTiAch5 codes for the enzyme responsible for cytokinin biosynthesis (*cyt*) and the lack of recognition of certain plants by a particular biotype is mainly due to the absence or inactivity of a *cyt* gene in the T-region of the narrow host-range Ti plasmid (Hoekema et al., 1984). Abscisic acid has been reported as a microbial metabolite of the plant pathogenic fungus *Cercospora rosicola* (Assante et al., 1977; Neill et al., 1982). Also, the structurally related trisporic acids are products formed during sexual mating of mucoraceous fungi (Austin et al., 1969). The instability of these compounds makes effects on roots difficult to determine (Lynch and White, 1977).

The rather slender evidence that growth regulators with known chemical structures are formed by rhizosphere microorganisms, and the lack of evidence for any growth-stimulatory effects of growth regulators applied exogenously to roots, puts into perspective the frequent claims that microorganisms stimulate plant growth by such mechanisms. For example, the inoculation experiments first with *Azotobacter paspali* and then with *Azospirillum* spp. were shown to increase the yields of tropical grasses. However, as much experimental evidence (Giller and Day, 1985) shows that the fixation of dinitrogen was not the mechanism, focus switched to the production of growth regulators as the mechanism. Hadas and Okon (1987) produced some interesting scanning electron micrographs in support of this theory. Root hair formation and/or branching were consistent with the effects that *might* be expected of growth regulators. However, no candidate chemicals have been extracted from roots and then reapplied to test their effects. As *Azospirillum* spp. can colonize the endorhizosphere (Patriquin and Dobereiner, 1978) and are well-characterized at the molecular level (Emerlich, 1984), proof of their metabolic potential in growth enhancement would open major opportunities for crop growth improvement. The opportunity for the critical experiments is enormous.

Fungi, notably *Trichoderma* spp. (Baker, 1988), can also stimulate plant growth and advance flowering of horticultural species. But although it is claimed that a chemical metabolite is likely to be involved (Windham et al., 1986), this has not yet been characterized. Certainly the effects appear to be real and in my laboratory (Lynch et al., unpublished) we have repeated Baker's observations. Again, there appears to be a major opportunity to characterize the factors involved.

3. PHYTOTOXINS

Accepting the principle that most substances can be toxic to plants if supplied at sufficient dose rate, it follows that the size of accumulated metabolite pools is a critical factor in determining microbial effects in the root region. Some phytotoxic metabolites are illustrated in Fig. 2.

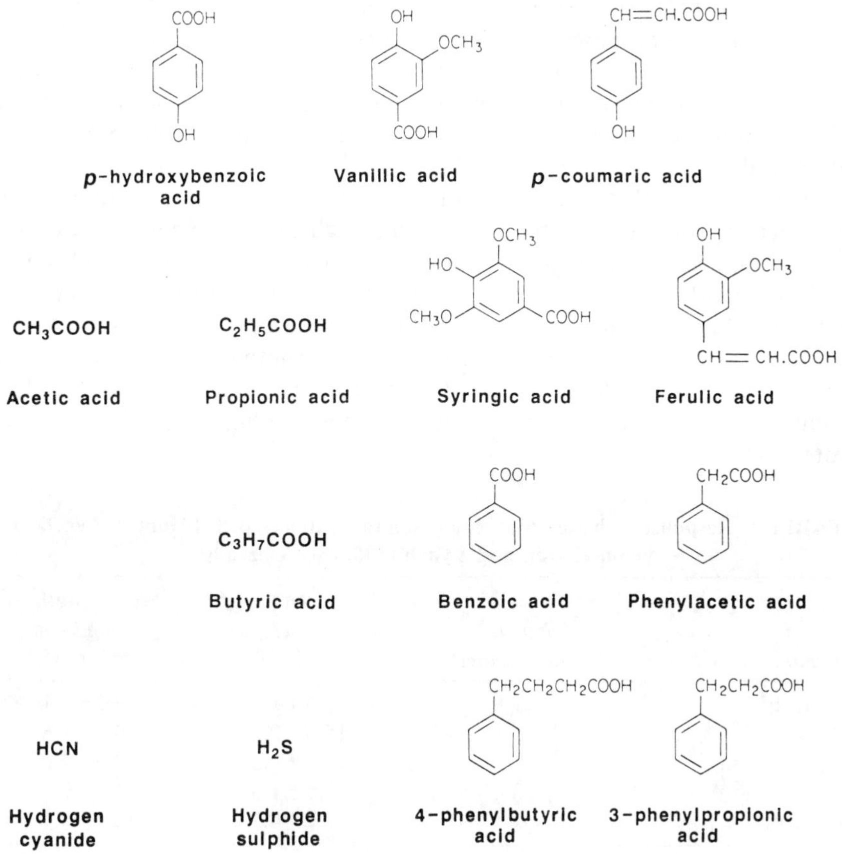

Fig. 2. Some phytotoxins.

3.1. Aliphatic Acids

One of the most extensively studied pools is that of the short-chain aliphatic acids (acetic, propionic and butyric), which are steam-volatile (for a review see Drew and Lynch, 1980). These are fermentative metabolites of soil bacteria and as such their formation is promoted in flooded soils with a large proportion of anaerobic microsites. Furthermore, under these conditions their catabolism is inhibited because this generally occurs aerobically. Cellulose and hemicellulose, which are major constituents of plant residues, are the major substrates for the biosynthesis of these organic acids. They are, therefore, most likely to accumulate around plant residues in soil; they either diffuse to the root or the root will come into contact with them. The diffusion properties have been approximated (Lynch et al., 1980), but it can be difficult to determine true diffusion characteristics because catabolism of the product occurs in aerated zones of the diffusion path. When roots grow through regions of plant residues, unless a continuous layer of residues is present it is unlikely that all the root system will experience the "toxic solution" and, therefore, the length of the root system exposed to the toxicity is critical in determining the phytotoxic effect (Gussin and Lynch, 1982). If a proportion of one root is exposed to the acid, there can be compensatory growth in other roots (Table 1). Potentially, this phenomenon can occur irrespective of the type of phytotoxin being produced.

It is assumed that these aliphatic acids would be common fermentative products of soil bacteria, but species with high rates of metabolite excretion do not appear to have been sought. Chemostat studies with soil slurries indicate that low redox potential is the critical factor governing accumulation and that the acids cause the soil solution pH to drop and Fe^{2+} and Mn^{2+} to become solubilized (Lynch and Gunn, 1978). Another factor to consider in organic acid accumulation is that it may promote plant growth indirectly by inhibiting harmful organisms; this is an area which has received little, if any, attention.

TABLE 1. Response of barley root elongation to treatment with 10 mol m^{-3} acetic acid. From Gussin and Lynch (1982) with permission

Region treated	*No. of roots treated*	*Mean length of non-treated root ± S.E.M.*	*Mean length of treated root ± S.E.M.*
2 cm tip	1	11.6 ± 1.7	8.2 ± 1.1
2 cm tip	3	15.8 ± 0.2	9.1 ± 0.6
2 cm section	1	16.5 ± 0.5	12.9 ± 2.6
2 cm section	3	17.5 ± 1.3	7.4 ± 0.8
Control[a]	—	11.2 ± 0.7	12.4 ± 0.7

[a]Means of plants where either one or three tips or lengths were treated with plant culture solution or where no roots were treated. Difference between control plants was not statistically different.

3.2. Phenolic Acids

Phenolic acids are generally more phytotoxic than aliphatic acids (Lynch, 1980) but phenolics seldom accumulate in such large concentrations. The accumulation of phenolic acids in soils has been reviewed by Hartley and Whitehead (1985). Ferulic, *p*-coumaric, *p*-hydroxybenzoic and vanillic acids are of widespread occurrence in soil but they occur mainly in chemically bound forms in plant cell wall materials. As such, their biological activity is limited. Although they can be released easily as their sodium salts with alkali, microorganisms producing cellulases and other lytic enzymes can also release them. Thus they should not really be reported as microbial biosynthetic metabolites in soil; they should more strictly be viewed as the products of biotransformation.

By contrast to the aliphatic acids, the role of phenolics against plant pathogens is well characterized (Friend, 1981), but this nevertheless relates primarily to the antimicrobial action occurring in the living plant cell wall rather than in the soil. Phenolics have also frequently been implicated in allelopathic interactions (Rice, 1984; Waller, 1987). Again this primarily relates to the direct phytotoxicity of one plant tissue on another but it also includes the microbial involvement in release of the toxin. The microbial action can also be indirect. For example, Rice and Pancholy (1974) showed that seven phenolic compounds (caffeic, chlorogenic, isochlorogenic and ferulic acids, isoquercitrin, myricetin and quercetin) could inhibit nitrification by *Nitrosomonas* and *Nitrobacter* in extremely low concentrations, thereby potentially changing the form of availability of plant nitrogen.

3.3. Hydrogen Sulphide

At the extremely low redox potentials which can occur in flooded soils, hydrogen sulphide can be produced by the obligately anaerobic bacteria *Desulfotomaculum*, *Desulfovibrio*, *Desulfomonas* and *Desulfuromonas* from sulphate, sulphite, thiosulphate, tetrathionate, and sulphur with a variety of organic substrates. Whereas sulphide is toxic, it usually reacts with iron, or is oxidized or is converted to the non-toxic sulphydryl ion (Gambrell and Patrick, 1978).

Sulphide can accumulate in the rhizosphere and spermosphere (Dommergues and Jacq, 1972; Garcia et al., 1974), even in rice where oxygen diffuses from roots, but it only accumulates in wet acidic soils. Under these conditions it can inhibit respiration and poison the cytochrome oxidase system (Allam and Hollis, 1972), causing a disorder in rice known in different parts of the world as browning, bronzing, bruzome, akiochi, hie-imochi, suffocation and straighthead diseases (Hollis et al., 1975).

3.4. Cyanide

The whole area of microbial cyanide metabolism has been reviewed by Knowles and Bunch (1986). "Cyanide" loosely refers to both the cyanide anion (CN^-)

and undissociated hydrogen cyanide (HCN). As the pK_a of cyanide is 9.3, it is largely present as HCN at physiological pH values and hence it can rapidly be volatilized. *Chromobacterium violaceum, Pseudomonas aeruginosa, P. fluorescens, P. chloraphis* and *P. aureofaciens*, the fungus *Marasmius oreades* and the green alga *Chlorella vulgaris* all form cyanide. However, it can be expected that in reality the list is much longer. Glycine, serine and threonine all act as substrates for cyanogenesis. Methionine can serve as an activator. Cyanogenesis is affected by iron and phosphate concentration. The cyanide could be toxic to roots if formed in the rhizosphere. Catabolism of cyanide can occur by five enzyme systems: β-cyanoalanine synthase (CAS), rhodanase, cyanide hydratase (formamide hydrolyase), nitrogenase and cyanide oxygenase with cyanase (Harris et al., 1987).

Several plant diseases can progress with the liberation of cyanide in host plant tissue, and over 2000 plant species produce cyanogenic glycosides (Conn, 1980). *M. oreades*, for example, is the fairy-ring fungus and cyanide production by the fungus could be involved in causing the ring by phytotoxicity where fructification occurs in a ring around the inoculum source. Some fungi that are pathogenic to cyanogenic plants form cyanide hydratase to detoxify cyanide by conversion into formamide (Fry and Myers, 1981). In bacteria, cyanide resistance is a taxonomic feature; *Enterobacter* spp., for example, are usually cyanide resistant because they contain the CAS enzyme.

Whereas little is known about cyanogenesis in the rhizosphere, sensitive detection techniques have recently been developed (Dartnall and Burns, 1987) and it seems that it could be a possible mechanism of plant growth inhibition (Alström and Burns, 1989) (see also next section). Interestingly, in carnation flowers, CAS activity and ethylene production are associated and it has been hypothesized that CN^- is formed during the conversion of 1-amino-cyclopropane-1-carboxylic acid to ethylene (Manning, 1986). On this basis, ethylene could become toxic to roots indirectly by allowing CN^- to accumulate, if CAS activity is absent.

Pathogen-resistant varieties of flax are more cyanogenic than susceptible varieties; of pathogenic and non-pathogenic fungi which colonize the roots, *Trichoderma* is tolerant of HCN (Timonin, 1941). This could be a factor in its biocontrol activity. As another dimension of the role of cyanide in biocontrol, *Pseudomonas fluorescens* CHA_0 produces cyanic acid which is toxic to *Thielaviopsis basicola*, the causative agent of black root rot of tobacco (Ahl et al., 1986). A cyanide-negative mutant (*hcn*$^-$) CHA_5 has been constructed by a gene replacement technique and gave less biocontrol activity than the wild type (Voisard et al., 1989). Complementation of CHA_5 by the cloned wild-type *hcn*$^+$ genes restored its ability to suppress disease. The genes were also inserted into another strain, P3, to induce cyanogenesis and disease suppressiveness. Clearly, cyanide could have a significant role in biological control mechanisms.

4. IONOPHORES

In recent years, a great deal of attention in rhizosphere studies has been focused on the production of ionophores, notably siderophores which chelate iron. It has been recognized for a long time that bacteria and fungi produce a range of siderophores (Fig. 3, Table 2) (Neilands, 1984). Yield increases by inoculation of *Pseudomonas fluorescens* and *P. putida* onto potato seedpieces (Burr et al., 1978) were subsequently attributed to the production of an extracellular siderophore, pseudobactin (Teintze and Leong, 1981; Teintze et al., 1981). These bacteria were termed "plant growth-promoting rhizobacteria" (PGPR) but this was probably a misnomer because their action was claimed to be as antagonists of "deleterious rhizobacteria" or other minor pathogens in the rhizosphere, rather than as plant growth promoters as such. The action was supported by the observations of Kloepper et al. (1980) which showed that addition of ferric iron—Fe(III)—abolished the antagonistic action *in vitro*. Mutations induced by chemicals or ultraviolet light (Kloepper and Schroth, 1981) or transposon mutagenesis (Marugg et al., 1985) were deficient in siderophore production and were ineffective as antagonists (Kloepper and Schroth, 1981). Pure pseudobactin could mimic the action of the bacteria (Kloepper et al., 1980).

The great excitement generated by these observations has been tempered by the failure to get consistent responses to inoculation in field trials (Suslow, 1982).

Why should iron chelation be so critical in soil/plant/microbial ecology? Iron is usually present in soil as $Fe(OH)_3$, the solubility constant of which is $10^{-38.7}$;

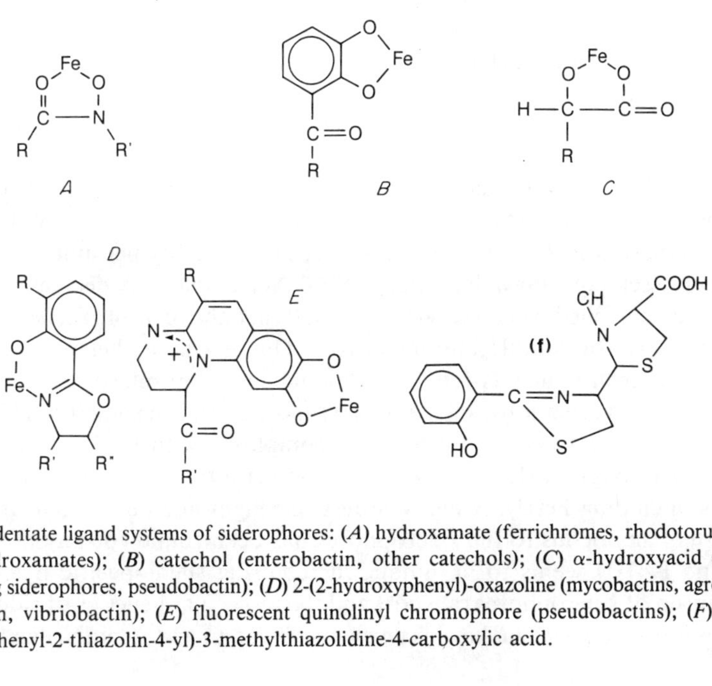

Fig. 3. Bidentate ligand systems of siderophores: (*A*) hydroxamate (ferrichromes, rhodotorulic acid, other hydroxamates); (*B*) catechol (enterobactin, other catechols); (*C*) α-hydroxyacid (citrate-containing siderophores, pseudobactin); (*D*) 2-(2-hydroxyphenyl)-oxazoline (mycobactins, agrobactin, parabactin, vibriobactin); (*E*) fluorescent quinolinyl chromophore (pseudobactins); (*F*) 2-(2-*o*-hydroxyphenyl-2-thiazolin-4-yl)-3-methylthiazolidine-4-carboxylic acid.

TABLE 2. Partial list of sidereophores in microbial species. From Neilands (1984)

Organism	*Siderophore(s)*
Bacteria	
Enteric spp.	Enterobactin, aerobactin
Paracoccus denitrificans	Parabactin
Vibrio cholerae	Vibriobactin
Bacillus megaterium	Schizokinen
Anabaena spp.	Schizokinen
Azotobacter vinelandii	α ϵ-*bis*-2,3-dihydroxybenzoyllysine
Pseudomonas spp.	Pyochelin, pyoverdine, pseudobactin, ferribactin
Actinomyces spp.	Ferrioxamines
Mycobacteria	Mycobactins
Fungi	
Widely distributed in Ascomycetes and Basidiomycetes	Ferrichromes, coprogen
Rhodotorula spp.	Rhodotorulic acid
Ectomycorrhizal spp.	Hydroxymate type

thus the concentration of soluble iron in soil is usually very low. The concentration of Fe(III) can be greater than 10^{-6} M, for example in anaerobic or low pH soils; here the Fe transport system is considered to be sufficient, low iron-affinity systems only being necessary for ferric ion uptake to occur. At lower concentrations the uptake is specific and high-affinity uptake systems are necessary to dissolve and transport Fe to microbial cells (Neilands, 1984; Swinburne, 1986). This can be achieved by complexing to siderophores. These are large molecules (1000–1500 D) and cannot pass normal membrane pores without the special high-affinity system. Siderophore production is usually only switched on when the Fe(III) concentration is less than 20 μM, and then its affinity constant and binding is heavily pH-dependent. Hydroxymate siderophores are products of fungi and bacteria and are widespread in soils, but catecholate siderophores are products only of bacteria (Crowley et al., 1987). Without siderophores there is 1000-fold decrease in Fe(III) availability per unit pH increase, making it essentially unavailable above pH 4. Fe(II) solubility decreases similarly but is also controlled by redox, with the result that above pH 4, Fe(II) is the most available form of soluble inorganic iron. This is where chelating agents are important. Agents such as fulvic and humic acids can bind with iron but are not specific and other ions, such as Cu(II) and Ca(II), can sometimes bind more readily. Siderophores also form stable complexes with Cr(III) and the rare Ga(III). Although Al(III) is abundant, the affinity for this and all divalent cations, including Fe(II), is many orders of magnitude below that of Fe(III). No system analogous to the siderophores has been found for any other metal ion, and Fe(III) seems to be unique in requiring such specific ligands. The mechanism of siderophore transport and the molecular genetics of siderophore systems in *Escherichia coli* have been characterized to an extent (Neilands, 1984).

In rhizospheres under iron deficiency, effective ion uptake mechanisms are crucial to plant root systems. Römheld and Marschner (1986) have described two strategies for iron acquisition. Type I occurs in most non-graminaceous species and is characterized by an inducible plasma membrane-bound reductase with enhancement of proton release. Type II occurs in grasses and is characterized by enhanced release of phytosiderophores and by a highly specific uptake system of Fe(III) phytosiderophores. Type II is at an ecological advantage over Type I because the solubilization of sparingly soluble inorganic Fe(III) compounds in the rhizosphere is influenced less by a high pH. Barley takes up Fe(III) phytosiderophores at a rate 100–1000 times faster than from synthetic Fe(III) chelates (e.g. FeEDTA) or microbial Fe(III) siderophores (e.g. ferrichrome). However, cucumber (*Cucumis sativus* L.) takes up synthetic chelates at the same rate as phytosiderophores. These processes are all strongly influenced by pH, and Marschner et al. (1986) have investigated the pH of the soil and rhizoplane by infiltrating soil with bromocresol purple. This showed that the rhizosphere pH can be two units higher or lower than the soil pH but this is strongly influenced by the form of N fertilization. With iron deficiency, several-fold increased Mn reduction and uptake can be induced, leading to higher Mn concentrations in Fe-deficient plants.

Bearing in mind the studies on Fe(III) uptake by microorganisms and plants, it is perhaps surprising how little attention has been addressed to competitive uptake. Becker et al. (1985a) showed that purified pseudobactin at concentrations as low as 0.1 μM and the siderophore from *Pseudomonas aureofaciens* ATCC 15926 inhibit the uptake of ferric iron from $^{59}FeCl_3$ by the roots of pea and maize sufficiently to reduce the synthesis of chlorophyll. They interpreted this as competitive binding. By contrast they found that aerobactin from *Klebsiella pneumoniae* had no effect on the uptake of ^{59}Fe by maize roots. Agrobactin, a linear catechol siderophore from *Agrobacterium tumefaciens* strain B6, enhances the uptake of Fe(III) into young pea and bean plants with a consequent increase in the synthesis of chlorophyll (Becker et al., 1985b). Clearly more studies to determine adsorption isotherms, in the presence or absence of microorganisms, of the type carried out for P uptake (Barber, 1978) could shed further light on competition for this critical rhizosphere nutrient.

Since the initial studies by Schroth's group in California, several groups have investigated the potential of siderophore producers as rhizosphere inoculants. Most study has been by Schippers' group in the Netherlands (Schippers et al., 1987). A range of *Pseudomonas* isolates have been shown to be effective in stimulating potato growth (presumably by controlling minor pathogens) and it seems that siderophore production is a prerequisite for the growth stimulation by *P. putida* WCS358 in the field (Bakker et al., 1986). However, responses are variable; for example, trials between 1981 and 1984 only showed a positive response in 1981, although the baseline yield was highly variable (Geels et al., 1986). This strain and others were effective in protecting carnation against *Fusarium* wilt in hydroponics but transposon mutants deficient in siderophore production did not colonize the roots and were ineffective in disease control (Xu et al., 1987).

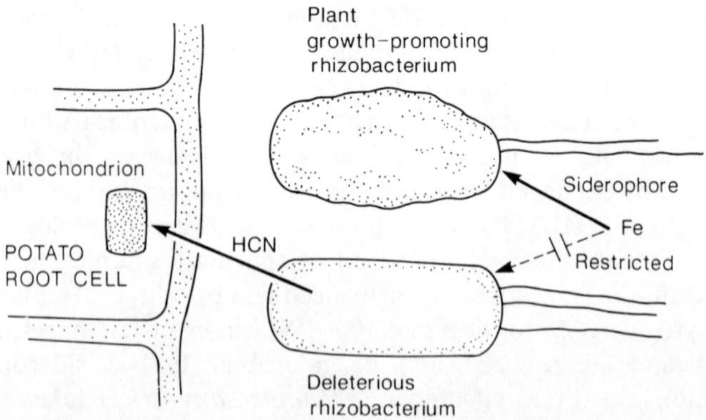

Fig. 4. The role of bacteria in the inhibition of root respiration by cyanide.

Bakker and Schippers (1987) suggested that PGPRs were primarily active by outcompeting cyanide-producing rhizobacteria (Fig. 4). They showed that 50% of potato rhizosphere pseudomonads produce HCN *in vitro* and that 5 μM HCN inhibits plant cytochrome oxidase respiration by 40% (again *in vitro*). Cyanide production by inhibitory rhizobacteria depends on the concentration of Fe(III) in growth media. Growth-promoting bacteria do not produce cyanide *in vitro* and their capacity for siderophore production enables them to compete with the inhibitory bacteria. When cyanide inhibits the cytochrome pathway in roots which store energy as ATP, it stimulates the alternative cyanide-resistant pathway, which is less energy-effective and can be inhibited by salicylhydroxamic acid (SHAM) (Fig. 5). Operation of the latter pathway causes depressed root cell energy metabolism, which in turn could impair nutrient uptake. The 'cyanide

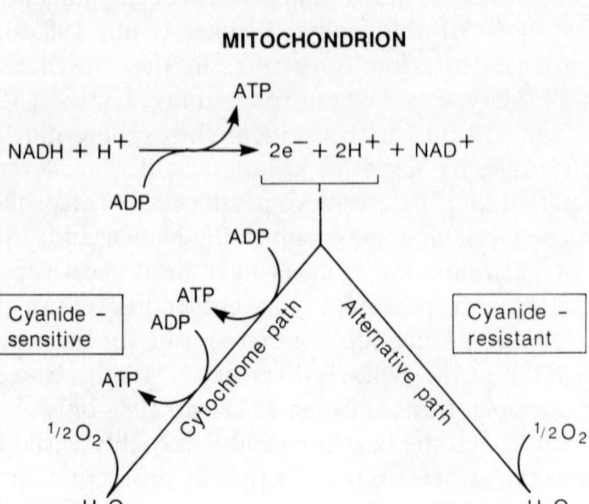

Fig. 5. Pathways of root respiration.

hypothesis' is interesting but based largely on *in vitro* observations. It should be evaluated *in vivo* and the active concentrations of cyanide in the rhizosphere should be estimated. It also seems that based on the evidence outlined above, presence of β-cyanoalanine synthase may be a more useful property than siderophore production in relieving the potential cyanide inhibition of root growth.

Ahl et al. (1986) demonstrated that iron-free siderophores from *Pseudomonas fluorescens* (pyoverdine) did not inhibit germination of endoconidia or chlamydospores or mycelial growth of *Thielaviopsis basicola*, but reduced the production of endoconidia. By contrast when the siderophore was complexed to Fe(III), there was a strong inhibition of germination and free Fe(III) was less toxic. It, therefore, seems that siderophores are not toxic by depletion but by elevation of Fe(III) to levels which are toxic to pathogens. This hypothesis has been supported by carrying out gnotobiotic experiments in clays of different iron contents (Keel et al., 1989). Furthermore a pyoverdine-negative mutant gave some protection. Using fluorescent siderophore negative mutants of *P. fluorescens* Migula str. 3551, Loper (1988) showed that siderophore production could only partially account for the antagonistic activity to *Pythium ultimum*. Gill and Warren (1988) constructed mutants to show that the production of the fungistatic agent from *Pseudomonas* sp. strain NZ 130, which is also active against *P. ultimum*, is antagonized by iron and the agent is not required for iron assimilation.

Iron can affect other metabolic processes in the rhizosphere, making the overall assessment of siderophore production even more difficult. For example Fe(III) at 0.2 mg ml^{-1} in the culture medium of the fungus *Mucor hiemalis* increases the production of the plant growth regulator ethylene by 2820% and Fe(II) increases it by 24 300% (Lynch, 1974). Similar results have been obtained for bacteria (Primrose, 1979). As ethylene is generally considered to be inhibitory to root growth, a scenario could develop where the siderophore-producers acquire Fe but in so doing inhibit root growth by producing ethylene.

During the microbial production of acetic acid as a phytotoxin at low redox potentials, Mn(II) and Fe(II) are solubilized and thus made potentially available to plants (Lynch and Gunn, 1978). Waterlogging damage is always associated with low soil redox potentials and is often associated with the accumulation of Fe and Mn in aerial tissues (Drew and Lynch, 1980). As a further complication in assessing the ecological significance of siderophore production in the rhizosphere, Focht and Verstraete (1977) have reviewed the role of hydroxamic acid siderophores as products of nitrifying bacteria using ammonium or amino N substrates.

Despite the many studies in recent years on siderophore-producing bacteria, nobody has yet convincingly identified their role in plant growth. Certainly in crops as diverse as wheat (Weller and Cook, 1986) and mushrooms (Henry et al., unpublished), bacterial siderophore production is not always a prerequisite for disease biocontrol, but there is clearly a need for evaluation of its direct and indirect antibiotic roles. It should also be noted that the effectiveness of the PGPR siderophore-producing bacteria as inoculants has been highly variable

(Suslow, 1982). One of the major factors likely to be responsible for this inconsistency is that production and biological activity (Fe competition) of siderophores of the fluorescent pseudomonads are influenced by pH, which of course can vary greatly between soils. Action in the field may be limited to a few cases where the soil pH, or more precisely the rhizosphere pH, is alkaline as Fe availability is inversely correlated with increased pH (Scher and Baker, 1982; Mishaghi et al., 1988). It would be unwise, however, to overstate the situation because despite the compelling evidence for antibiosis in the action of fluorescent pseudomonads in the rhizosphere, siderophores probably make a contribution to the biocontrol activity of bacteria (Thomashow and Weller, 1988).

It seems unlikely that the Fe(III) uptake system is plasmid encoded (Weger et al., 1986). If siderophore production was eventually proven to be a major factor in PGPR action, it might then be useful for the expression of gene product to be increased using molecular biology techniques (Marugg et al., 1985).

Nearly all the effort on rhizosphere ionophore production has concerned siderophores. However, investigation of other ionophores in the rhizosphere is likely to pay dividends. For example, a manganese-chelating agent was produced by bacteria growing on the roots of solution-grown barley plants, and this promoted Mn uptake by roots (Barber and Lee, 1974).

5. ANTIBIOTICS

One of the first experiments an undergraduate studying microbiology is likely to undertake is to demonstrate antibiosis on an agar plate by measuring a zone of inhibition around one organism against another. This is generally assumed to be evidence of antibiosis, although it is conceivable that in some situations the inhibition zone could be created by the secretion of a diffusible lytic enzyme by the inhibitory organism. Following the classical demonstration of penicillin as an antibacterial agent of *Penicillium notatum* by Alexander Fleming, attention in the 1940s was focused on the antibacterial agents (especially streptomycin) produced from soil *Streptomyces* spp. by Selman Waksman and his students. As with Fleming's work, this led to a Nobel Prize for Physiology and Medicine (Wainwright, 1988). The soil has continued since then to be a major source of antibiotic-producing organisms for the pharmaceutical industry, with the major focus of attention on antibacterial agents (Turner, 1971; Turner and Aldridge, 1983). It is, therefore, perhaps surprising that this has not also been a hunting ground for agrochemicals. Nevertheless, one antibacterial antibiotic has been found in the rhizosphere which has had a major impact on the development of modern plant molecular biology, while also yielding one of the few proven commercial examples of a biocontrol agent.

Crown gall of fruit trees (almond, peach, rose, cherry) is caused by *Agrobacterium tumefaciens, A. rhizogenes* and *A. rubi* (all biotypes of *A. radiobacter*). Virulence genes which induce tumours are located on a large

(200 kb) plasmid (the Ti-plasmid) and the T-DNA (24 kb) is transferred to the plant cell (Kerr and Tate, 1984) (see also Chapter 6). *A. rhizogenes* strain 84 was isolated as a non-pathogenic antagonist of the pathogens; it produces an antibiotic, agrocin 84 (Fig. 6), which is coded for on another plasmid, pAg 396, and taken up by the pathogen. The Ti-plasmid of the pathogens induces the synthesis by the plant of opines—octopine, nopaline and agropine—which can be catabolized by the inciting bacteria, but not by the plant or other microorganisms. Strain 84 kills the pathogen because agrocin 84 is actively transported into the pathogens via a permease which also transfers other opines (agropinopines) into its own cells. Thus toxin and food base are taken up at the same time. However, the genes on plasmid pAg 396 controlling agrocin 84 production (and resistance) can be transferred from the antagonist to a pathogenic recipient in the presence of nopaline. This has resulted in some failures of the biocontrol action where large populations of the pathogen in soil have left large residual pools of nopaline to promote plasmid exchange. However, an antagonist strain has now been engineered with a deficient plasmid-transfer system (Jones et al., 1988) and this is working very satisfactorily in the field (A. Kerr, personal communication).

The detailed chemical, physiological and molecular biological investigations of the *Agrobacterium* antibiosis have not only yielded a major proven commercial biocontrol agent but have also provided two of the major plasmid vectors used in plant molecular biology. This work should provide a sound basis for stimulating further scientific studies on plant disease biocontrol agents.

Most of the other studies on plant disease biocontrol have concentrated on fungal diseases, and some of the antibiotics so traced are shown in Fig. 6. They are likely to have a role in the interactions of microbial communities in the rhizosphere (Lynch, 1987a; Fravel, 1988; Weller, 1988).

Patulin was found to be an antibiotic metabolite of *Penicillium patulum, Aspergillus clavatus* and *A. terreus*, and to have some, but variable, biocontrol activity against damping-off diseases of tomato (Grossbard, 1952). This was a landmark study because it was one of the first to invoke a definitely identified chemical in a biocontrol action. It was also important because it clearly recognized that energy sources in the form of wheat straw, lawn mowings, bracken, timothy grass and sugar-beet pulp could provide suitable substrates for fungal growth and thus elevated levels of antibiotic. However, it did not recognize the potential toxicological properties of patulin. Even though it had been used in clinical trials to cure the common cold, it is a mycotoxin and mildly carcinogenic (Scott, 1974). This emphasizes the caution which should be exercised in handling any microbial metabolite which has not been evaluated toxicologically.

Pseudomonas fluorescens strain Pf-5 produces pyoluteorin and it is effective in controlling damping-off in cotton caused by *Pythium ultimum* (Howell and Stipanovic, 1980). The antibiotic is inactivated in soil and, therefore, its action can only be effective as a seed treatment; moreover, it does not act against a range of other damping-off diseases. However, the same investigators found

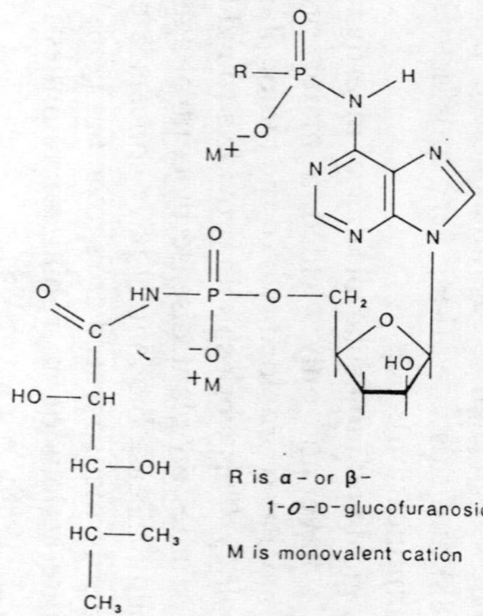

R is α- or β-
1-*O*-D-glucofuranoside

M is monovalent cation

Agrocin 84

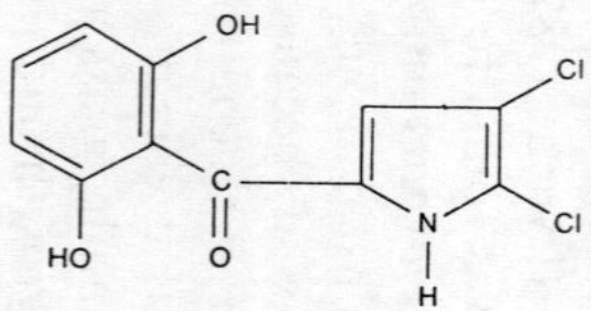

Pyoluteorin

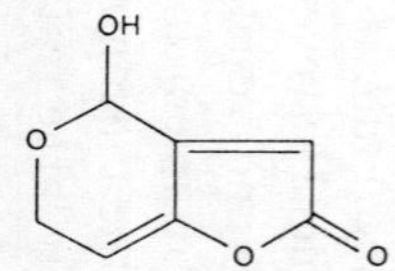

Patulin

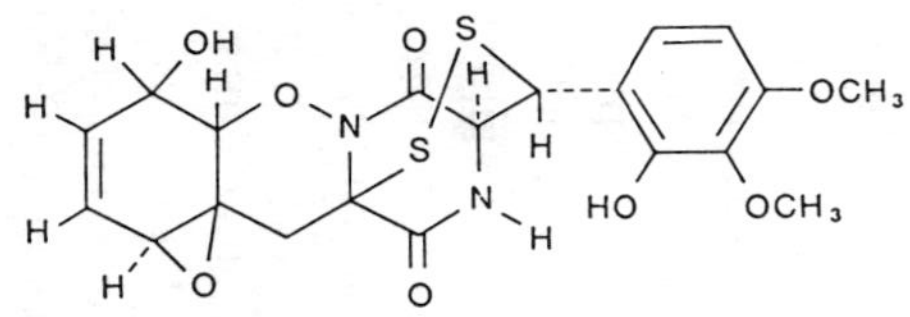

Gliovirin

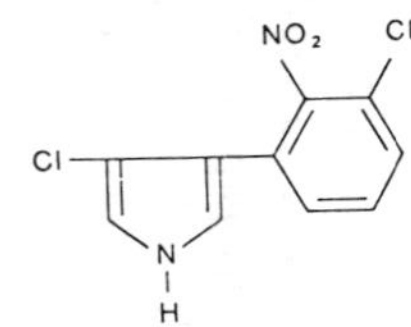

Pyrrolnitrin

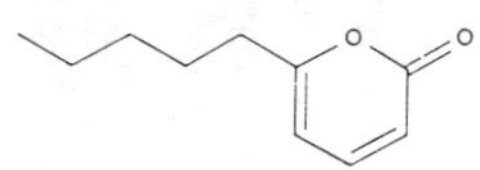

6n-pentyl-2H-pyran-2-one

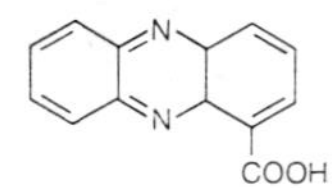

Phenazine-1-carboxylic acid

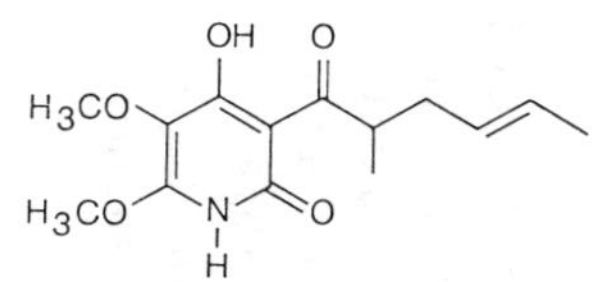

Harzianopyridone

Fig. 6. Some antibiotic microbial metabolites.

another strain of *P. fluorescens* which was inactive against *P. ultimum* but inhibited the growth of *Verticillium dahliae, Thielaviopsis basicola* and *Alternaria* spp. *in vitro* and *Rhizoctonia solani* on cotton seedlings *in vivo*. The action was attributed to the production of pyrrolnitrin (Howell and Stipanovic, 1979).

Trichoderma spp. produce a wide range of biologically active metabolites (Turner, 1971; Turner and Aldridge, 1983) and whereas the genus has yielded many strains with biocontrol activity, the critical correlation between biological activity and metabolite production has not usually been made. Dennis and Webster (1971) reported that biocontrol activity of *T. harzianum* against *R. solani* appeared to be associated with a coconut smell. Claydon et al. (1987) have shown this to be 6n-pentyl-2H-pyran-2-one and also demonstrated its biological activity. Subsequently, Simon et al. (1988) characterized chemically and biologically the same product from *T. koningii*. The substance is particularly interesting because, being volatile, it could be expected to have considerable ecological impact by readily diffusing to target pathogens. The broad range of activity included sclerotial-forming diseases and take-all (*Gaeumannomyces graminis* var *tritici*). However, the strain investigated by Claydon et al. (1987), which was first isolated from straw (Lynch, 1987b), also produces non-volatile harzianopyridone which also has a wide range of activity against soil-borne plant pathogens (Claydon et al., 1989). Perhaps there are more interesting finds in the *Trichoderma* metabolite pool! Even if not, the existing metabolites already provide the organic chemist with targets for synthesis of novel fungicides of agrochemical or medical importance.

Trichoderma is very closely related to *Gliocladium*, and whereas there have been fewer investigations of the *Gliocladium* spp. antibiotic pools, the potential is clearly present. *G. virens* is active against *P. ultimum* as the causative agent of damping-off in cotton; gliovirin appears to be involved in the action (Howell, 1982; Howell and Stipanovic, 1979, 1980, 1983). Support for this was provided by showing extra activity by a mutant strain of the antagonist with enhanced gliovirin production. However, by contrast in another study, a gliotoxin-producing isolate of *G. virens* did not limit damping-off or root rot of peanut caused by *Rhizoctonia solani* as effectively as a non-gliotoxin-producing isolate (Jones et al., 1984).

Various strains of *Pseudomonas fluorescens* have for a long time been suggested as potential agents against take-all of wheat (Cook, 1988). However, only recently has an antibiotic, phenazine-1-carboxylic acid, been characterized in studies in the USA (Thomashow and Weller, 1988) and Australia (Brisbane and Rovira, 1988) as being the active factor. Whereas both studies characterized the activity *in vitro*, the former in particular provided compelling evidence in favour of this hypothesis *in vivo*. Mutants defective in phenazine synthesis (*phz*$^-$) were generated by transposon (Tn5) mutagenesis and then compared with the parental strain. Six independent, prototrophic *phz*$^-$ mutants were non-inhibitory to the disease *in vitro* and *in vivo* (on wheat seedlings). Restoration of antibiotic synthesis and activity were coordinately restored in two mutants complemented with cloned DNA from a genomic library of the parent 2-79 strain containing the sequences necessary for phenazine-1-carboxylic acid production. These observations provide a clear example of the power of molecular biological approaches in identifying the mode of action of microbial antagonisms in the rhizosphere (see also Chapter 15).

6. LECTINS AND AGGLUTININS

Cell agglutinating proteins, notably haemagglutinins, have been recognized since the turn of the century and are widely distributed in nature (Mirelman et al., 1975). In plants they are termed phytohaemagglutinins or phytoagglutinins. Generally they are sugar specific and as such are now generally called lectins. They are usually oligomeric by binding to several sugar binding sites. Their physical and chemical properties vary enormously, with molecular weights ranging from about 11 000 to 335 000; often, but not always, they are glycoproteins. The lectins combine with lectin receptors (mono- or oligosaccharides) of the host cell walls (plant and microbial), usually by hydrophobic and hydrogen bonding. The specificity of lectin binding is determined by finding the monosaccharide(s) or oligosaccharide which most effectively inhibits binding. Clearly lectins have a considerable potential role in mediating adherence to surfaces and most of the interest in rhizosphere studies has focused on their possible roles in the *Rhizobium* association with roots, and the pathogenesis and biological control of diseases.

The initial studies on legumes were by Dazzo and co-workers (Dazzo, 1980; Dazzo and Gardiol, 1984). In the early "recognition" stage, the formation of a unique cross-reactive antigen (receptor) on the cell walls of *Rhizobium trifolii* is critical and this binds to the lectin of clover. In the second stage, firmer binding probably involves the production of cellulose microfibrils. Although substantial experimental evidence has been produced to support this hypothesis, some investigators have not always observed it, probably in part because there are two phases—attachment (Phase 1) and adhesion (Phase 2) (Dazzo et al., 1984). The current position has been reviewed by Dazzo et al. (1986). In particular it presents evidence for fimbriae (non-flagellar filamentous appendages) in *Rhizobium* being involved in the attachment. In addition the role of fimbriae-mediated bacterial adhesion to plants by other nitrogen-fixing bacteria (*Klebsiella* and *Enterobacter*) is considered (Duguid and Old, 1980).

Some of the classic studies of the role of lectins in pathogen adhesion to roots concerned *Agrobacterium tumefaciens* (Lippincott and Lippincott, 1969). The O-antigen from the outer envelope of these Gram-negative bacteria effectively inhibited tumorigenesis, the genetic information specifying adhesion being both chromosomal and plasmid-borne (Whatley et al., 1978).

Lectins are present in the fruiting bodies of higher fungi but they are also present in the lower fungi (Nordbring-Hertz and Chet, 1986). Plant lectins may be involved in defence reactions against pathogenic and other fungi. For example Mirelman et al. (1975) showed that wheat germ agglutinin could bind the hyphal tips of *Trichoderma viride*, and Hinch and Clarke (1980) showed that a fungal lectin was involved in the binding of *Phytophthora cinnamomi* zoospores to the root surface of maize (*Zea mays*).

Nordbring-Hertz and Chet (1986) have also reviewed the evidence of lectin involvement in fungal–nematode interactions.

The hyphal coiling around and binding by *Trichoderma* spp. to pathogen hyphae is referred to in Chapter 10. It seems that lectins of the pathogens *Rhizoctonia solani* (Elad et al., 1983) and *Sclerotium rolfsii* (Barak et al., 1984) are involved in the initial steps of attachment and coiling, which then lead to further steps generated by lytic enzymes bringing about the penetration of cell walls of the pathogen. Erythrocytes and *Escherichia coli* have been employed in determining the binding phenomena of *Trichoderma*.

Enterobacter cloacae is, like *Trichoderma* spp., effective in the biocontrol of damping-off fungi, especially *Pythium* spp. (Hadar et al., 1983) and again lectin binding appears to be involved (Nelson et al., 1986). Treating cucumber seeds with sugars (D-glucose, D-galactose, sucrose, N-acetyl-D-glucosamine or β-methyl-D-glycoside) that interfered with binding eliminated the biological control action.

There must be many examples of lectin binding being involved in root–microbe or microbe–microbe interactions in the rhizosphere. Opportunities should arise for regulating the processes phenotypically and genotypically to the benefit of the plant when the metabolites are characterized.

7. SOIL-AGGREGATING AGENTS

Soil aggregates are formed mainly by physical forces, while stabilization is effected by several factors, including organic materials, iron and aluminium oxides, and clays. However, as aggregation and stabilization are concurrent, the various stabilizing agents may act in conjunction with each other. It is generally regarded that rhizodeposition provides a major substrate to soil stabilizing microorganisms and this contributes, for example, to the rapid improvement in soil structures under grass leys (Lynch and Bragg, 1985). Chapter 2 has a picture of rhizosphere bacteria surrounded by an electron-transparent region (probably polysaccharide) causing clays to aggregate. Such stabilizing action can optimize the soil physical environment for root growth, especially in terms of controlling soil water availability, allowing adequate root aeration and minimizing the erosive forces of water causing run-off. Thus the production of soil stabilizing agents by rhizosphere microorganisms can have an indirect effect on crop growth and yield.

Because of their relative sizes, it is generally considered that bacteria provide the cementing action between clay particles to form micro-aggregates and that fungal hyphae provide the binding action to stabilize an aggregate. Some organisms act mechanically by cell/particle binding but others act by providing extracellular products.

Much debate has surrounded the role of microbial polysaccharides. Sodium periodate oxidizes sugars by clearing the bonds between adjacent carbon atoms containing —OH groups. The partially oxidized polymers no longer provide a bridge between soil particles. Variable results have been obtained for different soils. For example, in a Swiss Braunerde, Mehta et al. (1960) found that the stability of soil aggregates was not affected by periodate, whereas Greenland et al. (1962) found that a Red-Brown Earth soil owed most of its stability to periodate-sensitive materials. The bulk of available evidence points to polysaccharides generally being important and the detailed mechanisms for this have been discussed before (Lynch and Bragg, 1985).

Many of the experimental investigations have arisen from studies of the importance of crop residues providing substrates for the soil-stabilizing organisms (Chapman and Lynch, 1985a) (Table 3). With wheat straw as substrate, Chapman and Lynch (1985b) found that *Enterobacter cloacae* strain C2/4 was an effective soil stabilizer, especially if selected strains grew in association with *Trichoderma harzianum* and *Sordaria alcina*. This appears to be especially significant to the rhizosphere because the bacterium is known to be a dominant colonizer of the endorhizosphere of wheat (Kleeberger et al., 1983).

It should also be recognized that algae are very effective polysaccharide producers and their role in soil stabilization is discussed in Chapter 13.

8. ENZYMES

A wide range of free enzymes bound to inorganic particles, organic substances, microorganisms, plants and animals occur in the rhizosphere. Some of these can be located by ultracytochemical tests; for example, Martin and Foster (1985) located acid phosphatase and catalase in a model soil (calcined attapulgite, a non-swelling clay mineral) containing wheat roots. With the development of newer immunocytochemical techniques employing, for example, colloidal gold, it should be possible to locate a wide range of rhizosphere enzymes with precision using both transmission and scanning electron microscopy in natural soils. More than 50 enzymes are active in soil, including oxidoreductases, hydrolases, lyases and transferases, which are immobilized on or in soil clays and humates and constitute a persistent extracellular biocatalyst (Burns, 1983) (Table 4). Most products of the rhizosphere, including those involved in mineral cycling as well as those derived from organic substrates, are formed enzymically and are also subject to enzymic degradation. With greater substrate availability and microbial activity in the vicinity of the roots, enzyme activity is generally greater in the rhizosphere than in the bulk soil. Some enzymes, such as the cellulases and ureases, which release simple sugars and ammonia respectively to the

TABLE 3. Proportions (%) of neutral sugars and uronic acid (as D-glucuronic acid), molecular weight and intrinsic viscosity in the extracellular polysaccharide of some soil microorganisms. After Chapman and Lynch (1985a)

								Molecular weight		
	L-rhamnose	*L-fucose*	*D-xylose*	*D-mannose*	*D-glucose*	*D-galactose*	*Uronic acid*	*Viscometry*	*Gel filtration*	*Intrinsic viscosity* ($cm^3\ g^{-1}$)
Enterobacter cloacae (C2/4)		29.4			20.8	34.5	15.3	1 200 000	1 100 000	3500
E. cloacae (L1/14)					55.2	32.5	12.3	190 000	<40 000	530
Pseudomonas sp.	22.2			41.6	24.3	12.0		590 000	1 800 000	1600
P. fluorescens				12.0	57.4	30.6		1 600 000	1 400 000	4500
Rhodotorula rubra				28.6	29.7	25.8	16.1	1 000 000	1 800 000	2800
Candida humicola			55.0	45.0				200 000	1 700 000	610
Lipomyces starkeyi				53.5		29.1	17.4	610 000	1 800 000	1700

TABLE 4. Enzymes immobilized by soil clays and/or humates. From Burns (1983) with permission

Class	*EC No.*	*Recommended name*	*Substrate*
Oxidoreductases	1.7.3.3	Urate oxidase	Uric acid
	1.10.3.1	Catechol oxidase	Catechol
	1.10.3.2	Laccase	Phenylenediamine
	1.11.1.7	Peroxidase	Pyrogallol, chloroanilines[a]
	1.14.18.1	Monophenol monooxygenase	Catechol, pyrogallol, hydroquinone
Transferases	2.4.1.5	Dextransucrase	Sucrose
	2.4.1.10	Levansucrase	Sucrose
	2.8.1.1	Thiosulphate sulphur-transferase	Thiosulphate + cyanide
Hydrolases	3.1.1.1	Carboxylesterase	Hydroxy-methylcoumarin butyrate, Malathion[a]
	3.1.1.2	Arylesterase	Phenyl acetate
	3.1.1.3	Triacylglycerol lipase	4-methyl umbelliferone nonanoate
	3.1.3.1	Alkaline phosphatase	*p*-nitrophenyl phosphate
	3.1.3.2	Acid phosphatase	*p*-nitrophenyl phosphate
	3.1.6.1	Arylsulphatase	*p*-nitrophenyl sulphate
	3.2.1.1	α-amylase	Starch
	3.2.1.1	β-amylase	Starch
	3.2.1.4	Cellulase	Celluose, carboxymethylcellulose
	3.2.1.6	Endo-1,3(4)-β-D glucanase	Laminarin
	3.2.1.8	Xylanase	Xylan
	3.2.1.2	β-glucosidase	*p*-nitrophenyl β-D glucoside, cellobiose
		Peptidases	N-benzoyl L-arginine amide, benzyloxycarbonyl phenylalanyl leucine
		Proteinases	Casein, gelatine
	3.5.1.1	Asparaginase	Asparagine
	3.4.1.2	Glutaminase	Glutamine
	3.5.1.4	Amidase	Formamide, acetamide
	3.5.1.5	Urease	Urea
	3.5.1.13	Aryl acylamidase	Propanil[a]
Lyases	4.1.1.14	Glutamate decarboxylase	Aspartic acid
	4.1.1.25	Tyrosine decarboxylase	Tyrosine
	4.1.1.28	Aromatic-L-amino acid decarboxylase	Tryptophan, DOPA

[a]Xenobiotic substrates (pesticides).

rhizosphere organisms and roots, can thus regulate rhizosphere activity. Other enzymes can occur in the endorhizosphere, for example the "in-fill" between cells of the root cortex has a large pectin content and, therefore, pectinase is a particularly useful trait of the endorhizosphere organism.

Although it is assumed that enzyme activity is greater in the rhizosphere than in the bulk soil, few experiments have quantified this. One preliminary study was reported in Burns (1985) where the activity of various phosphatases was studied in the inner and outer rhizosphere of maize, barley and wheat. The inner rhizosphere (the fraction of soil removed from roots by fierce agitation) always had greater enzyme activity than the outer rhizosphere (the fraction of soil removed from roots by mild agitation). The rhizosphere to non-rhizosphere enzyme ratio was greatest in barley, with values of 3.1:1 (acid phosphatase, pH 5.2), 2.7:1 (neutral phosphatase, pH 6.9), 1.7:1 (alkaline phosphatase, pH 10.5) and 2.3:1 (phosphodiesterase). These activities, although hotly debated, especially in relation to mycorrhizas (Chapter 11), are likely to solubilize the phosphate in mineral particles around roots and may make this available to roots. One of the complications in interpreting activity in the soil environment itself is that there can be substantial interaction between the enzymes and soil mineral and organic colloids which can affect enzyme activity (Burns, 1983).

In the above discussions on microbial products, it is clear that much current interest focuses on products which govern microbial interactions and are involved in disease biocontrol. Besides antibiosis, pathogens can be inhibited by the action of lytic enzymes. Glucans and chitin are contained in most fungal cell walls and, therefore, it can be expected that β-1, 3-D-glucanases and chitinases would be involved in the action of antagonistic fungi. Such activity has been detected by Elad et al. (1982, 1984) and Artigues and Davet (1984). The enzymes can be induced by growing *Trichoderma* spp. on cell walls of *Rhizoctonia solani* (Ridout et al., 1986). However, when the exoenzymes were separated using a variety of protein separation techniques, a large number of proteases were found (Ridout et al., 1988). Protein is present in fungal cell walls and, therefore, the proteases could be involved in the biocontrol activity, a conclusion which has also been reached for another mycoparasite, *Pythium oligandrum* (Lewis et al., 1989). As yet there is no evidence that biocontrol activity of antagonistic fungi is correlated with exoenzyme production, which is not surprising because of the various modes of action possible, some of which will be in concert. However, there may be the opportunity to elevate the activity of lytic enzymes in strains which already have useful biocontrol properties.

9. CONCLUSION

The above examples have catalogued some chemicals as gene products of rhizosphere microorganisms which can influence plant growth and crop yield directly or indirectly. For some of them the biosynthetic pathway is known.

As such there should be the potential to identify suitable markers to exploit genetic engineering techniques now available with bacteria to amplify pathways leading to products beneficial to plants (Chapter 6; Davison, 1988). Suitable techniques such as transformation, are also being rapidly developed for fungi. Another strategy might be to introduce a beneficial pathway into a rhizosphere organism which already has a different pathway of benefit to plants, thus developing the concept of the "super-rhizosphere organism"! Clearly *in vitro* studies of rhizosphere microbial metabolites are crucial, but ecological studies *in vivo* are also required to ensure that the biosynthetic pathways are sustained. There is the possibility that amplification of the pathways could be achieved more readily phenotypically than genotypically, for example by modification of soil conditions. The opportunity for natural product chemists, microbiologists, molecular biologists and soil scientists to co-operate in rhizosphere studies has never been greater.

REFERENCES

Ahl, P., C. Voisard, and G. Defago (1986). Iron-bound siderophores, cyanic acid, and antibiotics involved in suppression of *Thielaviopsis basicola* by a *Pseudomonas fluorescens* strain. *J. Phytopathol.*, **116**:121–134.

Allam, A. I. and J. P. Hollis (1972). Sulfide inhibition of oxidases in rice roots. *Phytopathology*, **62**:634–639.

Alström, S. and R. G. Burns (1989). Cyanide production as a possible mechanism of plant growth inhibition. *Biol. Fert. Soils*, **7**:232–238.

Andel, O. M. van and A. Fuchs (1972). Interference with plant growth by microbial metabolites. In *Phytotoxins in Plant Disease* (R. K. S. Wood, A. Ballio, and A. Graniti, eds.). Academic Press, London, pp. 227–249.

Artigues, M. and P. Davet (1984). Activities β(1–3) glucanasique et chitinisique de quelques champignons, en relation avec leur aptitude a detruire les sclerotes de *Corticum rolfsii* dans la terne sterile. *Soil Biol. Biochem.*, **16**:527–528.

Assante, G., L. Merlini, and G. Nasini (1977). (+)-Abscisic acid, a metabolite of the fungus *Cercospora rosicola. Experientia*, **33**:1556–1557.

Austin, D. J., J. D. Bu'lock, and G. W. Gooday (1969). Trisporic acids: sexual hormones from *Mucor muscedo* and *Blakeslea trispora. Nature*, **223**:1178–1179

Baker, R. (1988). *Trichoderma* spp. as plant growth stimulants. *CRC Crit. Revs. Biotechnol.*, **7**:97–106.

Bakker, A. W. and B. Schippers (1987). Microbial cyanide production in the rhizosphere in relation to potato yield reduction and *Pseudomonas* spp.-mediated plant growth stimulation. *Soil Biol. Biochem.*, **19**:451–457.

Bakker, P. A. H. M., J. G. Lamers, A. W. Bakker, J. D. Marugg, P. J. Weisbeek, and B. Schippers (1986). The role of siderophores in potato tuber yield increase by *Pseudomonas putida* in a short rotation of potato. *Neth. J. Pl. Path.*, **92**:249–256.

Barak, R., Y. Elad, D. Mirelman, and I. Chet (1984). Lectins: a possible basis for specific recognition in the interaction of *Trichoderma* and *Sclerotium rolfsii. Phytopathology*, **75**:458–462.

Barber, D. A. (1978). Nutrient uptake. In *Interactions Between Non-Pathogenic Soil Micro-organisms and Plants* (Y. R. Dommergues and S. V. Krupa, eds.). Elsevier, Amsterdam, pp. 131–162.

Barber, D. A. and R. B. Lee (1974). The effect of micro-organisms on the absorption of manganese by plants. *New Phytol.*, **73**:47–106.

Becker, J. O., R. W. Hedges, and E. Messens (1985a). Inhibitory effect of pseudobactin on the uptake of iron by higher plants. *Appl. Environ. Microbiol.*, **49**:1090–1095.

Becker, J. O., E. Messens, and R. W. Hedges (1985b). The influence of agrobactin on the uptake of ferric iron by plants. *FEMS Microbiol. Ecol.*, **31**:171–175.

Brisbane, P. G. and A. D. Rovira (1988). Mechanisms of inhibition of *Gaeumannomyces graminis* var. *tritici* by fluorescent pseudomonads. *Plant Pathol.*, **37**:104–111.

Burns, R. G. (1983). Extracellular enzyme–substrate interactions in soil. In *Microbes in Their Natural Environments* (J. H. Slater, R. Whittenbury, and J. W. T. Wimpenny, eds.). Cambridge University Press, Cambridge, pp. 249–298.

Burns, R. G. (1985). The rhizosphere: microbial and enzymic gradients and prospects for manipulation. *Pedologie*, **25**:283–295.

Burr, T. J., M. N. Schroth, and T. Suslow (1978). Increased potato yields by treatment of seed pieces with specific strains of *Pseudomonas fluorescens* and *P. putida*. *Phytopathology*, **68**:1377–1383.

Chapman, S. J. and J. M. Lynch (1985a). Some properties of polysaccharides of micro-organisms from degraded straw. *Enzyme Microbiol. Technol.*, **7**:161–163.

Chapman, S. J. and J. M. Lynch (1985b). Polysaccharide synthesis by capsular micro-organisms in co-culture with cellulolytic fungi on straw and stabilization of aggregates. *Biol. Fert. Soils*, **1**:161–166.

Claydon, N., M. Allan, J. R. Hanson, and A. G. Avent (1987). Antifungal alkyl pyrones of *Trichoderma harzianum*. *Trans. Brit. Mycol. Soc.*, **88**:503–513.

Claydon, N., J. Dickinson, J. C. Hanson, and P. Hitchcock (1989). The structure and biosynthesis of harzianopyridone, an antifungal metabolite of *Trichoderma harzianum*. *J. Chem. Soc. Perkin Transactions*, **I**:(in press).

Conn, E. E. (1980). Cyanogenic compounds. *Ann. Rev. Pl. Physiol.*, **31**:433–451.

Cook, R. J. (1988). Management of the environment for the control of pathogens. *Phil. Trans. R. Soc. Lond.*, **B318**:171–182.

Crowley, D. E., C. P. P. Reid, and P. J. Szaniszlo (1987). Microbial siderophores as iron sources for plants. In *Iron Transport in Microbes, Plants and Animals* (G. Winkelmann, R. van der Helm, and J. Neilands, eds.). VCH Verlagsgesellschaft mbH, Weinheim, pp. 375–386.

Dartnall, A. M. and R. G. Burns (1987). A sensitive method for measuring cyanide and cyanogenic glucosides in sand culture and soil. *Biol. Fert. Soils*, **5**:141–147.

Davison, J. (1988). Plant beneficial bacteria. *Bio/Technology*, **6**:282–286.

Dazzo, F. B. (1980). Microbial adhesion to plant surfaces. In *Microbial Adhesion to Surfaces* (R. C. W. Berkeley, J. M. Lynch, J. Melling, P. R. Rutter, and B. Vincent, eds.). Ellis Horwood, Chichester, pp. 311–328.

Dazzo, F. B. and A. E. Gardiol (1984). Host specificity in *Rhizobium*-legume interactions. In *Genes Involved in Microbe–Plant Interactions* (D. P. S. Verma and T. Hohn, eds.). Springer-Verlag, New York, pp. 3–31.

Dazzo, F. B., G. L. Truchet, J. E. Sherwood, E. M. Hrabak, M. Abe, and S. H. Pankratz (1984). Specific phases of root hair attachment in the *Rhizobium trifolii*-clover symbiosis. *Appl. Env. Microbiol.*, **48**:1140–1150.

Dazzo, F. B., J. W. Kijne, K. Haahtela, and T. K. Korhoren (1986). Fimbrial lectins, and agglutinins of nitrogen fixing bacteria. In *Microbial Lectins and Agglutinins: Properties and Biological Activity* (D. Mirelman, ed.). John Wiley, New York, pp. 237–254.

Dennis, C. and J. Webster (1971). Antagonistic properties of species groups of *Trichoderma*. II. Production of volatile antibiotics. *Trans. Brit. Mycol. Soc.*, **57**:363–369.

Dommergues, Y. and V. Jacq (1972). Microbiological transformations of sulphur in the rhizosphere and spermosphere. *Ann. Agron.*, **23**:201–215.

Drew, M. C. and J. M. Lynch (1980). Soil anaerobiosis, micro-organisms and root function. *Ann. Rev. Phytopathol.*, **18**:37–67.

Duguid, J. P. and D. C. Old (1980). Adhesive properties of Enterobacteriaceae. In *Bacterial Adherence, Receptors and Recognition, Ser. B, Vol. 6* (E. H. Beachy, ed.). Chapman and Hall, London, pp. 185–217.

Elad, Y., I. Chet, and Y. Henis (1982). Degradation of plant pathogenic fungi by *Trichoderma harzianum*. *Can. J. Microbiol.*, **28**:719–725.

Elad, Y., R. Barak, and I. Chet (1983). The possible role of lectins in mycoparasitism. *J. Bacteriol.*, **154**:1431–1435.

Elad, Y., R. Barak and I. Chet (1984). Parasitism of sclerotia of *Sclerotium rolfsii* by *Trichoderma harzianum*. *Soil Biol. Biochem.*, **16**:381–386.

Emerlich, C. (1984). Molecular biology and ecology of diazotrophs associated with non-leguminous plants. *Biotechnology*, **2**:967–978.

Focht, D. D. and W. Verstraete (1977). Biochemical ecology of nitrification and denitrification. *Adv. Microb. Ecol.*, **1**:135–214.

Fravel, D. R. (1988). Role of antibiosis in the biocontrol of plant diseases. *Ann. Rev. Phytopathol.*, **26**:75–91.

Friend, J. (1981). Plant phenolics, lignification and plant disease. *Recent Adv. Phytochem.*, **7**:197–261.

Fry, W. E. and D. F. Myers (1981). Hydrogen cyanide metabolism by fungal pathogens of cyanogenic plants. In *Cyanide in Biology* (B. Vennesland, E. E. Conn, C. J. Knowles, J. Westley, and F. Wissing, eds.). Academic Press, London, pp. 321–324.

Gambrell, R. P. and W. H. Patrick (1978). Chemical and microbiological properties of anaerobic soils and sediments. In *Plant Life and Anaerobic Environments* (D. D. Hook and R. M. M. Crawford, eds.). Ann Arbor, Michigan, pp. 375–423.

Garcia, J. L., M. Raimbault, V. Jacq, C. Rinando, and P. Roger (1974). Activities microbiennes dans les sols de rizieres due Senegal: Relation avec les characteristiques physio-chemiques et influence de la rhizosphere. *Rev. Ecol. et Biol. Sol*, **11**:169–185.

Geels, F. P., J. G. Lamers, O. Hoekstra, and B. Schippers (1986). Potato plant responses to seed tuber bacterization in the field in various rotations. *Netherlands J. Plant Pathol.*, **92**:257–272.

Gill, P. R. and G. J. Warren (1988). An iron-antagonized fungistatic agent that is not required for iron assimilation from a fluorescent rhizosphere pseudomonad. *J. Bacteriol.*, **170**:163–170.

Giller, K. E. and J. M. Day (1985). Nitrogen fixation in the rhizosphere: significance in natural and agricultural systems. In *Ecological Interactions in Soil: Plants, Microbes and Animals* (A. H. Fitter, ed.). Blackwell Scientific Publications, Oxford, pp. 127–147.

Greenland, D. J., G. R. Lindstrom, and J. P. Quirk (1962). Organic materials which stabilise natural soil aggregates. *Soil Sci. Soc. Am. Proc.*, **26**:366–371.

Grossbard, E. (1952). Antibiotic production by fungi on organic manures and in soil. *J. Gen. Microbiol.*, **6**:295–310.

Gussin, E. J. and J. M. Lynch (1982). Effect of local concentrations of acetic acid around barley root on seedling growth. *New Phytol.*, **92**: 345–348.

Hadar, Y., G. E. Harman, A. G. Taylor, and J. M. Norton (1983). Effects of pregermination of peat and cucumber seeds and of seed treatment with *Enterobacter cloacae* on roots caused by *Pythium* sp. *Phytopathology*, **71**:569–572.

Hadas, R. and Y. Okon (1987). Effect of *Azospirillum brasilense* inoculation on root morphology and respiration in tomato seedlings. *Biol. Fert. Soils*, **5**:214–247.

Harris, R. E., A. W. Bunch, and C. J. Knowles (1987). Microbial cyanide and nitrite metabolism. *Sci. Prog. Oxf.*, **71**:243–304.

Hartley, R. D. and D. C. Whitehead (1985). Phenolic acids in soils and their influence on plant growth and soil microbial processes. In *Soil Organic Matter and Biological Activity* (D. Vaughan and R. E. Malcolm, eds.). Martinus Nijhoff/Dr W. Junk, Dordrecht, pp. 109–149.

Hinch, J. M. and A. E. Clarke (1980). Adhesion of fungal zoospores to root surfaces is mediated by carbohydrate determinants of root slime. *Physiol. Plant Pathol.*, **16**:303–307.

Hoekema, A., B. S. de Pater, A. J. Fellinger, P. J. J. Hooykaas, and R. A. Schilperoort (1984). The limited host range of an *Agrobacterium tumefaciens* strain extended by a wide host range T-region. *EMBO J.*, **3**:3043–3047.

Hollis, J. P., A. I. Allam, G. Pitts, M. M. Joshi, and I. K. A. Irrahim (1975). Sulfide diseases of rice on iron-excess soils. *Acta Phytopathol. Acad. Sci. Hung.*, **10**:329–341.

Howell, C. R. (1982). Effect of *Gliocladium virens* on *Pythium ultimum, Rhizoctonia solani* and damping-off of cotton seedlings. *Phytopathology*, **72**:496–498.

Howell, C. R. and R. D. Stipanovic (1979). Control of *Rhizoctonia solani* on cotton seedlings by *Pseudomonas fluorescens* and an antibiotic produced by the bacterium. *Phytopathology*, **69**:480–482.

Howell, C. R. and R. D. Stipanovic (1980). Suppression of *Pythium ultimum*-induced damping-off of cotton seedlings by *Pseudomonas fluorescens* and its antibiotic pyoluteorin. *Phytopathology*, **70**:712–715.

Howell, C. R. and R. D. Stipanovic (1983). Gliovirin, a new antibiotic from *Gliocladium virens*, and its role in the biological control of *Pythium ultimum*. *Can. J. Microbiol.*, **29**:321–324.

Jones, D. A., M. H. Ryder, B. G. Clare, S. K. Farrand, and A. Kerr (1988). Construction of a Tra^- deletion mutant of pAgK84 to safeguard the biological control of crown gall. *Mol. Gen. Genet.*, **212**:207–214.

Jones, R. W., R. E. Pettit, and R. A. Taber (1984). Lignite and stillage: carrier and substrate for application of fungal biocontrol agents to soil. *Phytopathology,* **74**:1167–1170.

Keel, C., C. Voisard, C. H. Berling, G. Kahr, and G. Defago (1989). Iron sufficiency, a prerequisite for suppression of tobacco black root rot by *Pseudomonas fluorescens* strain CHAO under gnotobiotic conditions. *Phytopathology*, **79**:584–589.

Kerr, A. and M. E. Tate (1984). Agrocins and the biological control of crown gall. *Microbiol. Sci.*, **1**:1–4.

Kleeberger, A., H. Castorph, and W. Klinmüller (1983). The rhizosphere microflora of wheat and barley with special reference to Gram-negative bacteria. *Arch. Microbiol.*, **136**:306–311.

Kloepper, J. W. and M. N. Schroth (1981). Relationship of *in vitro* antibiosis of PGPR to plant growth and displacement of root microflora. *Phytopathology*, **71**:1020–1024.

Kloepper, J. W., J. Leong, M. Teintze, and M. N. Schroth (1980). Enhanced plant growth by siderophores produced by plant growth-promoting rhizobacteria. *Nature*, **286**:885–886.

Knowles, C. J. and A. W. Bunch (1986). Microbial cyanide metabolism. *Adv. Microb. Physiol.*, **27**:73–111.

Kurosawa, E. (1926). Experimental studies on the secretion of *Fusarium heterosporum* on rice plants. *J. Nat. Hist. Soc. Formosa*, **16**:213–227.

Lewis, K., J. M. Whipps, and R. C. Cooke (1989). Mechanisms of biological disease control with special reference to the use of *Pythium oligandrum* as an antagonist. In *Biotechnology of Fungi for Improving Plant Growth* (J. M. Whipps and R. D. Lumsden, eds.) Cambridge University Press, Cambridge (in press).

Lippincott, B. B. and J. A. Lippincott (1969). Bacterial attachment to a specific wound site as an essential stage in tumour initiation by *Agrobacterium tumefaciens. J. Bacteriol.*, **97**:620–628.

Loper, J. E. (1988). Role of fluorescent siderophore production in biological control of *Pythium ultimum* by a *Pseudomonas fluorescens* strain. *Phytopathology*, **78**:166–172.

Lynch, J. M. (1974). Mode of ethylene formation by *Mucor hiemalis. J. Gen. Microbiol.*, **83**:407–411.

Lynch, J. M. (1976). Products of soil micro-organisms in relation to plant growth. *CRC Crit. Revs. Microbiol.*, **5**:67–107.

Lynch, J. M. (1980). Effects of organic acids on the germination of seeds and growth of seedlings. *Plant Cell & Environ.*, **3**:255–259.

Lynch, J. M. (1985). Origin, nature and biological activity of aliphatic substances and growth hormones found in soil. In *Soil Organic Matter and Biological Activity* (D. Vaughan and R. E. Malcolm, eds.). Martinus Nijhoff/Dr W. Junk, Dordrecht, pp. 151–174.

Lynch, J. M. (1987a). Biological control within microbial communities of the rhizosphere. In *Ecology of Microbial Communities* (M. Fletcher, T. R. G. Gray, and J. G. Jones, eds.). Cambridge University Press, Cambridge, pp. 55–82.

Lynch, J. M. (1987b). *In vitro* identification of *Trichoderma harzianum* as a potential antagonist of plant pathogens. *Curr. Microbiol.*, **16**:49–53.

Lynch, J. M. and E. Bragg (1985). Micro-organisms and soil aggregate stability. *Adv. Soil Sci.*, **2**:133–171.

Lynch, J. M. and K. B. Gunn (1978). The use of the chemostat to study the decomposition of wheat straw in soil slurries. *J. Soil Sci.*, **29**:551–556.

Lynch, J. M. and White, N. (1977). Effect of some non-pathogenic microorganisms on the growth of gnotobiotic barley plants. *Pl. Soil*, **47**:161–174.

Lynch, J. M., K. B. Gunn, and L. M. Panting (1980). On the concentration of acetic acid in straw and soil. *Pl. Soil*, **56**:93–98.

Manning, K. (1986). Ethylene production and β-cyanoalanine synthase activity in carnation flowers. *Planta*, **168**:61–66.

Marschner, H., V. Römheld, W. J. Horst, and P. Martin (1986). Root-induced changes in the rhizosphere: importance for the mineral nutrition of plants. *Z. Pflanzenernaehr. Bodenk.*, **149**:441–456.

Martin, J. K. and R. C. Foster (1985). A model system for studying the biochemistry and biology of the root–soil interface. *Soil Biol. Biochem.*, **17**:261–269.

Marugg, J. D., M. van Spanje, W. P. M. Hoekstra, B. Schippers, and P. J. Weisbeek (1985). Isolation and analysis of genes involved in siderophore biosynthesis in the plant growth-stimulating *Pseudomonas putida* strain WCS 358. *J. Bacteriol.*, **164**:563–570.

Mehta, N. C., H. Streuli, M. Muller, and H. Deuel (1960). Role of polysaccharides in soil aggregation. *J. Sci. Food Agr.*, **11**:40–47.

Mirelman, D., E. Galun, N. Sharon, and R. Lotan (1975). Inhibition of fungal growth by wheat germ agglutinin. *Nature*, **256**:414–416.

Mishaghi, I. J., M. W. Olsen, P. J. Cotty, and C. R. Donndelinger (1988). Fluorescent siderophore-mediated iron deprivation—a contingent biological control mechanism. *Soil Biol. Biochem.*, **20**:573–574.

Neilands, J. B. (1984). Siderophores of bacteria and fungi. *Microbiol. Sci.*, **1**:9–14.

Neill, S. J., R. Horgan, D. C. Walton, and T. S. Lee (1982). The biosynthesis of abscisic acid in *Cercospora rosicola. Phytochemistry*, **21**:61–65.

Nelson, E. B., W.-L. Chao, J. M. Norton, G. T. Nash, and G. E. Harman (1986). Attachment of *Enterobacter cloacae* to *Pythium ultimum* hyphae: possible role in the biological control of *Pythium* pre-emergence damping-off. *Phytopathology*, **76**:327–335.

Nordbring-Hertz, B. and I. Chet (1986). Fungal lectins and agglutinins. In *Microbial Lectins and Agglutinins: Properties and Biological Activity* (D. Mirelman, ed.). John Wiley, New York, pp. 393–408.

Patriquin, D.-G. and J. Dobereiner (1978). Light microscopy observations of tetrazolium-reducing bacteria in the rhizosphere of maize and other grasses. *Can. J. Microbiol.*, **24**:734–742.

Primrose, S. B. (1979). Ethylene and agriculture: the role of the microbe. *J. Appl. Bacteriol.*, **46**:1–25.

Rice, E. L. (1984). *Allelopathy*, 2nd edn. Academic Press, London.

Rice, E. L. and S. K. Pancholy (1974). Inhibition of nitrification by climax ecosystems. III. Inhibitors other than tannins. *Amer. J. Bot.*, **61**:1095–1103.

Ridout, C. J., J. R. Coley-Smith, and J. M. Lynch (1986). Enzyme activity and electrophoretic profile of extracellular protein induced in *Trichoderma* spp. by cell walls of *Rhizoctonia solani. J. Gen. Microbiol.*, **132**:1181–1187.

Ridout, C. J., J. R. Coley-Smith, and J. M. Lynch (1988). Fractionation of extracellular enzymes from a mycoparasitic strain of *Trichoderma harzianum. Enzyme Microbiol. Technol.*, **10**:180–187.

Römheld, V. and H. Marschner (1986). Evidence for a specific uptake system for iron phytosiderophores in roots of grasses. *Plant Physiol.*, **80**:175–180.

Scher, F. M. and R. Baker (1982). Effect of *Pseudomonas putida* and a synthetic iron chelator on induction of suppressiveness to *Fusarium* wilt pathogens. *Phytopathology*, **72**:1567–1573.

Schippers, B., B. Lutenberg, and P. J. Weisbeek (1987). Plant growth control by fluorescent pseudomonads. In *Innovative Approaches to Plant Disease Control* (I. Chet, ed.). John Wiley & Sons, New York, pp. 19–39.

Scott, P. M. (1974). Patulin. In *Mycotoxins* (I. F. H. Purchase, ed.). Elsevier, Amsterdam, pp. 383–403.

Simon, A., R. W. Dunlop, E. L. Ghisalberti, and K. Sivasithamparam (1988). *Trichoderma koningii* produces a pyrone compound with antibiotic properties. *Soil Biol. Biochem.*, **20**:263–264.

Suslow, T. V. (1982). Role of root-colonizing bacteria in plant growth. In *Phytopathogenic Prokaryotes, Vol. 1*. (M. S. Mount and G. H. Lacy, eds.). Academic Press, New York, pp. 187–223.

Swinburne, T. R. (ed.) (1986). *Iron, Siderophores and Plant Diseases*. Plenum Press, New York.

Teintze, M. and J. Leong (1981). Structure of pseudobactin A, a second siderophore from plant growth promoting *Pseudomonas* B10. *Biochemistry*, **20**:6446–6457.

Teintze, M., M. B. Hossain, M. B. Barnes, J. Leong, and D. Van der Helm (1981). Structure of ferric pseudobactin, a siderophore from a plant growth-promoting *Pseudomonas. Biochemistry*, **20**:6457–6462.

Thomashow, L. S. and D. M. Weller (1988). Role of a phenazine antibiotic from *Pseudomonas fluorescens* in biological control of *Gaeumannomyces graminis* var. *tritici. J. Bacteriol.*, **170**:3499–3508.

Timonin, M. I. (1941). The interaction of higher plants and soil microorganisms: III. Effect of by-products of plant growth on activity of fungi and actinomycetes. *Soil Sci.*, **2**:395–415.

Turner, W. B. (1971). *Fungal Metabolites*. Academic Press, London.

Turner, W. B. and D. C. Aldridge (1983). *Fungal Metabolites* II. Academic Press, London.

Vaughan, D. and R. E. Malcolm (1985). Influence of humic substances on growth and physiological processes. In *Soil Organic Matter and Biological Activity* (D. Vaughan and R. E. Malcolm, eds.). Martinus Nijhoff/Dr W. Junk, Dordrecht, pp. 37–75.

Voisard, C., C. Keel, D. Haas, and G. Defago (1989). Cyanide production by *Pseudomonas fluorescens* helps suppress black root rot of tobacco under gnotobiotic conditions. *EMBO J.*, **8**:351–358.

Wainwright, M. (1988). Selman A. Waksman and the streptomycin controversy. *SGM Quarterly*, **15**:90–92.

Waller, G. R. (ed.) (1987). *Allelochemicals: Role in Agriculture and Forestry*. American Chemical Society, Washington, D.C.

Weger de, L. A., R. van Boxtel, B. van der Burg, R. Gruters, F. P. Geels, B. Schippers, and B. Lutenberg (1986). Siderophores and outer membrane proteins of antagonistic, plant growth-stimulating root-colonizing *Pseudomonas* spp. *J. Bacteriol.*, **165**:585–594.

Weller, D. M. (1988). Biological control of soilborne plant pathogens in the rhizosphere with bacteria. *Ann. Rev. Phytopathol.*, **76**:379–407.

Weller, D. M. and R. J. Cook (1986). Suppression of root diseases of wheat by fluorescent pseudomonads and mechanisms of action. In *Iron, Siderophores and Plant Diseases* (T. R. Swinburne, ed.). Plenum Press, New York, pp. 99–107.

Whatley, M. H., J. B. Margot, J. Schell, B. B. Lippincott, and J. A. Lippincott (1978). Plasmid and chromosomal determination of *Agrobacterium* adherence specificity. *J. Gen. Microbiol.*, **107**:395–398.

Windham, M. T., Y. Elad, and R. Baker (1986). A mechanism for increased plant growth induced by *Trichoderma* spp. *Phytopathology*, **76**:518–521.

Xu, T., R. van Peer, H. Rattink, and B. Schippers (1987). The potential use of fluorescent pseudomonads in the protection of carnations against *Fusarium* wilt in hydroponics. *Act. Hort.*, **216**:93–100.

8

MATHEMATICAL MODELS OF INFECTION

C. A. GILLIGAN*

Department of Applied Biology, University of Cambridge, Pembroke Street, Cambridge CB2 3DX, UK

1. INTRODUCTION

Mathematical models underlie much of the expanding theory of epidemics of soil-borne disease, of which infection is an integral component. The density, location, viability and efficiency of inoculum to cause infection are, therefore, essential considerations in the construction of realistic models for these diseases. These considerations lead to hypotheses concerning the proportion of inoculum that is effective, the furthest distance a propagule can be from a host and still cause infection, and the effects of age and environmental factors, such as soil water potential, on the efficiency of inoculum. More recently, hypotheses for the effects of clustering of inoculum and the deployment of microbial antagonists on the probability of infection (Gilligan, 1985a) have also been considered. Other problems concern the identification of thresholds of infections for the expression of disease and the relationships between levels of infestation and root growth.

Numerous reviews have been published on the epidemiological analysis and modelling of soil-borne plant pathogens. These cover an extensive range and are classified elsewhere (Gilligan, 1990). They range from descriptive reviews of the biological bases of disease dynamics (Baker, 1978; Bloomberg, 1985; Gilligan, 1987) to largely mathematical reviews (Gilligan, 1983, 1985c; Campbell, 1986; Campbell et al., 1988). Many of these relate to the mathematical models

*Present address: Botany School, University of Cambridge, Downing Street, Cambridge CB2 3EA, UK

of infection. Some are restricted to spatial aspects (Campbell and Noe, 1985; Gilligan, 1988) or to temporal aspects (Gilligan, 1985c). Others specifically consider nematodes (Ferris, 1981, 1985; Freckman and Caswell, 1985) or the related topics of mycorrhizal fungi (Harley and Smith, 1983) and host growth (Huisman, 1982). These are not duplicated here, although some overlap is inevitable if a balanced picture is to be presented. Instead a synthesis is provided of recent and formative work on the processes and dynamics of infection of roots from inoculum in soil. Attention is focused initially, therefore, on the construction, testing and fitting of models for the infection process. Recent work on the dynamics of infection is then summarized and discussed.

A glossary of the principal random variables and parameters referred to in the text is given in the Appendix.

2. MODELLING THE INFECTION PROCESS: ESTIMATION OF PATHOZONE DIMENSIONS

The infection process is restricted here to primary infection (Gilligan, 1983) in which infection is initiated from inoculum in soil. This involves a sequence of events from survival of inoculum in soil to germination and growth or movement of propagules through soil to contact the host, and to the initiation of infection. Three objectives can be identified in modelling the infection process (Gilligan, 1985c):

(1) Estimation of the probability that an infection court, such as a root, will be infected, given certain densities of inoculum and hosts.
(2) Estimation of the mean or maximum distance from the surface of a host from which infection can be initiated.
(3) Construction of a mathematical summary of the primary infection component of the infection chain.

Most emphasis has been given to the estimation of rhizosphere or, more generally, pathozone dimensions (Gilligan, 1979, 1985c; Ferriss, 1981, 1982, 1983; Wilkinson et al., 1985a; Walker and Smith, 1988) and latterly to the infection efficiency of inoculum (Gilligan, 1985c; Reynolds et al., 1985; Gilligan and Simons, 1987).

The pathozone is the volume of soil surrounding subterranean plant organs within which the centre of a propagule must occur if it is to have any chance of infecting the organ (Gilligan, 1985a). Propagules outside this volume either do not respond to the presence of the host or, if they do, have insufficient reserves of nutrients to reach and infect the host (Gilligan and Simons, 1987). The pathozone is a more general term than rhizosphere or spermosphere and can be applied to any subterranean organ, such as a root, seed, hypocotyl, epicotyl or mesocotyl. Exceptionally, it can also be applied to composite organs, when, for example, a seed and an epicotyl comprise an infection court. The pathozone may also be regarded as surrounding propagules of inoculum, such as sclerotia. Strictly, the term cannot be applied to mycorrhizal infection because of the connotation of pathogenesis. An alternative term that enjoys the generality of the pathozone but which relates to infection, might be "molunozone"—literally "region of infection".

The efficiency of inoculum for infection has been variously defined, depending upon the pathogen under consideration (Gilligan, 1987). Essentially, however, efficiency concerns the ratio of input of inoculum to output of infection. It encompasses, therefore, the proportion of inoculum that is capable of causing infection and the amount of infection that results. Thus, Tomimatsu and Griffin (1982) effectively defined efficiency of propagules of *Cylindrocladium crotalariae* for infection of roots of peanut as the ratio of the number of discrete infections to the number of propagules within the pathozone. Implicit in this was an assumption that all propagules that were close enough to the host were capable of causing infection. The relationship between pathozone dimensions and the efficiency of inoculum for infection is discussed in Section 3.

Three approaches to the estimation of pathozone dimensions have been devised. The original method was based on theories of solid packing and surface density (Baker et al., 1967; Baker, 1971; Baker and Drury, 1981) as developed, for example, in physical chemistry (Gyani, 1945). The method permitted only qualitative distinction between rhizoplane and rhizosphere effects. For a rhizoplane effect, contact between inoculum and the host was adduced to be essential for infection to occur, whereas for a rhizosphere effect, location of viable inoculum within the rhizosphere (or pathozone) was sufficient. The surface density models have now been largely superseded by models that permit quantitative estimates of the pathozone dimensions. Of the two alternatives, currently the most widely applied method is that based on elementary probability theory for the occurrence of zero (unoccupied) cells (Gilligan, 1979, 1985a). These models have the advantage of requiring scores only for the presence or absence of infection on host targets. By contrast, the second alternative, which is derived from the theory for nearest neighbour distances (between two biological entities) (Wilkinson et al., 1985b), requires data for the number of infections on roots. Most discussion is given below to the so-called probability models.

2.1. Probability Models

Suppose that P propagules of inoculum occur in a fixed volume of soil, V. The probability that any one propagule occurs within the pathozone and infects the host is given by $\phi = v/V$ where v is the volume of the pathozone. For any one infection court, say a root, the probability that the root escapes infection is $1-\phi$. If there are N roots within the volume, V, the total number of roots that are uninfected (N_u) is given by:

$$N_u = N(1-\phi)^P \tag{1}$$

If v is small, i.e. $v \ll V$ and P large, the model can be rephrased using the first term of the Poisson, rather than the binomial, distribution to estimate the probability that a root escapes infection. Thus,

$$N_u = N \exp(-\phi P), \tag{2}$$

in which ϕP is the mean number of propagules per pathozone. Rearrangement of equations (1) and (2) allows solution for ϕ,

$$\phi = 1 - (N_u/N)^{1/P} \tag{3}$$

and

$$\phi = v/V = \ln(N/N_u)/P \tag{4}$$

Of more interest in many practical studies is the estimation of v, and in particular the width of the pathozone, under different experimental conditions. Thus Gilligan (1979) showed that, for data from Rouse and Baker (1978) on infection of seeds of *Raphanus sativus* by *Rhizoctonia solani*, amendment of soil with chitin effectively led to a reduction in pathozone width around the seed from 0.74 ± 0.153 to 0.23 ± 0.091. Rouse and Baker (1978), using surface density models, were unable to demonstrate any significant effect, but see Ferriss (1981) for criticisms of the calculations made by Gilligan (1979). More recently, Reynolds et al. (1985) defined a quantitative relationship between pathozone width and soil matric potential for infection of radicles of Fraser fir by *Phytophthora cinnamomi*. They also showed that the estimated width of the pathozone was independent of inoculum density.

Probability models have been further refined (Gilligan, 1985a) to allow for non-random occurrence of inoculum in soil, by substituting the expession for the zero cell of the negative binomial distribution in the equation for the proportion of roots that escape infection. The models have also been extended to allow for thresholds for the expression of disease and for changing probabilities of occurrence of infection with distance of inoculum across the pathozone. The shape of the pathozone (Ferriss, 1981, 1982) has been separately considered.

2.1.1. Shape of the Pathozone

The model originally proposed by Gilligan (1979) included the host volume in the pathozone on the assumption that the soil displaced by the growing root would occur within the pathozone. Thus, for a radially symmetrical pathozone surrounding a root, the radius of the pathozone is the sum of the radii of the root (r_r) and inoculum (r_i) plus a distance w across which the pathogen can grow or move to cause infection. For most host–symbiont systems the experimental dimension of interest is $w + r_i$. This has sometimes been redefined as w (Ferriss, 1981, 1982) when r is small compared to w. For exactness, we prefer to define the pathozone as $z = w + r_i$, with the outer radius of the pathozone $R = r_h + w + r_i$, where r_h is the radius of the host (Gilligan and Simons, 1987). The volume of the pathozone uniformly susceptible along its length L is therefore,

$$v = \pi L(r_r + z)^2$$

and

$$z = [\,v/(\pi L)\,]^{1/2} - r_r \tag{5}$$

Leonard (1980) and Ferriss (1981) challenged this model and Ferriss (1981) adjusted the model from a displacement to a shell form (Gilligan, 1985a), in which

$$v = \pi L\,[\,(r_r + z)^2 - r_r^2\,]$$

giving

$$z = [\,v/(\pi L) + r_r^2\,]^{1/2} - r_r \tag{6}$$

Selection of model (5) or (6) depended upon *a priori* hypotheses concerning the extent of changes in bulk density with distance from the root. If the furthest extent of change in bulk density is less than the outer limit of the pathozone, then the displacement model applies. If not, the shell model is more appropriate. Recent work by Walker and Smith (1988), based on a model by Dexter (1987), has provided compelling arguments in favour of the shell model.

Dexter (1987) proposed an empirical model for the change in porosity, $\eta(r)$, with radial distance (r) from the root surface:

$$\eta(r) = \eta_r + (\eta_i - \eta_r)\{1 - \exp[\,-(r - r_r)/(qr_r)\,]\} \tag{7}$$

in which η_r is the porosity at the root surface and η_i is the initial porosity; q is a parameter that is assumed to be independent of root radius. The final pore volume v_f out to some large but finite distance, $r = mr_r$, is therefore (Dexter, 1987):

$$v_f = 2\pi \int_{r_r}^{mr_r} r\eta(r)\mathrm{d}r, \tag{8}$$

and

$$\eta_r = \eta_i - [\,(1 - \eta_i)/2qA\,] \tag{9}$$

where,

$$A = q\,[\,1 - (X + 1)\exp(-X)\,] + [\,1 - \exp(-X)\,], \tag{10}$$

and

$$X = (m - 1)/q$$

As soil is compressed the local density of propagules can be modelled by recasting equations (7–9) in terms of the dry bulk density (ϱ). The pore size volume (η) is related to dry bulk density by

$$\eta = 1 - \varrho/\varrho_p, \tag{11}$$

in which ϱ_p is the particle density of soil, equivalent to 2.65 g cm^{-3} for most soils. Values for ϱ range from <1 g cm^{-3} for soils high in organic matter, to 1.0–1.4 g cm^{-3} for well-aggregated loamy soils, and to 1.2–2.0 g cm^{-3} for sands and compacted horizons in clay soils (White, 1979). Substitution for ϱ into equation (7) yields:

$$\varrho(r) = \varrho_i + (\varrho_r - \varrho_i)\exp[(r - r_r)/(qr_r)] \tag{12}$$

and from equations (9) and (11),

$$\varrho_r = \varrho_i(1 + 1/2qA) \tag{13}$$

The effective volume of the pathozone (v), after compression, is obtained by integration of equation (8) (Dexter, 1987; Walker and Smith, 1988), after substitution for bulk densities, within the limits ($r = r_r$) and ($r = r_r + z$) yielding:

$$v = \pi L\{(2r_r z + z^2) + 2r_r^2(\varrho_r/\varrho_i - 1)qA'\} \tag{14}$$

in which A' is given by equation (10), but with

$$X = z/qr_r$$

The first term in equation (14) is equivalent to the volume of the pathozone, for the shell model, in the absence of compression. The second term is the increase in volume due to compression of soil by the root.

The models in equations (7–14) assume that $\eta_r < \eta_{\min}$, where $\eta_{\min}$ is the minimum porosity for a given soil. Dexter (1987) argued that when the expected value of equation (7) was less than $\eta_{\min}$, the soil would be compressed to the minimum porosity out to some transition radius (r_l) and that beyond this, the porosity decreases according to equation (7), with $\eta_{\min}$ replacing η_r. The transition radius can be evaluated iteratively as:

$$r_l = r_r\left[\frac{1 - \eta_{\min}}{(\eta_i - \eta_{\min})(1 + 2qA'')}\right]^{1/2} \tag{15}$$

in which A'' is given by equation (10) in which,

$$X = (mr_r - r_l)/qr_l$$

The corresponding expressions for the effective volumes of the pathozone are:

$$v = \pi L(2r_r z + z^2) \qquad (r_r + z) \leq r_l \qquad (16a)$$

$$v = \pi L\{(r_l^2 - r_r^2) + [(z + r_r)^2 - r_l^2] + 2r_l^2(\varrho_{\min}/\varrho_i - 1)qA'''\} \qquad (r_r + z) > r_l, \qquad (16b)$$

in which A''' is given by the general form of equation (10) but with

$$X''' = (z - r_l)/qr_l$$

Details for the calculation of $\eta_{\min}$ are given by Dexter (1987). Note that Dexter derives the models for the numbers of propagules within the pathozone, assuming inoculum densities are expressed as propagules g^{-1}.

Walker and Smith (1988) used Dexter's model to compare the relative performance of the shell and displacement models for given values of pathozone width (z). Some examples, presented by Walker and Smith (1988), are shown in Fig. 1 for selected values of q. It may be seen that the shell model routinely did better than the displacement model in estimation of values for z: predictably the shell model led to overestimation and the displacement to underestimation of the true values. The bias of the models is a function of the value of z relative to the radius of the root. We have found little difference between estimated values for either model for pathozone dimensions for the infection of roots of wheat by *Gaeumannomyces graminis* (Gilligan and Simons, unpublished).

The effects of changing values of q on the relationship between bulk density and scaled distance are shown in Fig. 2. The value for q is critical in its effect on $\varrho(r)$. Partial differentiation of equation (12) with respect to the parameters reveals the sensitivity of the model to small changes in q, since

$$\partial\varrho(r)/\partial q = (x - q)e_i/2A \exp(-x/q)q^3,$$

$$x = r/r_r - 1$$

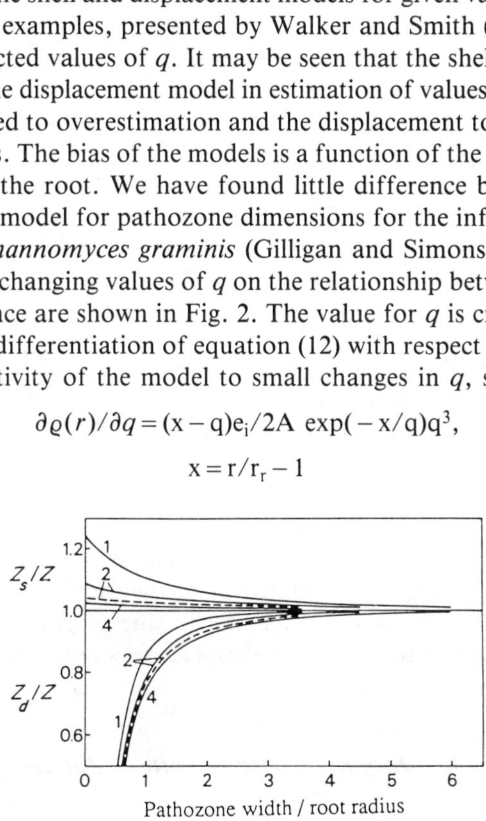

Fig. 1. Effects of soil compression on estimates of the pathozone widths, z_s and z_d for the "shell" and "displacement" models, respectively. The performance of the models is shown as ratios of calculated to given values (z) for selected values of q. Pathozone width is scaled relative to root radius. The solid lines refer to soils that do not reach the minimum compressed porosity. The dashed lines refer to a soil in which the minimum porosity was reached. Data reprinted with permission from Walker and Smith (1988). For other parameter values see the original paper.

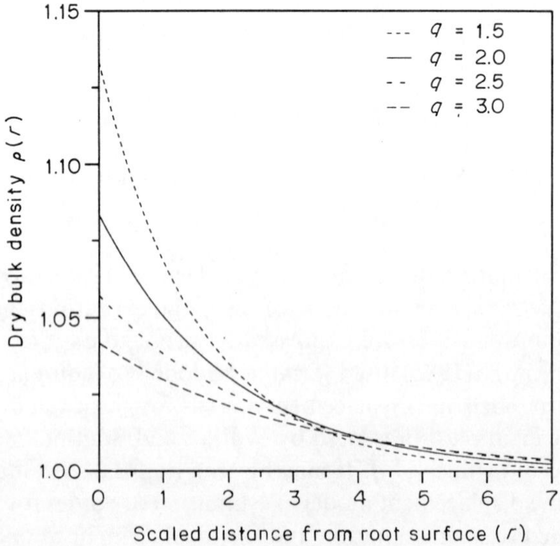

Fig. 2. Effects of changing values for the parameter q on the relationship between dry bulk density and scaled distance from the root surface (actual distance/root radius), due to soil compression as roots grow through soil. The values for ϱ_i and m were 1.0 and 100, respectively.

noting that ϱ_r is also a function of q. Dexter (1987) cites two examples (Graecen et al., 1968; Dexter and Tanner, 1972) which suggest that the value of q lies in the approximate range of 1.5 to 3.0, while a value for q of 2.0 is recommended in the absence of further information (Dexter, 1987). Scope for variation of initial bulk density within soils is comparatively limited: increasing the value for ϱ_i principally alters the vertical displacement of the relationship. The parameter m is relatively unimportant, unless the radius of a pot is less than approximately eight times the radius of the root, when the bulk density rises sharply towards the root surface. Such values, however, might obtain as root density increases within a pot. The displacement model would then be more accurate than the shell model.

2.1.2. Allowance for Varying Success with Distance

The assumption that the probability of infection is constant for all propagules within the pathozone was implicit in the derivation of the models above. There is considerable evidence, however, for a decrease in the probability of germination with distance from the root surface (see, for example, Griffin, 1969; Short and Lacy, 1974). More importantly this has also been shown for the probability of infection (Henis and Ben-Yephet, 1970; Punja and Grogan, 1981;

Punja, 1986; Gilligan and Simons, 1987). Some allowance for varying success in infection with distance across the pathozone is therefore necessary. A more realistic expression for the probability of a root escaping infection is given by:

$$P[\text{root is uninfected}] = (1 - \theta\psi) \tag{17}$$

in which θ is the probability that a propagule occurs within the pathozone and ψ is the conditional probability that the propagule is capable of infecting the root given that it occurs within the pathozone. The parameter θ is numerically equivalent to ϕ and is estimated by the ratio v/V. Unlike ϕ, however, it is used only to represent the probability of occurrence within the pathozone and not the probability of infection given occurrence.

Assuming a shell model, and ignoring changes in bulk density due to compression, the probability of infection of a root exposed to inoculum is given by:

$$\psi = \int_{r_r}^{R} (\psi_1|\theta)(\psi_2)dr, \tag{18}$$

where $(\psi_1|\theta)$ is the density function for the probability that the centre of a propagule occurs between say r and $r + \Delta r$ from the surface of the root; ψ_2 is the density function for the probability that a propagule can infect the host conditional upon its occurrence between r and $r + \Delta r$. Taking an empirical power law model for ψ_2,

$$\psi_2 = \beta\exp(-\alpha r^2) \tag{19}$$

the probability of infection has an analytical solution (Gilligan, 1985a):

$$\psi = \frac{\beta}{\alpha(R^2 - r_r^2)}[\exp(-\alpha r_r^2) - \exp(-\alpha R^2)]. \tag{20}$$

Recalling that $\phi = \theta\psi$, then for a root of length L,

$$\phi = \frac{\pi L\beta}{\alpha V}[\exp(-\alpha r_r^2) - \exp(-\alpha R^2)] \tag{21}$$

and since $R = r_r + z$, the extent of the pathozone is therefore calculated by:

$$z = \left\{\frac{1}{\alpha}\ln\left[\frac{\pi L\beta}{\pi L\beta\exp(-\alpha r_r^2) - \alpha V\phi}\right]\right\}^{1/2} - r_r \tag{22}$$

The parameter ϕ is estimated from experimental data using equations (3) or (4). The rationale for incorporation of parameters for varying success in infection is analogous to the models incorporating changes in bulk density. The two models could be combined, though at the expense of accumulating yet more parameters in the final model.

2.1.3. Estimation of Parameters

Little formal consideration has been given in the literature to the estimation of parameters for the probability models. Although the parameter of usual interest is the extent of the pathozone, z, it is more appropriate to estimate some simpler function of z, such as ϕ. Early work (Gilligan, 1979, 1980; Ferriss, 1981, 1982) involved the estimation of the mean number of infections or "hits" per root. This was obtained from the proportion of uninfected roots after exposure to inoculum for a fixed period, by use of the multiple infection transformation (Gregory, 1948). This is equivalent to an implicit calculation of ϕ from equation (4). Hence for the shell model and substituting for inoculum density per unit volume, $I = P/V$,

$$z = \left[\ln \left(\frac{N}{N_u} \right) \frac{1}{\pi LNI} + r_r^2 \right]^{1/2} - r_r \tag{23}$$

Hence for a single inoculum density, z can be calculated using the logarithm of the ratio of total to uninfected roots as an input variable. Smith et al. (1986), working with mycorrhizal systems, were able to count individual infections on roots and used these instead of the transformation to estimate z.

Variability in the estimation of the pathozone width is introduced by the inclusion of the mean root length and the inoculum density in the models. Inoculum density is usually estimated by most probable number method (Cochran, 1950) and root length by direct measurement or by one of the grid intersect methods (after Newman, 1966). Variation in root length can be reduced by restriction of inoculum to a layer, or removed completely by restriction to a plane of inoculum. The probability of infection is then defined in terms of relative areas rather than volumes (Gilligan, 1980). Baker and Drury (1981) and Drury et al. (1983) criticized the use of estimates of inoculum density in the estimation of z. The use of estimates of I, derived from recovery on selective media, clearly introduces bias into the estimate, thereby underestimating z. The problem is minimized by the use of bioassays in which I is estimated by the presence of infection and/or disease on bait plants (see, for example, Nicot and Rouse, 1987). The value for I used in equation (23), therefore, takes implicit account of the average ability of propagules to infect. The model has alternatively been amended to include a parameter for the "competence" (Grogan et al., 1980) of propagules to cause infection, given occurrence within the pathozone (Ferriss, 1981). Independent estimation of the parameter for competence also creates problems, requiring estimation of the ratio of the densities of infective to viable propagules. Additional sources of bias that result in underestimation of the pathozone dimension include the use of disease rather than infection as a criterion for successful infection, and the assumption that the root is uniformly susceptible over the entire surface. Overestimation of z can occur if there is time for secondary infection to develop (Smith et al., 1986). Failure to take account of this results in inflated estimates of the number of primary infections arising from the soil inoculum.

When, as is frequently the case, data are available for a range of inoculum densities, z can be estimated separately for each inoculum density or by optimization over the range of densities. Ferriss (1981) recast equation (23), for infection of a spherical seed of radius r_s, in the form:

$$z = \left(\frac{kS}{I} + r_s^3 \right)^{1/3} - r_s \tag{24}$$

in which $k = 3/4\pi$ and $S = \phi P$ is the estimated mean number of infections per seed. Several linear and non-linear models were proposed for the relationship between S and I (Ferriss, 1982). Two representative examples are given below:

$$S = \beta_1 I + \beta_2 I^2 \tag{25a}$$

$$S = \beta_3 [1 - \exp(-\beta_4 I)] \tag{25b}$$

After estimation of the parameters of the models (25a) and (25b), the ratio of S/I as I tends to zero was estimated by β_1 for the linear model and $\beta_3 \cdot \beta_4$ for the non-linear model. Calculated values for pathozone dimensions for infection of seeds, seed coats and colonization of dead seeds of soyabean by sporangia of *Pythium ultimum* are shown in Table 1 (after Ferriss, 1982). Inspection of the data shows good agreement between the methods. The calculated dimensions are of the order of 8, 130 and 60 times the average radius of sporangia of *P. ultimum*. For further discussion of the comparative susceptibility of the infection courts see Ferriss (1982).

An alternative, perhaps simpler, and more robust method for estimation of z is to estimate the parameter ϕ by maximum likelihood (Gilligan and Simons, unpublished). This method has the additional advantage of taking explicit account of the error structure of the data. The errors for the numbers of uninfected roots for a comparatively small number of roots exposed to inoculum are likely to be binomially rather than normally distributed. The effects of non-normality on parameter estimates become more important as the number of roots decreases and as the number of trials (roots) varies amongst treatments. The effects of plant density on the estimation of ϕ and z for infection and disease of seminal roots of wheat by inoculum of *G. graminis* are given in Table 2 (Gilligan and Simons, unpublished). The pathozone was estimated for the shell model from a range of inoculum densities restricted to a plane. The estimates were not independent of host density, suggesting some synergistic mechanism promoting more infection at the higher densities.

Additional complexity, as in the allowance for varying success with distance, can be allowed for in the estimation of parameters. However, although this satisfies the desire for biological reality, it is done at the expense of statistical efficiency in fitting the models. Increasing the numbers of parameters within non-linear models creates problems in finding unique solutions, while correlations amongst parameters increase. It is preferable, therefore, to obtain

TABLE 1. Calculated widths (z mm) of pathozone for infection of seeds, seed coats and colonization of dead seeds of soyabean by sporangia of *Pythium ultimum* using a linear and a non-linear model to relate the mean number of infections/colonizations per host unit to inoculum density. After Ferriss (1982)

Variable	z *(Linear model)*	z *(Non-linear model)*
Seed infection	0.084	0.114
Coat infection	1.342	1.365
Dead seed colonization	0.605	0.732

TABLE 2. Estimates of the probability of infection and disease (ϕ) and of the pathozone width (z mm) for seminal roots of wheat exposed to inoculum of *Gaeumannomyces graminis* comprising colonized grains of millet[a]

Host density (seedlings/pot)	d.f.	*Residual deviance*	$\phi \pm$ *S.E.*	z $(\psi = 1)$	z $(\psi \neq 1)$
5	4	26.24	0.019 ± 0.0014	3.96	4.21
10	4	36.17	0.035 ± 0.0019	5.47	5.83
15	4	113.65	0.043 ± 0.0016	6.13	6.55
Combined	14	280.14	0.035 ± 0.0009	5.52	5.89

[a]The probability of infection and disease (ϕ) was estimated by maximum likelihood, from equation (1) under an assumption of binomial errors. The pathozone width was calculated from the estimated values of ϕ, with ($\psi \neq 1$) and without ($\psi = 1$) allowance for varying success with distance from the root surface, using constants derived from independent experiments (Gilligan and Simons, 1987).

independent estimates of the parameters for efficiency, by separate experiment (Gilligan and Simons, 1987) and to use these as constants in the calculation of z. An example is shown for *G. graminis* in Table 2. In this case, allowance for varying success of infection with distance had little effect on the estimates of z.

2.2. Nearest Neighbour Models

Wilkinson et al. (1985b) estimated the mean value of z for the infection of seminal roots of wheat by artificial propagules of *G. graminis*, using a model derived from distances between nearest neighbours (Thompson, 1956). They assumed that the distribution of propagules around a root followed a Poisson process, whence for a density of I propagules per unit area, the expectation of the mean distance (r_n) to the nth nearest neighbour is:

$$\mathrm{E}(r_n) = \frac{(2n)!n}{I^{1/2}(2^n n!)^2} \tag{26}$$

Rearranging the equation and defining $E(r_n)$ as the mean distance for infection (EDI, *sensu* Wilkinson et al., 1985b) gives,

$$\mathrm{EDI} = \frac{(2n-1)!}{I^{1/2}2^{2n-1}[(n-1)!]^2} \tag{27}$$

The equation is used by assuming that the number of propagules, n, within the pathozone is equal to the number of lesions per plant, i.e. that propagules are capable of producing only one lesion. Whence if there are five lesions on roots exposed to inoculum, of horizontal density 0.75 cm^{-2}, the mean distance for infection is 1.42 cm. Certain additional assumptions are necessary in implementing the model. Thus Wilkinson et al. (1985b) treated the total number of lesions per plant as if they occurred on one root. In addition, they assumed that the three-dimensional experimental systems could be reduced to two dimensions. This was justified by the assumption that, for inoculum in narrow containers containing only one plant with an average of 4.2 roots per plant, each propagule was likely to interact with only one root and to do so in a plane perpendicular to the axis of the root (Wilkinson et al., 1985b).

The model was applied to data for infection of seminal roots of wheat by *G. graminis*, for propagules that were small enough to ensure an average of only one lesion per propagule. These yielded values ranging from 1.7 to 6.0 mm (Wilkinson et al., 1985b). The method was extended for propagules that produced more than one lesion. This involved substitution for n by the number of recorded lesions divided by the mean number of lesions produced per propagule. Presumably the latter was independently estimated by the placement of inoculum adjacent to roots, although the authors do not specify this. The variable, n, was no longer an integer and equation (27) was adjusted according to an assumption that the squared distance to the nth propagule from the root has a gamma distribution:

$$\mathrm{EDI} = \frac{\Gamma(2n)}{I^{1/2}2^{2n-1}\Gamma(n)^2} \tag{28}$$

Application of equation (28) to data for propagules approximately 1.0–2.0 mm in diameter yielded estimates of EDI between 10.8 and 11.6 mm for *G. graminis* (Wilkinson et al., 1985b). Fumigation of soil increased the estimates of EDI. This was most notable in a suppressive soil containing propagules between 0.1 and 1.0 mm in diameter, when the estimated distances increased from 2.8 to 7.9 mm. Suppressive soil *per se*, however, did not affect the estimated distances when compared to conducive soil (Wilkinson et al., 1985b).

Wilkinson et al. (1985b) note that the simple probability method used by Gilligan (1980) gave similar estimates but these were consistently slightly larger than their values for EDI. They argued that distances for z are based on squared distances to the rth inoculum propagule, whereas the EDI is determined as the expected linear distance to the rth propagule. More theoretical and experimental work is needed on the comparative approaches.

2.3. Surface Density Models

Baker et al. (1967) first proposed models based upon surface density considerations that distinguished between rhizoplane and rhizosphere infection. For "rhizoplane infection", direct contact between propagules of inoculum and root surfaces was essential, whereas "rhizosphere infection" occurred from propagules located at some distance from the root surface. The summary of the approach given here follows that of Gilligan (1983). For further details see the original papers of Baker et al. (1967), Baker (1971) and the vigorous and stimulating defences of the methods by Baker and Drury (1981) and Drury et al. (1983). Baker et al. (1967) originally proposed four models to deal with combinations of stationary and moving infection courts and motile and non-motile inoculum. These were later reduced to two (Baker, 1971) because of the artificiality of the distinction between motile and non-motile inoculum, in the context of distinguishing rhizoplane from rhizosphere infection. Thus inoculum may reach the infection court either by motility of propagules, such as zoospores, or by germination and growth of mycelium. Although most infection courts move through soil, they may be expected to grow past propagules of inoculum before infection occurs. The infection courts can be considered as stationary relative to the inoculum, thereby reducing the models to one.

The models take the form:

$$S = aI^b \tag{29}$$

in which S is the number of infections per root, I is the inoculum density and a and b are constants. For a stationary infection court the expected value of b is 1.0, if inoculum can infect from a distance, and 2/3, if contact between inoculum and root is essential for infection. The expectation of 2/3 reflects the action of a surface-density rather than a volume-density phenomenon. It is derived by considering the mean distances between propagules (d). Thus $I = k_1 d^{-3}$ and $S = k_2 d^{-2}$, which on dividing and rearranging gives equation (29), with $a = k_2/(k_1)^{2/3}$. Logarithmic transformation of equation (29) gives:

$$\ln S = a' + b \ln I \tag{30}$$

In practice, S is usually computed from the proportion of uninfected roots, using the multiple infection transformation. Thus $S = -\ln(N_u/N)$, yielding a complementary log-log transformation for the dependent variable in equation (30).

The model in equation (30) was widely used to infer distinction between rhizosphere and rhizoplane infection, by estimation of the parameter b from data for inoculum density experiments. Numerous statistical criticisms have been advanced against the surface-density models. These concern: the misuse of ordinary least-squares regression to estimate the parameters, without taking account of non-normal errors and unequal variances; the use of the multiple infection transformation to estimate S, when infection and disease may be non-random; and selection of portions of inoculum density–disease curves. The latter is a particularly important source of error, since the model only applies to the ascending portion of a typical inoculum density–disease response curve. Inclusion of points on the asymptote leads to underestimation of the value of b. The outstanding biologically-based criticism of the method concerns the existence of a rhizoplane effect as envisaged by Baker et al. (1967) and defended by Baker and Drury (1981). The term rhizoplane was defined by Clark (1949) as the external surfaces of plant roots together with any adhering particles of soil and debris. This definition suggests that the rhizoplane is not truly planar. Moreover, it has been noted previously (Gilligan, 1983) that Griffin (1969) equated germination of *Fusarium oxysporum* within 0.5 mm of the surface of peanut roots as a rhizoplane effect. Studies of the ultrastructure of the soil–root interface (Campbell and Rovira, 1973; Foster, 1981) indicate that the rhizoplane is dynamic, highly sculptured and intimately associated with the soil. Notwithstanding these criticisms, it must be noted that values for b of 0.67 have been frequently recorded. However, the dichotomous nature of the model in allowing only a qualitative distinction between ability or otherwise to infect from a distance is rather restrictive. Probability and nearest neighbour models offer greater flexibility. Baker and Drury (1981) and Drury et al. (1983) contend, however, that the probability models are rendered useless for practical purposes without concomitant estimates of the efficiency of inoculum for infection.

3. EFFICIENCY OF INOCULUM

Consideration of the efficiency of inoculum to cause infection is fundamental to models for the infection process. The simplest means of including efficiency is to incorporate a single parameter for the proportion of propagules within the pathozone that are capable of causing infection (Gilligan, 1979; Ferriss 1981; Reynolds et al., 1985). The parameter can be estimated from experiments in which inoculum is placed in contact with roots (Henis and Ben-Yephet, 1970; Gilligan, 1980; Gilligan and Simons, 1987) and the proportion of trials, say ψ, that result in successful infection, counted. Then the equation for the number of roots that escape infection (cf. equations 2 and 17) becomes:

$$N_u = N \exp(-\theta\psi P) \tag{31}$$

This was effectively done by Reynolds et al. (1985), although they used a formulation involving the estimate of numbers of infections. Routine calculation of the efficiency from independent experiments would greatly reduce the underestimation of z or the EDI. The estimates are still subject to bias, however, because of the failure to take account of the decline in probability of successful infection with increasing distance of the propagule from the infection court. Examples for the profiles of infection efficiency of propagules of *G. graminis*

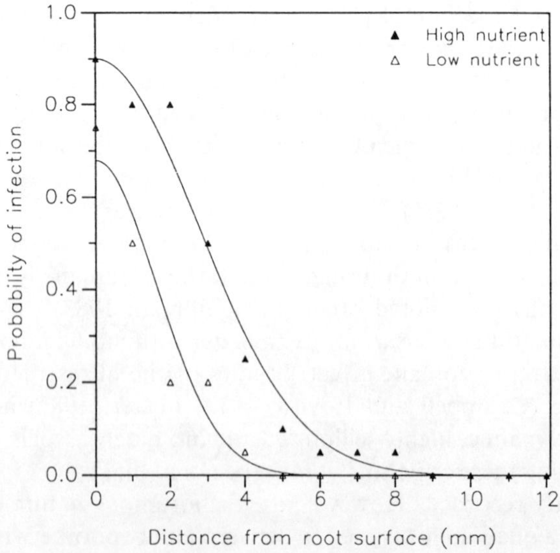

Fig. 3. Effects of high and low nutrient content in propagules of *Gaeumannomyces graminis* on the probability of infection (Y) of seminal roots of wheat with distance of placement (r) of the propagules from the root surface. Fitted curves were given by the model, $Y=\beta \exp(-\alpha r^2)$, after Gilligan and Simons (1987). High nutrient corresponded with colonized millet and low nutrient with colonized sand as inoculum.

are illustrated in Fig. 3, for two different sources of inoculum. The full model (equations 17–22), in which this decline is taken into account, has three parameters: z, the extent of the pathozone, measured from the root surface; and α and β. The latter derive from the function for ψ_2 (equation 19) that defines the change in inoculum efficiency across the pathozone (Gilligan and Simons, 1987). Hence α is a measure of the decline in infection efficiency with distance from the root, and β is an estimate of the asymptotic infection efficiency achieved when propagules are close to or contacting the root. The particular power of r in the equation used to describe the profile in infection efficiency was selected to permit analytical solution of the combined equation (Gilligan, 1985a). Because of the variability of experimental data, there is considerable flexibility in selecting non-linear models to describe the profile (Gilligan and Simons, 1987). The results of fits by maximum likelihood for the millet inoculum in Fig. 3 to three models are shown in Table 3. All three models gave similar estimates for the asymptotic parameter but the rate parameters differed considerably. One problem in fitting non-linear models to the profile of infection efficiency is the absence of a vertical asymptote to correspond with z, which must, therefore, be estimated by inspection in placement experiments.

Radial symmetry has been assumed for the pathozone surrounding roots as well as uniform susceptibility along the root axis. Reynolds et al. (1985) tested the latter assumption for infection of 2-week-old radicles of Fraser fir by *P. cinnamomi*. They concluded that the radicles were uniformly susceptible. In

TABLE 3. Summary of non-linear models used to describe the relationship between the probability of successful infection of seminal roots of wheat and distance (r mm) of placement of high nutrient propagules of *Gaeumannomyces graminis*

Model	*Parameters*[a] *(±S.E.)*	*Residual deviance (11 d.f.)*
$Y=\beta\exp(-\alpha r)$	α 0.34±0.035 β 0.95±0.035	20.33
$Y=\beta\exp(-\alpha r^2)$	α 0.07±0.010 β 0.90±0.050	4.56
$Y=\beta\exp[1-\exp(\alpha r)]$	α 0.20±0.015 β 0.94±0.040	8.21

[a]Parameter estimation was by maximum likelihood under an assumption of binomially distributed errors.

practice, it is probable that for more mature tissue, the emergence of lateral roots from main root axes will alter the susceptibility to infection. Some compromise is necessary. Although radial symmetry and uniform susceptibility have been assumed to apply also to seeds, for convenience in estimation of parameters (Gilligan and Simons, 1987), Baker and Drury (1981) caution against this for geometrically complex infection courts. Recent work by Hunter (1988) illustrates some of the complexity in relation to efficiency of inoculum for germination. The response surface for the average efficiency of germination of macroconidia of *F. culmorum* around germinating seeds of barley is shown in Fig. 4. The data were obtained by placement of seeds on soil-slides immersed in pots of soil and incubated at 10 °C. The slides were removed and germination of conidia that had been treated with a fluorescent stain was assessed in a grid of contiguous quadrats beneath the seed (Hunter, 1988). Maximum germination occurred at the embryo end of the seeds. Independent experiments using soil columns (Short and Lacy, 1974) showed that the furthest extent of germination around the seed surface varied from 0.7 to 2 mm (Hunter, 1988).

In considering the profile of infection efficiency across the pathozone, the author used a quantal variable for presence or absence of infection, rather than the number of infections as a measure of efficiency (Gilligan and Simons, 1987). Cook et al. (1986) defined the efficiency of inoculum of *G. graminis* for infection as the number of lesions per unit weight of colonized particles of inoculum. They also defined a threshold particle size as the minimum particle size required for maximum efficiency. Clearly quantitative variables such as these may be related to distance from the infection court by linear or non-linear models.

4. MODELLING THE DYNAMICS OF INOCULUM AND INFECTION

4.1. Standard Growth Curves

Standard growth curves have been widely used to describe the progress of epidemics of plant disease over time—see, for example, the reviews in Leonard

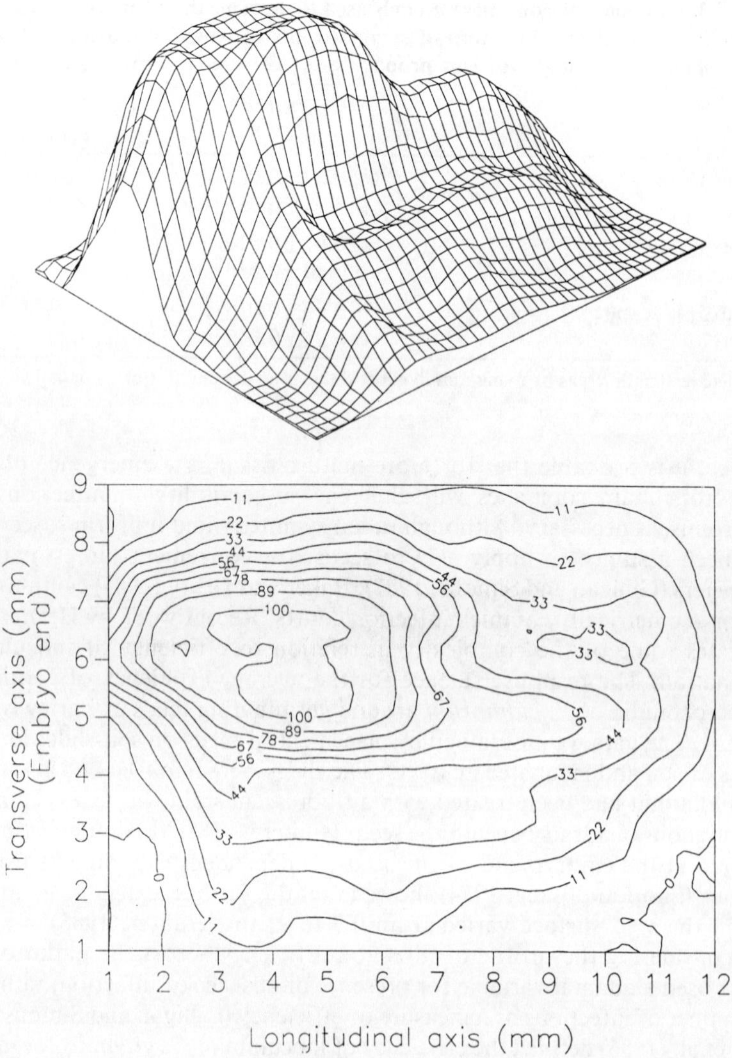

Fig. 4. Surface and contour plots to illustrate the mean percentage germination of macroconidia of *Fusarium culmorum* with respect to the longitudinal and transverse axes of germinating seedlings of barley (Hunter and Gilligan, unpublished). Distances are relative: the mean maximum seed dimensions extended from 1 to 11.5 mm on the abscissa and 2.75 to 7.25 mm on the ordinate.

and Fry (1986) and Gilligan (1985b). Certain aspects of the following discussion are expanded in Gilligan (1985c, 1990). One of the earliest applications of growth curve analysis to subterranean pathogens was that of Last et al. (1969), who used a multiple regression equation to describe the build-up of *Pyrenochaeta* brown rot on roots of tomatoes. Most attention has been given, however, to non-linear models. These have the advantage of having parameters to which biological interpretations can be attached. The general form of the standard non-linear models for growth processes may be expressed, after Turner et al. (1976) as:

$$\frac{dY}{dt} = rf(0,Y) \, . \, g(Y, \varkappa) \tag{32}$$

in which r is a rate constant, Y is the amount of disease (or infection) at time t and $\varkappa$ is the asymptotic level of disease. It may be seen that the function, f, tends to speed up epidemics whereas the function, g, has a decelerating effect. The common standard non-linear models are summarized in Table 4. Initially most attention was given to the monomolecular and logistic models, following the pioneering work of Van der Plank (1963). In the monomolecular model, in which $f(0,Y)$ is set to a constant, disease was considered to arise exclusively from primary infection from a reservoir of inoculum in soil. The model was thought to apply, therefore, to monocyclic pathogens. Polycyclic organisms were typified by the logistic model in which the epidemic is driven by overlapping cycles of secondary infection. This could be approximated during the early stages of an epidemic by the exponential function (Table 4). The cumulative curve for the monomolecular model has a rising exponential form. The logistic model is sigmoidal and symmetrical. The point of inflexion, when the epidemic switches from acceleration to deceleration, is located half-way between the upper and lower asymptotes. Dissatisfaction with the restrictive symmetry of the logistic model led to consideration of the Gompertz model, in which the point of inflexion is reached earlier (Berger, 1981). Search for still more flexibility, particularly

TABLE 4. Principal, standard non-linear models used to describe the progress of infection or disease in plant disease epidemics[a]

Model	*Rate equation*	*Integrated form*
Exponential	$\frac{dY}{dt} = rY$	$Y = \exp\,[r(t-\delta)]$
Mono-molecular	$\frac{dY}{dt} = r(\varkappa - Y)$	$Y = \varkappa\{1-\exp[-r(t-\delta)]\}\quad t \geq \delta$
Logistic	$\frac{dY}{dt} = rY\frac{(\varkappa - Y)}{\varkappa}$	$Y = \frac{\varkappa}{1+\exp[-r(t-\delta)]}$
Gompertz	$\frac{dY}{dt} = rY(\ln\varkappa - \ln Y)$	$Y = \varkappa\exp\{-\exp[-r(t-\delta)]\}$
Richards	$\frac{dY}{dt} = rY\frac{(\varkappa^n - Y^n)}{\varkappa^n}$	$Y = \frac{\varkappa}{\{1+\exp[-r_n(t-\delta)]\}^{1/n}}$
Weibull	$\frac{dY}{dt} = rn(\varkappa - y)\{\ln[\varkappa/(\varkappa - y)]\}^{1-1/n}$	$Y = \varkappa\{1-\exp[-(r(t-\delta))^n]\}\quad t \geq \delta$

[a]The random variable, Y, is used to denote amount of infection or disease. The parameters comprise rates (r), upper asymptotic levels ($\varkappa$), shapes (n) and locational parameters (δ).

with aerial pathogens, led to the exploration and use of the Weibull and Richards models, which each have an extra shape parameter (Table 4). These include the other models as exact or approximate special cases (see Waggoner, 1986).

Increasing the numbers of parameters in the models promotes flexibility. There are a number of disadvantages, however: the appeal of intuitive biological interpretation of the parameters may be compromised with the inclusion of shape or scale parameters; and the cost of collecting sufficient data points to allow comparison of treatments, with respect to individual parameters, may be too great for all but exceptional use. The selection of standard epidemiological models for disease progress is discussed by Madden (1980, 1986), Rouse (1985) and Waggoner (1986). The selection and fitting of models specifically for soil-borne pathogens and the interpretation of parameters is discussed by Campbell (1986), Gilligan (1985c) and Huisman (1982).

4.2. Incorporation of Inoculum and Host Density

Despite the flexibility of the standard non-linear growth models, they impose considerable restrictions that deviate from what is known about most systems involving root infection. Principal amongst these restrictions are the assumptions of constant host and inoculum density. Because of the assumption of constant host density, the models are frequently applied to data for whole plants, rather than roots. The growth of roots and the spread of infection within plants is therefore ignored. When the root is taken as the unit of population, changes in root density can be modelled over time and the diseased root population modelled relative to this. This can easily be done by selecting linked differential equations of the form

$$\frac{dN}{dt} = r_{rt} f_1(0,N).g_1(N,\varkappa) \tag{33a}$$

$$\frac{dN_i}{dt} = r_{irt} f_2(0,N_i).g_2(N_i,N) \tag{33b}$$

in which N is the total root density per unit area, N_i is the density of infected roots, r_{rt} and r_{irt} are the corresponding rates of increase of total and infected root populations, while $\varkappa$ is now the asymptotic root density. The models in equations (33a) and (33b) can be solved numerically. An analytical solution is given, however, by Turner et al. (1969) and Waggoner (1986), for the case in which both the total and infected root population densities increase logistically. Waggoner (1986) provides several analyses, in relation to aerial disease, of the relative importance of the initial and final sizes of host population and the rate of host to pathogen increase. The double logistic model, however, essentially relates to epidemics in which primary infection from inoculum in soil is of negligible importance relative to spread of infection from infected to healthy roots.

By expanding the rate parameter of the monomolecular model, Jeger (1987) allowed for an increase in mean root length and a declining reservoir of inoculum. The model applied to a system in which all infections were derived from soil inoculum. The components of the rate parameter were, a rate constant, k, a measure of inoculum, P, and a measure of root density, v. Then, taking Y as the proportion of infected plants, and an asymptote of 1.0, the model is given by:

$$\frac{dY}{dt} = kPv(1 - Y) \tag{34}$$

In practice the product Pv is a measure of the net effect of declining inoculum and growth of roots. Jeger (1987) obtained analytical solutions for cases in which the product took the form of a linear, exponential, logistic or monomolecular curve with time. Of particular importance was the demonstration of the variation in shape of the resulting curves. Hence, monocyclic systems may give rise to sigmoid curves. Jeger (1987) also suggested that the classical monomolecular model is unlikely to apply to subterranean systems, unless the rate of root growth is qualitatively balanced by the rate of loss of inoculum. Sigmoidal models have also been developed for the estimation of egg density by the monocyclic nematode *Globodera rostochiensis* (Jones and Kempton, 1978).

Two further developments in modelling of the dynamics of infection and disease concern the incorporation of components for primary and secondary infection in a single model and allowance for the growth of infections and lesions. Brassett and Gilligan (1988b) compounded rates of epidemic increase from monomolecular and logistic models to give:

$$\frac{dN_i}{dt} = (r_pP + r_sN_i)(N - N_i) \tag{35}$$

in which N and N_i are as defined previously, P is the density of inoculum, and r_p and r_s are rate parameters for primary and secondary infection, respectively. The model can be solved analytically for constant N and P. Thus, taking the transformation, $Y = N_i/N$, and reparameterizing for coefficients of primary and secondary infection, $\alpha = r_pP$ and $\beta = r_sN$ respectively, and introducing a parameter $\varkappa$, for an asymptotic proportion of infection yields:

$$\frac{dY}{dt} = (\alpha + \beta Y)(\varkappa - Y) \tag{36}$$

with solution:

$$Y = \frac{\alpha\varkappa\{1 - \exp[-(\alpha + \beta\varkappa)t]\}}{\alpha + \beta\varkappa \exp[-(\alpha + \beta\varkappa)t]} \tag{37}$$

A discrete version of the model, based upon probability models for an initial cycle of primary followed by successive cycles of secondary infection is given by Brassett and Gilligan (1988a). Gilligan (1990) has expanded the model, using numerical methods, to incorporate decay of inoculum and increase in density of the root population.

Growth of lesions or infections is an important aspect of infection. Smith and Walker (1981) and Walker and Smith (1984) have also used linked differential equations to model the increase in numbers of infections and the subsequent spread of infections of vesicular-arbuscular mycorrhizas, due to *Glomus mosseae*, on roots of *Trifolium subterraneum*. Two variables are modelled, the proportion of root length infected (l) and the frequency of infection units per unit length of root (u):

$$\frac{du}{dt} = A(1 - l) - gu \tag{38a}$$

$$\frac{dl}{dt} = Bu(1 - l) - gl \tag{38b}$$

in which g is the relative growth rate of roots, A is a measure of the frequency of infections per unit length per unit time and B is an estimate of the rate of spread of mycelium from a single infection point. Amongst numerous applications of the models to experimental data, Walker and Smith (1984) showed that, in the early stages of root infection, the mean length of an infection unit was independent of the density of infections. They also showed that the asymptotic level of infection would always be less than the total length of root available for infection. A similar conclusion was proposed by Jeger (1987) who extended his monomolecular model (equation 34), using concepts of pathozone volumes, to incorporate lesion growth.

As the complexity of models for infection increases, solution of models becomes sometimes more difficult, although numerical methods can always be used. Of more importance, however, are the problems created in the statistical validation of models by the proliferation of parameters. Much urgent and critical thought needs now to be given to these conflicting constraints of matching biological complexity and statistical fitting of models to data.

ACKNOWLEDGMENTS

I am grateful to my colleagues, S. A. Simons and I. G. Hunter, for permission to quote some of their results.

REFERENCES

Baker, R. (1971). Analyses involving inoculum density of soil-borne plant pathogens in epidemiology. *Phytopathology*, **61**:1280–1292.

Baker, R. (1978). Inoculum potential. In *Plant Disease: An Advanced Treatise, Vol. II.* (J. G. Horsfall and E. B. Cowling, eds.). Academic Press, New York, London, pp. 137–157.

Baker, R. and R. Drury (1981). Inoculum potential and soilborne pathogens: the essence of every model is in the frame. *Phytopathology*, **71**:363–372.

Baker, R., C. L. Maurer, and R. A. Maurer (1967). Ecology of plant pathogens in soil. VIII. Mathematical models and inoculum density. *Phytopathology*, **57**:662–666.

Berger, R. D. (1981). Comparison of the Gompertz and logistic equations to describe plant disease progress. *Phytopathology*, **71**:716–719.

Bloomberg, W. J. (1985). The epidemiology of forest nursery diseases. *Annu. Rev. Phytopathol.*, **23**:83–96.

Brassett, P. R. and C. A. Gilligan (1988a). A discrete probability model for polycyclic infection by soil-borne plant parasites. *New Phytol.*, **109**:183–191.

Brassett, P. R. and C. A. Gilligan (1988b). A model for primary and secondary infection in botanical epidemics. *Z. Pflanzenkr. Pflanzenschutz*, **95**:352–360.

Campbell, C. L. (1986). Interpretation and uses of disease progress curves for root diseases. In *Plant Disease Epidemiology: Population Dynamics and Management, Vol. I.* (K. J. Leonard and W. E. Fry, eds.). Macmillan, New York, pp. 38–54.

Campbell, C. L. and J. P. Noe (1985). The spatial analysis of soilborne pathogens and root diseases. *Annu. Rev. Phytopathol.*, **23**:129–148.

Campbell, C. L., K. M. Reynolds, and L. V. Madden (1988). Modelling epidemics of root diseases and development of simulators. In *Techniques in Plant Disease Epidemiology* (J. Kranz and J. Rotem, eds.). Springer-Verlag, Berlin, Heidelberg, New York, pp. 253–265.

Campbell, R. and A. D. Rovira (1973). The study of the rhizosphere by scanning electron microscopy. *Soil Biol. Biochem.*, **5**:747–752.

Clark, F. E. (1949). Soil microorganisms and plant roots. *Adv. Agron.*, **1**:241–288.

Cochran, W. G. (1950). Estimation of bacterial densities by means of the "most probable number". *Biometrics*, **6**:105–116.

Cook, R. J., H. T. Wilkinson, and J. R. Alldredge (1986). Evidence that micro-organisms in suppressive soil associated with wheat take-all disease do not limit the number of lesions produced by *Gaeumannomyces graminis* var. *tritici*. *Phytopathology*, **76**:342–345.

Dexter, A. R. (1987). Compression of soil around roots. *Plant Soil*, **97**:401–406.

Dexter, A. R. and D. W. Tanner (1972). Soil deformations induced by a moving cutting blade, an expanding tube and a penetrating sphere. *J. Agric. Engng. Res.*, **17**:371–375.

Drury, R. E., R. Baker, and G. J. Griffin (1983). Calculating the dimensions of the rhizosphere. *Phytopathology*, **73**:1351–1354.

Ferris, H. (1981). Dynamic action thresholds for diseases induced by nematodes. *Annu. Rev. Phytopathol.*, **19**:427–436.

Ferris, H. (1985). Population assessment and management strategies for plant parasitic nematodes. *Agric. Ecosyst. and Env.*, **12**:285–299.

Ferriss, R. S. (1981). Calculating rhizosphere size. *Phytopathology*, **71**:1229–1231.

Ferriss, R. S. (1982). Relationship of infection and damping-off of soyabean to inoculum density of *Pythium ultimum*. *Phytopathology*, **72**:1397–1403.

Ferriss, R. S. (1983). Calculating the dimensions of the rhizosphere—a response. *Phytopathology*, **73**:1355–1357.

Foster, R. C. (1981). The ultrastructure and histochemistry of the rhizosphere. *New Phytol.*, **89**:263–273.

Freckman, D. W. and E. P. Caswell (1985). The ecology of nematodes in agroecosystems. *Annu. Rev. Phytopathol.*, **23**:275–296.

Gilligan, C. A. (1979). Modeling rhizosphere infection. *Phytopathology*, **69**:782–784.

Gilligan, C. A. (1980). Dynamics of root colonisation by the take-all fungus, *Gaeumannomyces graminis*. *Soil Biol. Biochem.*, **12**:507–512.

Gilligan, C. A. (1983). Modeling of soilborne pathogens. *Annu. Rev. Phytopathol.*, **21**:45–64.

Gilligan, C. A. (1985a). Probability models for host infection by soil-borne fungi. *Phytopathology*, **75**:61–67.

Gilligan, C. A. (ed.) (1985b). *Mathematical Modelling of Crop Disease (Advances in Plant Pathology, Vol. 3)*, Academic Press, London, 255 pp.

Gilligan, C. A. (1985c). Construction of temporal models. III. Disease progress of soil-borne pathogens. In *Mathematical Modelling of Crop Disease* (C. A. Gilligan, ed.) *(Advances in Plant Pathology, Vol. 3)*. Academic Press, London, pp. 37–105.

Gilligan, C. A. (1987). Epidemiology of soil-borne plant pathogens. In *Populations of Plant Pathogens: Their Dynamics and Genetics* (M. S. Wolfe and C. E. Caten, eds.). Blackwell Scientific Publications, Oxford, pp. 119–133.

Gilligan, C. A. (1988). Analysis of the spatial pattern of soilborne pathogens. In *Techniques in Plant Disease Epidemiology* (J. Kranz and J. Rotem, eds.). Springer-Verlag, Berlin, Heidelberg, New York, pp. 85–98.

Gilligan, C. A. (1990). Mathematical modelling of soilborne plant pathogens. In *Epidemics of Plant Disease*, 2nd edn. (J. Kranz, ed.). Springer-Verlag, Berlin, Heidelberg, New York (in press).

Gilligan, C. A. and S. A. Simons (1987). Inoculum efficiency and pathozone width for two host–parasite systems. *New Phytol.*, **107**:549–566.

Graecen, E. L., D. A. Farrell, and B. Cockcroft (1968). Soil resistance to metal probes and plant roots. *Trans. 9th Int. Cong. Soil Sci.*, **1**. Angus and Robertson, Sydney, pp. 769–779.

Gregory, P. H. (1948). The multiple infection transformation. *Ann. Appl. Biol.*, **35**:412–417.

Griffin, G. J. (1969). *Fusarium oxysporum* and *Aspergillus flavus* spore germination in the rhizosphere of peanut. *Phytopathology*, **59**:1214–1218.

Grogan, R. G., M. A. Sall, and Z. K. Punja (1980). Concepts for modelling root infection by soilborne fungi. *Phytopathology*, **70**:361–363.

Gyani, B. P. (1945). Distribution law, adsorption and chemical reaction. *J. Phys. Chem.*, **49**:442–453.

Harley, J. L. and S. E. Smith (1983). *Mycorrhizal Symbiosis.* Academic Press, London, New York, 483 pp.

Henis, Y. and Y. Ben-Yephet (1970). Effect of propagule size of *Rhizoctonia solani* on saprophytic growth, infectivity and virulence of bean seedlings. *Phytopathology*, **60**:1351–1356.

Huisman, O. J. (1982). Interrelations of root growth dynamics to epidemiology of root-invading fungi. *Annu. Rev. Phytopathol.*, **20**:303–327.

Hunter, I. G. (1988). Epidemiological studies of *Fusarium culmorum* on winter barley. Ph.D. Thesis, University of Cambridge.

Jeger, M. J. (1987). The influence of root growth and inoculum density on the dynamics of root disease epidemics: theoretical analysis. *New Phytol.*, **107**:459–478.

Jones, F. G. W. and R. A. Kempton (1978). Population dynamics, population models and integrated control. In *Plant Nematology* (J. F. Southey, ed.). MAFF AS Publication GD/1 H. M. Stationery Office, London, pp. 333–361.

Last, F. T., M. H. Ebben, R. C. Hoare, E. A. Turner, and A. R. Carter (1969). Build-up of tomato brown root rot caused by *Pyrenochaeta lycopersicae* Schneider and Gerlach. *Ann. Appl. Biol.*, **64**:449–459.

Leonard, K. J. (1980). A reinterpretation of the mathematical analysis of rhizoplane and rhizosphere effects. *Phytopathology*, **70**:695–696.

Leonard, K. J. and W. E. Fry (1986). *Plant Disease Epidemiology: Population Dynamics and Management, Vol. I.* Macmillan, New York, 372 pp.

Madden, L. V. (1980). Quantification of disease progression. *Prot. Ecol.*, **2**:159–176.

Madden, L. V. (1986). Statistical analysis and comparison of disease progress curves. In *Plant Disease Epidemiology: Population Dynamics and Management, Vol. I.* (K. J. Leonard and W. E. Fry, eds.). Macmillan, New York, pp. 55–84.

Newman, E. I. (1966). A method of estimating the total length of root in a sample. *J. Appl. Ecol.*, **3**:139–145.

Nicot, P. C. and D. I. Rouse (1987). Precision and bias of three quantitative soil assays for *Verticillium dahliae. Phytopathology*, **77**:875–881.

Punja, Z. K. (1986). Relationships among soil depth, soil texture, and inoculum placement in infection of carrot roots by eruptively germinating sclerotia of *Sclerotium rolfsii. Phytopathology*, **76**:976–980.

Punja, Z. K. and R. G. Grogan (1981). Mycelial growth and infection without a food base by eruptively germinating sclerotia of *Sclerotium rolfsii. Phytopathology*, **71**:1099–1103.

Reynolds, K. M., D. M. Benson, and R. I. Bruck (1985). Epidemiology of *Phytophthora* root rot of Fräser fir: rhizosphere width and inoculum efficiency. *Phytopathology*, **75**:1010–1014.

Rouse, D. I. (1985). Construction of temporal models: I. Disease progress of air-borne pathogens. In *Mathematical Modelling of Crop Disease* (C. A. Gilligan, ed.) *(Advances in Plant Pathology, Vol. 3)*. Academic Press, London, pp. 11–29.

Rouse, D. I. and R. Baker (1978). Modeling and quantitative analysis of biological control mechanisms. *Phytopathology*, **68**:1297–1302.

Short, G. E. and M. L. Lacy (1974). Germination of *Fusarium solani* f.sp. *pisi* chlamydospores in the spermosphere of pea. *Phytopathology*, **64**:558–562.

Smith, S. E. and N. A. Walker (1981). A quantitative study of mycorrhizal infection in *Trifolium:* separate determination of the rates of infection and of mycelial growth. *New Phytol.*, **89**:225–240.

Smith, S. E., N. A. Walker, and M. A. Tester (1986). The apparent width of the rhizosphere of *Trifolium subterraneum* for vesicular-arbuscular mycorrhizal infection: effects of time and other factors. *New Phytol.*, **104**:547–558.

Thompson, H. P. (1956). Distribution of distance to *N*th neighbor in a population of randomly distributed individuals. *Ecology*, **37**:391–394.

Tomimatsu, G. S. and G. J. Griffin (1982). Inoculum potential of *Cylindrocladium crotalariae:* infection rates and microsclerotial density–root relationships on peanut. *Phytopathology*, **72**:511–517.

Turner, M. E., B. A. Blumenstein, and J. L. Sebaugh (1969). A generalization of the logistic law of growth. *Biometrics*, **25**:577–580.

Turner, M. E., E. L. Bradley, and K. A. Kirk (1976). A theory of growth. *Math. Biosci.*, **29**:367–373.

Van der Plank, J. E. (1963). *Plant Diseases: Epidemics and Control*. Academic Press, New York, London, 349 pp.

Waggoner, P. E. (1986). Progress curves of foliar diseases: their interpretation and use. In *Plant Disease Epidemiology: Population Dynamics and Management, Vol. I.* (K. J. Leonard and W. E. Fry, eds.). Macmillan, New York, pp. 3–37.

Walker, N. A. and S. E. Smith (1984). The quantitative study of mycorrhizal infection. II. The relation of rate of infection and speed of fungal growth to propagule density, the mean length of the infection unit and the limiting value of the fraction of the root infected. *New Phytol.*, **96**:55–69.

Walker, N. A. and S. E. Smith (1988). Effect of soil compression on estimates of rhizosphere width: comparing Ferriss's equation with Gilligan's. *Phytopathology*, **78**:253–255.

White, R. E. (1979). *Introduction to the Principles and Practice of Soil Science*. Blackwell Scientific Publications, Oxford, 198 pp.

Wilkinson, H. T., R. J. Cook, and J. R. Alldredge (1985a). Relation of inoculum size and concentration to infection of wheat by *Gaeumannomyces graminis* var. *tritici. Phytopathology*, **75**:98–103.

Wilkinson, H. T., J. R. Alldredge, and R. J. Cook (1985b). Estimated distances for infection of wheat roots by *Gaeumannomyces graminis* var. *tritici* in soils suppressive and conducive to take-all. *Phytopathology*, **75**:557–559.

6. APPENDIX: GLOSSARY OF SYMBOLS USED TO REPRESENT PRINCIPAL RANDOM VARIABLES AND PARAMETERS IN TEXT

A 1. Rate parameter for frequency of infections per unit length per unit time.
2. Intermediate variable used in computation of soil porosities.

a Arbitrary constant used in model of the form $y = ax^b$

α 1. Rate of decrease in infection efficiency with distance from a host surface.
2. Coefficient of primary infection.

B	Rate of spread of mycelium within a root from a single infection point.
b	Arbitrary constant used in model of the form $y = ax^b$
β	1. Asymptotic infection efficiency.
	2. Coefficient of secondary infection.
β_i	Regression parameters.
δ	Locational parameter in non-linear model.
η_i	Initial porosity of soil prior to penetration by root.
$\eta_{\min}$	Minimum porosity of a soil.
η_r	Porosity of soil at root surface.
$\eta(r)$	Porosity at radial distance r from root surface.
θ	Probability that a propagule occurs within the pathozone.
I	Inoculum density per unit volume.
k	Miscellaneous constant.
$\varkappa$	Asymptotic parameter in non-linear model.
l	Proportion of root length infected.
L	Length of single, main root axis.
m	Pot radius relative to radius of single root.
N	Number of hosts.
N_i	Number of infected hosts.
N_u	Number of uninfected hosts.
n	1. Number of propagules within the pathozone.
	2. Shape parameter for generic growth model.
P	Total number of propagules within a fixed volume of soil (V).
ϕ	Probability that a propagule of inoculum occurs in a pathozone and infects the host.
q	Measure of sensitivity of soil porosity to root growth.
R	Outer radius of the pathozone.
r	1. Rate parameter for non-linear model.
	2. Radial distance.
r_h	Radius of host.
r_i	Radius of inoculum.
r_{irt}	Rate of increase of infected root population.
r_l	Transition radius when soil is compressed by a root to the minimum porosity.
r_n	Mean distance to nth nearest neighbour.
r_P	Rate parameter for primary infection.
r_r	Radius of root.
r_{rt}	Rate of increase of total root population.
r_s	Rate parameter for secondary infection.
ϱ_i	Initial bulk density of soil prior to penetration by root.
ϱ_p	Particle density of soil.
ϱ_r	Bulk density of soil at root surface.
$\varrho(r)$	Bulk density of soil at radial distance (r) from root surface.
S	Number of infections per seed (root).
u	Number of entry points or infection units per unit length of root.
V	Volume of soil containing P propagules.
v	1. Volume of pathozone.
	2. Measure of root density.
w	Component of width of the pathozone.
X	Intermediate variable used in computation of soil porosities.
Y	Level of disease and/or infection.
ψ	Probability that propagule infects hosts, given that the propagule occurs in the pathozone.
z	1. Width of pathozone, including radius of inoculum.

9

ROOT DISEASES

D. HORNBY

AFRC Institute of Arable Crops Research, Rothamsted Experimental Station, Harpenden, Herts AL5 2JQ, UK

1. INTRODUCTION

The role of the rhizosphere in root disease is an expanding but diffuse subject without definitive texts. Comprehensive catalogues of root-infecting organisms and root diseases and the ways in which each is thought to be affected by the rhizosphere do not exist. The rhizosphere's involvements in disease have not all been studied to the same extent, so our knowledge is incomplete and uneven. Most investigations of the rhizosphere in relation to plant disease have focused on fungal pathogens (Curl and Truelove, 1986). This bias is present in the selection of root pathogens, often of arable crops, used here to illustrate (a) some of the ways in which the rhizosphere is known, or assumed, to influence root infection and disease progress, and (b) the sorts of association that are often present and that make interpretations of cause and effect difficult. The selection is further biased in that it presents the perspective of a field-orientated plant pathologist and is, therefore, concerned more with what is than with what might be.

2. SETTING THE SCENE

2.1. Historical

The significance of the rhizosphere in root disease was not recognized until the 1930s, when Garrett (1956) provided evidence that the activities of the rhizosphere microflora of wheat affected the take-all fungus. In flax the population of fungi in the rhizosphere of a variety resistant to wilt was found to differ from that in the rhizosphere of a susceptible variety (Timonin, 1941). This was attributed to a greater excretion of HCN by the resistant variety, although *Trichoderma viride*, a potential biological control agent, tolerated HCN better than many fungi (see also Chapter 7). The question of whether the difference in rhizospheres was the result, or the cause, wholly or partly, of resistance became a recurrent theme. Hornby and Ullstrup (1967) concluded that in root-rot-susceptible varieties of maize large populations of microorganisms around the roots were probably the result and not the cause of root decay.

A general picture emerged of a finely balanced equilibrium of harmless, beneficial and pathogenic microorganisms in the rhizosphere, susceptible to a whole range of edaphic and plant-mediated modifying influences. Soil amendments could shift this equilibrium to control certain root diseases (Katznelson and Richardson, 1948; Katznelson, 1960). Basic ideas about biological control arose, mostly without reference to the rhizosphere, but from the 1940s interest in the use of antagonists to control root diseases increased and after 1960 merged into a period dominated by the idea of biological control (Hornby, 1978).

2.2. Doctrine

Infection of roots by a soil-borne plant pathogen is influenced by the physical and chemical properties of the rhizosphere environment and interaction of the pathogen with other microorganisms in that environment. Most attention has been given to situations where the pathogen is affected adversely before it achieves the parasitic state. There is much less information about how the rhizosphere affects subsequent disease development.

Simple model systems have provided only a basis for speculation about natural root environments and, for the most part, current methodology to monitor and measure rhizosphere effects under field conditions is inadequate (Curl and Truelove, 1986). In nature, and often in agriculture, mycorrhizas and the mycorrhizosphere add yet another dimension of complexity (Chapter 11).

In root disease studies the importance of the rhizosphere is tacitly accepted. Intuitively this seems right because the pathogen, or its vector, has to cross through the rhizosphere, or is in the rhizosphere when contact is made with the host. Despite this, most purported involvements of the rhizosphere in root disease are so correlated with other factors that cause and effect are blurred.

2.3. Milieu

Greater microbial populations and activity than exist in soil away from the roots, or in unplanted soil, make the rhizosphere a formidable barrier (Malajczuk, 1983) and the plant's outermost defence against pathogens (Garrett, 1956). It manifests itself as a dynamic process initiated by root exudation and other release of organic nutrients, and its extent is determined by many factors, such as species, age and condition of the host plant, the type and condition of the soil, and cultural practices. Growing roots compact the soil and deplete nutrients (Rovira, 1985; Linderman, 1988). Mineral nutrition in the rhizosphere is affected by several factors; these include ionic depletion (e.g. of phosphorus or potassium) or accumulation (e.g. of calcium, sodium, or chloride) profiles; a different pH (by as much as ± 2 units) to the bulk soil; altered redox potential; and organic solutes of low molecular weight, mucilage, and sloughed-off cells (Marschner et al., 1986).

Control of root diseases and nutrient availability together with nitrogen fixation and the production of plant growth-promoting substances are the most important microbiological processes in the rhizosphere that affect plant growth (Rovira, 1985). Pathogen populations (inoculum density), growth and survival, and infection-pathogenesis, are all influenced by the rhizosphere (Curl and Truelove, 1986). The ecological relationships between the host, the pathogen and populations of soil microbes have been represented by a "triangle of interactions" (Park, 1963).

The current "dynamic era" of rhizosphere studies (Bowen, 1980) is concerned with the dynamics of microbial populations around roots, interactions, mechanisms and the use of simulation modelling (Chapter 5). Spatial and physiological relations between roots and microorganisms and the ultrastructure of mucilage have been examined directly, and histochemical tests have provided information about polysaccharides (Rovira, 1985). This new enlightenment is offset by a history of rhizosphere research using methods that are rarely standardized and conflicting results from apparently similar studies, making generalizations often hazardous, or impossible (Curl and Truelove, 1986).

3. POPULAR MECHANISMS

3.1. Exudates

Suitable substrates for microbes released by plant roots are exudates, secretions, plant mucilages, mucigel and lysates (Rovira, 1985; Linderman, 1988) (see also Chapter 4). Root exudates and secretions consist of water-soluble sugars, antibiotics, organic acids, nucleotides, flavones, enzymes, vitamins and volatile compounds (Rovira, 1965; Malajczuk, 1983). They play an important role in regulating interactions between plant hosts and pathogens in the rhizosphere by, for instance, stimulating or inhibiting spore germination and/or mycelial growth, causing chemotaxis and affecting binding to root surfaces. Their effects rarely extend beyond a few millimetres from the root surface and seem more general than specific.

Virus infection usually increases root exudation and infection by root-attacking fungi (Zeyen, 1979). The existence of quantitative and qualitative differences in root exudates between resistant and susceptible cultivars has sustained the idea that root pathogens may be controlled by changing root exudation (Funck-Jensen and Hockenhull, 1984). The host genotype has a major role in determining the characteristics of the microbial population in the rhizosphere of certain wheat cultivars resistant or susceptible to common root rot (Atkinson et al., 1975). The composition of rhizodeposition products was one of the selection criteria in the quest for resistance to several soil-borne diseases of cotton (Bird, 1982). It was assumed that cotton has the genetic potential (multi-adversity resistance genes) to alter the natural symbiotic mycoflora on its seeds, roots and in its tissues. This is achieved by changes in the quality and quantity of rhizodeposition such that it becomes unfavourable for pathogens and selectively favourable for organisms which compete with pathogens.

The rhizosphere, therefore, exerts its major influence on root diseases through rhizodeposition (Curl and Truelove, 1986), either directly by affecting changes in populations of pathogens, or indirectly by influencing the general population of microbes, nutrient availability and uptake, synthesis of growth factors and the initiation of antagonism phenomena. Both direct and indirect influences affect inoculum potential, which in turn is linked more or less closely with disease severity (Hornby, 1981). The expectations raised by these insights have yet to be realized in any new and significant way in the field.

3.2. Antagonism

Rhizosphere organisms capable of affecting the plant or other microorganisms operate in a chemically and physically dynamic environment of enhanced microbial activity, which may exhibit all principal forms of antagonism (antibiosis, competition, parasitism and predation). Relatively few attempts to utilize antibiosis and competition to control root diseases have been successful in the field. Lack of appreciation of the complexity of the root zone and its interactions explains the failure of introduced antagonists to control *Phytophthora* spp., except where soils have been pre-treated by steam or fumigants (Malajczuk, 1983).

The rhizosphere as a site where biological control occurs is much discussed and the brief summary in this section is based mostly on a review by Rovira (1985). Often there is no correlation between inhibition of pathogens on agar by biological control agents and the results obtained on plants (Chapter 10). *Bacillus mycoides* is, however, believed to control the take-all fungus by producing antibiotics *in situ*, and control of damping-off of cotton (caused by *Rhizoctonia solani* and *Pythium ultimum*) by strains of *Pseudomonas fluorescens* has been attributed to the antibiotics pyrrolnitrin and pyoluteorin (Chapter 7). Siderophores may be the mechanism whereby fluorescent pseudomonads decrease take-all, "minor pathogens" and *Fusarium* wilt pathogens. *Trichoderma harzianum* controls *R. solani*, *Sclerotium rolfsii*, *S. cepivorum* (white rot of onions) and *Rhizoctonia* root disease of strawberries by hyperparasitism. *Gaeumannomyces graminis* var. *graminis* competes with and may displace var. *tritici*, and/or increase resistance of the host. Non-pathogenic *Fusarium oxysporum* competes with the pathogen to control *Fusarium* wilt of sweet potato and seems to induce a resistance response in the host. Enzymic and mechanical penetration are the mechanisms of mycoparasitism, whereby *Trichoderma* spp. and *Gliocladium catenulatum* kill sclerotia of *Sclerotinia sclerotiorum*, the cause of basal stem rot of sunflower and other hosts (Zazzerini and Tosi, 1985).

4. CURRENT ISSUES AND STRATEGIES

These are listed with comments in Tables 1 and 2.

5. MANIPULATING ANTAGONISTS

5.1. The Problems

In the search for biological control there are those who exhort empiricism and those who feel little can be achieved without a comprehensive knowledge of microbial ecology (Hornby, 1978). The evidence for antagonistic organisms controlling disease in the field rests largely on correlations and, being presumptive, this tends to polarize views. Because the rhizosphere is so complex, it is often beyond our capabilities to influence its microflora in a controlled fashion. The balance between pathogen and the antagonist is a special case of the equilibrium between organisms that stimulate and those that inhibit plant growth. When disease develops strongly it is assumed the equilibrium favours the pathogen, but when the opposite is true despite the presence of the pathogen, the antagonist is assumed to be favoured. However, if the host, pathogen and antagonist each interact differently with the environment, a shift to favour the antagonist may be offset by an adverse effect on the host. When members of the *Gaeumannomyces/Phialophora* complex were used to cross-protect spring wheat against take-all in the field, unusual weather conditions suppressed disease, but favoured root colonization by one of the introductions (Hornby, 1978).

Much attempted manipulation of antagonists in the rhizosphere has been crude and has produced variable results (Hornby, 1978). In reference to control of *Phytophthora* by antagonistic microorganisms, Malajczuk (1983) drew attention to severe deficiencies in numbers of examples of the phenomenon and of our understanding of the subject. A popular impression has been created that we are near to introducing antagonists into the rhizosphere without them being rejected before disease is controlled and that we are getting better at stimulating beneficial, resident antagonists. The prosaic view is that much more needs to be known before we will be able to do this in the field as a matter of course.

5.2. Ways of Changing the Rhizosphere

The rhizosphere responds to additions of organic matter, fertilizers or pesticides to soil (Fig. 1).

TABLE 1. Issues relevant to root disease in current rhizosphere research

Issue	*Comments*	*References*
1. *Using, or manipulating microbial activity* in the rhizosphere to reduce infection and the incidence of root diseases	Much discussion of mechanisms of antagonism, utilizing genetic engineering and understanding rhizosphere ecology	Funck-Jensen and Hockenhull (1984) Rovira (1985)
2. *Microbially-induced starvation*	May explain low incidences of root diseases in modern hydroponics and probably results from good agricultural and horticultural practice	Funck-Jensen and Hockenhull (1984)
3. *Specific growth rates* of microorganisms in the rhizosphere	Change with age of root?	Bowen (1980)
4. Amount of *mycelium on the root surface*	Increased on ryegrass by low concentrations of available N and P	Turner and Newman (1984)
5. *Sources of inoculum* for roots		Bowen (1980)
6. *Mycorrhizal fungi*	Interest in quantitative relationships between populations, infection and plant response	Bowen (1980)
7. *Genetic transfer* between bacteria	High rhizosphere populations may increase chances of plasmid-promoted transfer	P. R. Hirsch, personal communication
8. *In situ* observation	Relatively recent use of borescope mini-rhizotron. Full potential yet to be demonstrated	Rush et al. (1984)

TABLE 2. Some rhizosphere-orientated strategies for the control of root diseases. After Bowen (1980), Rovira (1985)

Strategy	*Comments*
1. Using beneficial organisms where they occur naturally	E.g. symbionts and organisms suppressive to root diseases
a) Selecting host plants	For well-developed antagonistic rhizosphere microfloras, root exudates, or key components in exudates
b) Management techniques	Apply knowledge of epidemiology and microbial energy sources
2. Introduced organisms	Many reports, but how useful in agriculture? Mechanisms insufficiently understood to explain results and variability
a) Analysis of specific examples	E.g. successful biological control in the complex environment where it functions
b) Sites of action	Root, soil, rhizosphere?
c) Creation of favourable environment	Physical and chemical modification of whole soil (most feasible for protected and irrigated crops). Control of H_2O and forms of N applied. Soil sterilization: heat, fumigation and solarization
d) Modify root environment and biological control agent	*Trichoderma* spp.: UV mutants show resistance to fungicides, but retain biological control properties
e) Select appropriate host cultivars	
f) Select microorganisms	With competitive root-colonizing abilities and biological control properties

5.2.1. Fumigation

Some species of fungi occurred less frequently in the rhizosphere of wheat grown in soil fumigated by a 2:1 mixture of chloropicrin and methyl bromide, than in the rhizosphere of wheat in untreated soil, but *Fusarium merismoides*, *Trichoderma koningii* and *T. viride* were increased significantly (Sivasithamparam et al., 1987). Increase in *Trichoderma* spp. has often been associated with a reduction in root diseases in treated soils, or with impressive yield increases in the apparent absence of root disease, as in maize (Barnard and Hornby, 1982).

5.2.2. Other Agrochemicals

Urea and antibiotics applied to foliage will modify rhizodeposition, but this does not appear to have successfully controlled root diseases (Hornby, 1978).

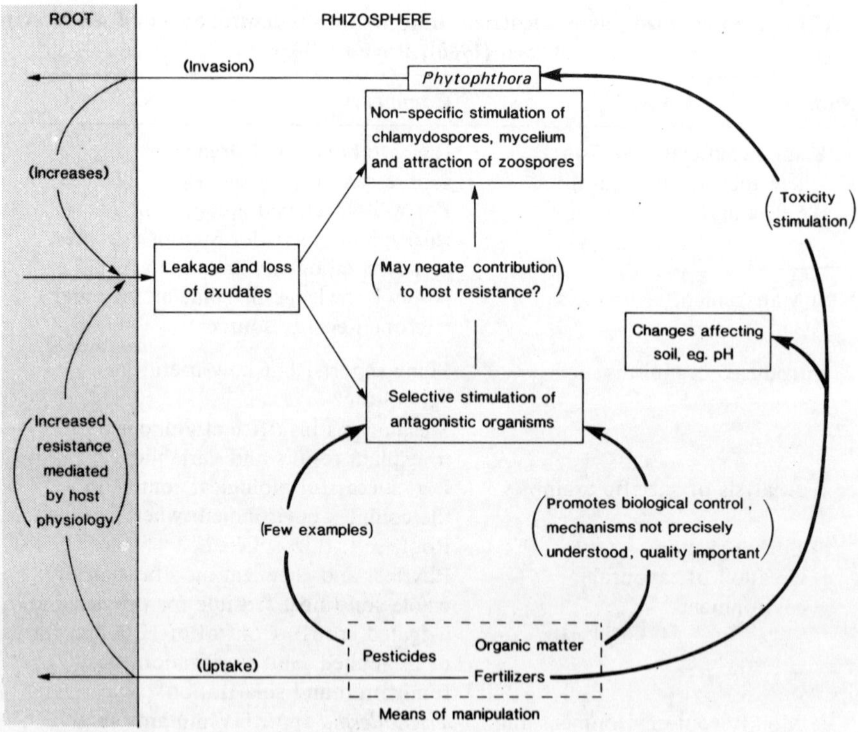

Fig. 1. Some involvements of the rhizosphere in diseases caused by *Phytophthora* spp. (based on information in Malajczuk, 1983). Thick lines represent processes that may also operate in non-rhizosphere soil.

Nutrient sprays and fungitoxicants may affect symbiotic relationships. Foliar applications to leeks of fosetyl-Al, a systemic, symplastic, anti-oomycete fungicide, increased colonization by fungi producing vesicular-arbuscular mycorrhizas (VAM), plant growth and soluble sugars in root exudates. Physiological changes involving amino acids and lipids occurred when VAM and fungicide interacted (Jabaji-Hare and Kendrick, 1985). The significance of host-mediated changes in carbohydrate metabolism to the expression of resistance to pathogens deserves further investigation (Cohen and Coffey, 1986).

6. OTHER CONSIDERATIONS

6.1. The Mycorrhizosphere

Rhizodeposition is greatest in the zone of elongation, which is also where mycorrhizal colonization is initiated when phosphorus reserves in the seed have been expended (Graham, 1988). Before P sufficiency is attained, there exists an increased risk from pathogens because phospholipid depletion makes cell membranes more permeable and exudation is greatest. Mycorrhiza-induced decreases in rhizodeposition have been correlated with reduction in some soil-borne disease (Graham, 1988). In high-input agriculture, fertilizer application may eliminate mycorrhizas. Cereals grown without fertilizer at Rothamsted had 33–36% of root length infected with mycorrhizas, but only 12–13% was infected where nitrogen was added and only 5% where both nitrogen and phosphorus had been used (Hayman, 1975). In Kansas, wheat planted in autumn failed to become mycorrhizal until just prior to harvest and mycorrhizal fungi neither increased yield nor decreased soil-borne diseases (Hetrick et al., 1984). In barley, in Yorkshire, infection depended more on cropping pattern and weather than on soil type or composition, and appreciable infection that developed late in the season was unlikely to have had any nutritional effects on the crop (Black and Tinker, 1979). However, high levels of infection in winter wheat in south-east England occurred during autumn and early spring, with up to 60% of root length infected in November–December (Dodd and Jeffries, 1986). Most VAM infections seem to have little or no effect on *Phytophthora* spp. (Malajczuk, 1983).

Ectomycorrhizas may protect plants against pathogens such as *Phytophthora cinnamomi*, *Rhizoctonia solani*, *Pythium* spp. and *Fusarium oxysporum*. The association formed between *Pinus sylvestris* and *Hebeloma crustuliniforme* brings about microbial change in the rhizosphere which is unfavourable to *Pythium*, whereas terpenes are increased in the association of *Laccaria laccata* with *Picea abies* and decrease root necrosis caused by *F. oxysporum* (Perrin, 1986).

6.2. Virulence of the Pathogen

Seed exudates play an important role in infection of seedlings by *Rhizoctonia solani* (Weinhold et al., 1972). Increased virulence through utilization of external nutrients is one element of the "multiple-component system" which a pathogen in the rhizosphere, or on the rhizoplane, is subject to and which determines its virulence (Curl and Truelove, 1986).

7. SELECTED EXAMPLES INVOLVING *FUSARIUM* SPP.

Many investigations into resistance to root diseases caused by *Fusarium* spp. have fuelled the controversy of whether resistance is due to rhizosphere microorganisms or biochemical constituents of the host. The use of antagonists and biological control has been much researched and has led to the concept that establishment and domination in the infection court may be prerequisites for biological control (Dupler and Baker, 1984). These and other aspects are summarized in Table 3.

TABLE 3. Some rhizosphere and root surface involvements with pathogenic *Fusarium* spp.

Interaction	*Topic*[a]	*Observation*	*Reference*
F. oxysporum wilt of flax	1	Resistance correlated with *Trichoderma viride* in rhizosphere	Timonin (1941)
	2	Wilt suppressed by added *Pseudomonas putida*, less so in biologically active soils	Dupler and Baker (1984)
on various plants	4	Previous cropping may increase or decrease rhizosphere populations	Besri (1975)
ff. sp.	2	Fluorescent pseudomonads may produce smaller amounts of siderophores in rhizospheres with little exudation—inhibition of chlamydospore germination less	Elad and Baker (1985)
f. sp. *vasinfectum* wilt cotton	1	Herbicides increased exudation, but not disease, in resistant varieties	Youssef and Heitefuss (1983)
ff. sp. cotton, wheat muskmelon	2	Diseases decreased when *T. harzianum* from cotton rhizosphere used as soil treatment or seed dressing. Effect persisted through 3 plantings	Sivan and Chet (1986)
f. sp. *melonis* musk melon	3	Competition more intense in rhizospheres of plants in suppressive soils	Alabouvette et al. (1984)
f. sp. *carthami* safflower	2	Antibiotic-producing microorganisms more in rhizospheres of healthy and partially recovered plants	Chakrabarti and Basuchaudhary (1982)
f. sp. *dianthi* carnation	2	*Serratia liquefaciens* from carnation rhizosphere used to protect root cuttings	Sneh et al., (1985)

TABLE 3. *(continued)*

Interaction	*Topic*[a]	*Observation*	*Reference*
F. udum wilt pigeon pea	1	No rhizosphere organisms screened were potent antagonists	Murthy and Bagyaraj (1978)
Fusarium spp.			
lucerne	5	*F. oxysporum* (true vascular pathogen): sparse growth on roots, hyphae orientated between cells. *F. avenaceum* and *F. culmorum* dense growth and hyphal aggregation	Parry and Pegg (1985)
maize	4	Greater rhizosphere populations where reduced cultivation used. Herbicides no effect	Skoglund et al. (1986)
on various plants	2	*Serratia plymuthica* common in rhizosphere. Antagonistic to *Fusarium culmorum* and *Pythium* sp.	Alström and Gerhardson (1987)

[a]Topic key: 1, host resistance; 2, antagonists; 3, suppressive soils; 4, crop sequence and cultivations; 5, root colonization.

8. SELECTED EXAMPLES INVOLVING *GAEUMANNOMYCES GRAMINIS*

8.1. Rhizosphere Involvements

Some of the ways in which the rhizosphere is used in explanations of take-all are given in Table 4. They raise questions of whether all these processes are effective and whether they all function and interact at the same time, or whether there are different mechanisms for different circumstances.

8.1.1. Pre-infection Growth

G. graminis requires nitrogen for germination, hyphal growth and penetration (Garrett, 1967). It produces proteolytic, pectinolytic and cellulolytic enzymes, but not extracellular toxin, and grows more densely and penetrates more frequently in meristematic regions than elsewhere on roots. First infections may occur through the root hairs, which may be deformed by polygalacturonase (Weste, 1972). Wheat roots make possible limited saprophytic growth outside host residues. The fungus may require a feeding phase, in which mycelium

TABLE 4. Some rhizosphere involvements in take-all

Occurrence	*Significance*	*Reference*
Host-specific rhizosphere effects	May have slight influence on susceptibility of host	Scott (1981)
Changed physicochemical environment due to uptake of NH_4^+-N	pH_r decreased. Antagonism by pseudomonads increased.	Huber (1981); Smiley (1978)
	Affects host, pathogen, or both.	Hornby and Goring (1972)
	Not likely to last to host maturity	Huber (1981)
Changed microbiota	Source of *Phialophora* spp.	Sivasithamparam (1975)
	Proliferation of suppressive bacteria	Rovira (1985)
Ectotrophic growth of the pathogen	1. Site for control 2. Affected by light: less spread on short-day plants	
	3. Feeding stage to increase inoculum potential	Brown and Hornby (1971)
Fungicides in the rhizosphere	Unable to affect fungus in root. Benomyl and nuarimol: no long-term protection	Bateman (1985)

increases by forming clumps and colonizing root hairs, before infecting seminal root axes (Brown and Hornby, 1971). The potential of infective host fragments is, therefore, enhanced by (1) the ability to respond at some distance from the host root and (2) using growth in the root region to increase inoculum potential.

Inoculum responds to roots at distances determined to a large extent by the suppressive condition of the soil, but usually at least 5 mm (Brown and Hornby, 1971; Sivasithamparam and Parker, 1978). In take-all decline (TAD) soils the distance was about 5 mm, compared to 8 mm in conducive soil (Pope and Jackson, 1973). Suppressive soil placed within 3 mm of the root decreased the density of runner hyphae arising from added inoculum in plant chambers (Wildermuth et al., 1979). Gilligan's (1980) model of rhizosphere infection, estimated w (the zone of potential infection) as 11.6 ± 0.88 mm, which compared favourably with infections from inocula placed at 10 mm, but not at 12.5 mm in experiments (see also Chapter 8). Statistical methods of estimating the width of the rhizosphere, or w, offer an interesting possibility for investigating some types of biological control. However, the site of suppression will first need to be clarified, for whereas differences in nutrients available for pre-penetration growth in the inoculum and in the rhizosphere might explain different infection efficiencies (the number of lesions produced per unit weight of inoculum) of different sources of inoculum, differences in suppressiveness between soils could not be explained by effects on the incidence of root infections in the experiments of Cook et al. (1986).

When cucumbers were used in a bioassay of *Pythium* spp. (Bouhot and Joannes, 1983; Section 9.6), only the collars of the plants were exposed initially to infested soil, whereas in bioassays of the take-all fungus (Hornby, 1981) the host's roots grew through infested soil. In the bioassay for *Pythium*, inoculum density increased during incubation, because the pathogen competed successfully with the soil microflora for space in the steamed soil in which the cucumbers were growing. The take-all fungus has little competitive saprophytic ability in soil, so increases in inoculum density or inoculum potential by saprophytic growth were severely limited. In standard assays of the take-all fungus there is usually no uncolonized soil, so competition for space is probably unimportant, whereas competition for nutrients near the roots may, as suggested above, be significant. In both systems, however, the relationship between initial inoculum density and disease will be more or less modified by inoculum changes during periods of saprophytic growth prior to host infection. In turn, saprophytic growth may be influenced by the microflora of the soil, rhizosphere, rhizoplane, or all three and this provides a link with studies of biological control mechanisms.

8.1.2. Significance of the Death of Cortical Cells in the Root

Cortical cell death is natural in cereals and is more rapid in wheat than barley. It starts in the outer cortex and progresses inwards. In the field half of the cortex in the proximal part of the first seminal root of spring wheat was dead after 35 days. Cell death may: (1) enhance infection through making nutrients available and decreasing host resistance; (2) increase competition between pathogens and weak parasites and soil saprophytes; and (3) relate to changes in the endodermis and adjacent cells that retard invasion of the stele (Holden, 1976; Deacon and Henry, 1978).

8.1.3. Significance of the Rhizosphere in Soil Bioassays

Microbial antagonism has long been accepted as modifying the relationship between take-all and temperature and other soil variables in bioassays (Hornby, 1981). Plants in sterile soil inoculated with the take-all fungus are most severely diseased near 25 °C, but in field soils the disease is most severe between 12 and 20 °C. The rhizosphere is likely to be involved in this, although saprophytic survival away from roots must also be a factor. Butler (1961) suggested that nutrients added with artificial inoculum exaggerated the role of microbial antagonism in many experiments and such ideas have prompted much work recently on suitable forms of inoculum for work with this disease (see Section 8.1.1). Fragmented and/or degraded inoculum becomes increasingly less able to infect and all that is detected in the favourable conditions of most bioassays is unlikely to function as inoculum in the field.

8.1.4. Mineral Nutrition in the Rhizosphere

The form of nitrogen applied in a fertilizer is not necessarily the form present in the bulk of the soil, or in the rhizosphere, or taken up by the plant (Hornby and Brown, 1977). The pH of the rhizosphere (pH_r) may affect soil microorganisms antagonistic to the ectotrophic growth of *Gaeumannomyces* on the rhizoplane (Smiley and Cook, 1971) (see also Chapter 15). Antagonistic fluorescent pseudomonads are favoured by media with low pH and control of take-all by these organisms may be more successful in plants receiving NH_4^+-N than in plants in unfertilized soils, or in soils receiving NO_3^--N (Smiley, 1978). An attempt to explain TAD by nutritional changes in the rhizosphere, including changes in the $NO_3^- : NH_4^+$ ratio (Brown et al., 1973), has not found support in subsequent field work. In wheat rhizospheres in SE England the concentration of NO_3^--N fell from 20 to 3 μg N g^{-1} between January and April, whilst NH_4^+-N remained fairly constant at 4–5 μg N g^{-1}. After a spring application of 'Nitro-chalk' they became 47.5 and 22 μg N g^{-1}, respectively, but then decreased rapidly. Before April and after June, ammonium was the dominant form of nitrogen in the rhizosphere, but changes in NH_4^+-N seemed to be unrelated to pH_r. All this is consistent with a preferential uptake of NO_3^--N and the movement of this ion down an assumed shallow concentration gradient from the bulk soil to the root surface. NH_4^+-N concentrations in the rhizosphere exceeded 20 μg N g^{-1} for only a few days after fertilizer application in the spring, and pH_r did not fall below 6. Thus the conditions described for the control of take-all by ammonium fertilizer in the USA never occurred in England and it was not established that any of the observed changes in mineral nitrogen concentration and pH_r affected the disease (Hornby and Brown, 1977).

Wheat plants given ammonium phosphate *in vitro* produced more and greater concentrations of amino acids, fewer nucleate cells in the cortex and larger take-all lesions than plants given calcium nitrate (Brown and Hornby, 1987). These effects were most strongly associated in the proximal region of the roots and were attributed to the different forms of nitrogen. Plants receiving NH_4^+-N produced root exudates that were less favourable to the growth of *Gaeumannomyces* than those from plants receiving NO_3^--N. Others (Deacon and Henry, 1980) found no evidence that differences in mineral nutrition, or in rhizosphere microorganisms, were implicated in the greater infection of 5-day-old roots, compared with 15-day-old roots.

P, K, S, Mg, Cl, Zn and Cu have all been shown to decrease the severity of take-all, and Ca and K to increase it (Huber, 1981; Reis et al., 1982). Environmental and cultural conditions (e.g. pH_r) modify the individual or collective effects of these elements. Enhancement of the activity of fluorescent pseudomonads antagonistic to *Gaeumannomyces* on the rhizoplane has been attributed to chloride fertilizers in the USA (Halsey and Powelson, 1981). In UK agriculture the appropriate adjustment of nutrients is unlikely on its own to solve the problem of take-all nationally, but "fine-tuning" is worthwhile and is an obvious consideration in any approach to disease control (Hornby, 1985).

8.1.5. Take-all Decline (TAD)

Most popular hypotheses for TAD still revolve around the idea of antagonistic rhizosphere microorganisms (Cook and Weller, 1987). There is confusion over the types of disease or pathogen suppression that have been encountered in artificial and in natural systems (Hornby, 1983). Specific antagonism is perhaps the most researched area of TAD, but like most microbial antagonism in soil, it is difficult to demonstrate in the field and leans heavily on presumptive evidence. I am not convinced that TAD is explained by the simple, single antagonist theory. The development of specific antagonism, changes in the pathogen population and changes in the microflora that affect the pathogen indirectly through the host were all part of a composite view which tried to accommodate contemporary knowledge (Hornby, 1979), and in which the rhizosphere played a central role. Of the antagonistic microbiota, bacteria, especially *Pseudomonas* spp., actinomycetes and fungi were prominent at times, occurring on the root, in the rhizosphere and associated with diseased host residues, affecting the pathogen in both its saprophytic and parasitic states. Infection, ectotrophic growth and perhaps other stages in disease development were inhibited, the pathogen was deprived of root exudates and its hyphae were lysed. Inhibitory factors were sensitive to heat, partly or wholly transmissible to other soils and sensitive to some forms of nitrogen. The pathogen was most virulent and least able to survive saprophytically when disease peaked in wheat monoculture, but then it became more variable and less pathogenic.

8.1.6. Antagonism

Four out of five trials with pseudomonads applied on seed sown in commercial fields with take-all in Washington State, USA, resulted in yields of 5–21% more than yields from untreated seed (Cook and Weller, 1987). Tests with pseudomonads and other bacteria in England have given less promising results (Hornby, 1987). Observations that as wheat roots aged, the numbers of bacteria and actinomycetes in the rhizosphere decreased (Sivasithamparam et al., 1987), may in part explain why biological control by bacteria in England occurred early and was only temporary (Hornby, 1987). Chapters 7 and 15 discuss recent studies on fluorescent pseudomonads for the control of take-all and how such bacteria might form part of an integrated programme of control.

Various fungi (*Trichoderma* sp., *Didymella exitialis* (Mor.) E. Müller) and even some mycetozoans parasitize the take-all fungus in experimental conditions, but the occurrence and significance of this in nature is not known. *D. exitialis*, for example, decreased take-all severity by 50% when inoculated simultaneously with the pathogen in sterile conditions in a glasshouse (Siegle, 1961). It penetrated and destroyed hyphae of the take-all fungus in the rhizosphere, and amino acids from it were apparently responsible for suppressing pathogenicity by interfering with nitrogen metabolism and cellulase production by *Gaeumannomyces*. Amoebae, actinomycetes and other fungi have each been considered responsible for the inhibition of take-all in suppressive soils (Brisbane and Rovira, 1988).

8.1.7. Resistance

The question of whether or not there is any significant difference amongst wheat cultivars to take-all has been a controversial topic for many years (Scott, 1981). The idea that the rhizosphere has a role in shaping the small differences in resistance detected in the field has been put forward, but using surface sterilized grains grown in steam sterilized sand does not test this (Penrose, 1985).

8.1.8. Mycorrhizosphere

Wheat plants in phosphorus-deficient soil and free from *Glomus fasciculatus* were more diseased than non-mycorrhizal plants given P, or plants with high levels of VAM infection. This was apparently due to improvements in the P status of roots and a decrease in root exudation before inoculation by the pathogen (Graham and Menge, 1982). In glasshouse tests NO_3^--N salts inhibited wheat growth and VAM development more than NH_4^+-N salts and applying them in bands was more inhibitory to root colonization by most VAM than incorporation (Davis and Young, 1985). *Glomus clarum* and *G. fasciculatum* appeared tolerant to incorporated ammonium sulphate or ammonium chloride. VAM colonization was not often indicative of growth response, and chloride and ammonium nitrogen did not suppress VAM as is claimed they do the take-all fungus. In-furrow infestation by *Glomus mosseae* or *G. epigaeum* in Kansas failed to increase winter wheat yields or decrease take-all or *Cephalosporium* stripe disease, but although these mycorrhizal fungi had little or no effect, cold-tolerant, highly efficient isolates should still be sought (Hetrick et al., 1984).

8.2. Connections

Often in take-all research the dominant factor in one view of rhizosphere involvement is later found to be associated with dominant factors in other views. For example, nitrogen was implicated in the mechanism whereby *Didymella exitialis* is said to suppress *Gaeumannomyces*; its form affected root exudates and death of cortical cells in the root *in vitro* (Brown and Hornby, 1987), and pH_r, and it interacted with antagonistic pseudomonads (Smiley, 1978). Many phenomena conspicuous in laboratory work may be of little relevance to crops in the field. Much field experience suggests there is still a considerable gap between theory and practice.

9. SELECTED EXAMPLES INVOLVING *PYTHIUM* AND *PHYTOPHTHORA* SPP.

A general scheme and examples illustrating the role of the rhizosphere in some infections caused by *Phytophthora* spp. are given in Fig. 1 and Tables 5 and 6. The principles and many of the presumed mechanisms are similar to those for other soil-borne pathogens and root diseases.

9.1. Host Resistance

Pythium uses nutrients from roots to infect and overcome the resistance of the host. Infection occurs at sites of intensive exudation such as the root tip and

TABLE 5. Some examples of the antagonism represented in Fig. 1. Information based on Malajczuk (1983)

Host	*Antagonists in rhizosphere*	*Comments*
Legumes	*Rhizobium* spp.	Produce antibiotics against *Phytophthora cactorum*. Lyse hyphae of *P. megasperma*. Not complete protection
Lupin	None	Infection by zoospores of *P. cinnamomi* greater in absence of rhizosphere microbes
Eucalyptus spp.	1) Antibiotic-producing *Penicillium, Aspergillus* and *Trichoderma* spp. and bacteria	In the region of roots infected by *Phytophthora*
	2) Bacteria and actinomycetes	More intensive populations and more antagonists around roots of resistant eucalypts. Antagonism not total
Loblolly pine	Ectomycorrhizal fungus, *Leucopaxillus cerealis* var. *piceina*	Produces a diatretyne nitrite antibiotic—kills zoospores of *P. cinnamomi* at 2 ppm. Altered exudation decreases chemotactic response to seedling roots. Mycorrhiza is also mechanical barrier

TABLE 6. Some examples of manipulation of diseases caused by *Phytophthora* spp. using organic matter. Based on information in Malajczuk (1983)

Host	*Pathogen*	*Organic matter*	*Results and comments*
Avocado	*P. cinnamomi*	1) Alfalfa meal	Root rot decreased (markedly in glasshouse, partially in field). Microbial antagonism presumed. Ammonia produced, increased microbial activity, meal phytotoxic to root growth
		2) "Organic farming" legumes and forage grown under trees: incorporated with straw mulches, fowl manure and dolomite	No root rot. Enhanced microbial activity. Antagonistic microflora. (Best example of *Phytophthora* control in the field?)
Avocado and Citrus	*P. cinnamomi* *P. parasitica*	Chicken manure, alfalfa meal, processed sewage sludge	Increased microbial activity. Soil becomes suppressive. Increased lysis and sporangial abortion
Nursery-grown ornamentals	*P. cinnamomi*	Hardwood bark (composted)	*P. cinnamomi* and other fungi controlled: lysis of hyphae, zoospores and germinating cysts. Rapid build up of mycophagous antagonistic microorganisms (e.g. actinomycetes, fluorescent pseudomonads and *Trichoderma* spp.)

bases of lateral roots (Funck-Jensen and Hockenhull, 1984). Peas susceptible to damping-off (caused by *P. debaryanum*) release greater amounts of sucrose during germination than do resistant peas (Youssef and Heitefuss, 1983). Resistance to *Phytophthora cinnamomi* is determined by rhizosphere mycofloras (Yip and Weste, 1985). Microorganisms may interact with *P. cinnamomi* in the soil, the rhizosphere, or on the rhizoplane causing decline in inoculum and a reduced incidence of disease development through hyphal lysis, sporangium abortion and reduced germination of cysts (Malajczuk, 1988).

9.2. Suppressive Soils

Pythium graminicola is associated with "poor root syndrome" (PRS) of sugar cane. Many rhizosphere bacteria inhibit this fungus *in vitro*, but individually they were ineffective in pot and field trials, and using the suppressiveness of many different microbes for the economic control of PRS would be difficult (Birch, 1986).

9.3. Antagonists

Microbial antagonism of *Phytophthora* spp. has been most studied in *P. cinnamomi* (Malajczuk, 1983). Evidence for amensalism (production of antibiotics or toxic by-products by antagonists) is largely circumstantial and correlation between disease suppression and antibiotic production on agar may not be indicative of antibiosis in soil, or the rhizosphere. On roots of jarrah (*Eucalyptus marginata*) growing in litter, germ tubes of *P. cinnamomi* were trapped by an unknown fungus employing a constrictive ring method not unlike that used by nematode-trapping fungi. Certain ectomycorrhizal fungal symbionts seemed to contribute chemically to the antagonism of zoospores (Malajczuk, 1988). Antagonistic microorganisms may in many cases be the only means available for controlling *Phytophthora* (Malajczuk, 1983).

Loss of control of black shank in tobacco coincides with a rapid increase of propagules of *Phytophthora parasitica* var. *nicotianae* in the rhizosphere (Cohen and Coffey, 1986). Populations of the pathogen are extremely small in soil and highly aggregated at planting. The extensiveness and heterogeneity of the soil environment make contact between pathogen and antagonist unlikely and so the rhizosphere is a more appropriate region for seeking control, because both organisms will be active there in a considerably smaller volume of soil (English and Mitchell, 1988). However, no community antagonistic to *P. parasitica* var. *nicotianae* has been reported within the rhizosphere of tobacco, so to continue this approach there is a need to understand population dynamics and patterns of rhizosphere and rhizoplane colonization.

Certain indigenous ''deleterious rhizobacteria'' have been shown to interact with *Pythium* spp. to increase root damage, but fluorescent pseudomonads may inhibit these (Rovira, 1985). *Pseudomonas fluorescens* Migula isolated from the rhizosphere of cotton inhibited *Pythium ultimum in vitro*. It produced an extracellular phenolic polypetide (pyoluteorin) with antifungal properties (Bencini et al., 1983).

9.4. Crop Sequence

Phycomycete (Mastigomycotina) zoospores typically accumulate and encyst in large numbers immediately behind the root tip, and with few exceptions this is not host specific (Chang-ho and Hickman, 1970). *Pythium* zoospores are attracted to oat roots and lysed, a phenomenon attributed to avenacin (Deacon and Mitchell, 1985) and one which suggests oats could be grown to reduce populations of the pathogen.

9.5. Fungicides

Lysis of hyphae of *Phytophthora cinnamomi* was greater in unsterile leachates containing metalaxyl than in sterile treatments with the same amount of fungicide (Malajczuk, 1983). Experiments with *P. parasitica*, genetically modified root cultures of *Medicago sativa* and fosetyl-Al (Decor, 1985) revealed ethylphosphonate and phosphonate in the roots and the rhizosphere, thus providing evidence for absorption, transport and exudation of fungicide.

9.6. Modelling

Modelling the rhizosphere continues to be a vigorous, competitive topic, where models vie for general acceptability (Walker and Smith, 1988) (see also Chapters 5 and 8). The French concept, applied to damping-off of cucumbers by *Pythium* (Bouhot and Joannes, 1983 and Section 8.1.1), is that of infectious soil stores, and has available, a quantity of pathogenic energy called the "potentiel infectieux des sols" (soil infectivity), which is the sum of the metabolic activities of the pathogen necessary and sufficient to establish infection. The flow of pathogenic energy through a pathogen/soil/host system has been likened to the flow of electricity through a circuit (Bouhot, 1979), but the pathogenic energy cannot be expressed without an external energy flux provided by the plant and soil (Estivalet, 1984). The analogy with electrical flow is interesting and helps understanding, but the idea that there is some energy which is pathogenic is hard to grasp. It seems more a question of a pathogenic potential of inoculum (as genetic or cytoplasmic factors) that is expressed when energy is made available and a host is present. The concept of pathogenic energy (as inoculum potential) is not new (Garrett, 1956). Several workers have developed equations which purport to relate disease severity to inoculum potential (Hornby, 1981). The French concept also involves spatial relationships and increases in pathogen biomass, which may complicate the relationship between inoculum concentration and disease severity. "Potentiel infectieux des sols" is impossible to measure directly, but a unit weight, or volume, of soil that contains it ("le potentiel infectieux standard", or SIP) may be estimated in standard biological tests.

In the French work, principal component analysis has helped in formulating a biological explanation of the observations. This has been simulated in a computer using a simple deterministic model which applies a logistic equation to two competing populations. Even though some implicit assumptions may be false, it is not unusual for actual populations to behave like the simple models, because different conceptual models can lead to similar curves of population change. Data obtained by observing population sizes in a succession of times are not likely to be adequate for discriminating amongst possible models, but such knowledge is a necessary preliminary to the development of more elaborate models (Pielou, 1969). The French bioassay and model may be used to characterize soils, to determine the "receptivity" of soils to pathogens, to predict disease risk, to assess the importance of climatic factors on disease, and to study the direct and indirect effects of pesticides. [Soil receptivity to a pathogen, R_t, is measured as the quotient SIP_t/SIP_{t_0}, where t is time. Where log $R_t<0$ soils are suppressive, where log $R_t \approx 0$ soils are tolerant and where $R_t>0$ soils are conducive.]

9.7. Environmental Factors

Low root temperatures (less than 10 °C) restricted the attraction of *P. cinnamomi* to root tips of eucalypts, encystment and germ tube development. In susceptible eucalypts the interaction between temperature and effects of inoculation was significant (Halsall and Williams, 1984).

10. CONCLUSIONS

Plant pathologists studying root diseases instinctively include the rhizosphere in explanations of what they observe. Most rhizosphere–microbe interactions recorded in the literature in some way concern the efficacy of seed and soil treatments, biological control agents, bioenvironmental management, or breeding for resistance (Curl and Truelove, 1986). Words like "believed", "apparent" and "may have" often occur in these reports, yet the impetus is now so great that nobody doubts that the rhizosphere holds the key to a better understanding of host–parasite relationships in root diseases and in mechanisms responsible for biological and, in some cases chemical, control (Curl and Truelove, 1986).

There is remarkably little disagreement about the way forward. The needs for rhizosphere studies in general are: (1) basic studies to be continued and expanded into the future; (2) systems to utilize or stimulate natural conditions; (3) more knowledge of the synergistic actions of components of the soil solution; (4) new and better methods for extracting and growing rhizosphere microorganisms; and (5) new and improved methods of direct observation (Curl and Truelove, 1986). There are now new opportunities in modelling the rhizosphere and new challenges in biological control, but biological control by introduced antagonists requires a better understanding of both antagonist and pathogen in the soil, rhizosphere and root (Malajczuk, 1983). Theoretical models incorporating relationships among pathogen, antagonist and environment, and principles for predicting epidemics or control are needed, as are suppressive soils to formulate and test models.

A major problem is the number of apparently relevant differences between the rhizospheres of diseased and healthy plants and the number of rhizosphere processes that are believed to affect disease. For example, VAM require living cells and cortical cell death occurs naturally, yet the literature suggests both phenomena occur significantly in wheat in temperate climates. Macleod et al. (1986) found that the death of the cortex in wheat was most rapid in plants moderately deficient in phosphorus and that VAM delayed it in phosphorus-deficient plants and accelerated it in phosphorus-adequate plants. This kind of complexity may be the background to many root diseases in the field and must be understood if widespread, consistent management of the rhizosphere's ability to influence disease is ever to become commonplace.

In many reports it is not clear whether the rhizosphere influenced, or reflected, the outcome of a challenge by a root pathogen, and in many diseases the mechanisms of the rhizosphere's involvement remain presumptive. Much research gives great emphasis to antagonism in the rhizosphere, but it is not the only influential rhizosphere mechanism affecting root disease. The dominant mechanisms or agents are not the same for all root diseases, and they may differ for the same disease in different locations and at different stages of plant growth.

The resources brought to bear in rhizosphere research have rarely matched the complexity of the problem. Our researches reveal fragments of a gigantic jigsaw puzzle, which we painstakingly and enthusiastically continue to piece together. Creative thinking has outstripped methodological progress, which is slow, and practical progress in exploiting the rhizosphere to our advantage in the field is likely to be sporadic and fortuitous until all aspects of the problem are identified and approached with multidisciplinary expertise. A saying, Japanese I think, elegantly encapsulates this: "Look up to ideals, but do not run away from reality".

REFERENCES

Alabouvette, C., Y. Couteaudier, and J. Louvet (1984). Recherches sur la résistance des sols aux maladies. X. Comparaison de la mycoflore colonisant les racines de melons cultivés dans un sol résistant ou dans un sol sensible aux fusarioses vasculaires. *Agronomie*, **4**:735–740.

Alström, S. and B. Gerhardson (1987). Characteristics of a *Serratia plymuthica* isolate from plant rhizospheres. *Plant and Soil*, **103**:185–189.

Atkinson, T. G., J. L. Neal, and R. I. Larson (1975). Genetic control of the rhizosphere microflora of wheat. In *Biology and Control of Soilborne Plant Pathogens* (G. W. Breuhl, ed.). The American Phytopathological Society, St. Paul, Minnesota, pp. 116–122.

Barnard, A. J. and D. Hornby (1982). The effects of dazomet and nitrogen fertilizer on successive crops of maize (*Zea mays* L.) grown for either grain or forage. *J. Agr. Sci., Camb. (England)*, **98**:7–15.

Bateman, G. L. (1985). The effects of distribution of two soil-incorporated fungicides on the control of take-all (*Gaeumannomyces graminis* var. *tritici*) in wheat. *J. Plant Dis. Protect.*, **92**:194–203.

Bencini, D. A., C. R. Howell, and J. R. Wild (1983). Production of phenolic metabolites by a soil pseudomonad. *Soil Biol. Biochem.*, **15**:491–492.

Besri, M. (1975). Recherches sur les fusarioses. Influence du précédent cultural sur l'évolution de la population de *Fusarium oxysporum* dans la rhizosphère de quelques plantes. *Ann. Phytopathol.*, **7**:1–8.

Birch, R. G. (1986). Biological suppressiveness against sugarcane poor root syndrome in Queensland soils. *Australasian Plant Pathol.*, **15**:51–56.

Bird, L. S. (1982). The MAR (Multi-Adversity Resistance) system for genetic improvement of cotton. *Plant Dis.*, **66**:172–176.

Black, R. and P. B. Tinker (1979). The development of endomycorrhizal root systems. II. Effect of agronomic factors and soil conditions on the development of vesicular-arbuscular mycorrhizal infection in barley and on endophyte spore density. *New Phytol.*, **83**:401–413.

Bouhot, D. (1979). Estimation of inoculum density and inoculum potential: techniques and their value for disease prediction. In *Soil-borne Plant Pathogens* (B. Schippers and W. Gams, eds.). Academic Press, London, pp. 21–34.

Bouhot, D. and H. Joannes (1983). Potentiel infectieux des sols—concept et modèles. *EPPO Bull.*, **13**:291–295.

Bowen, G. D. (1980). Misconceptions, concepts and approaches in rhizosphere biology. In *Contemporary Microbial Ecology* (D. C. Ellwood, M. J. Latham, J. M. Hedger, J. M. Lynch and J. H. Slater, eds.). Academic Press, London, pp. 283–304.

Brisbane, P. G. and A. D. Rovira (1988). Mechanisms of inhibition of *Gaeumannomyces graminis* var. *tritici* by fluorescent pseudomonads. *Plant Pathol.*, **37**:104–111.

Brown, M. E. and D. Hornby (1971). Behaviour of *Ophiobolus graminis* on slides buried in soil in the presence or absence of wheat seedlings. *Trans. Brit. Mycol. Soc.*, **56**:95–103.

Brown, M. E. and D. Hornby (1987). Effects of nitrate and ammonium on wheat roots in gnotobiotic culture: amino acids, cortical cell death and take-all (caused by *Gaeumannomyces graminis* var. *tritici*). *Soil Biol. Biochem.*, **19**:567–573.

Brown, M. E., D. Hornby, and V. Pearson (1973). Microbial populations and nitrogen in soil growing consecutive cereal crops infected with take-all. *J. Soil Sci.*, **24**:296–310.

Butler, F. C. (1961). Root and foot rot diseases of wheat. *Sci. Bull. Dep. Agric. N. S. W.* No. 77: 98 pp.

Chakrabarti, D. K. and K. C. Basuchaudhary (1982). Effect of rhizosphere microflora of healthy and wilt-diseased *Carthamus tinctorius* L. plants. *Proc. Indian Natn. Sci. Acad.*, **B48**:649–651.

Chang-ho, Y. and C. J. Hickman (1970). Some factors involved in the accumulation of phycomycete zoospores on plant roots. In *Root Diseases and Soil-borne Pathogens* (T. A. Tousson, R. V. Bega, and P. W. Nelson, eds.). University of California Press, Berkeley, Los Angeles, pp. 103–108.

Cohen, Y. and M. D. Coffey (1986). Systemic fungicides and the control of oomycetes. *Ann. Rev. Phytopathol.*, **24**:311–338.

Cook, R. J. and D. M. Weller (1987). Management of take-all in consecutive crops of wheat or barley. In *Innovative Approaches to Plant Disease Control* (I. Chet, ed.). John Wiley & Sons, New York, pp. 41–66.

Cook, R. J., H. T. Wilkinson, and J. R. Alldredge (1986). Evidence that microorganisms in suppressive soil associated with wheat take-all decline do not limit the number of lesions produced by *Gaeumannomyces graminis* var. *tritici*. *Phytopathology*, **76**:342–345.

Curl, E. A. and B. Truelove (1986). *The Rhizosphere*. Springer-Verlag, Berlin, 288 pp.

Davis, E. A. and J. L. Young (1985). Endomycorrhizal colonization of glasshouse-grown wheat as influenced by fertilizer salts when banded or soil-mixed. *Can. J. Bot.*, **63**:1196–1203.

Deacon, J. W. and C. M. Henry (1978). Death of the cereal root cortex: its relevance to biological control of take-all. *Ann. Appl. Biol.*, **89**:100.

Deacon, J. W. and C. M. Henry (1980). Age of wheat and barley roots and infection by *Gaeumannomyces graminis* var. *tritici*. *Soil Biol. Biochem.*, **12**:113–118.

Deacon, J. W. and R. T. Mitchell (1985). Toxicity of oat roots, oat root extracts and saponins to zoospores of *Pythium* spp. and other fungi. *Trans. Brit. Mycol. Soc.*, **84**:479–487.

Decor, J. P. (1985). Stratégie d'étude de nouveaux fongicides systémiques: example de phoséthyl-Al. In *Fungicides for Crop Protection. 100 Years of Progress, Vol. 1* (I. M. Smith, ed.). Monograph No. 31. BCPC Publications, Croydon, UK, pp. 119–125.

Dodd, J. C. and P. Jeffries (1986). Early development of vesicular-arbuscular mycorrhizas in autumn-sown cereals. *Soil Biol. Biochem.*, **18**:149–154.

Dupler, M. and R. Baker (1984). Survival of *Pseudomonas putida*, a biological control agent, in soil. *Phytopathology*, **74**:195–200.

Elad, Y. and R. Baker (1985). The role of competition for iron and carbon in suppression of chlamydospore germination of *Fusarium* spp. by *Pseudomonas* spp. *Phytopathology*, **75**:1053–1059.

English, J. T. and D. J. Mitchell (1988). Development of microbial communities associated with tobacco root systems. *Soil Biol. Biochem.*, **20**:137–144.

Estivalet, D. (1984). Influence de quelques facteurs climatiques sur l'expression du potentiel infectieux des sols infestés par *Pythium* spp. DEA Thesis, Dijon, 35 pp. + appendices.

Funck-Jensen, D. and J. Hockenhull (1984). Root exudation, rhizosphere microorganisms and disease control. *Växtskyddsnotiser*, **48**:49–54.

Garrett, S. D. (1956). *Biology of Root-infecting Fungi*. Cambridge University Press, 293 pp.

Garrett, S. D. (1967). Effect of nitrogen level on survival of *Ophiobolus graminis* in pure culture on cellulose. *Trans. Brit. Mycol. Soc.*, **50**:519–524.

Gilligan, C. A. (1980). Zone of potential infection between roots and inoculum units of *Gaeumannomyces graminis*. *Soil Biol. Biochem.*, **12**:513–514.

Graham, J. H. (1988). Interactions of mycorrhizal fungi with soilborne plant pathogens and other organisms: an introduction. *Phytopathology*, **78**:365–366.

Graham, J. H. and J. A. Menge (1982). Influence of vesicular-arbuscular mycorrhizae and soil phosphorus on take-all disease of wheat. *Phytopathology*, **72**:95–98.

Halsall, D. M. and J. D. Williams (1984). Effect of root temperature on the development of *Phytophthora cinnamomi* root rot on *Eucalyptus* seedlings. *Aust. J. Bot.*, **32**:521–528.

Halsey, M. and R. Powelson (1981). The influence of NH_4Cl with KCl on antagonism between fluorescent pseudomonads (fp) and *Gaeumannomyces graminis* var. *tritici* (GGT) on the wheat rhizoplane. *Phytopathology*, **71**:105.

Hayman, D. S. (1975). The occurrence of mycorrhiza in crops as affected by soil fertility. In *Endomycorrhizas* (F. E. Sanders, B. Mosse, and P. B. Tinker, eds.). Academic Press, London, pp. 495–509.

Hetrick, B. A. D., W. W. Bockus, and J. Bloom (1984). The role of vesicular-arbuscular mycorrhizal fungi in the growth of Kansas winter wheat. *Can. J. Bot.*, **62**:735–740.

Holden, J. (1976). Infection of wheat seminal roots by varieties of *Phialophora radicicola* and *Gaeumannomyces graminis*. *Soil Biol. Biochem.*, **8**:109–119.

Hornby, D. (1978). Microbial antagonisms in the rhizosphere. *Ann. Appl. Biol.*, **89**:97–100.

Hornby, D. (1979). Take-all decline: a theorist's paradise. In *Soil-borne Plant Pathogens* (B. Schippers and W. Gams, eds.). Academic Press, London, pp. 133–156.

Hornby, D. (1981). Inoculum. In *Biology and Control of Take-all* (M. J. C. Asher and P. J. Shipton eds.). Academic Press, London, pp. 271–293.

Hornby, D. (1983). Suppressive soils. *Ann. Rev. Phytopathol.*, **21**:65–85.

Hornby, D. (1985). Soil nutrients and take-all. *Outlook Agric.*, **14**:122–128.

Hornby, D. (1987). Field testing putative biological controls of take-all: rationale and results. *EPPO Bull.*, **17**:615–623.

Hornby, D. and M. E. Brown (1977). Nitrate and ammonium in the rhizosphere of wheat crops and concurrent observations of take-all. *Plant and Soil*, **48**:455–471.

Hornby, D. and C. A. I. Goring (1972). Effects of ammonium and nitrate nutrition on take-all disease of wheat in pots. *Ann. Appl. Biol.*, **70**:225–231.

Hornby, D. and A. J. Ullstrup (1967). Fungal populations associated with maize roots. Quantitative rhizosphere data for genotypes differing in root rot resistance. *Phytopathology*, **57**: 76–82.

Huber, D. M. (1981). The role of nutrients and chemicals. In *Biology and Control of Take-all* (M. J. C. Asher and P. J. Shipton, eds.). Academic Press, London, pp. 317–341.

Jabaji-Hare, S. H. and W. B. Kendrick (1985). Effects of fosetyl-Al on root exudation and on composition of extracts of mycorrhizal and nonmycorrhizal leek roots. *Can. J. Plant Pathol.*, **7**:118–126.

Katznelson, H. (1960). Observations on the rhizosphere effect. In *The Ecology of Soil Fungi* (D. Parkinson and J. S. Waid, eds.). Liverpool University Press, pp. 192–201.

Katznelson, H. and L. T. Richardson (1948). Rhizosphere studies and associated microbiological phenomena in relation to strawberry root rot. *Scient. Agric.*, **28**:293–308.

Linderman, R. G. (1988). Mycorrhizal interactions with the rhizosphere microflora: the mycorrhizosphere effect. *Phytopathology*, **78**:366–371.

Macleod, W. J., A. D. Robson, and L. K. Abbott (1986). Effects of phosphate supply and inoculation with a vesicular-arbuscular mycorrhizal fungus on the death of the root cortex of wheat, rape and subterranean clover. *New Phytol.*, **103**:349–357.

Malajczuk, N. (1983). Microbial antagonism to *Phytophthora*. In *Phytophthora. Its Biology, Taxonomy, Ecology and Pathology* (D. C. Erwin, S. Bartnicki-Garcia, and P. H. Tsao, eds.). The American Phytopathological Society, St. Paul, Minnesota, pp. 197–218.

Malajczuk, N. (1988). Interaction between *Phytophthora cinnamomi* zoospores and micro-organisms on non-mycorrhizal and ectomycorrhizal roots of *Eucalyptus marginata*. *Trans. Brit. Mycol. Soc.*, **90**:375–382.

Marschner, H., V. Römheld, W. J. Horst, and P. Martin (1986). Root-induced changes in the rhizosphere: importance for the mineral nutrition of plants. *Z. PflErnähr. Bodenk.*, **149**:441–456.

Murthy, G. S. and D. J. Bagyaraj (1978). Rhizosphere microflora of *Cajanus cajan* in relation to Fusarium wilt resistance. *Plant and Soil*, **50**:485–487.

Park, D. (1963). The ecology of soil-borne fungal disease. *Ann. Rev. Phytopathol.*, **1**:241–258.

Parry, D. W. and G. F. Pegg (1985). Surface colonization, penetration and growth of three *Fusarium* species in lucerne. *Trans. Brit. Mycol. Soc.*, **85**:495–500.

Penrose, L. (1985). Evidence for resistance in wheat cultivars grown in sand culture to the take-all pathogen, *Gaeumannomyces graminis* var. *tritici*. *Ann. Appl. Biol.*, **107**:105–108.

Perrin, R. (1986). Maladies d'origine tellurique en pépinère forestière: problème du diagnostic et lutte intégrée. *EPPO Bull.*, **16**:553–560.

Pielou, C. (1969). *An Introduction to Mathematical Ecology*. Wiley Interscience, New York, 286 pp.

Pope, A. M. S. and R. M. Jackson (1973). Effects of wheatfield soil on inocula of *Gaeumannomyces graminis* (Sacc.) Arx and Olivier var. *tritici* J. Walker in relation to take-all decline. *Soil Biol. Biochem.*, **5**:881–890.

Reis, E. M., R. J. Cook, and B. L. McNeal (1982). Effect of mineral nutrition on take-all of wheat. *Phytopathology*, **72**:224–229.

Rovira, A. D. (1965). Plant root exudates and their influence upon soil microorganisms. In *Ecology of Soil-borne Plant Pathogens* (K. F. Baker and W. C. Snyder, eds.). University of California Press, Berkeley, pp. 170–184.

Rovira, A. D. (1985). Manipulation of the rhizosphere microflora to increase plant production. In *Biotechnology and Recombinant DNA Technology in the Animal Production Industries* (R. A. Leng, J. S. F. Barker, D. B. Adams, and K. J. Hutchinson, eds.). Reviews in Rural Science 6. University of New England Publishing Unit, Armidale, pp. 185–197.

Rush, C. M., D. R. Upchurch, and T. J. Gerik (1984). In situ observation of *Phymatotrichum omnivorum* with a borescope mini-rhizotron system. *Phytopathology*, **74**:104–105.

Scott, P. R. (1981). Variation in host susceptibility. In *Biology and Control of Take-all* (M. J. C. Asher and P. J. Shipton, eds.). Academic Press, London, pp. 219–236.

Siegle, H. (1961). Über Mischinfektionen mit *Ophiobolus graminis* und *Didymella exitialis*. *Phytopath. Z.*, **42**:305–348.

Sivan, A. and I. Chet (1986). Biological control of *Fusarium* spp. in cotton, wheat and muskmelon by *Trichoderma harzianum*. *J. Phytopathol.*, **116**:39–47.

Sivasithamparam, K. (1975). *Phialophora* and *Phialophora*-like fungi occurring in the root region of wheat. *Aust. J. Bot.*, **23**:193–212.

Sivasithamparam, K. and C. A. Parker (1978). Effect of infection of seminal and nodal roots by the take-all fungus on tiller numbers and shoot weight of wheat. *Soil Biol. Biochem.*, **10**:365–368.

Sivasithamparam, K., G. C. MacNish, C. S. Fang, and C. A. Parker (1987). Microflora of soil and wheat rhizosphere in a field following fumigation. *Aust. J. Soil Res.*, **25**:491–498.

Skoglund, L. G., W. M. Brown Jr., and R. R. Ruppel (1986). Assessment of rhizosphere population densities of *Fusarium* spp. associated with corn stalk rot and conservation tillage. *Phytopathology*, **76**:1130.

Smiley, R. W. (1978). Antagonists of *Gaeumannomyces graminis* from the rhizoplane of wheat in soils fertilized with ammonium or nitrate nitrogen. *Soil Biol. Biochem.*, **10**:169–174.

Smiley, R. W. and R. J. Cook (1971). Influence of nitrogen fertilizers on rhizosphere pH and take-all of wheat caused by *Ophiobolus graminis*. *Phytopathology*, **61**:911–912.

Sneh, B., O. Agami, and R. Baker (1985). Biological control of *Fusarium*-wilt in carnation with *Serratia liquefaciens* and *Hafnia alvei* isolated from rhizosphere of carnation. *Phytopath. Z.*, **113**:271–276.

Timonin, M. I. (1941). The interaction of higher plants and soil microorganisms: III. Effects of by-products of plant growth on activity of fungi and actinomycetes. *Soil Sci.*, **52**:395–413.

Turner, S. M. and E. I. Newman (1984). Fungal abundance on *Lolium perenne* roots: influence of nitrogen and phosphorus. *Trans. Brit. Mycol. Soc.*, **82**:315–322.

Walker, N. A. and S. E. Smith (1988). Effect of soil compression on estimates of rhizosphere width: comparing Ferriss's equation with Gilligan's. *Phytopathology,* **78**:253–255.

Weinhold, A. R., R. L. Dodman, and T. Bowman (1972). Influence of exogenous nutrition on virulence of *Rhizoctonia solani*. *Phytopathology*, **62**:278–281.

Weste, G. (1972). The process of root infection by *Ophiobolus graminis*. *Trans. Brit. Mycol. Soc.*, **59**:133–147.

Wildermuth, G. B., A. D. Rovira, and J. H. Warcup (1979). Mechanism and site of suppression of *Gaeumannomyces graminis* var. *tritici* in soil. In *Soilborne Plant Pathogens* (B. Schippers and W. Gams, eds.). Academic Press, London, pp. 157–164.

Yip, H. Y. and G. Weste (1985). Rhizoplane mycoflora of *Gahnia radula* and *Isopogon ceratophyllus* in soils infested and free from *Phytophthora cinnamomi*. In *Ecology and Management of Soilborne Plant Pathogens* (C. A. Parker, A. D. Rovira, K. J. Moore, P. T. W. Wong, and J. F. Kollmorgen, eds.). The American Phytopathological Society, St. Paul, Minnesota, pp. 92–98.

Youssef, B. A. and R. Heitefuss (1983). Side-effects of herbicides on cotton wilt caused by *Fusarium oxysporum* f. sp. *vasinfectum* II. Effect of herbicides on the quantitative and qualitative composition of sugars and amino acids in cotton seed and root exudates. *Z. PflKrankh. Pflschutz.*, **90**:36–49.

Zazzerini, A. and L. Tosi (1985). Antagonistic activity of fungi isolated from sclerotia of *Sclerotinia sclerotiorum*. *Plant Pathol.*, **34**:415–421.

Zeyen, R. J. (1979). Viruses. In *Ecology of Root Pathogens* (S. V. Krups and Y. R. Dommergues, eds.). Elsevier Scientific Publishing Company, Amsterdam, pp. 179–205.

10

BIOLOGICAL DISEASE CONTROL

G. E. HARMAN

Department of Horticultural Sciences, New York State Agricultural Experiment Station, Cornell University, Sturtevant Hall, Geneva, NY 14456, USA

R. D. LUMSDEN

Biocontrol of Plant Diseases Laboratory, Plant Protection Institute, USDA-ARS, Beltsville, MD 20705, USA

Biological control of soil-borne pathogens of plants has several potential advantages over chemical pesticides. Firstly, chemicals pose problems to the environment, and environmental concerns have resulted in removal of some chemicals from the market. For example, several chemical seed protectants have been banned from use in the USA, including mercury-containing fungicides. The commonly used broad-spectrum fungicide captan is being reconsidered for use, and the seed industry expects it to be banned within 5 years. Other widely used broad-spectrum chemicals, including soil fumigants and the seed protectant thiram, are also becoming increasingly questioned for use because of state and federal regulations regarding worker safety, ground water contamination, and disposal of surplus chemicals and excess treated seed. Secondly, biological control depends upon manipulation of the environment around a crop to favor organisms contributing to plant health, rather than the application of a pesticide that may eliminate a range of organisms that include the target pathogen. Broad-spectrum chemical fungicides often induce resistance in pests and bioprotectants are generally not thought to induce such resistance. Thus, biocontrol methods are less disruptive to ecosystems than are chemical pesticides (Cook and Baker, 1983). Thirdly, biocontrol may be economically feasible, whereas chemical control frequently is not. Because of the large mass of field soil—one hectare of soil 15 cm deep weighs about 1.7×10^7 kg—prohibitively large quantities of chemicals are needed even if they are effective at low concentrations. Biological control agents potentially can overcome the limitation of quantity of treatment material needed because they have the ability to proliferate. Thus a small amount of initial inoculum may increase and colonize a large volume of soil or plant surfaces.

In spite of these advantages, biological control agents, hereafter called "bioprotectants", are little used for control of plant pathogens, primarily because they have been less effective and more variable than competitive chemical pesticides (Kommedahl et al., 1981; Ruppel et al., 1983). This lower level of reliability must be overcome if bioprotectants are to become important tools of plant disease control. Additionally, bioprotectants may be used in conjunction with judicious use of chemicals to obtain enhanced disease control. This is considered an integrated approach. In this chapter, we shall examine several strategies to enhance the efficacy of biocontrol systems, with emphasis on protection of the spermosphere and rhizosphere.

1. BIOCONTROL STRATEGIES

1.1. Cropping System Management

Some root diseases of important crop plants may be controlled with appropriate cultural practices. Such systems no doubt evolved from experience and were practiced more intensively prior to the development of modern agricultural practices that feature soil tillage and extensive culture of only one or a few crops. These components of biological/cultural control will be considered elsewhere in this volume (Chapters 9 and 15).

1.2. Suppressive Soils

Disease-suppressive soils are defined as those soils in which disease development is suppressed when the pathogen is present or is introduced in the presence of a susceptible host (Schneider, 1982; Linderman et al., 1983). Wheat planted in take-all-infested soil initially is infected heavily, but after several successive wheat crops, the disease largely disappears (Cook, 1981). This phenomenon is called take-all decline. Unfortunately, this example of management of an agricultural system to encourage development of disease suppression is impracticable from a farmer's point of view. It usually would not be feasible for a farmer to endure several years of destructive disease levels to eventually attain a tolerable level of disease. A more realistic approach to inducement of suppressive soil is in the case of avocado trees in Australia which are highly susceptible to root rot caused by *Phytophthora cinnamomi* (Broadbent and Baker, 1975). Disease suppression is induced in infested soils by addition of large amounts of manure, plant material and lime. With this practice, the amount of organic matter, the soil structure, cation exchange capacity, and calcium content of the soil are maintained at high levels. These cultural practices encourage greater microbial activity and a general soil suppressiveness based on competition of the soil microbiota towards *P. cinnamomi*, thus less disease development. Another example of this type of general disease suppression induced by organic matter was developed by the Aztecs in Mexico (Lumsden et al., 1987). The chinampa system of agriculture was developed to its peak around modern-day Mexico City by the pre-Columbian Aztec empire. It has enhanced disease suppressive properties toward *Pythium* spp., described as having similar characteristics to the Australian system.

1.3. Addition of Specific Antagonists

Whereas biological control can be encouraged by ecological manipulations of the soil–plant environment (i.e. in applied terms, crop management systems) the greatest emphasis on research has been with the addition of specific antagonists to soil systems. Antagonists tested include bacteria, amoebae, actinomycetes and fungi. Genera of organisms frequently used have been summarized by Cook and Baker (1983) and total more than 40.

Antagonists may be added to ecosystems to protect roots in several ways, depending upon the specific substrate/infection court to be colonized. Strategies differ for selecting, manipulating and delivering bioprotectants to each area or substrate to be colonized. The various general cases are considered below.

1.3.1. Addition to Bulk Soil or Planting Mix

Bioprotectants may be added to bulk soil to establish a new population or augment natural populations (Papavizas and Lewis, 1981). Deliberate introduction of antagonists into the environment was thought at first to be futile (Garrett, 1956). Attempts to increase the population of antagonists were thought to be contrary to the ecological principle that ''the population reflects the habitat'' and that introduced changes are only transient. Although this concept is largely true (Papavizas and Lewis, 1981), notable exceptions exist. Lumsden and Locke (1989) introduced *Gliocladium virens* into soilless mixes in a glasshouse environment with successful control of damping-off of zinnia caused by *Pythium ultimum* and *Rhizoctonia solani*. In a similar case, the addition of an antagonist to pasteurized greenhouse soil led to successful control of chrysanthemum blight caused by *Fusarium solani* (Locke et al., 1985). These data indicate that application of bioprotectants to specialized environments where ecological and environmental variables can be controlled is likely to be successful.

Others have described field control of pathogens with in-furrow or broadcast applications. Wells et al. (1972) were among the first to report control in the field, in this case of *Sclerotium rolfsii* on tomato seedlings, with *Trichoderma harzianum*. Backman and Rodriguez-Kabana (1975) used a two-component carrier system consisting of diatomaceous earth impregnated with a molasses food base for control of *S. rolfsii* with *T. harzianum*. Generally, since nutrients are lacking in soil, the bioprotectants must be added at or near the level required for effective biocontrol, or adequate nutrients or food base must be supplied (Papavizas and Lewis, 1981). Further, bioprotectants will have to be produced or formulated in such a way that they will remain active in the soil or planting mix for the required period needed for protection.

1.3.2. Seed Treatment as a Delivery System for Root Protectants

Seed treatments are excellent delivery systems for root protectants. These treatments require a smaller amount of material than that required for bulk addition to soil volumes, and are an accepted practice in commercial agriculture. Moreover, biological seed treatment provides the significant economic advantage of protecting the seed against rots. For biological seed treatments to protect the roots, the bioprotectant first must grow well on the seed, then transfer to, and grow around, upon, or within the root. Good growth on the seed is a prerequisite for successful root colonization, and is also critical for successful seed protection. Therefore, any modification of treatment procedure or of the organism itself that enhances growth of the protectant upon the seed, increases its efficacy as a seed protectant. Seed treatments have been reviewed (Kommedahl and Windels, 1981).

1.3.3. Direct Application to Plant Roots

Most notable among examples of biological control of plant disease is the successful commercial application of the bacterium *Agrobacterium radiobacter* (Strain 84) for control of crown gall caused by *Agrobacterium radiobacter* var. *tumefaciens* (Kerr, 1980). The non-pathogenic strain of *A. radiobacter* produces an antibiotic, agrocin, which inhibits the pathogenic strains of *A. radiobacter* (Chapters 6 and 7). At transplant, bioprotectant is applied to the entire root system of young trees or other plants being protected and thus prevents invasion of the rhizosphere by the pathogen.

2. PROBLEMS ASSOCIATED WITH BIOLOGICAL CONTROL

Biological control has met with only limited success to date. In terms of overall agricultural pest control strategies biological control is at present an insignificant component of pest management systems. It is well recognized that protectants do not usually perform spectacularly and often do not work as quickly or as reliably as broad-spectrum chemicals (Cook and Baker, 1983). In order to establish biocontrol efficacy we must realize that there are many inherent obstacles to overcome. These factors include basic aspects of strain selection, efficient production of biomass, formulation, storage ability and methods of application. Also, we need a better understanding of mechanisms of action, nutrition and ecology of protectants and pathogens. We need to examine the potential for genetic manipulation of fungi to create genetically superior strains or hybrids that can perform better than wild types selected from the environment.

The soil is a complex and competitive environment for microorganisms. Yet this is the environment from which bioprotectants are often isolated and into which bioprotectants are usually introduced. Soil edaphic factors and biological components can vary markedly even within a single field. Moisture may vary from waterlogged conditions, where oxygen is deficient, to extreme dryness. Temperature may range from 0 °C to greater than 30 °C, and the pH may vary from pH 5 to pH 8. Levels of organic matter, available nutrients and ionic balance may differ.

Moreover, there are very large numbers of microorganisms in soil. Bacterial numbers range from $<10^6$ to $>10^9\,g^{-1}$ soil, may account for up to 0.2% of the soil weight, and occupy about 0.4% of the space available to them (Clark, 1967). Soil fungi, while lower in numbers than bacteria, may contribute even more to the total soil biomass (Anderson and Domsch, 1973). These, together with the soil fauna, make up an extremely diverse, densely populated and highly competitive microenvironment. This environment is so highly competitive that most soils exhibit fungistasis (which applies equally well to bacteria), which may be defined as exogenous dormancy imposed upon propagules of microorganisms. This stasis is readily broken by the addition of nutrients (Lockwood, 1977). When nutrients are added, the soil microflora will proliferate. These properties of soils make them extremely stable ecologically, and so special strategies are required to enable a bioprotectant to become established and/or function in sufficient numbers for biocontrol to be successful.

For biocontrol agents to be effective, they need to grow and proliferate (Harman and Hader, 1983) and thus compete with other soil microflora to protect infection sites or destroy pathogen propagules. Of particular and critical importance is rapid initial colonization of the specific plant part (e.g. newly planted seeds or newly formed roots) by the bioprotectant. A good example of problems that may arise in this regard is provided by the interactions of *Pseudomonas, Trichoderma* and *Pythium* spp. on and in freshly planted seeds. Sporangia of *Pythium* spp. rapidly germinate and infect seeds within 4–6 h of planting (Stasz et al., 1980), and soil-borne pseudomonads also colonize seeds rapidly, with levels of 10^6 per seed within 24 h of planting (Hubbard et al., 1983). Spores of the bioprotectant simply coated onto seeds do not germinate until 10–14 h after planting (Hubbard and Harman, unpublished; Lifshitz et al., 1986). The pseudomonads, particularly in low-iron soils, are strongly antagonistic to some *Trichoderma* strains and prevent their growth (Hubbard et al., 1983). Thus, the bioprotectant, even though it possesses reasonable ability to parasitize *Pythium* (Chet, 1987), is at a double disadvantage, i.e. (1) the pathogen colonizes and occupies the infection court before the bioprotectant becomes active, and (2) the bioprotectant is damaged by antagonistic soil microflora. Not suprisingly, seed protection is not obtained under these conditions. If, however, the *Trichoderma*-treated seeds are allowed to germinate for 12–24 h before planting, the ability of the bioprotectant to control seed rots is markedly improved. Even a short period where the biocontrol strain can proliferate in the absence of its naturally-occurring competitors and colonize the appropriate infection sites can markedly increase its apparent efficacy.

3. STRATEGIES FOR ENHANCING EFFICACY AND RELIABILITY

Recognizing that living systems are difficult to manipulate and that complex ecological environments compound this difficulty, we need to provide every possible advantage to the bioprotectant for biocontrol to be successful. This means that we will require as much knowledge as possible about the biological control agent to encourage its growth, metabolism and survival, so that it may successfully overwhelm pathogen activity. We also need to learn more about the behavior of the pathogens themselves. The following are strategies for improving the efficacy of biocontrol agents.

3.1. Selection and Screening of Strains

Many past efforts to isolate bioprotectants have been done empirically, ranging from random selection from plant surfaces and soil to chance encounters on culture plates where zones of growth inhibition were detected. A better approach to selection is to examine natural biological niches where direct effects on plant pathogens would be expected to occur. As mentioned earlier, disease-suppressive soils are those where biological control occurs naturally. It is in these soils where we might hope to locate effective strains of biocontrol agents. Lumsden et al. (1987) examined *Pythium* damping-off suppression in soils from a traditional chinampa agroecosystem in the valley of Mexico compared to soil from a nearby agroecosystem. Strains of antagonists isolated from the suppressive soil were capable of reducing disease caused by *Pythium* spp. in the greenhouse.

Disease-suppressive soils are often difficult to identify. Therefore, in the absence of identifiable disease-suppressive soils, an alternative is to examine the rhizospheres of healthy plants in otherwise heavily infested agricultural sites. Isolation of microorganisms from these disease-free root systems may enhance the chances of obtaining effective strains.

Once strains of potential bioprotectants have been isolated, systems must be designed to assess biological control ability to eliminate the vast majority that are ineffective. This assessment should be done in an environment as similar as possible to the environment where the antagonist will be applied. If the target is a pathogen of annual crops or potted plants in greenhouse production, then assessment in greenhouses or growth chambers is appropriate (Lumsden and Locke, 1989). If the target is a pathogen of annual or perennial crops in field production then outdoor beds or microplots are best used (Adams and Ayers, 1982). Natural soils should be used (Linderman et al., 1983) or, if appropriate, commercially used soil mixes or soilless potting products (Lumsden and Locke, 1989). The assays should be standardized with a known amount of antagonist inoculant, and an amount and type of pathogen inoculum that will give a level of disease so as not to overwhelm the antagonist. Pathogen inoculum should be as ''natural'' as possible. In some cases, indirect screening assays may yield useful information. Most of these methods are *in vitro* assays which may give possible insights into mechanisms of action. These assays include studies of mycoparasitism; competition for space and nutrients; production of antibiotics, metabolites or lytic enzymes; induction of resistance; prevention of pathogen propagule germination; and ability of antagonists to proliferate in soil, colonize organic substrates, and affect pathogen survival (Cook and Baker, 1983; Papavizas and Lumsden, 1980). However, it is important to remember that indirect screening methods may only infer potential for biocontrol and may be misleading. To date, nothing can replace the results obtained from pathogen–antagonist–host interactions in a natural ecosystem.

3.2. Modification of the Soil Environment

As discussed earlier in this chapter, the soil environment is extremely complex and competitive and introduced microorganisms may have great difficulty competing. For this reason, it is imperative to reduce the amount of competition imposed on the introduced bioprotectant. Several modifications to the soil environment have been described in the past, including modification of the pH of soil, partial sterilization, and addition of numerous organic substrates (Papavizas and Lewis, 1981). To date, however, none of these has been used extensively to control soil-borne plant pathogens successfully. We need new approaches to solve old problems. One such approach is to modify our concept of the agricultural environments in which to apply bioprotectants to enable them to perform more effectively.

3.2.1. Selection of the Environment for Application of Biocontrol

Attempts at biocontrol in the past have been done without regard for where the biocontrol agents perform best. Therefore, instead of attempting to change a well-buffered, complex soil environment, we should attempt to select the environment where biocontrol might work best. The level of ecological complexity where the antagonist–pathogen–host interaction will occur may have more bearing on the success of biocontrol than our attempts to modify the environment. For example, biocontrol may be expected to work best in a simple environment, such as the surface of freshly cut tree stumps (Rishbeth, 1975), less well in glasshouse environments (Lumsden and Locke, 1989), and least well in complex field crops or undisturbed ecosystems (Papavizas, 1985). For application of biocontrol, we perhaps should turn away from looking for an all-encompasing panacea to specific, specialized applications.

3.2.2. Improving the Soil Microenvironment for Bioprotectants

A recent approach has been introduced which may greatly enhance the possibilities of tailoring the microenvironment in soil for improving bioprotectant performance (Fravel et al., 1985; Lewis and Papavizas, 1985). This approach is based on the concept of entrapping propagules of antagonists within pellets formed with an alginate matrix. The process of making these pellets consists of mixing biomass (fermentor biomass or harvested cells) with a bulking agent. The bulking agent may be inert, or may provide a food base (e.g. powdered wheat bran) for rapid and non-competitive growth of the antagonist in soil. The mixture is dripped into a calcium chloride solution to form insoluble gelatinous beads which dry into stable, hard pellets or granules of uniform size. Alginate pellets have been prepared with bacterial cells, spores of *Talaromyces flavus*, and fermentor biomass of *Trichoderma* and *Gliocladium* isolates and of other fungi (Fravel et al., 1985; Lewis and Papavizas, 1985, 1987). Addition to soils of pellets containing various biocontrol fungi resulted in the reduction of various diseases (Lewis and Papavizas, 1987; Lumsden and Locke, 1989), and allowed proliferation of the biocontrol fungi in soil (Lewis and Papavizas, 1985). This approach enables incorporation of the bioprotectant into a discrete particle that may also include other additives that can give the biocontrol agent enhanced capability to overcome competition from the soil microbiota.

3.3. Modification of Seed Treatment Systems

Seed treatments may be used as a delivery system for root protectants; however, their first function is to protect seeds. Two main strategies have been employed to improve biological seed treatment systems. The first of these is to use modifications of various physiological seed treatment systems originally designed to enhance seed vigor or performance, while the second is to add various amendments to seeds specifically to enhance activity of the bioprotectant.

3.3.1. Seed Priming

Seed priming is defined by Bradford (1986) as controlled hydration of seeds to a level that permits pre-germinative metabolic activity to proceed, but that prevents actual emergence of the radicle. Seeds are placed at a controlled temperature in an aerated aqueous solution of salts, or polyethylene glycol, to provide sufficient osmotic potential to prevent sprouting. Thus the critical components of such systems are a non-phytotoxic solution of high osmotic strength, a source of aeration, and controlled temperature. Seeds are placed in such systems for several days. As a consequence of activation of cellular machinery, seeds are more resistant to various biotic and abiotic stresses and emerge more rapidly in the field than do non-treated seeds (Bradford, 1986; Taylor et al., 1988). However, such systems are expensive and not readily adaptable to treatment of large volumes of seeds. Taylor et al. (1985) showed that, as in the aerated water systems, bacterial proliferation resulted in a relatively high measure of seed protection, and thus such systems could be used for delivery of biocontrol strains.

Recently, a modification designated solid matrix priming (SMP) was developed by John Eastin, Kamterter Inc., and utilized by Harman and Taylor (1988) to improve markedly the efficacy of biological seed treatments. In this process, seeds are moistened, and bioprotectants are added. To this is added porous solid material, such as powdered coal or peat moss, with high water-holding capacity which permits priming to occur (Harman and Taylor, 1988).

Efficacy of *Trichoderma* and *Enterobacter cloacae* were markedly enhanced by SMP with cucumber seed in soils heavily infested with *Pythium ultimum* (Fig. 1). Without SMP, non-treated seeds did not emerge, since all of them were rotted by the pathogen. With SMP, the bioprotectants were very effective for control of pre-emergence damping-off. Initial stands with seeds treated with *E. cloacae* increased from 0 to 70% as a consequence of SMP, and those treated with *T. harzianum* increased from 30 to 90%. However, post-emergence damping-off was not as effectively controlled. With *E. cloacae*, stand was reduced to 0, and with *T. harzianum*, to about 60%. SMP did not improve performance of the fungicide thiram. Consequently, overall performance of biological seed treatments using *T. harzianum* was considerably better than treatment with thiram (Fig. 1).

With tomato, results differed in that *T. harzianum* was as effective as thiram even in the absence of SMP. Further, activity of *E. cloacae* dropped to nearly zero after SMP treatment. The differences in the results with tomato and cucumber can be attributed to the fact that the pH of the tomato seeds was strongly acidic (pH 2.8), since tomato seeds are commercially extracted with HCl (Harman and Taylor, 1988). This pH favors *Trichoderma* (Weindling and Fawcett, 1936; Wright, 1956) but is harmful to bacteria. Similarly, the pH of the solid matrix was critical; the matrix material could be obtained at either pH 4.1 or 6.6. The pH 4.1 material was highly effective in enhancing activity of *T. harzianum*, but less effective with *E. cloacae*, while the converse was true with the pH 6.6 material (Harman and Taylor, 1988).

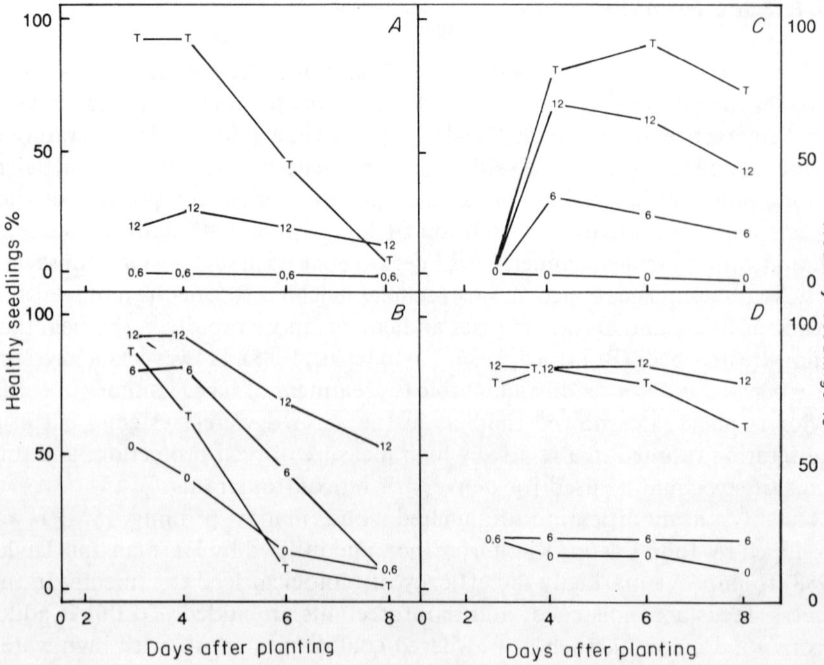

Fig. 1. Percent healthy cucumber (*A* and *B*) and tomato (*C* and *D*) seeds at various times after planting. *A* and *C* provide percentages of seedlings produced from seeds in the absence of solid matrix (SMP) while *C* and *D* provide percentages in the presence of SMP. Symbols on graphs are for seeds that are not treated (0), treated with thiram (T), treated with *T. harzianum* strain T12 (12), or with *E. cloacae* strain E6 (6). The pH of cucumber seeds was near 7, while the pH of tomato seeds was about 3, since they had been treated by the supplier with HCl. Reprinted with permission from *Phytopathology*, **78**:520–525.

The mechanisms of the enhancement of the bioprotectant ability are related to growth of the microorganisms during the priming process. Numbers of colony-forming units of either microorganism increased 10-fold during priming, when they were applied in systems providing favorable pH levels. Increased levels of both organisms persisted when seeds were dried and stored (Harman and Taylor, 1988). This growth permits full colonization of the seedling infection sites, and allows utilization of seed exudates prior to planting the seed in the competitive soil environment.

3.3.2. Fluid Drilling

Fluid drilling is a second system for enhancing performance of seed. In this procedure, seeds are germinated under controlled conditions prior to planting. Usually, seeds are added to an aerated water mixture at a controlled temperature and removed after emergence of the radicle. These sprouted seeds are then transferred to a gel to avoid damage from handling the fragile sprouted seeds.

The seed–gel mix is added to a planter equipped with a peristaltic pump, and this mix is delivered into a planting furrow (Taylor, 1977). There are two opportunities in such a system to add bioprotectants. First, the microorganisms can be added to the aerated water solution in which the seeds are sprouted. Exudates from the seeds leach into the water in sufficient quantities to support microbial growth. Thus, actively growing bioprotectants (e.g. germlings of *Gliocladium* or *Trichoderma*) could be produced this way, or actively growing bacteria may be obtained. Secondly, various propagules, either actively growing or quiescent, can be added to the gel in which the seeds are suspended, and this used as a delivery system. In either case, bioprotectants can be delivered both to the seed and to the planting furrow.

Hadar et al. (1984) added conidia of *Trichoderma* strains to various gels used in fluid drilling, and found that protection was obtained. The level of protection was dependent on the choice of gel: *Trichoderma* in a silica-based material at pH 9 was largely ineffective, while the same strains in a cellulosic gel at lower pH were more effective. Regardless of the gel used, however, most conidia did not germinate, and were therefore ineffective in biocontrol. Conway (1986) used similar cellulosic gels to deliver sclerotia of *Laetisaria arvalis* to control *Rhizoctonia solani* on pepper, and sclerotia of *L. arvalis* or chlamydospores of *T. harzianum* to control *Sclerotium rolfsii* on apple. Control was obtained with both systems.

3.3.3. Seed Pelleting

Seed pelleting is a process in which seeds are coated with successive layers of a mixture of a binder and a solid filler. Usually, irregularly shaped seeds are pelleted to aid in precision planting. Seeds frequently are encased in layers of pelleting material that may be several millimetres thick.

Seed pellets, because of their bulk, offer the potential of delivering bioprotectants with amendments to enhance their activity. Wu (1976) obtained improved stands of small grains with seeds pelleted with bioprotectants, when compared to non-treated seeds. Pelleting alone improved stands. Harman and Hadar (1983) similarly found that pelleting alone reduced seed rots caused by *Pythium* spp., but that addition of *Trichoderma* did not further reduce seed rots. Much remains to be done to exploit the full potential of pelleted seeds as a delivery system to augment activity of bioprotectants.

3.3.4. Seed Treatment Amendments

Another strategy to enhance activity of bioprotectants is to add specific amendments to seed treatments to modify the pH of the seed coating. The pH of any seed treatment should be strongly buffered to a region optimal for the microbe in question. In addition, specific nutrients may be added that will favor the organism. Nelson et al. (1988) tested 88 compounds for their ability to support growth of *Trichoderma harzianum* or *T. koningii* while not supporting growth of *Pythium ultimum*. Such compounds were evaluated further as seed treatment amendments. Activity of *T. koningii* was stimulated primarily by fatty or organic acids, while most of the compounds enhancing activity of *T. harzianum* were carbohydrates (Table 1). Earlier, Harman et al. (1981) found that activity of *T. hamatum* was stimulated by chitin or peat, while cellulose reduced seed protection using this organism. Both studies utilized a single concentration of the compounds in question; Nelson et al. (1988) used a concentration of 1 mg ml^{-1}, while Harman et al. (1981) used 30 mg ml^{-1}. The concentration of the compounds used is critical, and a great deal of testing of combinations and permutations of compounds and concentrations will have to be done. It seems obvious, however, that appropriate augmentation should give markedly improved seed treatment systems.

TABLE 1. Compounds promoting biological control activity of *Trichoderma koningii* and *T. harzianum*[a]. Reprinted with permission from *Soil Biol. Biochem.*, 20: 145–150

T. koningii		*T. harzianum*	
Compound	*Increase in biocontrol activity[b] (%)*	*Compound*	*Increase in biocontrol activity (%)*
Myristic acid	48	Arabitol	44
Linolenic acid	48	Starch	44
Caprylic acid	36	Inulin	36
p-hydroxybenzoic acid	32	Ribose	32
Maleic acid	32	Adonitol	32
Tannic acid	32	Malonic acid	32
Oleic acid	32	Mannitol	28
Palmitic acid	28	Laminarin	20
Chloroform	28	Xylan	20
Stearic acid	28	Sorbic acid	24
Tween 40	24	Butyric acid	20
Lauric acid	24		
Caproic acid	24		
p-hydroxybenzaldehyde	20		

[a]Includes only those compounds providing significant (p = 0.05) increases in activity over the *Trichoderma* seed treatments alone. Compounds incorporated into a 2% methylcellulose (Methocel A4C) suspension (1 mg ml^{-1}) containing 10^8–10^9 *Trichoderma* conidia ml^{-1}.
[b]Means represent % increases in activity calculated using the formula: % increase = [(TC) − C] − (T − M), where TC = % emergence from seeds treated with *Trichoderma* + compound; C = % emergence from seeds treated with compound only; T = % emergence from seeds treated with *Trichoderma* only; and M = % emergence from methylcellulose-treated seeds (control).

3.3.5. Seed Treatment/Rhizosphere Competence

A distinctly different strategy is to apply the bioprotectant to seed or other plant parts, and for it to grow onto the developing roots, proliferate on or around the surface of the root, and protect against root pathogens. The ability of an organism to transfer and to colonize the root surface is designated rhizosphere (or rhizoplane) competence (Ahmad and Baker, 1987). Bacteria with rhizosphere competence have been known since at least 1978 (Burr et al., 1978) and include several genera (e.g. *Pseudomonas, Enterobacter* and *Azospirillum*) (Burr et al., 1978; Okon, 1985; Chao et al., 1986; Lynch, 1987).

Rhizosphere competence among fungi was more recently discovered. It appeared that most fungi lacked this characteristic (Chao et al., 1986). However, through mutation, protoplast fusion, and even by careful selection, strains of *Trichoderma* have been developed or discovered that are highly rhizosphere competent (Fig. 2) (Ahmed and Baker; 1987; Sivan et al., 1987; Harman et al., 1989).

3.3.6. Seed Treatment/Internal Plant Colonization

Research and development for the introduction of bioprotectants into the innermost plant spaces is only in its infancy. Of course, some beneficial microbes (e.g. *Rhizobium*, mycorrhizal fungi) do colonize the outer portions of roots. In this section, we wish, however, to consider the endophytic microbes that colonize the xylar vessels and fluids.

Gardner et al. (1982) examined the microbial composition of xylar fluids or the xylem itself from citrus roots and isolated a total of 556 bacteria. In order of frequency of isolation, *Pseudomonas* (primarily non-fluorescent strains, particularly *P. cepacia*), *Enterobacter, Bacillus* and *Serratia* spp. were most frequently isolated. A strain of *P. cepacia* was studied further, and its ability to persist in the growing plant in the xylem or in the rhizosphere was compared. Greater numbers of the bacterium were present in the xylem 11 months after inoculation than in the rhizosphere. These data clearly demonstrate that bacterial species previously identified as effective in biocontrol of root surfaces (e.g. *Pseudomonas, Enterobacter*; cf. Cook and Baker, 1983) may proliferate in the xylem of the plant. It seems reasonable that strains might be found that persist in both the xylem and rhizosphere, and that such strains might provide both an internal and an external layer of protection.

4. GENETIC IMPROVEMENT

Most biological control agents that have been described are indigenous, naturally-occurring microorganisms. These will continue for some time to be a

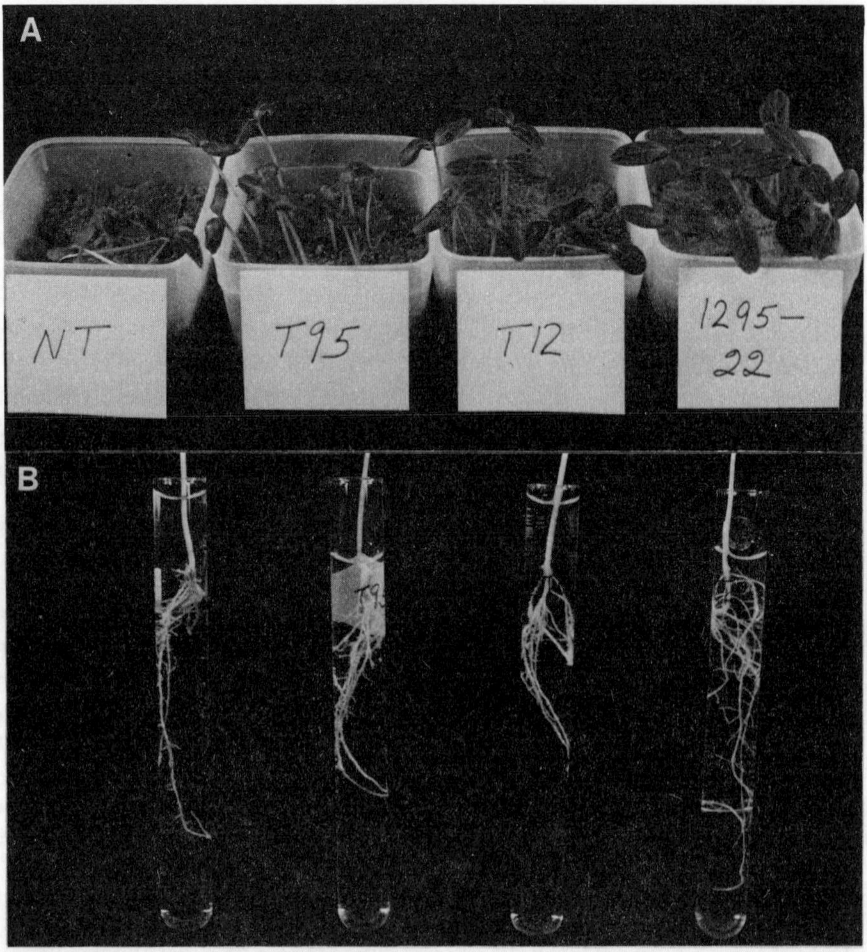

Fig. 2. Appearance of cucumbers 9 days after planting from seeds treated with Pelgel alone, or with *T. harzianum* strains T95, T12, or 1295-22. Strain 1295-22 was produced by protoplast fusion of the other two strains, and is strongly rhizosphere competent. Reprinted with permission from Harman et al. (1989).

main source of potentially effective biocontrol agents. However, in the future we will undoubtedly rely on genetic manipulation and improvement of strains of antagonists isolated from the wild for major advances to be made in biocontrol. The following are strategies for improving the biological control ability of bioprotectants.

4.1. Strategies of Genetic Modification of Antagonists

Once effective antagonists are selected, it is possible to produce superior strains by genetic modifications. The choices for such modification are non-specific mutation (as induced by irradiation, chemical mutagens or plasmid insertion), sexual or asexual recombination of genomes of two parental strains, or site-specific transformation.

Sexual or asexual recombination (i.e. through anastomosis, protoplast fusion or sexual processes) permits recombination of the genomes of two strains. When strains are identified with specific complementary activities (e.g. ability to control a different range of pathogens or improved ability to compete effectively with competitive soil microflora), a useful strategy may be to combine these valuable traits into a single superior organism. Such a procedure may be the method of choice if the traits to be combined are undefined or if their expression is controlled by several genes. Recombinant systems are useful for addition of specific genes. Such techniques also permit specific deletion and addition of specific alleles and their products; therefore, they can be used to determine and quantify the role of specific genes in the biocontrol process via a molecular equivalent of Koch's Postulates. This permits a more precise evaluation of mechanisms than heretofore has been possible. In Chapter 6 there is a discussion of plant and microbial genetics, including modern techniques of genetic engineering.

4.1.1. Mutation

Mutation using chemical mutagens or radiation has been used to produce improved strains of biocontrol fungi. Papavizas and his co-workers (Abd-El Moity et al., 1982; Papavizas et al., 1982; Papavizas and Lewis, 1983) used selection and mutation to develop strains resistant to benomyl and related fungicides. These differed in their antibiotic production and in their ability to control soil-borne plant pathogens. Some were improved in their biocontrol capability, while others were poorer in this attribute relative to the parents. Antibiotic production was not correlated with biocontrol ability. Ahmad and Baker (1987) similarly prepared mutants of *T. harzianum* that were resistant to benomyl. Most were stronger rhizosphere colonizers than the non-mutated strains. As mentioned earlier, this ability is critical if biocontrol strains are to protect roots successfully when seed treatments are used as the delivery system.

Insertional mutation has been used to inactivate specific genes in bacteria. Such mutation can be shown to occur only at a single site, and therefore effects can be demonstrated to occur as a consequence of the deletion of activity of a single gene. In this way Weger et al. (1987) showed that *Pseudomonas fluorescens* required flagella to colonize potato roots. However, in similar experiments Scher et al. (1988) found no character, including motility, that was required for rhizosphere competence. Thomashow and Weller (1988) demonstrated that phenazine antibiotics were required for control of take-all on wheat by *Pseudomonas* spp. specifically deleting genes coding for these antibiotics (see also Chapters 7 and 15). Such studies are among the first to utilize insertional mutation to begin unequivocal studies on the mechanisms of biocontrol.

4.1.2. Sexual and Asexual Hybridization

Biocontrol ability is a property of specific strains, rather than being widely distributed among most strains of a taxon. For example, Lumsden and Locke (1989) showed that some strains of *Gliocladium virens* have good ability to control *Pythium* spp., others were effective against *R. solani*, while others were effective against both, and some were ineffective against either pathogen. Still others may possess other desirable traits (e.g. ability to grow and protect at low temperatures, ability to compete well against soil microflora, good rhizosphere competence, etc.). Such traits are likely to be polygenic in nature, and certainly their genetic bases are undefined. Therefore, it seems desirable to have systems permitting the combination of the entire genome of strains with desirable complementary traits.

Either sexual or asexual processes can accomplish such recombination. With many bioprotectants, the sexual stage is lacking, and therefore sexual recombination has been little used for strain improvement. Katan et al. (1984) used sexual recombination to elucidate the genetics of fungicide resistance in the bioprotectant *Talaromyces flavus*, but we know of no reports on strain improvement by this procedure. More has been done with asexual recombination in fungal bioprotectants. Fungi have a parasexual cycle in which plasmogamy (cell fusion without nuclear fusion) occurs via anastomosis. This gives rise to a mixture of nuclei and other organelles from both parents. Such heterokaryons by definition contain all the genetic information from both parents. Occasionally, dissimilar nuclei fuse (karyogamy) and give rise to a diploid strain. With time some chromosomes may be lost, while (rarely) crossing over may occur during mitosis. Aneuploid strains frequently result, with traits derived from both parents (Pontecorvo, 1956; Peberdy and Ferenczy, 1985).

Heterokaryons may be produced in large numbers by protoplast fusion. Typically, in this process, nutritional auxotrophs are produced by chemical or radiation mutagenesis. Cell walls are then removed by enzymatic digestion to release single protoplasts. Protoplasts from the two different nutritional auxotrophs are mixed and fused by addition of polyethylene glycol or high pH plus calcium. Electroporation also can be used for fusion. The mixture of fused and non-fused protoplasts can then be placed on a medium deficient in the nutrients required for growth of either auxotrophic parental strain. Only those fusion products with nuclei and genes from both parents can grow, and therefore, heterokaryons are selected. Such heterokaryons can be maintained on these selective media, and their continued growth as a single thallus forced by nutrient pressure. Other methods of selection and forced growth are possible (Mellon, 1985). Reliable protoplast preparation and fusion systems have been developed for several fungal genera, including *Penicillium, Aspergillus* and *Saccharomyces* (Peberdy and Ferenczy, 1985), *Cephalosporium* (Minuth and Esser, 1983), and *Trichoderma* and *Gliocladium* (Stasz et al., 1988; Harman and Stasz, 1989).

Protoplast fusion among some closely related strains of *Penicillium* and *Aspergillus* results in progeny that immediately are fully prototrophic heterokaryons, and that grow as rapidly as the parental strains. Some nuclei in these strains undergo kayrogamy and the parasexual cycle is completed. In other cases, however, fusions between less closely related strains are less likely to occur. Progeny initially grow very slowly, and they sector to give rise to more rapidly growing prototrophic strains. Such rapidly growing sectors have been described as "hybrids" and presumed to result from classic parasexuality (Anne and Peberdy, 1985).

Fusions among strains of *Trichoderma* give rise to progeny similar to the latter class (Stasz et al., 1988; Harman et al., 1989). Following fusion, a very few progeny give rise to weakly prototrophic strains. Many grow poorly even on permissive media. When their isozymic pattern is analyzed, they invariably are similar to one parent or the other. In any given cross, all or nearly all of the progeny strains will be identical to one (the prevalent) parental strain, while few or none will be identical to the other (non-prevalent) parental strain. These strains are unstable, and sectors develop. These sectors frequently grow more rapidly than the original progeny strain, and are more strongly prototrophic. They express a very great range of colony morphologies, and may be isozymically identical with either parental strain. A discussion of the post-fusion genetic events in *Trichoderma* is given elsewhere (Harman et al., 1989).

Protoplast fusion in *Trichoderma* gives rise to a wide range of variation in progeny strains. When coupled with effective screening procedures, superior biocontrol strains can be obtained. For example strain 1295-22, which is a selection from a fusion between *T. harzianum* strains T12 and T95, grows more rapidly than either parental strain, and is a more efficient seed protectant. Perhaps more importantly, it is very strongly rhizosphere competent and can protect plant roots (Fig. 2; Harman et al., 1989). In field trials with peas, nearly all (>95%) of root segments sampled from mature plants grown from treated seeds were colonized by the new strain, while many fewer (<50%) of root segments of plants produced from seeds treated with the parental strains were colonized (Harman et al., 1989).

4.1.3. Transformation Systems

The recent advances in molecular biology will make an important impact on strain improvement in bioprotectants. It will be possible to delete specific genes, and to add any identifiable gene from any source to any organism. With this technology, it will be possible to elucidate very specifically the mechanisms of biocontrol at the gene level. This will make it possible to prove the role of a specific gene in biocontrol activity via Koch's Postulates. Specific genes can be found to be associated with a specific biocontrol property, the gene can be isolated, and inserted into an organism that does not possess it. Moreover, the activity of specific genes (i.e. quantity of gene product) may be enhanced by linking them with specific, highly active promoter regions. Finally, critical genes can be attached to regulator regions that will allow production of the specific gene product only when it is needed, and not at other times. Such regulation will be important to minimize accumulation of undesirable products of bioprotectants in foodstuffs, and to minimize detrimental effects on the bioprotectant caused by constant overproduction of products of added genes.

The first steps already have been taken; e.g. the gene for the toxin (delta-endotoxin) to Lepidoptera from *Bacillus thuringiensis* has been cloned into the endophytic bacterium *Clavibacter xyli* subsp. *cynodotis* (Fahey, 1988) and into rhizosphere competent *Pseudomonas fluorescens* (U.S. Congress, Office of Technology Assessment, 1988). These are the first of numerous new strains prepared through recombinant DNA systems that will be available in the near future.

There are limitations to this technology, however. The most serious of these will be that no doubt many biocontrol processes are polygenic (for example, mycoparasitism involves tropic growth, attachment, production of extracellular enzymes, formation of haustoria-like structures and eventual invasion of the target hypha) (Chet, 1987). Such polygenic characters will be difficult to manipulate using recombinant DNA systems, since these depend upon isolation, cloning and insertion of specific gene systems. For manipulating such complex characters, sexual or asexual recombination offers a more straightforward approach.

5. CONCLUSION

Biological control of plant pathogens with bioprotectants has not yet materialized to a large extent for practical solutions of agricultural problems. We are at a significant turning point in biotechnology that may nevertheless enable us to make appreciable advances in the development and application of biological control agents. Some of the strategies that we will employ in the development of microorganisms for disease control are as follows:

(1) Improved methods of isolating and screening potential biocontrol agents, including searching for them in naturally suppressive soils and microsites.
(2) Improving the procedures for producing and formulating bioprotectants.
(3) Modifying goals and approaches for disease control by selecting specific diseases and agricultural sites for applying bioprotectants.
(4) Improving strategies for favoring bioprotectant performance with amendments that stimulate their activity, whether in bulk soil formulations or as root or seed protectants.
(5) Determining factors which encourage their ability to colonize plant spermospheres and rhizospheres.

(6) Exploitation of new technologies, especially genetic manipulation and hybridization, to induce new and improved biotypes.

Our ability to manage plant disease problems with microbial approaches will be improved with implementation of these new strategies to favor bioprotectants in a hostile ecological environment over other competing microorganisms and the plant pathogens themselves.

REFERENCES

Abd-El Moity, T. H., G. C. Papavizas, and M. N. Shatla (1982). Introduction of new isolates of *Trichoderma harzianum* tolerant to fungicides and their experimental use for control of white rot of onion. *Phytopathology*, **72**:396–400.

Adams, P. B. and W. A. Ayers (1982). Biological control of Sclerotinia lettuce drop in the field by *Sporidesmium sclerotivorum*. *Phytopathology*, **72**:485–488.

Ahmad, J. S. and R. Baker (1987). Rhizosphere competence of *Trichoderma harzianum*. *Phytopathology*, **77**:182–189.

Anderson, J. P. E. and K. H. Domsch (1973). Quantification of bacterial and fungal contributions to soil respiration. *Arch. Mikrobiol.*, **93**:113–127.

Anne, J. and J. F. Peberdy (1985). Protoplast fusion interspecies hybridization in *Penicillium*. In *Fungal Protoplast, Applications in Biochemistry and Genetics* (Peberdy, J. F. and L. Ferenczy, eds.). Marcel Dekker Inc., New York, pp. 259–278.

Backman, P. A. and R. Rodriguez-Kabana (1975). A system for the growth and delivery of biological control agents to the soil. *Phytopathology*, **65**:819–821.

Bradford, K. J. (1986). Manipulation of seed water relations via osmotic priming to improve germination under stress conditions. *Hort. Sci.*, **21**:1105–1112.

Broadbent, P. and K. F. Baker (1975). Soils suppressive to *Phytophthora* root rot in eastern Australia. In *Biology and Control of Soilborne Plant Pathogens* (G. W. Bruehl, ed.). American Phytopathological Society, St. Paul, Minnesota, pp. 152–157.

Burr, T. J., M. N. Schroth, and T. Suslow (1978). Increased potato yields by treatment of seed pieces with specific strains of *Pseudomonas fluorescens* and *P. putida*. *Phytopathology*, **68**:1377–1383.

Chao, W. L., E. B. Nelson, G. E. Harman, and H. C. Hoch (1986). Colonization of the rhizosphere by biological control agents applied to seeds. *Phytopathology*, **76**:60–65.

Chet, I. (1987). *Trichoderma* application, mode of action, and potential as a biocontrol agent of soilborne plant pathogenic fungi. In *Innovative Approaches to Plant Disease Control* (Chet, I. ed.). John Wiley & Sons, New York, pp. 137–160.

Clark, F. E. (1967). Bacteria in soil. In *Soil Biology* (A. Burges and F. Raw, eds.). Academic Press, New York, pp. 16–49.

Conway, K. E. (1986). Use of fluid-drilling gels to deliver biological control agents to soil. *Plant Dis.*, **70**:835–839.

Cook, R. J. (1981). Biological control of plant pathogens: overview. In *Biological Control of Crop Production* (G. C. Papavizas, ed.), BARC Symposium 5, Allanheld & Osmun, Totowa, New Jersey, pp. 23–44.

Cook, R. J. and K. F. Baker (1983). *The Nature and Practice of Biological Control of Plant Pathogens*. American Phytopathological Society, St. Paul, Minnesota, 539 pp.

Fahey, J. W. (1988). Endophytic bacteria for the delivery of agrichemicals to plants. In *Biologically Active Natural Products: Potential Use in Agriculture* (H. G. Cutler, ed.). American Chemical Society Symposium Series 380, Washington D.C., pp. 120–128.

Fravel, D. R., J. J. Marois, R. D. Lumsden, and W. J. Connick, Jr. (1985). Encapsulation of potential biocontrol agents in alginate-clay matrix. *Phytopathology*, **75**:774–777.

Gardner, J. M., A. W . Feldman, and R. M. Zablotowicz (1982). Identity and behavior of xylem-residing bacteria in rough lemon roots of Florida citrus trees. *Appl. Environ. Microbiol.*, **43**:1335–1342.

Garrett, S. D. (1956). *Biology of Root-Infecting Fungi.* Cambridge University Press, Cambridge, 294 pp.

Hadar, Y., G. E. Harman, and A. G. Taylor (1984). Evaluaton of *Trichoderma koningii* and *T. harzianum* from New York soils for biological control of seed rot caused by *Pythium* spp. *Phytopathology*, **74**:106–110.

Harman, G. E. and Y. Hadar (1983). Biological control of *Pythium* species. *Seed Sci. Technol.*, **11**:893–906.

Harman, G. E. and T. E. Stasz (1989). Protoplast fusion for the production of superior biocontrol fungi. In *Microbial Weed Control* (Te Best D. O., ed.). Chapman and Hall, New York.

Harman, G. E. and A. G. Taylor (1988). Improved seedling performance by integration of biocontrol agents at favorable pH levels with solid matrix priming. *Phytopathology*, **78**:520–525.

Harman, G. E., I. Chet, and R. Baker (1981). Factors affecting *Trichoderma hamatum* applied to seeds as a biocontrol agent. *Phytopathology*, **71**:569–572.

Harman, G. E., A. G. Taylor, and T. E. Stasz (1989). Combining effective strains of *Trichoderma harzianum* and solid matrix priming to provide improved biological seed treatment systems. *Plant Dis.*, **73**:631–637.

Hubbard, J. P., G. E. Harman, and Y. Hadar (1983). Effect of soilborne *Pseudomonas* spp. on the biological control agent, *Trichoderma hamatum*, on pea seeds. *Phytopathology*, **73**:655–659.

Katan, T., M. T. Dunn, and G. C. Papavizas (1984). Genetics of fungicide resistance in *Talaromyces flavus*. *Can. J. Microbiol.*, **30**:1079–1087.

Kerr, A. (1980). Biological control of crown gall through production of agrocin 84. *Plant Dis.*, **64**:24–30.

Kommedahl, T. and C. E. Windels (1981). Introduction of microbial antagonists to specific courts of infection: seeds, seedlings, and wounds. In *Biological Control in Crop Protection* (G. C. Papavizas, ed.). Allanheld & Osmun, Totawa, New Jersey, pp. 227–248.

Kommendahl, T., C. E. Windels, G. Sarbini, and H. B. Wiley (1981). Variability in performance of biological and fungicidal seed treatments in corn, peas, and soybeans. *Protec. Ecol.*, **3**:55–61.

Lewis, J. A. and G. C. Papavizas (1985). Characteristics of alginate pellets formulated with *Trichoderma* and *Gliocladium* and their effect on the proliferation of the fungi in soil. *Plant Pathol.*, **34**:571–577.

Lewis, J. A. and G. C. Papavizas (1987). Application of *Trichoderma* and *Gliocladium* in alginate pellets for control of Rhizoctonia damping-off. *Plant Pathol.*, **36**:438–446.

Lifshitz, R., M. T. Windham, and R. Baker (1986). Mechanism of biological control of preemergence damping-off of pea by seed treatment with *Trichoderma* spp. *Phytopathology*, **76**:720–725.

Linderman, R. G., L. W. Moore, K. F. Baker, and D. A. Cooksey (1983). Strategies for detecting and characterizing systems for biological control of plant pathogens. *Plant Dis.*, **67**:1058–1064.

Locke, J. C., J. J. Marois, and G. C. Papavizas (1985). Biological control of fusarium wilt of greenhouse-grown chrysanthemums. *Plant Dis.*, **69**:167–169.

Lockwood, J. L. (1977). Fungistasis in soils. *Biol. Rev.*, **52**:1–43.

Lumsden, R. D. and J. C. Locke (1989). Biological control of *Pythium ultimum* and *Rhizoctonia solani* damping-off with *Gliocladium virens* in soilless mix. *Phytopatholgy*, **79**:361–366.

Lumsden, R. D., R. García, J. A. Lewis, and G. A. Frías (1987). Suppression of damping-off caused by *Pythium* spp. in soil from the indigenous Mexican chinampa agricultural system. *Soil Biol. Biochem.*, **19**:501–508.

Lynch, J. M. (1987). Microbial interactions in the rhizosphere. *Soil Microorg.*, **30**:33–41.

Mellon, F. M. (1985). Protoplast fusion and hybridization in *Penicillium*. In *Molecular Genetics of Filamentous Fungi* (W. E. Timberlake ed.). Alan R. Liss, New York.

Minuth, W. and K. Esser (1983). Intraspecific, interspecific, and intergeneric recombination in β-lactam producing fungi via protoplast fusion. *Eur. J. Appl. Biotechnol.*, **18**:38–46.

Nelson, E. B., G. E. Harman, and G. T. Nash (1988). Enhancement of *Trichoderma*-induced biological control of *Pythium* seed root and preemergence damping-off of peas. *Soil Biol. Biochem.*, **20**:145–150.

Okon, Y. (1985). Azospirillum as a potential inoculant for agriculture. *Trends Biotechnol.*, **3**:223–228.

Papavizas, G. C. (1985). Soilborne plant pathogens: new opportunities for biological control. 1984. British Crop Protection Conference—Pests and Disease, BCPC Publications, Croydon, pp. 371–378.

Papavizas, G. C. and J. A. Lewis (1981). Introduction and augmentation of microbial antagonists for the control of soilborne plant pathogens. In *Biological Control in Crop Production* (G. C. Papavizas, ed.), Allanheld & Osmum, Totowa, New Jersey, pp. 305–322.

Papavizas, G. C. and J. A. Lewis (1983). Physiological and biocontrol characteristics of stable mutants of *Trichoderma viride* resistant to MBC fungicides. *Phytopathology*, **73**:407–411.

Papavizas, G. C. and R. D. Lumsden (1980). Biological control of soilborne fungal propagules. *Ann. Rev. Phytopathol.*, **18**:389–413.

Papavizas, G. C., J. A. Lewis, and T. H. Abd-El Moity (1982). Evaluation of new biotypes of *Trichoderma harzianum* for tolerance to benomyl and enhanced biocontrol capabilities. *Phytopathology*, **72**:126–132.

Peberdy, J. F. and L. F. Ferenczy (eds.) (1985). *Fungal Protoplasts, Applications in Biochemistry and Genetics*. Marcel Dekker Inc., New York.

Pontecorvo, G. (1956). The parasexual cycle in fungi. *Ann. Rev. Microbiol.*, **10**:393–400.

Rishbeth, J. (1975). Stump inoculation: a biological control of *Fomes annosus*. In *Biology and Control of Soil-borne Plant Pathogens* (G. W. Bruehl, ed.). American Phytopathological Society, St. Paul, Minnesota, pp. 158–162.

Ruppel, E. G., R. Baker, G. E. Harman, J. P. Hubbard, R. J. Hecker, and I. Chet (1983). Field tests of *Trichoderma harzianum* Rifai aggr. as a biocontrol agent of seedling disease in several crops and *Rhizoctonia* root rot of sugar beet. *Crop Protect.*, **2**:399–408.

Scher, F. M., J. W. Kloepper, C. Singleton, I. Zaleska, and W. Laliberte (1988). Colonization of soybean roots by *Pseudomonas* and *Serratia* species: relationship to bacterial motility, chemotaxis, and generation time. *Phytopathology*, **78**:1055–1059.

Schneider, R. W. (1982). *Suppresive Soils and Plant Disease*. American Phytopathological Society, St. Paul, Minnesota, 96 pp.

Sivan, A., O. Ucko, and I. Chet (1987). Biological control of *Fusarium* crown rot of tomato by *Trichoderma harzianum* under field conditions. *Plant Dis.*, **71**:587–592.

Stasz, T. E., G. E. Harman, and G. A. Marx (1980). Time and site of infection of resistant and susceptible germinating pea seeds by *Pythium ultimum*. *Phytopathology*, **70**:730–733.

Stasz, T. E., G. E. Harman, and N. F. Weeden (1988). Protoplast preparation and fusion in two biocontrol strains of *Trichoderma harzianum*. *Mycologia*, **80**:141–150.

Taylor, A. G. (1977). Comparative performance of pregerminated, high moisture content and dry vegetable seed in greenhouse and field studies. *J. Seed Technol.*, **2**:52–61.

Taylor, A. G., Y. Hadar, J. M. Norton, A. A. Kahn, and G. E. Harman (1985). Influence of presowing seed treatments of table beets on the susceptibility to damping-off caused by *Pythium*. *J. Am. Soc. Hort. Sci.*, **110**:516–519.

Taylor, A. G., D. E. Klein, and T. H. Whitlow (1988). SMP: Solid matrix priming of seeds. *Sci. Hort.*, **37**:1–11.

Thomashow, L. S. and D. M. Weller (1988). Role of a phenazine antibiotic from *Pseudomonas fluorescens* in biological control of *Gaeumannomyces graminis* var. *tritici*. *J. Bacteriol.*, **170**:3499–3508.

U.S. Congress, Office of Technology Assessment (1988). *New Developments in Biotechnology—Field-testing Engineered Organisms: Genetic and Ecological Issues*. OTA-BA-350, U.S. Government Printing Office, Washington, D.C.

Weger de, L. A., C. I. M. Van der Vlugt, A. H. M. Wijfjes, P. A. H. M. Bakker, B. Schippers, and B. Lugtenberg (1978). Flagella of a plant-growth-stimulating *Pseudomonas fluorescens* strain are required for colonization of potato roots. *J. Bacteriol.*, **169**:2769–2773.

Weindling, R. and H. S. Fawcett (1936). Experiments in the control of Rhizoctonia damping-off of citrus seedlings. *Hilgardia*, **10**:1–16.

Wells, H. D., D. K. Bell, and C. A. Jaworski (1972). Efficacy of *Trichoderma harzianum* as a biocontrol for *Sclerotium rolfsii*. *Phytopathology*, **62**:442–447.

Wright, J. M. (1956). The production of antibiotics in soil. III. Production of gliotoxin in wheatstraw buried in soil. *Ann. Appl. Biol.*, **44**:461–466.

Wu, W. (1976). Biological control of seed- and soil-borne fungi associated with wheat and oats. *Bot. Bull. Acad. Sinica*, **17**:161–168.

11

MYCORRHIZAS

C. P. P. REID

Department of Forestry, 118 Newins-Ziegler Hall, University of Florida, Gainesville, FL 32611, USA

1. INTRODUCTION

Mycorrhizas provide an intimate link between the soil environment and the functional nutrient absorbing system of the plant. The modification of plant roots by symbiotic fungi into the distinct structures characteristic of mycorrhizas results in a unique and intriguing component of the rhizosphere. Since the first published description of a mycorrhizal association by A. B. Frank in 1885, scientists continue to be challenged as to the role of mycorrhizas in the ecological and physiological context of plants. Although most research and observations on mycorrhizas have been concerned with nutrient uptake by mycorrhizas, especially immobile elements such as phosphorus, there is an increasing awareness of their potential importance in many diverse aspects of a plant's ability to grow and survive in natural and man-altered environments.

In this chapter, I will not attempt to present an exhaustive review of the large collection of literature that now exists on mycorrhizas, but rather I will highlight those aspects of structure and function that relate specifically to the root–soil interface, i.e. the rhizosphere. First, I will provide a brief description of the various groupings of mycorrhizas with emphasis on their structure and morphology, followed by sections on how mycorrhizas affect the root–soil interface and their interactions with other organisms in the rhizosphere, both spatially and temporally. I will also examine the current views on functional characteristics of mycorrhizas, such as nutrient and water uptake, physical protection of roots, and alteration of the chemical and physical environment of the rhizosphere. It will be necessary to briefly discuss carbon utilization by mycorrhizal fungi in the context of other functions, although the carbon economy of the rhizosphere is discussed in greater detail in Chapter 4. Finally, some of the interesting work on connectivity of plants by mycorrhizal fungi and the influence of mycorrhizas in broader ecological terms will be discussed.

2. GROUPINGS OF MYCORRHIZAL SYMBIOSES

Five broad groupings of mycorrhizas have come into general use over the years on the basis primarily of morphology and anatomy but also of either host plant taxonomy or fungal taxonomy. These generally accepted groupings are: the ectomycorrhizas, the vesicular-arbuscular mycorrhizas, the ericaceous mycorrhizas, the ectendomycorrhizas and the orchidaceous mycorrhizas. These groupings are not necessarily definitive categories but they have proven useful as a means for examining structure and function. One is referred to the excellent review of mycorrhizal symbioses by Harley and Smith (1983) for a more exhaustive treatment of each of these groups. All of these groupings do share the commonality of a host plant having a fungal associate in intimate, mutualistic symbiosis where both partners are living and exchange of substances can occur in both directions, i.e. a biotrophic association (Lewis, 1973).

As the characteristics of these mycorrhiza types are described it is useful to keep in mind how these features might influence function in the rhizosphere—e.g. surface area for nutrient absorption, microbe colonization, source of carbon for microbes, physical contact with the soil, and protection from pathogens or grazing soil fauna.

2.1. Ectomycorrhizas (ECM)

The ectomycorrhizas (ECM) are sometimes referred to as "sheathing" mycorrhizas because of the distinctive feature of the presence of a sheath or mantle of fungal mycelium that covers the absorbing root. ECM are found almost exclusively on woody perennials. The ECM found on *Pinus* in the Pinaceae and on *Fagus* in the Fagaceae are often considered archetypal of the group.

The presence of a Hartig net in the cortex is a key diagnostic feature of this association (Marks and Foster, 1973). This net of hyphae extends into the cortex but is confined to the intercellular spaces and never penetrates the wall into individual cortical cells, although moribund root cap cells may be invaded (Kottke and Oberwinkler, 1986). Further, the fungus never penetrates beyond the endodermis. A general review of ECM can be found in Marks and Kozlowski (1973) and in Harley and Smith (1983).

Features of the mantle and of the extra-matrical mycelium are particularly relevant to function in the rhizosphere because of surface area considerations and the likely influence on carbon use. Fungal mantles vary considerably in thickness—from a few layers of hyphae to a radial thickness equal to or much greater than the radial diameter of the root itself. The thickness of the mantle directly affects the surface area of the absorbing organ that is in contact with the soil through change in circumference, and may be an indication of the magnitude of carbon utilized in the symbiosis. The packing or tightness of the mantle may be important to the colonization by other microorganisms. This, coupled with changes in quantity and quality of rhizodeposition products, undoubtedly influences the population of rhizosphere microorganisms. Bacteria are intimately associated with the ECM sheath of pine although primarily in a zone immediately outside the sheath (Foster and Marks, 1967; Foster et al., 1983).

The degree to which hyphae or mycelial strands extend out into the soil from the mantle is another important characteristic that varies considerably. The presence of mycelial strands and an extensive extra-matrical system would be very advantageous in nutrient uptake in terms of surface area and soil exploitation (Bowen, 1973; Reid, 1984). Mycelial strands seem to be well suited to long-distance transport of phosphate, water and other substances (Skinner and Bowen, 1974; Duddridge et al., 1980). Other aspects of soil binding and proliferation in nutrient-rich sites will be discussed below (see Section 3.1.1).

Root growth patterns of the host plant are often altered by the ECM development on the root system. For example, in *Pinus* an impressive proliferation of short roots is stimulated by fungal colonization and dichotomy (bifurcation) of short roots is very prominent in pine ECM. The longevity of short roots is often increased if they develop into ECM (see Section 3.1.1 below), and this certainly has implications for the length of time a functional absorbing organ remains at one location in the soil, and for the carbon cost due to loss and replacement.

A very large number of fungal genera have been identified or strongly implicated in the ECM association, with most being basidiomycetes or ascomycetes.

2.2. Vesicular-arbuscular Mycorrhizas (VAM)

The broader term endomycorrhizas, parallel in development to the term ectomycorrhizas, has been replaced to a great extent by the term vesicular-arbuscular mycorrhizas (VAM), albeit with a more restrictive meaning. The term refers to the presence of intracellular structures—vesicles and arbuscules—that form in the root during various phases of development. These mycorrhizas are the most commonly occurring group since they occur on a vast taxonomic range of plants, both herbaceous and woody. Although VAM were recognized and described in the late 1800s, and their wide distribution on plant species was known in the early 1900s, little importance was given to this "curiosity" until the mid-1950s. Whereas ECM may be formed by a range of fungi, the grouping of VAM is delimited to those mycorrhizas that are formed by the single family, Endogenaceae.

The VAM are in great contrast to the ECM in both anatomical features and morphology. There is no formation of a sheath of mycelium, and more importantly, intracellular penetration of cortical cells takes place and is the norm. As with ECM, penetration of the fungus does not go beyond the endodermis. A number of excellent reviews are available which describe in detail current knowledge on VAM (Sanders et al., 1975; Harley and Smith, 1983; Powell and Bagyaraj, 1984; Safir, 1987).

It seems that root structure is rarely altered by the development of VAM to the degree that is obvious with ECM, and microscopic examination of the root is often necessary to determine if VAM are present. Internally, arbuscules, vesicles or coils can form within individual cortical cells. Internal structures are contiguous with a network of hyphae that develops externally to the root and serves as a nutrient-absorbing system. Initial infection of a root system involves a hypha encountering the root surface and penetrating the root, either by direct penetration (after formation of an appressorium) of an epidermal cell (including root hair cells) or by intercellular penetration and eventual cell wall penetration of a cortical cell in an internal position in the cortex (see Bonfante-Fasolo, 1984). The infecting hypha can arise from a germinated spore in the soil or by extra-matrical hyphae arising from previously infected roots. A normal sequence of events in development appears to be an internal spread of the fungus in the cortex following penetration, and then a delayed proliferation of hyphae outside the root, presumably in response to a ready supply of carbohydrate from the host (Bethlenfalvay et al., 1982a; Schubert et al., 1987; Sylvia, 1988). As discussed for ECM, extent of the external mycelial system, its longevity, and its biomass will influence nutrient uptake, carbon use and the physical and chemical environment of the rhizosphere.

2.3. Ericaceous Mycorrhizas

This particular grouping of mycorrhizas is defined as those mycorrhizas occurring only in the order Ericales but whose characteristics may have similarity to some of those in both the ECM and VAM. The ericaceous mycorrhizas are further classified into three types which are generally related to the type of root system of the host: (1) ericoid; (2) arbutoid; and (3) monotropoid. The "ericoid" type is generally associated with diffuse root systems, with no root hairs, and with very small diameter roots (1–3 layers of cortical cells). Most genera in the family Ericaceae form this type of mycorrhiza. The "arbutoid" and "monotropoid" are associated with coarser root systems and thereby appear to occur in specific families of the Ericales. Reviews on ericaceous mycorrhizas include Harley and Smith (1983) and Read (1983). Ericaceous plants, especially those that form the ericoid-type mycorrhiza, are often prevalent in nutrient-poor soils of low pH and high organic matter content (Read, 1983). Mycorrhizal development has been shown to improve nutrient uptake by ericaceous plants, in particular nitrogen uptake (Read, 1987), although the enhancement of other nutrients has been demonstrated as well.

2.3.1. Ericoid Mycorrhizas

The ericoid mycorrhizas are endomycorrhizas in the general sense, since the fungal symbiont penetrates the cortical cell wall and invaginates the plasmalemma (as do all endomycorrhizal types). Infection of each cortical cell takes place from the outer cortical wall; lateral spread from cell to cell does not occur. Infected cells appear to be fully packed with fungal hyphae. The mycorrhizas do not form a sheath although a loose weft of hyphae around the root can sometimes be observed. The functional life of the association in epidermal cells may be short-lived, being only a matter of weeks in *Rhododendron* (Duddridge and Read, 1982). Duddridge and Read (1982) point out that the functional relationship of the fungus to the infected cortical cells is quite different from that of VAM. In the ericoid mycorrhiza, the host cell dies as the association disintegrates thereby restricting the functional life (i.e. nutrient absorption) of these epidermal cells to the period prior to breakdown of the infected cell. It has been conclusively shown for some associations that the fungal endophyte is an ascomycete (Read, 1983).

2.3.2. Arbutoid Mycorrhizas

The arbutoid mycorrhizas have characteristics found in both ECM and other endomycorrhizas. Intracellular penetration of cortical cells can occur, as can the formation of a sheath, and a Hartig net is present. These mycorrhizas, as found in such genera as *Arbutus* and *Arctostaphylos*, can resemble the ECM formed in pine except for the intracellular penetration. A feature distinguishing them from ericoid mycorrhizas is the presence of dolipore septae in internal hyphae. It appears from most reports that the fungal associate in arbutoid mycorrhizas is a basidiomycete (Read, 1983). The importance of the various characteristics of these mycorrhizas to the rhizosphere should be similar to those discussed above. Although little experimental work has specifically examined the function of arbutoid mycorrhizas, it is likely they are functionally comparable to ECM.

2.3.3. Monotropoid Mycorrhizas

This term is applied specifically to mycorrhizas that are observed on the achlorophyllous plants in the family Monotropaceae. These mycorrhizas are very similar to the ECM and form a distinct sheath and Hartig net. However, they exhibit a distinctive type of intracellular penetration in cortical cells that is unlike other endomycorrhizal types (Duddridge and Read, 1982). Rather than penetrating the cell wall and invaginating the plasmalemma, the fungus forms a fungal peg which invaginates the cell wall. Early studies on interplant transfer of substances via mycorrhiza fungal hyphae were done in systems where the mycorrhizas of *Monotropa* shared the same mycorrhizal fungi as did the ECM of spruce (Bjorkman, 1960). The movement of both ^{32}P and ^{14}C was shown to occur from the autotrophic spruce to the heterotrophic *Monotropa*.

2.4. Ectendomycorrhizas

The term used for this particular type of mycorrhizal association has presented some confusion in the literature and is used here in reference to a particular group of mycorrhizas that are only formed with genera in the Pinaceae. Readers are referred to the reviews by Wilcox (1983) and Harley and Smith (1983) concerning details on this group. These mycorrhizas form a Hartig net in the cortex of the root but develop little or no sheath. Additionally, intracellular penetration of cortical cells takes place, and thus they are similar to the arbutoid type. Ectendomycorrhizas in Pinaceae seem to be limited to forest nurseries and are formed by a group of fungi called E-strain. These fungi are most likely to be the imperfect stage of ascomycetes; they may cause ectendomycorrhiza in some tree species and ECM in other tree species. Very often the ectendomycorrhizas formed on nursery seedlings are eventually replaced by ECM.

2.5. Orchidaceous Mycorrhizas

The orchid mycorrhizas are somewhat unusual in the sense that the "roles" are reversed in this symbiosis, at least in the early stages of the orchid's life cycle where the plant is not autotrophic. The fungal symbiont appears to provide carbon to the host plant. Considerable work has been done on the role of mycorrhizas during the seedling stages of orchids (Harley and Smith, 1983; Hadley, 1984), yet little is known about the role of mycorrhizas of mature plants, either achlorophyllous or chlorophyllous. The fungal association is of the endomycorrhizal type, where the fungus penetrates the cell wall and invaginates the plasmalemma and forms hyphal coils within the cell. Once the plant is invaded, spread of the fungus may occur from cell to cell internally. The internal hyphae eventually collapse or are digested by the host cell. Since the symbiosis forms an external network of hyphae, it would seem probable that the fungal hyphae function in nutrient uptake as with other mycorrhizas and that the coarse root system of orchids would be supplemented by the increased absorbing surface area of the hyphae (Harley and Smith, 1983). Indeed, enhanced uptake of phosphorus by mycorrhizal adult plants of the orchid *Goodyera repens* Br. was shown by Alexander et al. (1984). A number of basidiomycete genera have been shown to be involved in the symbiosis, although many reports on isolation of the symbiotic fungus from the roots of orchids have placed the symbiont in the form genus *Rhizoctonia* when the perfect stage was not known or the isolate was not induced to fruit in culture.

3. FUNCTION OF MYCORRHIZAS IN SOIL

In discussing the function of mycorrhizas in the rhizosphere, one should conceptualize mycorrhizas in the context of the importance of soil nutrient availability to the success of plants in their terrestrial habitats. In terrestrial ecosystems, the organic detritus-decomposer pathway accounts for the majority of energy flow and nutrient turnover. Further, microflora coupled with microfauna in the soil are the major components of both biomass and activity affecting nutrient availability (Coleman et al., 1983). Although soil bacteria and fungi generally immobilize mineral nutrients as carbon is consumed and thereby compete with plants for macronutrients, mycorrhizal fungi, because of their unique carbon strategies, can efficiently couple soil mineralization and nutrient uptake by the plant roots. In many cases the mycorrhizal system actually "bridges" across the rhizosphere, and provides an organic link between the root and the bulk soil. Fig. 1 adapted from Coleman et al. (1983) attempts to identify some of the characteristics and interactions of mycorrhizas in the soil.

There seems to be little doubt that mycorrhizal plants have an enhanced capability over non-mycorrhizal plants in acquiring nutrients from soil, particularly where the nutrient is poorly soluble and in low concentration. This efficiency in nutrient uptake, primarily through increased absorbing surface area, is likely to be one of the most important features of mycorrhizas; but there is evidence for a number of other important functions, and these are also presented below.

3.1. Nutrient Acquisition

3.1.1. The Nutrient-absorbing System

Regardless of mycorrhizal type, most of the beneficial effects attributed to nutrient uptake are related to increases in the surface area that is effective in ion uptake. Numerous reviews have explored in great detail the evidence which contributes to this conclusion. See, for example, Bowen (1973) and Harley (1978) for ectomycorrhizas; Powell and Bagyaraj (1984), Rhodes and Gerdemann (1980), Sanders et al. (1975) and Tinker (1975) for VA mycorrhizas; Read (1983) for ericaceous mycorrhizas; and Clarkson (1985), Coleman, et al. (1983), Harley and Smith (1983), Reid (1984) and Tinker and Gildon (1983), for discussions related to mycorrhizas in general.

The movement of inorganic ions to absorbing surfaces in the soil, whether it be to root surfaces or to hyphal surfaces, occurs by diffusion and massflow. The rate of uptake by the root will depend on the demand by the absorbing organ, the size and distribution of the absorbing system, and the mobility of the nutrient in soil (Nye, 1979). Generally, positive growth responses to mycorrhizal development can be expected where the concentration of some

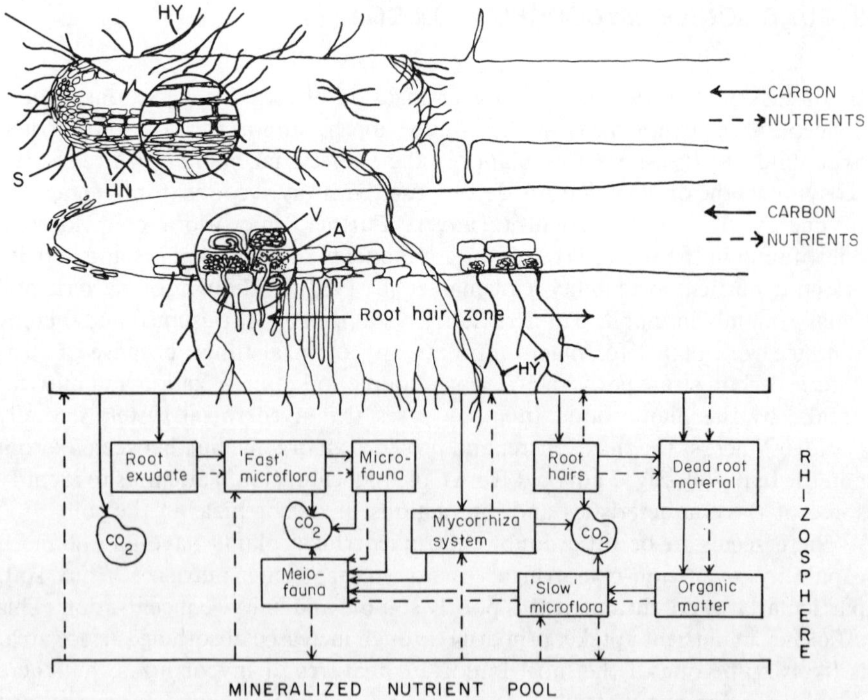

Fig. 1. Diagrammatic representation of ectomycorrhiza (top), vesicular-arbuscular mycorrhiza (middle), and important flows of nutrients and carbon in the rhizosphere (bottom). HY = hyphae, S = sheath, HN = Hartig net, V = vesicle, A = arbuscule. Solid arrows represent flows of carbon, dashed arrows flows of inorganic nutrients. (Adapted from Coleman et al., 1983.)

nutrient is extremely low in the aqueous phase but some solid or unavailable form exists in reserve (Tinker, 1975). The extension of hyphae from the mycorrhiza beyond the geometrically defined depletion zone of the root effectively allows the absorbing system to access the bulk soil solution concentration (Clarkson, 1985). This process is obviously more significant for ions with low diffusion coefficients. Based on models of diffusion for cylindrical geometry, a fibrous root system would have low mean inflow rates (I, uptake per unit length of root) and normal diffusion may be adequate for supply of nutrients to the root (Tinker, 1975). In coarse root systems (with higher I values), factors that increase the length of absorbing surface per unit volume of soil (L_v), such as root proliferation, mycorrhizal hyphae or mycelial strands, result in lower I values and the possibility that normal rates of diffusion may not be limiting to uptake (e.g. for poorly mobile nutrients such as phosphate).

If enhanced nutrient uptake by mycorrhizal plants is primarily related to the phenomenon of surface area increase then it is important to consider factors that influence this, such as physical extent of the hyphal system, length of time for which it is functional, soil conditions under which it is functional, absorbing power of the hyphae, and exploitation of nutrient-rich sites in the soil.

There are numerous observations on the extent of the extra-matrical system of hyphae in soil with various types of mycorrhizas, although the gathering of quantitative data is extremely difficult and laborious. Sanders and Tinker (1973) calculated that VAM on onion had 80 cm of hyphae per cm of infected root. Abbott and Robson (1985) reported over 3000 cm of hyphae per cm of infected root of subterranean clover at 5 weeks when inoculated with *Glomus fasciculatum*, a quantity that was more than four times greater than *Gigaspora calospora* under the same experimental conditions. In contrast, in a pot study with bahiagrass colonized with two different *Glomus* spp., Sylvia (1988) found only 1.6–3.2 cm of hyphae per cm of colonized root 6–13 weeks following seeding, but over 96% were considered active on the basis of indonitrotetrazolium staining. The peak in development of external hyphae of VAM following infection may vary considerably. Abbott and Robson (1985) observed peaks in maximum length of hyphae per unit root length at between 4 and 5 weeks, depending on the fungal symbiont, while Sylvia (1988) showed a peak and leveling off of active hyphae after 8 weeks as root colonization continued to increase up to the last date of harvest at 13 weeks. Bethlenfalvay et al. (1982b) observed a similar increase and peaking of external hyphae at 10 weeks, whereas internal fungal biomass continued to increase to 19 weeks. VAM have been shown to transport substances up to 8 cm distant from the root (Pearson and Tinker, 1975; Rhodes and Gerdemann, 1978), and the depletion zone around roots has been shown to be increased 1–2 mm by mycorrhizal infection (Owusu-Bennoch and Wild, 1979). Sylvia (1990) has recently reviewed the subject of VAM hyphae development.

The development of VAM on the root system with its associated hyphal system has been shown repeatedly to enhance the phosphorus nutrition of numerous plants, many of them cultivated crop plants (Harley and Smith, 1983; Powell, 1984). Most often these reports have been from experimental trials where the soil was previously sterilized and then reinoculated with mycorrhizal fungi, and under low phosphorus conditions. However, as emphasized by Powell (1984), the benefit of field inoculation of crop plants with VAM fungi is very much dependent on field conditions and in many experimental trials it has not been possible to conclusively determine a positive plant response to inoculation. The potential for increasing plant growth and yield by inoculation will very much depend on the probability of natural inoculation by the indigenous fungi and the level of available nutrients, especially phosphorus. Other factors that influence successful field inoculation will be the selection of the correct fungal isolate for the crop host, and inoculum type (e.g. spores, infected root pieces), formulation and placement (Powell, 1984). As indicated by Marschner (1986), the potential of using VAM fungi as "biofertilizers" in agricultural production is fascinating, but large-scale use in agriculture is still not feasible because of difficulties in economical production of large quantities of inoculum, competition from indigenous VAM fungi, production of pathogen-free inoculum, and the impracticality of large-scale soil sterilization for most crops.

The development of the hyphae and mycelial system connected to the sheath of the ECM is certainly variable and suggests that the characteristics of this external system should be reflected in the relative effectiveness of the mycorrhiza in nutrient uptake (Harley and Smith, 1983). The hyphae of ECM fungi often aggregate into collections of hyphae that form mycelial strands, strands that can become differentiated into a structure with an outer rim of dead, thick-walled cells that protect and enclose a core of living cells which may further differentiate into large empty "vessel hyphae" (Foster et al., 1983). Schramm (1966) observed that ECM mycelial strands in anthracite coal mine spoils could extend for metres. Bowen (1973) has discussed the apparent advantage of mycelial strands to the nutrient uptake capability of ECM, and Skinner and Bowen (1974) have demonstrated the importance of strands for long-distance transport of phosphorus and other substances. The transport of tritiated water through mycelial strands was demonstrated by Duddridge et al. (1980), as was the movement of ^{14}C in mycelial strands when connected to the host root, up to distances of 20 cm (Finlay and Read, 1986; Reid and Woods, 1969). Some combinations of host and fungus may never develop strands, while in other combinations they are quite prevalent.

In those mycorrhizas that form a sheath the thickness affects various features: surface area increases with the circumference of the root; there is possible exclusion of other microflora or microfauna; and the biomass utilizing carbon and the capacity for storage of nutrient elements both depend on sheath thickness. The sheath in ECM is typically 20–40 μm thick and may comprise 20–30% of the volume of the rootlet (Harley and Smith, 1983). A doubling of the root diameter by the sheath, which certainly has been observed, would double the surface area in contact with the soil, exclusive of the "roughness" of the sheath surface. This mode of increasing surface area contact, however, would seem to require a much greater carbon investment by the host plant to gain absorbing surface area than having an extensive extra-matrical hyphal system.

Mycorrhizal associations may actually displace certain characteristics of the root rhizosphere to a more distant location from the root surface, whether through displacement by the thickness of the ECM mantle or through extension of the nutrient-absorbing surfaces of the external mycelium of all the mycorrhiza types.

Much more information is needed on how long components of the mycorrhizal system remain functional. The length of life of ECM roots has been discussed by Harley (1969). It would appear that the infection and colonization of short apices, at least in *Pinus* and *Picea*, prolongs their life. In pines, ECM roots may live from several months to 1 year, but in spruce they may live as long as 4 years, although there is some uncertainty as to their metabolic activity (Harley and Smith, 1983). Studies by Fogel and Hunt (1979) with Douglas fir indicate that the average turnover time of mycorrhizal roots appears to be about 1.6 years. Turnover time was defined as the interval for an amount of substance to be transferred into and out of a biomass component to equal the amount present as the "standing crop".

Little information is available on the expected life of hyphae and mycelial strands in the soil, although observations have been made on the length of time that the fungus appears functional in the root. In VAM, various estimates indicate that an infection unit remains functional in the root from about 4 days (Cox and Tinker, 1976) to 15 days (Bowen et al., 1975). This functional period is based on the time that the arbuscule remains intact in the cortical cell—the internal phase that seems most appropriate for nutrient exchanges between host and fungus (Bonfante-Fasolo, 1984). Any given segment of a VAM root may have a much longer functional life as new cortical cells are infected thereby retaining an effective nutrient exchange interface between the root cortex and the extra-matrical hyphal system. In ericoid mycorrhizas the active life of the infection unit is probably not more than 5 to 6 weeks (Read, 1983).

The degree to which fungal hyphae remain functional in nutrient uptake and transport under extremes of soil moisture, temperature and other factors of the soil environment is an important factor in assessing their importance under different ecological circumstances, yet most experimental studies on nutrient uptake have been done under "optimal" conditions. However, some studies have examined the possible advantage of mycorrhizas to plants under low soil moisture conditions; this will be discussed later on. There are reports, however, on the impact of harsh environments on the survival and viability of VAM propagules, including spores, hyphae and infected root pieces (Hetrick, 1984). There are many reports on the advantage of mycorrhizal plants on mine spoils, where extremes in such factors as temperature, acidity, heavy metal concentrations, and moisture are likely to exist; this suggests that more attention needs to be given to the functioning of the external hyphal system. For example, Bradley et al. (1982) demonstrated that the ericoid mycorrhizas can offer resistance to heavy metals primarily by an exclusion mechanism in which the mycorrhizal fungus presumably increases the surface area in the root and thereby allows for adsorption/complexing of the metals.

It is clear that root systems respond to localized soil conditions and often show enhanced root proliferation in nutrient-rich zones (Weaver, 1926; Lyr and Hoffman, 1967; Rogers and Head, 1969). Because of the heterogeneity of soil, it would seem a useful strategy for the host/fungus association to "seek out" nutrient-rich sites in the rooting zone. Although it is unlikely that the fungal associate searches out such sites, evidence suggests that proliferation of hyphae does occur at microsites with relatively higher concentrations of nutrients. Observations on the association of mycorrhizal hyphae with organic matter particles in the soil (Dowding, 1959; Mosse, 1959; Nicolson, 1959; Nicolson and Johnston, 1979) have been supported by experimental evidence demonstrating proliferation of VAM fungal hyphae at such sites (St. John et al., 1983a). St John et al. (1983b) showed that hyphae grew randomly in the soil until they encountered a relatively richer site (organic matter) then proliferated at the organic matter microsite. Other microbes might be involved in the process since the attraction of other fungi (non-mycorrhizal) to decomposing organic material has been shown to be related to the presence of other microflora (Menzies and Gilbert, 1967). The presence of mycorrhizal hyphae in organic matter zones may provide a very efficient interface to microsites where active mineralization by other microbes is taking place. In forest systems with predominantly ECM species, the mycorrhizal roots are intensively developed in the surface horizons of the soil, especially in the humic layers (Harley and Smith, 1983). Ericoid mycorrhizal roots appear to be concentrated in humic surface horizons as well (Read, 1983).

3.1.2. Alteration of Nutrient Availability

It is a prevalent belief that the processes of absorption of nutrients by mycorrhizal fungi are similar to those of simple roots of higher plants, and that enhanced uptake by mycorrhizal roots is simply attributable to increased surface area (Bowen, 1973; Tinker and Gildon, 1983). Yet, various alternative hypotheses have been put forward. In the case of beech ECM it has been demonstrated that the absorbing power of mycorrhizas can be much greater than non-mycorrhizal roots. Phosphate absorption rate by ECM was shown to be five times greater than non-mycorrhizal roots on an area basis, and twice as great on a mass basis (Harley and McCready, 1950). The beech ECM also absorbed alkali metals at a rate 1.5 to 2 times greater than non-mycorrhizal roots on a mass basis (Harley, 1978). Studies such as this, performed in aerated nutrient solution, are difficult to extrapolate to the soil environment, where diffusion impedance in the soil may be the rate-limiting step (Clarkson, 1985; Tinker and Gildon, 1983). For similar reasons, experimental data which suggest that the fungal infection increases phosphate uptake of VAM roots by lowering the apparent K_m may be of little long-term importance unless it can be shown that the external hyphae, outside the root's depletion zone, also have a lower K_m (Clarkson, 1985).

From an ecological perspective, the lack of greater absorption efficiency by mycorrhizas would be consistent with ideas on nutrient strategies by plants under low fertility conditions (relative to agronomic standards). According to Chapin (1980), plants have not usually adapted to nutrient stress by evolving an enhanced capacity to extract nutrients (absorption per unit weight of root) from soil. Rather, plants have developed strategies that increase the root absorbing surface area in the soil for the least amount of energy expenditure.

Apart from absorption kinetics, several lines of evidence suggest that biochemical mineralization and chelation by the fungal component of mycorrhizas could be important in solubilizing certain nutrients in the rhizosphere. Bartlett and Lewis (1973) suggested that the presence of surface acid phosphatases in ECM and VAM could contribute to the effectiveness of mycorrhizas in utilizing complex phosphates, and studies have shown increases in rhizosphere activity of acid phosphatases of ECM when compared to non-mycorrhizal roots (Antibus et al., 1981; Williamson and Alexander, 1975). MacDonald and Lewis (1978) found acid phosphatase activity in the germ tubes and hyphae of the VAM fungus *Glomus mosseae*, and Dodd et al. (1987) showed higher levels of acid phosphatase activity in the root and rhizosphere of plants infected with *G. mosseae* and *G. geosporum* when compared with control plants. However, infection with *G. monosporum* caused no significant increase in acid phosphatase even though increases in plant growth occurred in response to mycorrhizas, and plant shoot concentrations of phosphorus were nearly double that of non-mycorrhizal winter wheat plants. Mitchell and Read (1981) showed that two mycorrhizal endophytes of ericaceous plants, when grown in pure culture, had the capacity to utilize Al and Fe salts of inositol hexaphosphate. Presumably the use of these insoluble phosphates was through the activity of an acid phosphatase, the occurrence of which had been demonstrated earlier for an endophyte of *Calluna* (Pearson and Read, 1975).

Although increased phosphatase activity has been demonstrated in a number of mycorrhizal symbioses, plant roots and other soil microorganisms can also produce acid phosphatases, making the situation very complex and giving no clear picture of the importance of mycorrhizas in enhancing phosphorus availability. Tinker and Gildon (1983) point out that several studies using isotopic labeling of the soil phosphate pool have failed to provide any evidence that VAM fungi can "solubilize" soil P that is unavailable to the non-mycorrhizal root. However, it still may be too soon to conclude that mycorrhizal fungi and non-mycorrhizal roots access the same sources of phosphorus in the soil (Stribley, 1987).

An additional factor to be considered in nutrient uptake is the relative ability of mycorrhizal roots (compared to non-mycorrhizal roots) to produce ligands to chelate ions in the rhizosphere (Harley, 1978). A number of authors have suggested that the plant root (Nye and Tinker, 1977; Nye, 1979; Kepert et al., 1979) or rhizosphere microorganisms (Woldendorp, 1978) can modify the availability of phosphorus by producing metabolites which release phosphorus from insoluble phosphates (Ca, Al, Fe) by formation of soluble complexes with metal ions. One potentially important class of compounds that is receiving increased attention from plant nutritionists and plant pathologists is the siderophore, produced by almost all aerobic microorganisms (see Chapter 7). Although the importance of these ligands in the iron nutrition of microorganisms has been recognized for some time (see Winkelmann et al., 1987) it is only recently that they have been implicated in biological control of root and foliar diseases and in the iron nutrition of plants. It is now well established that plants can utilize these compounds and that they occur in soil at significantly important concentrations (Reid et al., 1986; Crowley et al., 1987). Szaniszlo et al. (1981) demonstrated that a number of ECM fungi produce hydroxamate siderophores when grown in pure culture, and Reid et al. (1984) showed that the concentration of hydroxamate siderophores in the rhizosphere of pine, both nursery seedlings and mature trees in the field, was as much as 50 times greater than in bulk soil nearby. Although all the root systems had well-developed ECM, the elevated levels of siderophores could have been the result of the general rhizosphere microflora.

It may be possible for siderophores produced in the rhizosphere to release phosphorus from insoluble phosphates under certain conditions. Reid et al. (1985), using a simulated root system in soil, showed that the microbial siderophores, ferrioxamine B and ferrichrome, and the synthetic chelate ethylenediamine [di(*o*-hydroxyphenylacetic)acid] (EDDHA) were very effective in releasing ^{32}P from a clayey, kaolinitic, typic Hapludult soil of pH 5.2 (previously labeled with ^{32}P and ^{59}Fe), as ^{59}Fe was chelated and pulled into the solution phase. However, the possibility that siderophores enhance iron availability may be of more significance.

The suggestion that the mycorrhizal fungus plays an important role in the iron nutrition of its host, especially in high pH soils where iron solubility is extremely limiting, is intriguing. The intimate association of the fungal symbiont with the root and the known ability of other soil microorganisms to produce siderophores under iron stress makes this a complicated system for experimental study. Recently, Cress et al. (1986) showed that the iron nutrition of the semi-arid warm season grass, galleta (*Hilaria jamesii*), was enhanced by VAM formed by several species of *Glomus*. The mycorrhizal plants, which showed greater iron uptake than non-mycorrhizal plants, tested positively when bioassayed for the presence of hydroxamate siderophores.

Because it seems certain that mycorrhizal fungi can produce and utilize siderophores in the soil environment, and that the plant root can also utilize ferric iron from the microbial siderophores, the proposed role for certain rhizosphere bacteria in inhibiting root pathogenic fungi through competition for iron via siderophore production (see Chapter 7) must be viewed with some caution. Inhibition of infection and growth of basidiomycetous root pathogens by bacterially produced siderophores could have a similar impact on basidiomycetous mycorrhizal fungi. Although no evidence exists for the specific type of siderophore produced by the VAM fungi in the Endogenaceae, I would suspect that it will likely be similar to those characterized for other fungi, i.e. the hydroxamate type. This is supported by the circumstantial evidence of Cress et al. (1986). If this proves to be correct, the mechanism proposed for the role of bacterial siderophores in competitively inhibiting root pathogen growth should apply to VAM fungi in the same manner.

An additional mycorrhizal fungal product that has received some attention is oxalic acid. Graustein et al. (1977) have found calcium salts of oxalic acid to be very prevalent on the outer surface of fungal hyphae found in forest litter. They propose that the fungal-produced oxalic acid releases phosphate from insoluble metal phosphates. Large amounts of calcium oxalate in fungal mats of *Hysterangium crassum*, a probable ECM fungus with Douglas fir, have been shown to be associated with more intense weathering of clay minerals (Cromack et al., 1979). Knutson et al. (1980) isolated bacteria from the ECM of Douglas fir and from hyphal material of *H. crassum* which they believed important in utilizing calcium oxalate as a carbon source and possibly affecting microsite pH.

3.1.3. Transport and Storage of Nutrients

Although it has been demonstrated that many nutrients can be taken up by mycorrhizal hyphae and transported to the root, most experimental work has been concerned with phosphorus uptake, translocation and storage, and to a lesser extent nitrogen. Nitrogen, especially ammonium, is receiving increasing attention as more experimental studies are demonstrating that mycorrhizas can enhance nitrogen nutrition of the plant (Ames et al., 1983; Barea et al., 1987; Smith et al., 1985; Stribley and Read, 1980).

For example, the uptake and transport of P, S, Zn, Cu, Ca, N, K and Sr have been reported for VAM fungi; P, N, K, Ca, Cl and Zn for ECM (see reviews by Harley and Smith, 1983; Tinker and Gildon, 1983); P for orchidaceous mycorrhizal fungi (Alexander et al., 1984; Harley and Smith, 1983); and P and N for ericaceous mycorrhizal fungi (Read, 1983).

A discussion on uptake and transport of phosphorus should provide a general idea on how nutrients are moved by mycorrhizal fungi, and the reader is referred to the detailed description of these by Clarkson (1985), Tinker and Gildon (1983) and Sanders and Tinker (1973). Calculations by Tinker and Gildon (1983) based on "inflow"—uptake per unit length of root or hyphae—indicate that the mean uptake rate for phosphorus by VAM fungi is approximately 2×10^{-15}–10^{-17} mol cm^{-1} hyphae s^{-1} based on data of Sanders and Tinker (1973) and Cooper and Tinker (1978). Allowing for a mean hypha diameter of 8 μm, the higher rate implies a flux in the hyphae of about 10^{-12} mol cm^{-2} s^{-1}, similar to fluxes in roots (Tinker and Gildon, 1983). Sanders and Tinker (1973) indicated that the larger hyphae at the root entry point would have to generate a flux near 3.8×10^{-8} mol cm^{-2} s^{-1} to be consistent with the observed increase in uptake of phosphorus by mycorrhizal roots over non-mycorrhizal roots of onion. Ames et al. (1983) examined the ability of extra-matrical hyphae of *Glomus mosseae* of VAM celery to take up and transport ^{15}N when supplied with organic matter or ammonium-sulfate labeled with ^{15}N. Based on measures of the total cross-sectional area of hyphae that were presumed to be involved in transport of N, Ames et al. (1983) calculated a mean flux of nitrogen of between 1.7 and 7.4×10^{-8} mol N cm^{-2} s^{-1}, a value that seems reasonable when considering estimated fluxes of phosphorus, which range between about 10^{-8} and 10^{-10} mol P cm^{-2} s^{-1}, and an assumption that fluxes may have to be larger to meet the greater plant demand for nitrogen.

Considerable work is still needed on elucidating the mechanisms of transport in fungal hyphae, but it appears that cytoplasmic streaming coupled with massflow gives a reasonable explanation for many observed results (Tinker and Gildon, 1983). The presence of polyphosphate granules in vacuoles (Cox et al., 1980; Strullu et al., 1981) that can be carried by cytoplasmic streaming along the hyphae toward the root and then broken down in the hyphae within the host cortical cells, suggests a possible model for maintaining a proper concentration gradient for phosphorus movement through the hyphal system (Clarkson, 1985). Failure of cytoplasmic streaming inhibitors to completely disrupt phosphorus transport and phosphorus translocation responses to transpiration support a massflow component (Clarkson, 1985).

Although no strong evidence exists, movement of nitrogen in hyphae could be as amino acids, especially glutamic acid and glutamine, following the uptake of ammonium, as postulated in the model by France and Reid (1983) for pine ECM. Stribley and Read (1980) showed that ericoid mycorrhizal plants could readily utilize nitrogen when supplied as amino acids, in addition to ammonium. Smith et al. (1985) have concluded that the fungi of VAM are able to assimilate ammonium via glutamine synthetase and that this would have a significant influence on the ability of extra-matrical hyphae in the soil to exploit ammonium. This provides a mechanism in support of the demonstrated uptake and transport of ammonium by VAM hyphae from organic matter as shown by Ames et al. (1983). The absorption and assimilation of ammonium by extra-matrical hyphae of mycorrhizas would have further consequences in the rhizosphere through the extrusion of protons that must occur as a result of the uptake of the ammonium ion. This could then affect the pH profile around the root and hyphae, thereby affecting the availability of other ions such as phosphate (Smith et al., 1985). Rygiewicz et al. (1984) have also shown that ECM of several northwestern US conifers readily utilize ammonium. Interestingly, they showed that proton extrusion into the external root bathing solution by mycorrhizal roots of spruce and Douglas fir was significantly less per ammonium ion taken up than from non-mycorrhizal roots. Although no definitive reason was given for the above observation, they suggested that in soil, non-mycorrhizal roots would tend to reduce rhizosphere pH more than mycorrhizal roots.

The potential for storage of nutrients in the fungal components of the mycorrhizal structure would seem to be much greater for those mycorrhizas with well-developed sheaths, as in the ECM and monotropoid types. The mass of the fungal component of the ECM root may be 40% or more of the total rootlet biomass (Harley and Smith, 1983), whereas in VAM the fungal component of an infected root is nearer 10% (Cooper, 1984).

The importance of storage of phosphorus as polyphosphate and the possible storage of nitrogen as glutamine in ECM sheaths has been discussed in considerable detail by Harley and Smith (1983). Some storage of phosphorus as polyphosphate in the arbuscules of VAM also occurs and may be as great as 40% of the total phosphorus in the fungus (Cooper, 1984). Read (1983) suggests that ericoid mycorrhizas store seasonally released nitrogen in the cortical fungal elements, which can then be used later when environmental conditions are less favorable for nitrogen acquisition from the soil. An obvious advantage of this "capacitance" of the fungal tissue for nutrients is the ability for the mycorrhiza to accumulate nutrients when they are available in the soil environment and then continue to provide nutrients to the host plant during times when soil supply is reduced. This would tend to reduce the oscillations in the rhizosphere environment for other organisms as well, especially if they are dependent on active metabolism and growth of the plant.

3.2. Mycorrhizas and Water Use

A number of observations have indicated that the presence of mycorrhizas on the plant root system modifies the water relations of the plant when compared to non-mycorrhizal plants. However, considerable controversy exists as to the causes of changes in such parameters as transpiration, hydraulic conductivity of the plant, stomatal conductance, and avoidance of or recovery from wilting. Cooper (1984) has summarized the experimental evidence reported for VAM relative to their effects on resistance of the plant to water transport. Harley and Smith (1983) have also reviewed the water relations of VAM and ECM, while Reid (1979) examined mycorrhizas and water stress. Much of the difficulty in assessing the changes in water status of the host plant is related to separating secondary effects of better plant nutrition (e.g. phosphorus) from more direct effects resulting from increased surface area for water absorption. It is likely that improved water relations of mycorrhizal plants under well-watered soil conditions is the result of improved plant nutrition, since there would usually be little soil impediment to diffusion and massflow of water to the root under such conditions. Therefore much of my discussion will focus on mycorrhizas under conditions of low soil moisture.

Is it conceivable that the increased surface area provided by the extra-matrical hyphal system can provide a direct pathway for water uptake to the root? In the soil–plant–atmosphere continuum, the flow of water is driven by a potential gradient established by transpiring leaf surfaces, and modified by a series of resistances and capacitances in the continuum. If we examine just the soil–root part of the continuum, the major resistances to water movement are at the root–soil interface and in the soil. Since water moves in the soil by bulk or massflow, changes in water content have marked changes on conductivity. Nye and Tinker (1977) discussed the movement of water in the rhizosphere and concluded that it is reasonable to expect gradients of water content around individual roots under some conditions, especially if the bulk soil is dry. If I_w (water uptake per unit length of root per unit time) is small, it is unlikely that a significant gradient in water content (and thereby increased impedance) will occur in the rhizosphere. Large values of I_w may arise if plants have short lengths of active root, or their transpiration rates are high (Nye and Tinker, 1977). It would seem logical that an increase in absorbing surface area of the root system due to mycorrhizas would also reduce the value of I_w and thereby prevent large gradients of water content from occurring around individual absorbing surfaces. Critical to this supposition is whether the flux of water through the hyphae is sufficient to make a significant difference in water uptake.

Both Clarkson (1985) and Sanders and Tinker (1973) express some reservations about the importance of hyphae in the overall water economy of the plant. Assuming well-watered conditions, Sanders and Tinker (1973) and Graham and Syvertsen (1984) submit that calculations based on number of hyphal entry points, cross-sectional area of hyphae, and observed water uptake by plants suggest that hyphal transport alone cannot adequately account for observed increases in water uptake by mycorrhizal plants. On the other hand, Allen (1982) and Hardie (1985) maintain that the additional surface area provided by the extra-matrical hyphae does offer a reasonable explanation for increased water uptake by mycorrhizal plants of *Bouteloua gracilis*, and of *Trifolium pratense* and *Allium porrum*, respectively.

The aggregation of hyphae into differentiated strands by ECM fungi may provide the potential for greater fluxes of water to the root, although few studies have examined this. Duddridge et al. (1980) examined the transport of tritiated water through rhizomorphs formed by the ECM of *Pinus sylvestris* and found that the movement of water through the fungus to the host could be rapid, at least 27 cm h^{-1}.

Certainly, what remains unclear is the ability of hyphae and mycelial strands to conduct water at low soil moisture or in very coarse-textured soils where unsaturated hydraulic conductivity in the soil becomes limiting to water uptake. Are hyphae functional under these dry conditions where presumably they would offer the greatest advantage to water uptake? A number of the ECM fungi have been shown to be tolerant of low water potential, and some, like *Cenococcum geophilum*, can actively grow at potentials as low as −1.5 MPa (Mexal and Reid, 1973). Worley and Hacskaylo (1959) observed that black mycorrhizas formed by this fungus were more prevalent under drought conditions, and Read et al. (1977) suggested that the formation of mycorrhizas by *Cenococcum* with *Helianthemum* was an adaptation to dry soil conditions. Early reports by Cromer (1935) and Harley (1940) proposed that ECM were more resistant to desiccation than non-mycorrhizal roots. These types of study suggest that perhaps the extra-matrical hyphae would be tolerant of low soil moisture conditions. Hu (1977) (as cited by Harley and Smith, 1983) did demonstrate that the hyphae of red pine ECM were functional in uptake and transport of nutrients at a soil potential of −0.1 MPa, whereas non-mycorrhizal roots had reductions in uptake at −0.02 MPa.

In experiments with osmotica, others have found no difference in uptake of phosphorus between ECM and non-mycorrhizal roots of pine when subjected to nutrient solutions at potentials as low as −1.2 MPa (Reid and Bowen, 1979), indicating that uptake processes *per se* may not be changed by the fungus, at least over short periods. This may indicate that if the hyphae can maintain their physical integrity and allow long-distance transport in dry soil, uptake of nutrients at microsites may not be adversely affected by low soil water potential. Perhaps, observed benefits of mycorrhizas to plants under low soil moisture is more a factor of continued capacity to take up nutrients, where diffusivity of ions is markedly decreased under reduced water content, rather than any direct effect from water uptake.

Although it has been proposed that the sheath may provide a barrier to unimpeded water movement into the root, Sands et al. (1982) could show no significant difference in the radial resistance to water uptake between mycorrhizal and non-mycorrhizal roots, suggesting perhaps that the sheath has little impact on the conductivity of the water pathway between the soil at the sheath surface and the conducting center of the root.

Secondary effects of the symbiosis that might be important to water uptake are those that modify the physical environment of the rhizosphere. The external hyphae should serve to bind the soil near the root and prevent the development of gaps between the soil and the root as the root expands and contracts due to hydrostatic tension and as the soil shrinks and swells with changes in water content, thereby maintaining liquid continuity across the root–soil interface. This benefit has been suggested for root hairs (Barley, 1970; Newman, 1974; Weatherly, 1975; Tinker, 1976) and would seem to be even more appropriate to roots with extensive development of extra-matrical hyphae near the root. Hyphae, coupled with mucilaginous material produced in the rhizosphere (Foster et al., 1983), should provide an effective binding matrix in the soil. The ability of mycorrhizal hyphae to contribute to the aggregation of sand grains and stabilization of soil (Koske, 1975; Clough and Sutton, 1978; Forester and Nicolson, 1979) provides support for this suggestion.

3.3. Root Protection in the Rhizosphere

As Harley and Smith (1983) point out, the subject of biological antagonism in the rhizosphere is of long standing. The interactions between mycorrhizal fungi and other potentially deleterious organisms in the soil are complex. Ways in which mycorrhizal fungi might influence root pathogens include:

1. Competition for the uptake of essential nutrients in the rhizosphere and at the root surface.
2. Competition for actual sites of infections in and on the root.
3. Alteration of the physiology of the host plant.
4. Presence of a physical barrier to infection (in the case of sheathing mycorrhizas).
5. Selection for favorable microflora by changes in root exudate products.
6. Production of toxic or inhibitory compounds.
7. Nutrient absorption system compensation for damage to roots by disease (Harley and Smith, 1983; Marx, 1975).

The review by Bagyaraj (1984) indicates that most studies with VAM plants show that VAM decrease the severity caused by plant pathogenic fungi. Reports show that VAM can reduce damage from such pathogens as *Thielaviopsis basicola* on cotton (Schonbeck and Dehne, 1977), species of *Phytophthora* on soybean (Chou and Schmitthenner, 1974) and citrus (Schenck et al., 1977), *Rhizoctonia* on several plant species, and other plant/pathogen combinations. Explanations for beneficial effects vary and often are not known. However, enhanced nutrient status of the plant for VAM citrus (Davis, 1980; Davis and Menge, 1980), high arginine levels in VAM tobacco that were inhibitory to pathogen chlamydospores (Baltruschat and Schonbeck, 1975), and cell wall thickenings in VAM onion which restricted pathogen penetration (Becker, as cited by Bagyaraj 1984), have been proposed.

Marx (1969) was one of the first to demonstrate that the ECM fungus could provide protection to roots from the pathogen *Phytophthora cinnamomi*. Both antibiotic production by the fungus (such as shown with *Leucopaxillus cerealis*) and the physical barrier of the sheath (as shown with *Pisolithus tinctorius*) were effective in protecting the root. More recently, Chakravarty and Unestam (1987) showed almost complete inhibition of the rhizospheric fungus *Clindrocarpon destructans* by the ECM fungi *Laccaria laccata* and *P. tinctorius* on *Pinus sylvestris*. Although the specific mechanism was not determined, the authors suggested that utilization of root-produced substrate by the mycorrhizal fungi may have limited the growth of *C. destructans*. Sylvia and Sinclair (1983) also reported that *L. laccata* could protect primary roots of Douglas fir seedlings from the root rot *Fusarium oxysporum*. They attributed inhibition of the pathogen to phenolic osmiophilic substances produced in the root cortical cells in response to the mycorrhizal fungus.

The only other group of plant pathogenic organisms that has received considerable attention in its interactions with mycorrhizas is nematodes. Generally, nematodes are grouped on the basis of their feeding habits into: (a) sedentary endoparasites; (b) migratory endoparasites; and (c) ectoparasites. In a recent review, Ingham (1988) concludes that endoparasitic nematodes and VAM are generally mutually inhibitory. Most VAM studies have been conducted with sedentary endoparasitic nematodes in the genus *Meloidogyne* (root-knot nematodes), although a few studies with migratory endoparasites indicate that VAM may inhibit this group of nematodes as well. A diversity of interactions between VAM and plant parasitic nematodes gives an indication that each nematode–fungus–plant combination may have unique characteristics. However, the single most common effect of VAM fungi on nematode-susceptible plants is to promote tolerance to nematodes (Hussey and Roncadori, 1982). Little information is available on ECM protection against nematodes. The possible mechanisms by which VAM and ECM reduce the damage by nematodes is similar to those proposed mechanisms given above for other plant pathogens. Certainly, better nutrition status of the host and spatial competition by the fungus and the parasite have been suggested (Hussey and Roncadori, 1982).

The third group of nematodes, the ectoparasites, may be more important as grazers of mycorrhizal fungi and are discussed in Section 3.4 below.

3.4. Interactions between Mycorrhizas, Other Microbes, and Soil Fauna

It appears clear from examination of the morphological and anatomical changes that take place in mycorrhizal roots, the development of the extra-matrical fungal system, and the physiological changes that occur in the plant as a result of the symbiosis, that the rhizosphere of the mycorrhizal plant root is different from the non-mycorrhizal plant root. A recent paper by Linderman (1988) discusses this uniqueness of the rhizosphere of mycorrhizal roots (or "mycorrhizosphere" as termed by some) and stresses the need to evaluate plant responses to microbial inoculations in the light of the complex of factors strongly influenced by the mycorrhizal association. The altering of both the mineral nutrition and the carbon allocation of the host plant undoubtedly alters the chemical and physical environment for root-associated microbes in the soil. Nevertheless, experimental studies on mycorrhiza–microbe interactions are relatively few.

Coleman (1985) emphasizes that the root–microbe interactions are the key to understanding ecosystem function, and places mycorrhizas in perspective with the many other complex interactions taking place in the rhizosphere. The model of trophic interactions among plants, substrates, microflora and microfauna presented by Hunt et al. (1987) attempts to place mycorrhizas in the context of labile and resistant substrates, primary decomposers and various levels of microbial predators (Fig. 2). The authors discuss the relative importance of bacteria and fungi as primary mineralizers of N and then, in turn, how the transfer of N in the detrital food web is quantitatively influenced by fauna grazing upon the microflora.

Most published work related to mycorrhizal roots and associated microorganisms has been concerned with nitrogen-fixing organisms, phosphate-solubilizing bacteria, changes in populations of general microflora, or root pathogens (as discussed earlier). As reviewed by Bagyaraj (1984), most of the studies on the tripartite association of the leguminous plant, VAM fungus and *Rhizobium* suggest that mycorrhizas improve nodulation and nitrogen fixation by the legume bacteria by primarily improving phosphorus nutrition of the plant. However, Barea et al. (1987) suggest that the formation of VAM in legumes may contribute to increases in nitrogen uptake from the soil in addition to improving phosphorus nutrition. By using ^{15}N, Barea et al. (1987) demonstrated that the increased nitrogen content of VAM legumes was from both increased N_2 fixation and enhanced nitrogen uptake from the soil. Studies by Smith et al. (1985) with mycorrhizal and non-mycorrhizal subterranean clover and onion indicate that the mycorrhizal fungus contributes to the production of glutamine synthetase in the VAM root system and thereby increases the uptake of ammonium from soil. In a legume symbiosis with *Rhizobium* and VAM, this would presumably lead to additional nitrogen uptake beyond that resulting from improved phosphorus nutrition and enhanced N_2 fixation.

Barea et al. (1983) proposed that the combination of VAM and associative nitrogen-fixing microorganisms such as *Azospirillum*, which colonizes cortical cells, would provide a means by which graminaceous plants, lacking symbiotic nodulating nitrogen-fixers, could compensate for nutrient-deficient conditions. Although in studies with maize and ryegrass they were unable to demonstrate a synergistic effect of VAM and *Azospirillum*, there was an indication that *Azospirillum* in combination with VAM could mimic the response of VAM plants fertilized with nitrogen but without *Azospirillum*.

Later, Pacovsky et al. (1985) showed that inoculation with VAM fungus or inoculation with *Azospirillum* improved the shoot and root dry weight of sorghum. Inoculation with both organisms together produced an additive effect on shoot dry weight. Of particular interest is that in some experiments, inoculation with both endophytes resulted in less total nitrogen in the plant than if inoculated only with *Azospirillum*. They attributed this reduction in nitrogen

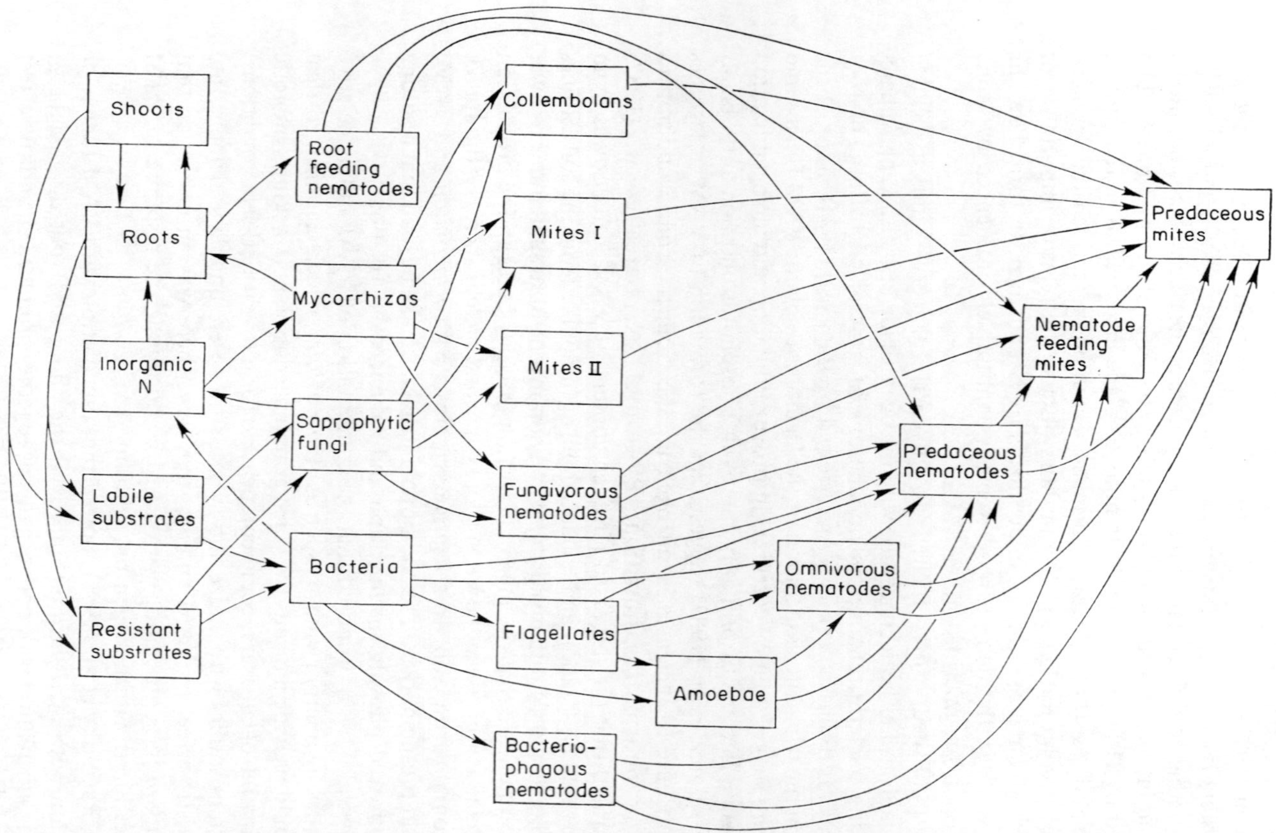

Fig. 2. Representation of detrital food web in a shortgrass prairie. Flows omitted for clarity include transfers from every organism to the substrate pools (death) and transfers from every animal to the substrate pools (defecation) and to inorganic N (ammonification). Note grazing of mycorrhizas by collembolans, mites and nematodes. (From Hunt et al., 1987.)

content to a competition for carbohydrates or other root metabolites between the two endophytes. Shunting of carbohydrate to the VAM fungus could result in nitrogen stress if the plant is dependent on nitrogen fixed by *Azospirillum* (Pacovsky et al., 1985). Brown and Carr (1984) also suggested that the decrease in the populations of *Azotobacter* in the rhizosphere of lettuce when inoculated with VAM fungi was possibly the result of competition between the VAM fungus and *Azotobacter* for both phosphorus and carbon. In their experiments they showed that a combination of inoculation of both VAM fungus and *Azotobacter* resulted in greater plant weights than either alone. No particular mechanism was shown to be responsible for the positive response to the interaction, although others have suggested that the plant growth responses to *Azotobacter* are more likely to be related to growth regulators than nitrogen fixation (see Brown, 1982; Bagyaraj, 1984).

The effectiveness of mycorrhizas in enhancing plant growth in phosphorus-deficient soils has naturally led to an interest in the role of phosphate-solubilizing bacteria in the rhizosphere of mycorrhizal plants. A number of reports suggest that phosphate-solubilizing bacteria, because they may be able to release phosphate ions in soil microsites, interact with VAM for a better exploitation of poorly-soluble phosphorus sources, yet the production of growth hormones and other metabolites by these bacteria confounds the interpretation of observed responses of VAM plants with these bacteria (Bagyaraj, 1984; Azcon-Aguilar et al., 1986). Raj et al. (1981) reported that VAM finger millet produced more dry matter and removed more ^{32}P from the soil when inoculated with the phosphate-solubilizing bacteria, and Piccini and Azcon (1987) indicated that dual inoculations of VAM fungi and phosphate-solubilizing bacteria significantly increased alfalfa growth. Their results showed that rock phosphate was transformed into available forms in the presence of the bacteria and was utilized by the plants. Piccini and Azcon (1987) suggested that it was likely that the bacteria's main effect was through solubilization of the rock phosphate since neither the bacteria nor the VAM fungus, *Glomus mosseae*, when supplied singly, had any effect on the alfalfa in the absence of rock phosphate. However, phosphate-solubilizing bacteria may not always improve the utilization of phosphorus by mycorrhizal plants. Azcon-Aguilar et al. (1986) reported that inoculation of soybean with phosphate-solubilizing bacteria did not improve the utilization of tricalcium phosphate by either roots or mycorrhizas, and no synergistic interaction could be shown. However, they did observe an increase in mycorrhizal infection as a result of the bacteria and suggested that this may have been a result of the production of hormones by the bacteria.

As stated earlier, mycorrhizal development undoubtedly causes alteration of the rhizosphere, and changes in the quantity and quality of root exudates, at least in VAM plants, has been linked to the effect of phosphorus nutrition on the membrane permeability of roots. Under low phosphorus nutrition, root cell membranes have higher permeability thereby allowing increased loss of root metabolites, which may be beneficial to colonization by VAM fungi. Improved phosphorus nutrition of the plant by the VAM subsequently leads to reduced membrane permeability and a reduction in root metabolites (Ratnayake et al., 1978; Graham et al., 1981; Schwab et al., 1983). Such changes potentially impact other microbes in the rhizosphere. In the ECM association, Ling-Lee et al. (1977) were able to demonstrate that the development of the sheath on roots of *Eucalyptus* substantially reduced the amount of acidic polysaccharide (root cap slime) that accumulated around the outer root cap when compared to non-mycorrhizal roots. The reduction was considered to be the result of diversion of carbohydrate to the fungal partner.

Several studies have examined the effect of mycorrhizal development on the populations of rhizosphere microflora, presumably through alterations in the physical and chemical environment. Some of these studies have been initiated in response to a need to determine if the experimental results from mycorrhizal fungi inoculum treatments are being confounded by the introduction and development of a quite different microflora. Ames et al. (1984) looked at both the rhizosphere and non-rhizosphere soil of mycorrhizal and non-mycorrhizal blue grama grass grown in pots. No difference in bacterial or actinomycete colony-forming units could be shown between the non-rhizosphere soil of the two groups of plants. But in soil assayed from the rhizosphere, one species of bacterium consistently had higher values of colony-forming units from mycorrhizal roots, whereas three other bacterial species and actinomycetes (as a group) showed no statistically significant differences. In later studies, Ames et al. (1987) concluded that similar functional groups of microorganisms can become established in both VAM plants and non-mycorrhizal plants if the same VAM-fungal inoculum wash is applied to both treatments. However, in studies with maize, Berthelin and Leyval (1982) indicated that the population of rhizospheric bacteria was significantly different between mycorrhizal and non-mycorrhizal plants, and the absence of non-symbiotic rhizospheric bacteria reduced the plant uptake of K, Ca and Mg.

Although mycorrhizal roots were not compared with non-mycorrhizal roots, Summerbell (1987) showed that *Trichoderma* species associated with the ECM of spruce could vary considerably in their ability to inhibit fungal colonization under axenic conditions. It has also been suggested that particular bacteria and actinomycetes that might be associated with ECM fungi could affect the development of mycorrhizas through cellulolytic and pectolytic enzymes (Dahm et al., 1987).

The possibility that rhizosphere microorganisms affect the iron nutrition of plants and mycorrhizal fungi through production of iron-chelating siderophores has been previously discussed in Section 3.1.2 above.

Study of interactions with soil fauna has been focused primarily on nematodes and arthropods, but some work has been done with mycophagous amoebae (Old and Chakraborty, 1986). Ectoparasitic nematodes could be effective in reducing the nutrient-absorbing area of the mycorrhizas by grazing on extra-matrical hyphae. Fungal-feeding nematodes have been reported to feed on both VAM and ECM fungi (see Sutherland and Fortin, 1968; Riffle, 1971; Ingham, 1988). Impacts of fungal grazing on phosphorus uptake

(Salawu and Estey, 1979) and nitrogen (Ingham et al., 1986) by VAM plants has been indicated.

Several studies, as reviewed by Moore and Walter (1988), suggest that collembola or mites can consume significant quantities of mycorrhizal fungal hyphae. In laboratory studies, Finlay (1985) demonstrated that the yield of VAM plants was reduced when collembola densities were in excess of 150 dm^{-3}. In field plots treated with the insecticide chlorfenvinphos, Finlay (1985) observed that collembola densities were lower and plant shoot weights were correspondingly greater than in control plots. This suggested that higher levels of collembola were deleterious to plant growth through their impact on the mycorrhizas. Shaw (1985) also observed that it is very likely that collembola in stands of *Pinus contorta* graze on various species of ECM fungi, although the significance of this on mineral nutrition of the host plant is unknown.

Grazing of mycorrhizal fungal hyphae by soil fauna may not always be detrimental. Low to moderate amounts of grazing could be beneficial, perhaps by stimulating further hyphal exploration of soil, but this possibility needs to be examined much more thoroughly (Coleman, 1985).

4. IMPACT OF MYCORRHIZAS ON CARBON BALANCE

Until recently, there appears to have been a preoccupation with the role of mycorrhizas in plant yield improvement through enhanced nutrient uptake under various soil fertility regimes, often with crop plants under modern cultural practices. However, it should be obvious that the development of mycorrhizal associations over evolutionary time has not necessarily resulted in features that immediately translate into yield or productivity. The benefit of mycorrhizas to the host plant in the broadest sense might be described in terms of "fitness" (see Ricklefs, 1979), a concept originally developed for animals, i.e. a measure of the success of a species in maintaining itself (in an evolutionary sense), involving a combination of competitive adaptation in growth and reproduction. In plants, the idea of fitness is applied in an even more narrow context, i.e. the ability of an individual to be successful through the most appropriate scheme in allocating carbon to photosynthetic and non-photosynthetic tissues. Another way of stating this is "What are the marginal returns on investment of carbon into specific tissues?". Consideration of the carbon utilization by the mycorrhizal system with its fungal associate, and even by free-living microorganisms in the rhizosphere, seems to be extremely relevant to determining the competitive abilities of different plants. The logical extension of this idea is that the allocation of carbon to the fungal symbiont, rather than to the production of additional root tissue, contributes to the fitness of the host plant. Understanding the utilization of carbon in the symbiotic association is vital to fully understanding the benefits of mycorrhizas. Here I will only briefly discuss general relationships between carbon use by mycorrhizas and mineral uptake.

Plant physiological ecologists have discussed at great length the importance of carbon allocation between shoots and roots to survival and competitive ability (see Schulze, 1982). The development of mycorrhizas undoubtedly has an impact on the carbon balance of the plant, especially in the sheathing type, yet our knowledge of the physiological mechanisms involved in maintaining the symbiosis is extremely limited. It seems essential we better understand the trade-offs between nutrient acquisition and the carbon requirements of the mycorrhizal fungal symbiont.

Over the years, studies in forest trees and ecosystems have given us some indications that the allocation of carbon to ECM fungi can be substantial. Early on, J. L. Harley pointed this out by calling our attention to Rommel's observations on fruiting in spruce forests, and to the calculations of Tranquillini (Harley and Smith, 1983). Vogt et al. (1982) calculated that the mycorrhizal fungal component comprised approximately 14–15% of the net primary production in Pacific silver fir forests ($\simeq$3300–3700 kg ha^{-1} yr^{-1}) and represented 2800–4100 kg ha^{-1} of dry matter. This was 1% (180-year-old stand) to 3.5% (23-year-old stand) of total living tree biomass. Fogel and Hunt (1979) indicated that the annual fungal throughput in a second-growth Douglas-fir stand was about 50% of the total annual stand throughput of 30 300 kg ha^{-1}. This included both mycorrhizal fungi and fungal hyphae in the litter and soil. The mycorrhizal sheath and sclerotia biomass (12 800 kg ha^{-1}) was estimated to be about 4.0% of the total standing crop of tree components.

I provide these examples simply to indicate that fungal cost can be substantial in mature forest systems with ECM. The above estimates are also very conservative since they do not include the cost of respiration of the fungal system and may not include all of the turnover of fungal structures. Admittedly, the significance of carbon diversion to VAM is likely to be much less. There are a number of compelling reasons why a greater understanding of carbon allocation in mycorrhizas is needed. At minimum, a substitution of the fungal component for root biomass in nutrient absorption function means that root biomass alone does not represent the carbon needed to maintain a nutrient uptake system (*à la* conventional shoot/root concepts).

Relative to the broad definition of nutrient cost given by Clarkson (1985), the following would be relevant to consider in comparing mycorrhizal plants to non-mycorrhizal plants:

1. Direct energy cost of substrate transfer to fungus (also fruiting), including respiration to support growth of fungal biomass (relative to the growth respiration of root tissue), and the respiration to support maintenance of the fungal absorbing system (relative to maintenance cost of root tissue for the same absorbing capacity).
2. Energy content of rhizodeposition relative to possible change in net loss by presence of the mycorrhizal fungus (and the change in quantity and quality of these products to other rhizosphere microorganisms).
3. Difference in actual C utilized by root biomass and fungal biomass to provide similar absorptive capacity.

Even if we could determine fungal biomass and respiration costs accurately, the act of forming the mycorrhizal association can result in an increase of the C fixation rates, as has been shown with both VAM and ECM (see Chapter 4). Although the reasons for the stimulation of photosynthesis are not unequivocal, the simplest explanation, consistent with research on nutrient uptake by mycorrhizas, is that an increase in phosphorus uptake removes limitations on photosynthesis. However, other more complex interactions cannot be entirely ruled out, such as source-sink and hormonal influences. A better understanding of the carbon cost of mycorrhizas is also necessary to fully assess the degree of benefit of the association in older and mature plants. Most of the convincing evidence for "mutualism" and positive benefit to the host plant is derived from extensive studies with seedlings in pot cultures, or in similar restrictive experimental conditions. The observation that mycorrhizas occur on mature plants in natural settings is not evidence for a positive benefit in itself. Much more documented experimental evidence is required.

5. INTERCONNECTIVITY OF MYCORRHIZAL ROOT SYSTEMS

In the broader ecological context of mycorrhizas and the rhizosphere, the possibility that mycorrhizas provide a means by which plant root systems are interconnected could profoundly alter our approach to studying plant community dynamics and alter our view of the extent of the rhizosphere. Observations on intraspecific root grafting and unexplained interspecific below-ground transfer of radioisotopes between donor plants and numerous receiver plants, led Woods and Brock (1964) to suggest that root systems in plant communities are not independently functioning units but may be interconnected by shared mycorrhizal fungi. Later, Reid and Woods (1969) attempted to demonstrate that substances could be transferred between root systems of pine by shared ECM fungus, but with only limited success. However, it is now clear that such transfer can occur in both ECM and VAM systems (Hirrel and Gerdemann, 1979; Heap and Newman, 1980; Chiariello et al., 1982; Whittingham and Read, 1982; Francis and Read, 1984; Finlay and Read, 1986), with the transfer of both carbon and phosphorus having been demonstrated. Consideration of the quantity and the ecological significance of the transfer of specific substances is certainly required. One consequence of such interconnection is the possibility that seedlings or understory plants under severe competition for light and nutrients have access to phosphorus reserves or carbon from neighboring plants that are in a better position to exploit light or nutrients (Whittingham and Read, 1982; Finlay and Read, 1986). This may allow the suppressed plant to survive during critical stages of development when competition is especially severe. Studies with ^{14}C seem to leave little doubt that carbon can be transferred from one plant to another via connecting hyphae (Francis and Read, 1984; Finlay and Read, 1986). There is also considerable evidence to indicate that hyphal transfer of ^{32}P between plants also occurs (Heap and Newman, 1980; Chiariello et al., 1982; Whittingham and Read, 1982; Francis et al., 1986), although results are somewhat circumstantial. Newman and Ritz (1986) suggest that the movement of ^{32}P between donor and receiver mycorrhizal plants can be adequately explained by the ''soil pool pathway'' where phosphorus is lost from the donor root into the soil solution and then absorbed by the receiver plant root or mycorrhizal hyphae.

The prevalence of interconnectivity of root systems due to mycorrhizal fungi at best remains uncertain. St. John and Coleman (1983) discuss circumstances where hyphal interconnections may be few or infrequent between different species, perhaps as a result of allelopathy.

For a more in-depth discussion of mycorrhizas in the context of ecological understanding the reader is referred to Miller (1987) for VAM, and to St. John and Coleman (1983) for mycorrhizas in general.

6. SUMMARY

The rhizosphere is truly a dynamic region of immense importance to the functioning and ecological success of the plant. Under most soil conditions, the rhizosphere is, in reality, the ''envelope'' of activities that provides the interface between the composite root–fungus structure and the soil. The utilization of root carbon by the fungal symbiont, and the impact of the fungus on nutrient uptake from the mineralized nutrient pool of the soil, suggest only the more obvious influences of mycorrhizas on the characteristics of the rhizosphere. The ability of mycorrhizas to enhance the uptake of poorly mobile nutrients from soil is well documented. However, as we learn more about how mycorrhizas alter the rhizosphere in biological, chemical and physical terms, we will begin to fully appreciate the advantage and necessity of considering the nutrition of plants in soil in a more holistic manner.

REFERENCES

Abbott, L. K. and A. D. Robson (1985). Formation of external hyphae in soil by four species of vesicular-arbuscular mycorrhizal fungi. *New Phytol.*, **99**:245–255.

Alexander, C., I. J. Alexander, and G. Hadley (1984). Phosphate uptake by *Goodyera repens* in relation to mycorrhizal infection. *New Phytol.*, **97**:401–411.

Allen. M. F. (1982). Influence of vesicular-arbuscular mycorrhizae on water movement through *Bouteloua gracilis* (H.B.K.) Lag ex Steud. *New Phytol.*, **91**:191–196.

Ames, R. N., C. P. P. Reid, L. K. Porter, and C. Cambardella (1983). Hyphal uptake and transport of nitrogen from two ^{15}N-labelled sources by *Glomus mosseae*, a vesicular-arbuscular mycorrhizal fungus. *New Phytol.*, **95**:381–396.

Ames, R. N., C. P. P. Reid, and E. R. Ingham (1984). Rhizosphere bacterial population responses to root colonization by a vesicular-arbuscular mycorrhizal fungus. *New Phytol.*, **96**:555–563.

Ames, R. N., K. L. Mihara, and G. J. Bethlenfalvay (1987). The establishment of microorganisms in vesicular-arbuscular mycorrhizal and control treatments. *Biol. Fert. Soils,* **3:**217–223.

Antibus, R. K., J. G. Croxdale, O. K. Miller, and A. E. Linkins (1981). Ectomycorrhizal fungi of *Salix rotundifolia*. III. Resynthesized mycorrhizal complexes and their surface phosphatase activities. *Can. J. Bot.*, **59:**2458–2465.

Azcon-Aguilar, R., V. Gianinazzi-Pearson, J. C. Fardeau, and S. Gianinazzi (1986). Effect of vesicular-arbuscular mycorrhizal fungi and phosphate-solubilizing bacteria on growth and nutrition of soybean in a neutral-calcareous soil amended with ^{32}P-^{45}Ca-tricalcium phosphate. *Plant Soil*, **96:**3–15.

Bagyaraj, D. J. (1984). Biological interactions with VA mycorrhizal fungi. In *VA Mycorrhiza* (C. Ll. Powell and D. J. Bagyaraj, eds.). CRC Press, Boca Raton, FL, pp. 131–153.

Baltruschat, H. and F. Schonbeck (1975). The influence of endotrophic mycorrhiza on the infestation of tobacco by *Thielaviopsis basicola*. *Phytopathol. Z.*, **84:**172–188.

Barea, J. M., A. F. Bonis, and L. Olivares (1983). Interactions between *Azospirillum* and VA mycorrhiza and their effects on growth and nutrition of maize and ryegrass. *Soil Biol. Biochem.*, **15:**705–709.

Barea, J. M., C. Azcon-Aguilar, and R. Azcon (1987). Vesicular-arbuscular mycorrhiza improve both symbiotic N_2 fixation and N uptake from soil as assessed with a ^{15}N technique under field conditions. *New Phytol.*, **106:**717–725.

Barley, K. P. (1970). The configuration of the root system in relation to nutrient uptake. *Adv. Agron.*, **22:**159–201.

Bartlett, E. M. and D. H. Lewis (1973). Surface phosphate activity of mycorrhizal roots of beech. *Soil Biol. Biochem.* **5:**249–257.

Berthelin, J. and C. Leyval (1982). Ability of symbiotic and non-symbiotic rhizosphere microflora of maize (*Zea mays*) to weather micas and to promote plant growth and plant nutrition. *Plant Soil*, **68:**369–377.

Bethlenfalvay, G. J., M. S. Brown, and R. S. Pacovsky (1982a). Relationships between host and endophyte development in mycorrhizal soybeans. *New Phytol.*, **90:**537–543.

Bethlenfalvay, G. J., R. S. Pacovsky, M. S. Brown, and G. Fuller (1982b). Mycotrophic growth and mutualistic development of host plants and fungal endophyte in an endomycorrhizal symbiosis. *Plant Soil*, **68:**43–54.

Bjorkman, E. (1960). *Monotropa hypopitys* L., an epiparasite on tree roots. *Physiol. Plant.*, **13:**308–327.

Bonfante-Fasolo, P. (1984). Anatomy and morphology of VA mycorrhizae. In *VA Mycorrhiza* (C. Ll. Powell and D. J. Bagyaraj, eds.). CRC Press, Boca Raton, FL, pp. 5–33.

Bowen, G. D. (1973). Mineral nutrition of ectomycorrhizae. In *Ectomycorrhizas: Their Ecology and Physiology* (G. C. Marks and T. T. Koslowski, eds). Academic Press, New York, pp. 151–205.

Bowen, G. D., D. I. Bevege, and B. Mosse (1975). Phosphate physiology of vesicular-arbuscular mycorrhizas. In *Endomycorrhizas* (F. E. Sanders, B. Mosse, and P. B. Tinker, eds.). Academic Press, London, pp. 242–260.

Bradley, R., A. J. Burt, and D. J. Read (1982). The biology of mycorrhiza in the Ericaceae. VIII. The role of mycorrhizal infection in heavy metal resistance. *New Phytol.*, **91:** 197–209.

Brown, M. E. (1982). Nitrogen fixation by free-living bacteria associated with plants—fact or fiction. In *Bacteria and Plants* (M. E. Rhodes-Roberts and I. A. Skinner, eds.). *Soc. Appl. Bacteriol. Symp. Ser. No. 10*, Academic Press, London, pp. 26–41.

Brown, M. E. and G. R. Carr (1984). Interactions between *Azotobacter chroococcum* and vesicular-arbuscular mycorrhiza and their effects on plant growth. *J. Appl. Bacteriol.*, **56:**429–437.

Chakravarty, P. and T. Unestam (1987). Mycorrhizal fungi prevent disease in stressed pine seedlings. *J. Phytopathol.*, **118:**335–340.

Chapin, F. S. III (1980). The mineral nutrition of wild plants. *Annu. Rev. Ecol. Syst.*, **11:**233–260.

Chiariello, N., J. C. Hickman, and H. A. Mooney (1982). Endomycorrhizal role for interspecific transfer of phosphorus in a community of annual plants. *Science*, **217:**941–943.

Chou, L. G. and A. F. Schmitthenner (1974). Effect of *Rhizobium japonicum* and *Endogone mooseae* on soybean root rot caused by *Pythium ultimum* and *Phytophthora megasperma* var. *sojae*. *Plant Dis. Rep.*, **58**:221–225.

Clarkson, D. (1985). Factors affecting mineral nutrient acquisition by plants. *Annu. Rev. Plant Physiol.*, **36**:77–115.

Clough, K. S. and J. C. Sutton (1978). Direct observation of fungal aggregates in sanddune soil. *Can. J. Microbiol.*, **24**:326–333.

Coleman, D. C. (1985). Through a ped darkly: an ecological assessment of root-soil-microbial-faunal interactions. In *Ecological Interactions in Soil* (A. H. Fitter, A. H. Atkinson, D. J. Read, and M. B. Usher, eds.). Blackwell Scientific Publications, Oxford, pp. 1–22.

Coleman, D. C., C. P. P. Reid, and C. V. Cole (1983). Biological strategies of nutrient cycling in soil systems. *Adv. Ecol. Res.*, **13**:1–55.

Cooper, K. M. (1984). Physiology of VA mycorrhizal associations. In *VA Mycorrhiza* (C. Ll. Powell and D. J. Bagyaraj, eds.). CRC Press, Boca Raton, FL, pp. 155–186.

Cooper, K. M. and P. B. Tinker (1978). Translocation and transfer of nutrients in vesicular-arbuscular mycorrhizas. II. Uptake and translocation of phosphorus, zinc, and sulphur. *New Phytol.*, **81**:43–52.

Cox, G. and P. B. Tinker (1976). Translocation and transfer of nutrients in vesicular-arbuscular mycorrhizas. I. The arbuscule and phosphorus transfer: A quantitative ultrastructural study. *New Phytol.*, **77**:371–378.

Cox, G., K. J. Moran, F. Sanders, C. Nockols, and P. B. Tinker (1980). Translocation and transfer of nutrients in vesicular-arbuscular mycorrhiza. III. Polyphosphate granules and phosphorus translocation. *New Phytol.*, **84**:649–659.

Cress, W. A., G. V. Johnson, and L. L. Barton (1986). The role of endomycorrhizal fungi in iron uptake by *Hilaria jamesii*. *J. Plant Nutr.*, **9**:547–556.

Cromack, K., Jr., P. Sollins, W. C. Graustein, K. Speidel, A. W. Todd, G. Spycher, C. Y. Li, and R. L. Todd (1979). Calcium oxalate accumulation and soil weathering in mats of the hypogeous fungus *Hysterangium crassum*. *Soil Biol. Biochem.*, **11**:463–468.

Cromer, D. A. N. (1935). The significance of the mycorrhiza of *Pinus radiata*. *Bull. For. Bur. Aust.*, **16**:1–19.

Crowley, D. E., C. P. P. Reid, and P. J. Szaniszlo (1987). Microbial siderophores as iron sources for plants. In *Iron Transport in Microbes, Plants and Animals* (G. Winklemann, R. van der Helm, and J. Neilands, eds.). Verlagsgesellschaft mbH, Weinheim, pp. 375–386.

Dahm, H., E. Strzelczyk, and L. Majewska (1987). Cellulolytic and pectolytic activity of mycorrhizal fungi, bacteria and actinomycetes associated with the roots of *Pinus sylvestris*. *Pedobiologia*, **30**:73–80.

Davis, R. M. (1980). Influence of *Glomus fasciculatus* on *Thielaviopsis basicola* root rot of citrus. *Plant Dis.*, **64**:839–840.

Davis, R. M. and J. A. Menge (1980). Influence of *Glomus fasciculatus* on *Phytophthora* root rot of citrus. *Phytopathology*, **70**:447–452.

Dodd, J. C., C. C. Burton, R. G. Burns, and P. Jeffries (1987). Phosphatase activity associated with the roots and the rhizosphere of plants infected with vesicular-arbuscular mycorrhizal fungi. *New Phytol.*, **107**:163–172.

Dowding, E. W. (1959). Ecology of *Endogone*. *Trans. Br. Mycol. Soc.*, **42**:449–457.

Duddridge, J. and D. J. Read (1982). An ultrastructural analysis of the development of mycorrhizas in *Rhododendron ponticum*. *Can. J. Bot.*, **60**:2345–2356.

Duddridge, J. A., A. Malibari, and D. J. Read (1980). Structure and function of mycorrhizal rhizomorphs with special reference to their role in water transport. *Nature* (Lond.), **287**:834–836.

Finlay, R. D. (1985). Interactions between soil microarthropods and endomycorrhizal associations of higher plants. In *Ecological Interactions in Soil* (A. H. Fitter, A. H. Atkinson, D. J. Read, and M. B. Usher, eds.). Blackwell Scientific Publications, Oxford, pp. 319–331.

Finlay, R. D. and D. J. Read (1986). The structure and function of the vegetative mycelium of ectomycorrhizal plants. I. Translocations of ^{14}C-labelled carbon between plants interconnected by a common mycelium. *New Phytol.*, **103**:143–156.

Fogel, R. and G. Hunt (1979). Fungal and arboreal biomass in a western Oregon Douglas-fir ecosystem: distribution patterns and turnover. *Can. J. For. Res.*, **9**:245–256.

Forester, S. M. and T. H. Nicolson (1979). Microbial aggregation of sand in a maritime dune succession. *Soil Biol. Biochem.*, **13**:205–208.

Foster, R. C. and G. C. Marks (1967). Observations on the mycorrhizas of forest trees. II. The rhizosphere of *Pinus radiata* D. Don. *Aust. J. Biol. Sci.*, **20**:915–926.

Foster, R. C., A. D. Rovira, and T. W. Cock (1983). *Ultrastructure of the Root–Soil Interface.* American Phytopathology Society, St. Paul, MN.

France, R. C. and C. P. P. Reid (1983). Interactions of nitrogen and carbon in the physiology of ectomycorrhizae. *Can. J. Bot.*, **61**:964–984.

Francis, R. and D. J. Read (1984). Direct transfer of carbon between plants connected by vesicular-arbuscular mycorrhizal mycelium. *Nature* (Lond.), **307**:53–56.

Francis, R., R. D. Finlay, and D. J. Read (1986). Vesicular-arbuscular mycorrhiza in natural vegetation systems. IV. Transfer of nutrients in inter- and intra-specific combinations of host plants. *New Phytol.*, **102**:103–111.

Graham, J. H. and J. P. Syvertsen (1984). Influence of vesicular-arbuscular mycorrhiza on the hydraulic conductivity of roots of two citrus rootstocks. *New Phytol.*, **97**:277–284.

Graham, J. H., R. T. Leonard, and J. Menge (1981). Membrane mediated decrease in root exudation responsible for phosphorus inhibition of vesicular-arbuscular mycorrhiza formation. *Plant Physiol.*, **68**:548–552.

Graustein, W. C., K. Cromack, and P. Sollins (1977). Calcium oxalate: occurrence in soils and effect on nutrient and geochemical cycles. *Science*, **198**:1252–1254.

Hadley, G. (1984). Uptake of [^{14}C]-glucose by asymbiotic and mycorrhizal orchid protocorms. *New Phytol.*, **96**:263–273.

Hardie, K. (1985). The effect of removal of extraradical hyphae on water uptake by vesicular-arbuscular mycorrhizal plants. *New Phytol.*, **101**:677–684.

Harley, J. L. (1940). A study of the root system of the beech in woodland soils with special reference to mycorrhizal infection. *J. Ecol.*, **28**:107–117.

Harley, J. L. (1969). *The Biology of Mycorrhiza*, 2nd edn. Leonard Hill, London.

Harley, J. L. (1978). Ectomycorrhizas as nutrient absorbing organs. *Proc. R. Soc. Lond., B. Biol. Sci.*, **203**:1–21.

Harley, J. L. and C. C. McCready (1950). The uptake of phosphate by excised mycorrhizal roots of the beech. *New Phytol.*, **49**:388–397.

Harley, J. L. and S. E. Smith (1983). *Mycorrhizal Symbiosis.* Academic Press, London.

Heap, A. J. and E. I. Newman (1980). The influence of vesicular-arbuscular mycorrhizas on phosphorus transfers between plants. *New Phytol.*, **85**:173–179.

Hetrick, B. A. D. (1984). Ecology of VA mycorrhizal fungi. In *VA Mycorrhiza* (C. Ll. Powell and D. J. Bagyaraj, eds.). CRC Press, Boca Raton, FL, pp. 35–55.

Hirrel, M. C. and J. W. Gerdemann (1979). Enhanced carbon transfer between onions infected with a vesicular-arbuscular mycorrhizal fungus. *New Phytol.*, **83**:731–738.

Hu, H. T. (1977). The influences in soil moisture on survival, growth, and historical change in ectomycorrhizal red pine seedlings. *Bull. Exp. For. Natl. Taiwan Univ.*, **120**:13–37.

Hunt, H. W., D. C. Coleman, E. R. Ingham, R. E. Ingham, E. T. Elliot, J. C. Moore, C. P. P. Reid, and C. R. Morley (1987). The detrital food web in a shortgrass prairie. *Biol. Fert. Soil*, **3**:57–68.

Hussey, R. S. and R. W. Roncadori (1982). Vesicular-arbuscular mycorrhizae may limit nematode activity and improve plant growth. *Plant Dis.*, **66**:9–14.

Ingham, R. E. (1988). Interactions between nematodes and vesicular-arbuscular mycorrhizae. In *Biological Interactions in Soil* (C.A. Edwards, B.R. Stinner, D. Stinner, and S. Rabatin, eds.). Elsevier Science Publishers, Amsterdam, pp. 169–182.

Ingham, E. R., J. A. Trofymow, R. N. Ames, H. W. Hunt, C. R. Morley, J. C. Moore, and D. C. Coleman (1986). Trophic interactions and nitrogen cycling in a semiarid grassland soil.

II. System responses to removal of different groups of soil microbes or fauna. *J. Appl. Ecol.*, **23**:615–630.

Kepert, D. G., A. D. Robinson, and A. M. Posner (1979). The effect of organic root products and the availability of phosphorus to plants. In *The Soil–Root Interface* (J. L. Harley and R. S. Russell, eds). Academic Press, London, pp. 115–124.

Knutson, D. M., A. S. Hutchins, and K. Cromack, Jr. (1980). The association of calcium oxalate-utilizing *Streptomyces* with conifer ectomycorrhizae. *Antonie Leeuwenhoek J. Microbiol.*, **46**:611–619.

Koske, R. E. (1975). *Endogone* spores in Australian sand dunes. *Can. J. Bot.*, **53**:668–672.

Kottke, I. and F. Oberwinkler (1986). Root–fungus interactions observed on initial stages of mantle formation and Hartig net establishment in mycorrhizas of *Amanita muscaria* on *Picea abies* in pure culture. *Can. J. Bot.*, **64**:2348–2354.

Lewis, D. H. (1973). Concepts in fungal nutrition and the origin of biotrophy. *Biol. Rev. Camb. Philos. Soc.*, **48**:261–278.

Linderman, R. G. (1988). Mycorrhizal interactions with the rhizosphere microflora: The mycorrhizosphere effect. *Phytopathology*, **78**:366–370.

Ling-Lee, M., A. E. Ashford, and G. A. Chilvers (1977). A histochemical study of polysaccharide distribution in eucalypt mycorrhizas. *New Phytol.*, **78**:329–335.

Lyr, H. and G. Hoffman (1967). Growth rates and growth periodicity of tree roots. *Int. Rev. Microbiol.*, **5**:67–107.

MacDonald, R. M. and M. Lewis (1978). The occurrence of some acid phosphatases and dehydrogenases in the vesicular-arbuscular mycorrhizal fungus *Glomus mosseae*. *New Phytol.*, **80**:135–141.

Marks, G. C. and R. C. Foster (1973). Structure, morphogenesis and ultrastructure of ectomycorrhizae. In *Ectomycorrhizae: Their Ecology and Physiology* (G. C. Marks and T. T. Kozlowski, eds.). Academic Press, New York, pp. 1–41.

Marks, G. C. and T. T. Kozlowski (eds.) (1973). *Ectomycorrhizae: Their Ecology and Physiology*. Academic Press, New York.

Marschner, H. (1986). *Mineral Nutrition of Higher Plants*. Academic Press, London.

Marx, D. H. (1969). The influence of ectotrophic mycorrhizal fungi on the resistance of pine roots to pathogenic fungi. *Phytopathology*, **59**:153–163.

Marx, D. H. (1975). Role of ectomycorrhizae in the protection of pine from root infection by *Phytophthora cinnamomi*. In *Biology and Control of Soil-borne Plant Pathogens* (J. W. Bruehl, ed.). American Phytopathology Society, St. Paul, MN, pp. 112–115.

Menzies, J. D. and R. G. Gilbert (1967). Responses of the soil microflora to volatile components in plant residues. *Soil Sci. Am. Proc.*, **31**:495–496.

Mexal, J. and C. P. P. Reid (1973). The growth of selected mycorrhizal fungi in response to induced water stress. *Can. J. Bot.*, **51**:1579–1588.

Miller, R. M. (1987). The ecology of vesicular-arbuscular mycorrhizae in grass- and shrublands. In *Ecophysiology of VA Mycorrhizal Plants* (G. Safir, ed.). CRC Press, Boca Raton, FL, pp. 136–170.

Mitchell, D. T. and D. J. Read (1981). Utilization of inorganic and organic phosphates by the mycorrhizal endophytes of *Vaccinium macrocarpon* and *Rhododendron ponticum*. *Trans. Br. Mycol. Soc.*, **76**:255–260.

Moore, J. C. and D. E. Walter (1988). Arthropod regulation of micro- and mesobiota in below-ground detrital food webs. *Annu. Rev. Entomol.*, **33**:419–439.

Mosse, B. (1959). Observations on the extramatrical mycelium of a vesicular-arbuscular endophyte. *Trans. Br. Mycol. Soc.*, **42**:439–448.

Newman, E. I. (1974). Root and soil water relations. In *The Plant Root and Its Environment* (E. W. Carson, ed.). University Press of Virginia, Charlottesville, pp. 363–440.

Newman, E. I. and K. Ritz (1986). Evidence on the pathways of phosphorus transfer between vesicular-arbuscular mycorrhizal plants. *New Phytol.*, **104**:77–87.

Nicolson, T. H. (1959). Mycorrhiza in the gramineae. I. Vesicular-arbuscular endophytes with special reference to the external phase. *Trans. Br. Mycol. Soc.*, **42**:421–438.

Nicolson, T. H. and C. Johnston (1979). Mycorrhiza in the gramineae. III. *Glomus fasciculatus* as the endophyte of pioneer grasses in a maritime sand dune. *Trans. Br. Mycol. Soc.*, **72**:261–268.

Nye, P. H. (1979). Soil properties controlling the supply of nutrients to the root surface. In *The Soil–Root Interface* (J. L. Harley and R. S. Russell, eds.). Academic Press, London, pp. 39–49.

Nye, P. H. and P. B. Tinker (1977). Solute movement in the soil–root system. University of California Press, Berkeley.

Old, K. M. and S. Chakraborty (1986). Mycophagous soil amoebae: their biology and significance in the ecology of soil borne plant pathogens. *Progr. Protistol.*, **1**:163–194.

Owusu-Bennoch, E. and A. Wild (1979). Autoradiography of the depletion zone of phosphate around onion roots in the presence of vesicular-arbuscular mycorrhiza. *New Phytol.*, **84**:133–140.

Pacovsky, R. S., G. Fuller, and A. E. Paul (1985). Influence of soil on the interactions between endomycorrhizae and *Azospirillum* in sorghum. *Soil Biol. Biochem.*, **17**:525–531.

Pearson, D. P. and D. J. Read (1975). The physiology of the mycorrhizal endophyte of *Calluna vulgaris*. *Trans. Br. Mycol. Soc.*, **64**:1–7.

Pearson, V. and P. B. Tinker (1975). Measurement of phosphorus fluxes in the external hyphae of endomycorrhizas. In *Endomycorrhizas* (F. E. Sanders, B. Mosse, and P. B. Tinker, eds.). Academic Press, London, pp. 275–287.

Piccini, D. and R. Azcon (1987). Effect of phosphate-solubilizing bacteria and vesicular-arbuscular mycorrhizal fungi on the utilization of Bayovar rock phosphate by alfalfa plants using a sand–vermiculite medium. *Plant Soil*, **101**:45–50.

Powell, C. Ll. (1984). Field inoculation with VA mycorrhizal fungi. In *VA Mycorrhiza* (C. Ll. Powell and D. J. Bagyaraj, eds.). CRC Press, Boca Raton, FL, pp. 205–222.

Powell, C. Ll. and D. J. Bagyaraj (eds.) (1984). *VA Mycorrhiza*, CRC Press, Boca Raton, FL.

Raj, J., D. J. Bagyaraj, and A. Manjunath (1981). Influence of soil inoculation with vesicular-arbuscular mycorrhiza and a phosphate-dissolving bacterium on plant growth and ^{32}P-uptake. *Soil Biol. Biochem.*, **13**:105–108.

Ratnayake, M., R. T. Leonard, and J. A. Menge (1978). Root exudation in relation to supply of phosphorus and its possible relevance to mycorrhizal formation. *New Phytol.*, **81**:543–552.

Read, D. J. (1983). The biology of mycorrhiza in the Ericales. *Can. J. Bot.*, **61**:985–1004.

Read, D. J. (1987). The biology of mycorrhiza in heathland ecosystems with special reference to the nitrogen nutrition of the Ericaceae. In *Microbiol Ecology* (M. W. Loutit and J. A. R. Miles, eds.). Springer Verlag, New York, pp. 324–328.

Read, D. J., H. Kianmehr, and A. Malibari (1977). The biology of mycorrhiza in *Helianthemum* Mill. *New Phytol.*, **78**:305–312.

Reid, C. P. P. (1979). Mycorrhizae and water stress. In *Root Physiology and Symbiosis, Vol. 6* (A. Riedacker and J. Gagnaire-Michard, eds.). CFNR, Nancy, France, pp. 392–408.

Reid, C. P. P. (1984). Mycorrhizae: a root–soil interface in plant nutrition. In *Microbial–Plant Interactions*. Soil Science Society of America, Madison, WI, pp. 29–50.

Reid, C. P. P. and G. D. Bowen (1979). Effect of water stress on the uptake of ^{32}P from solution by mycorrhizae and nonmycorrhizal roots of *Pinus radiata*. *New Phytol.*, **83**:103–107.

Reid, C. P. P. and F. W. Woods (1969). Translocation of C-14 labeled compounds in mycorrhizae and its implications in interplant nutrient cycling. *Ecology*, **50**:179–187.

Reid, C. P. P., P. J. Szaniszlo, and D. E. Crowley (1986). Siderophore involvement in plant iron nutrition. In *Iron, Siderophores, and Plant Diseases* (T. R. Swinburne, ed.). Plenum, New York, pp. 29–42.

Reid, R. K., C. P. P. Reid, P. E. Powell, and P. J. Szaniszlo (1984). Comparison of siderophore concentrations in aqueous extracts of rhizosphere and adjacent bulk soils. *Pedobiologia*, **26**:263–266.

Reid, R. K., C. P. P. Reid, and P. J. Szaniszlo (1985). Effects of synthetic and microbially produced chelates on the diffusion of iron and phosphorus to a simulated root in soil. *Biol. Fert. Soils*, **1**:45–52.

Rhodes, L. H. and J. W. Gerdemann (1978). Hyphal translocation and uptake of sulphur by vesicular-arbuscular mycorrhizae of onion. *Soil Biol. Biochem.*, **10**:355–360.

Rhodes, L. H. and J. W. Gerdemann (1980). Nutrient translocation in vesicular-arbuscular mycorrhizae. In *Cellular Interactions in Symbiosis and Parasitism* (C. B. Cook, P. W. Pappas, and E. D. Rudolph, eds.). Ohio State University Press, Columbus, Ohio, pp. 173–195.

Ricklefs, R. E. (1979). *Ecology*, 2nd edn. Chiron Press, New York.

Riffle, J. W. (1971). Effect of nematodes on root-inhabiting fungi. In *Mycorrhizae* (E. Hacskaylo, ed.). USDA For. Serv. Misc. Publ. 1189, pp. 97–113.

Rogers, W. S. and G. C. Head (1969). Factors affecting the distribution and growth of roots of perennial woody species. In *Root Growth* (W. J. Whittington, ed.). Plenum Press, New York, pp. 280–295.

Rygiewicz, P. T., C. S. Bledsoe, and R. J. Zasoski (1984). Effects of ectomycorrhizae and solution pH on [^{15}N] ammonium uptake by coniferous seedlings. *Can. J. For. Res.*, **14**:885–892.

Safir, G. R. (ed.) (1987). *Ecophysiology of VA Mycorrhizal Plants*. CRC Press, Boca Raton, FL.

Salawu, E. O. and R. H. Estey (1979). Observations on the relationships between a vesicular-arbuscular fungus, a fungivorous nematode and the growth of soybeans. *Phytoprotection*, **60**:99–102.

Sanders, F. E. and P. B. Tinker (1973). Phosphate flow into mycorrhizal roots, *Pestic. Sci.*, **4**:385–395.

Sanders, F. E., B. Mosse, and P. B. Tinker (eds.) (1975). *Endomycorrhizas*, Academic Press, London.

Sands, R., E. L. Fiscus, and C. P. P. Reid (1982). Hydraulic properties of pine and bean roots with varying degrees of suberization, vascular differentiation and mycorrhizal infection. *Aust. J. Plant Physiol.*, **9**:559–569.

Schenck, N. C., W. H. Ridings, and J. A. Cornell (1977). Interactions of two vesicular-arbuscular mycorrhizal fungi, and *Phytophthora parasitica* on two citrus root stocks. *3rd N. Am. Conf. Mycorrhizae, Athens, GA, Abst.*

Schonbeck, F. and H. W. Dehne (1977). Damage to mycorrhizal and non-mycorrhizal cotton seedlings by *Thielaviopsis basicola*. *Plant Dis. Rep.*, **61**:266–267.

Schramm, J. R. (1966). Plant colonization studies on black wastes from anthracite mining in Pennsylvania. *Trans. Am. Philos. Soc.*, **56**:1–194.

Schubert, A., C. Marzachi, M. Mazzitelli, M. C. Cravero, and P. Bonafante-Fasolo (1987). Development of total and viable extraradical mycelium in the vesicular-arbuscular mycorrhizal fungus *Glomus clarum* Nicol. & Schenck. *New Phytol.*, **107**:183–190.

Schulze, E. D. (1982). Plant life forms and their carbon, water, and nutrient relations. In *Physiological Plant Ecology II* (O. L. Lange et al., eds.). Springer-Verlag, Berlin, pp. 615–676.

Schwab, S. M., J. A. Menge, and R. T. Leonard (1983). Quantitative and qualitative effects of phosphorus on extracts and exudates of sudangrass roots in relation to vesicular-arbuscular mycorrhiza formation. **73**:761–765.

Shaw, P. J. A. (1985). Grazing preferences of *Onychiurus armatus* (Insecta: Collembola) for mycorrhizal and saprophytic fungi of pine plantations. In *Ecological Interactions in Soil* (A. H. Fitter, A. H. Atkinson, D. J. Read, and M. B. Usher, eds.). Blackwell Scientific Publications, Oxford, pp. 333–337.

Skinner, M. F. and G. D. Bowen (1974). The uptake and translocation of phosphate by mycelial strands of pine mycorrhizas. *Soil. Biol. Biochem.*, **6**:53–56.

Smith, S. E., B. J. St. John, F. A. Smith, and J. D. Nicholas (1985). Activity of glutamine synthetase and glutamate dehydrogenase in *Trifolium subterraneum* L. and *Allium cepa* L.: effects of mycorrhizal infection and phosphate nutrition. *New Phytol.*, **99**:211–227.

St. John, T. V. and D. C. Coleman (1983). The role of mycorrhizae in plant ecology. *Can. J. Bot.*, **61**:1005–1014.

St. John, T. V., D. C. Coleman, and C. P. P. Reid (1983a). Association of vesicular-arbuscular mycorrhizal hyphae with soil organic particles. *Ecology*, **64**:957–959.

St. John, T. V., D. C. Coleman, and C. P. P. Reid (1983b). Growth and spatial distribution of nutrient-absorbing organs: selective exploitation of soil heterogeneity. *Plant Soil*, **71**:487–493.

Stribley, D. P. (1987). Mineral nutrition. In *Ecophysiology of VA Mycorrhizal Plants* (G. Safir, ed.). CRC Press, Boca Raton, FL, pp. 59–70.

Stribley, D. P. and D. J. Read (1980). The biology of mycorrhiza in the Ericaceae. VIII. The relationship between mycorrhizal infection and the capacity to utilize simple and complex organic nitrogen sources. *New Phytol.*, **86**:365–371.

Strullu, D-G., J-P. Gourret, and J-P. Garrec (1981). Microanalyse des granules vacuolaires des ectomycorhizes, endomycorhizes, et endomycothalles. *Physiol. Veg.*, **19**:367–378.

Summerbell, R. C. (1987). The inhibitory effect of *Trichoderma* species and other soil microfungi on formation of mycorrhiza by *Laccaria bicolor in vitro*. *New Phytol.*, **105**:437–448.

Sutherland, J. R. and A. Fortin (1968). Effect of the nematode *Aphelenchus avenae* on some ectotrophic mycorrhizal fungi and on a red pine mycorrhizal relationship. *Phytopathology*, **58**:519–523.

Sylvia, D. M. (1988). Activity of external hyphae of vesicular-arbuscular mycorrhizal fungi. *Soil Biol. Biochem.*, **20**:39–43.

Sylvia, D. M. (1990). Distribution, structure, and function of external hyphae of vesicular-arbuscular mycorrhizal fungi. In *Rhizosphere Dynamics* (J. E. Box and L. Hammond, eds.). Westview Press, Boulder, CO (in press).

Sylvia, D. M. and W. A. Sinclair (1983). Phenolic compounds and resistance to fungal pathogens induced in primary roots of Douglas-fir seedlings by the ectomycorrhizal fungus *Laccaria laccata*. *Phytopathology*, **73**:390–397.

Szaniszlo, P. J., P. E. Powell, C. P. P. Reid, and G. R. Cline (1981). Production of hydroxamate siderophore iron chelators by ectomycorrhizal fungi. *Mycologia*, **73**:1158–1174.

Tinker, P. B. H. (1975). Effects of vesicular-arbuscular mycorrhizas on higher plants. *Symp. Soc. Exp. Biol.*, **29**:325–349.

Tinker, P. B. H. (1976). Roots and water. Transport of water to plant roots in soil. *Phil. Trans. R. Soc. Lond. B. Biol. Sci.*, **273**:445–461.

Tinker, P. B. and A. Gildon (1983). Mycorrhizal fungi and ion uptake. In *Metals and Micronutrients: Uptake and Utilization by Plants* (D. A. Robb and W. S. Pierpont, eds.). Academic Press, New York, pp. 21–32.

Vogt, K. A., C. C. Grier, C. E. Meier, and R. L. Edmonds (1982). Mycorrhizal role in net primary production and nutrient cycling in *Abies amabilis* ecosystems in western Washington. *Ecology*, **63**:370–380.

Weatherly, P. E. (1975). Water relations of the root system. In *The Development and Function of Roots* (J. G. Torrey and D. T. Clarkson, eds.). Academic Press, New York, pp. 397–413.

Weaver, J. E. (1926). *Root Development and Field Crops*. McGraw-Hill, New York.

Whittingham, J. and D. J. Read (1982). Vesicular-arbuscular mycorrhiza in natural vegetation systems. III. Nutrient transfer between plants with mycorrhizal interconnections. *New Phytol.*, **90**:277–284.

Wilcox, H. E. (1983). Fungal parasitism of woody plant roots from mycorrhizal relationships to plant disease. *Annu. Rev. Phytopathol.*, **21**:221–242.

Williamson, B. and I. J. Alexander (1975). Acid phosphatase localized in the sheath of beech mycorrhiza. *Soil. Biol. Biochem.*, **7**:195–198.

Winkelmann, G., R. van der Helm, and J. Neilands (eds.) (1987). *Iron Transport in Microbes, Plants, and Animals*. Verlagsgesellschaft mbH, Weinheim.

Woldendorp, J. W. (1978). The rhizosphere as part of the plant–soil system. In *Structure and Functioning of Plant Populations* (A. H. J. Freysen and J. W. Woldendorp, eds.). Konink. Nederl. Akad. Wetens. Verh., Afdeling Natuurkunde, Tweede. Reeks, deel 70 North-Holland, New York, pp. 237–267.

Woods, F. W. and K. Brock (1964). Interspecific transfer of ^{45}Ca and ^{32}P by root systems. *Ecology*, **45**:886–889.

Worley, J. F. and E. Hacskaylo (1959). The effect of available soil moisture on the mycorrhizal association of Virginia pine. *For. Sci.*, **5**:267–268.

12

SOIL AND RHIZOSPHERE ASPECTS OF N_2-FIXING PLANT-MICROBE ASSOCIATIONS

D. A. ZUBERER

Department of Soil and Crop Sciences, Texas A & M University, College Station, Texas 77843-2474, USA

1. INTRODUCTION

Interest in the potential for increasing the yields of non-leguminous grains and the productivity of pastures through manipulation of nitrogen fixation in the soil and rhizosphere has remained strong for the past 15–20 years since reports of high rates of N_2-fixation associated with tropical grasses (Döbereiner et al., 1972a). These reports stimulated a widespread, renewed interest in the possibilities for using free-living N_2-fixing bacteria as well as symbiotic associations to enhance plant growth in a variety of cropping systems. The high price of oil and the linked high prices of nitrogenous fertilizers during the early 1970s gave a strong impetus to studies of the abilities of non-legumes to fix nitrogen through root-associated bacteria. Much information has been gained regarding the intricacies of the many plant–bacteria associations demonstrating some degree of nitrogen fixation capacity, yet there are still substantial gaps in our collective knowledge of the establishment, functioning and manipulation of these systems for practical use in agricultural systems throughout the world. Interest in N_2-fixing associations remains keen and there is research activity throughout the world at levels ranging from the molecular genetics of the bacteria to field application of bacteria to crops.

This chapter is a review of some of the soil and rhizosphere aspects of various N_2-fixing plant–bacteria associations, including non-nodulating plants (mainly grasses) as well as the *Rhizobium/Bradyrhizobium* and *Frankia* root nodule symbioses. Since the root nodule symbioses begin with the arrival of the microsymbiont at the rhizoplane of a suitable host, soil and rhizosphere aspects of these associations are included. Detailed information regarding recognition in these plant–microbe systems can be found in Chapter 6.

2. SOIL AND RHIZOSPHERE ASPECTS OF ACTINORHIZAL SYMBIOSES

2.1. Growth and Survival of *Frankia* spp. in Soil

Members of the genus *Frankia*, filamentous actinomycetes, are able to nodulate a large variety of woody non-leguminous trees and shrubs of economic and ecological importance. Notable among these are trees of the genera *Alnus* (the alders) and *Casuarina* (the "Australian pine" common to the tropics), which are commercially valued for wood and for fuel. Many of the lesser shrubs are of value as nitrogen-fixing nurse crops for timber species, soil improvement on impoverished or disturbed sites, and in windbreak plantings. In addition many of these species are valued for their role in wildlife refuges, where they provide browse and protective cover (Fessenden, 1979; Klemmedson, 1979). Currently, the list of actinorhizal plants encompasses 22 genera among eight families of woody dicots, including more than 170 species (Torrey and Tjepkema, 1983).

Because of the importance of actinorhizal species as nitrogen-fixing plants with a wide range of uses, considerable research efforts have been devoted to gaining a fuller understanding of the nature of the symbiosis between the actinomycete and its hosts (Figs. 1–3). These efforts were thwarted for a long time because of an inability to readily isolate the nodule endophyte and grow it in pure culture. Prior to 1977 few confirmed reports of successful isolation, culture and reinfection by the isolate appeared in the literature. Thus Koch's postulates remained unfulfilled for the putative symbiotic actinomycete observed in root nodules. In 1977, a group of researchers led by Dr. John Torrey at Harvard University succeeded in culturing the endophyte of *Comptonia peregrina* (L.) Coult. (sweet fern) and in repeatedly reinfecting the host in sand or aeroponic culture systems (Callaham et al., 1978). Following the success of the group at Harvard many other researchers succeeded in culturing the endophytes from a wide variety of actinorhizas and the way was cleared for detailed study of the endophyte in pure cultures (Baker et al., 1979; Berry and Torrey, 1979; Lalonde et al., 1981; Benson, 1982; Baker and O'Keefe, 1984). A great deal of information has been amassed in the 10 years since the isolation and confirmation of reinfection by the *Comptonia* isolate. The reader is referred

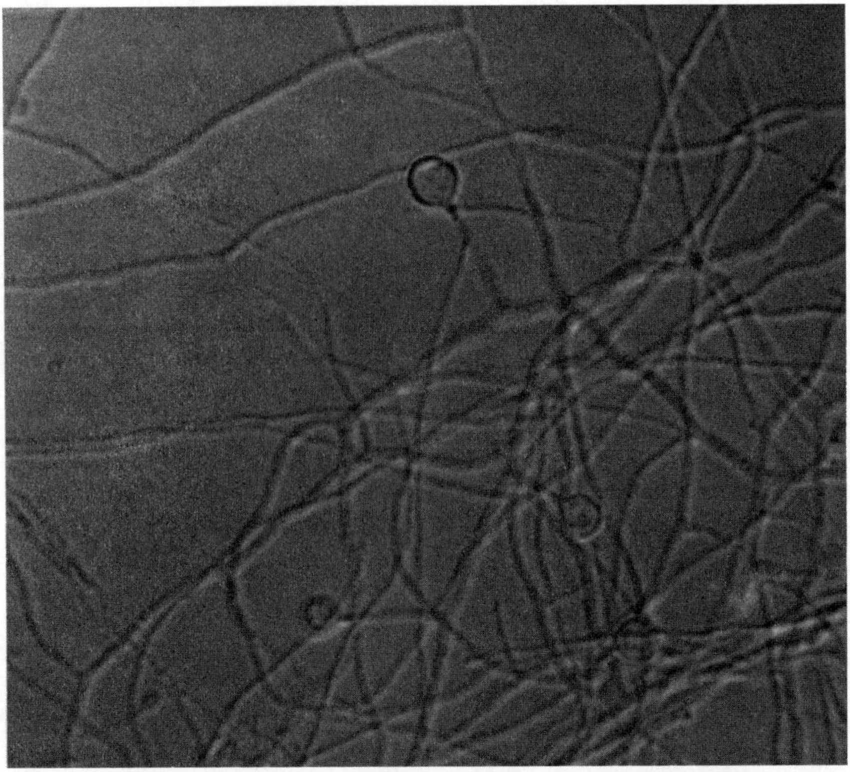

Fig. 1. Hyphae and vesicles of *Frankia* EuI1 from *Elaeagnus umbellata* Thunb. The vesicles confer protection for the nitrogenase enzyme complex against damage by oxygen. Photomicrograph courtesy of Dwight Baker, Yale University.

to the many reviews of this subject before and after the report of Callaham et al. (1978) for detailed accounts of the biology of *Frankia* and its symbiotic association with non-legume hosts (e.g. Becking, 1970, 1975; Tjepkema et al., 1986; Benson, 1988).

Knowledge that the actinomycete could be isolated and grown in pure culture and that it was confirmed as the nodule endophyte led quickly to the discovery that the organism could fix N_2 independently of a host plant. Nitrogen fixation by the free-living endophyte was demonstrated by acetylene reduction (Tjepkema et al., 1980; Gauthier et al., 1981) and by $^{15}N_2$ incorporation (Torrey et al., 1981). Thus in a short period of time, what was previously regarded as an obligate symbiont proved to be an organism capable of free-living growth and N_2 fixation under culture conditions appropriate for the induction of formation of vesicles containing nitrogenase (Tjepkema et al., 1980; Meesters et al., 1987). These studies demonstrated that *Frankia* grows as a free-living bacterium, yet knowledge of the behavior of the organism in soil is still quite sparse.

The paucity of information on the non-symbiotic phases of *Frankia* in soil stems from the difficulties in isolating the bacterium in the face of competition

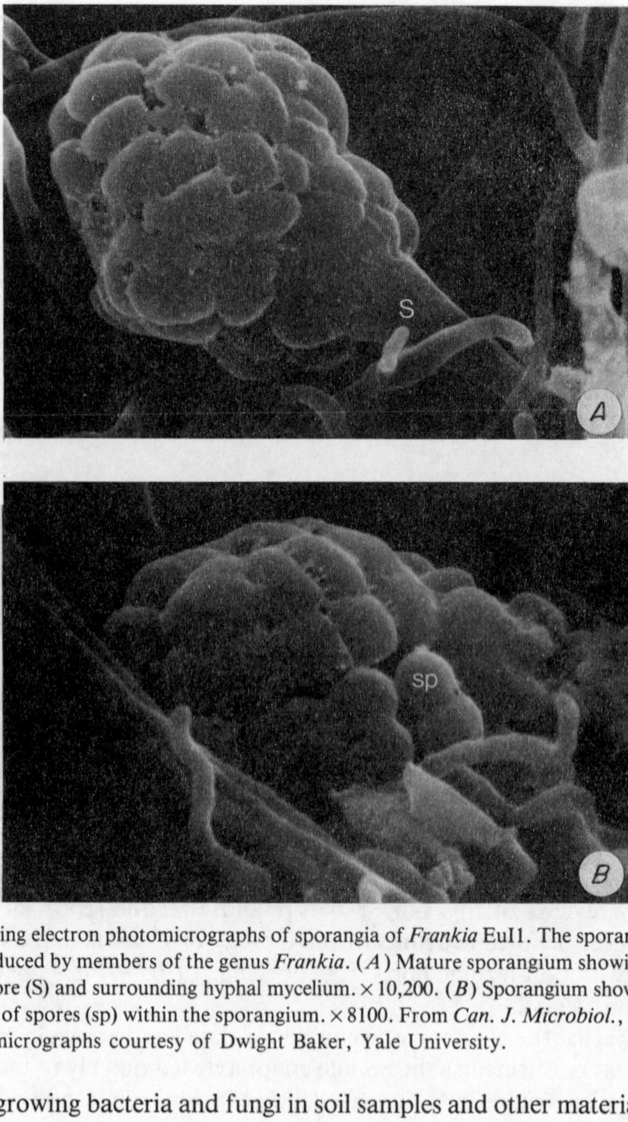

Fig. 2. Scanning electron photomicrographs of sporangia of *Frankia* EuI1. The sporangia are typical of those produced by members of the genus *Frankia*. (*A*) Mature sporangium showing the swollen sporangiophore (S) and surrounding hyphal mycelium. × 10,200. (*B*) Sporangium showing the paired arrangement of spores (sp) within the sporangium. × 8100. From *Can. J. Microbiol.*, **26**:1072–1089; SEM photomicrographs courtesy of Dwight Baker, Yale University.

from fast-growing bacteria and fungi in soil samples and other materials. Because of the slow growth rate extreme measures to decontaminate nodule materials or to separate the organism from other organisms are required. Berry and Torrey (1979) used microdissection techniques to separate the endophyte from nodules. Lalonde et al. (1981) reported a high rate of success using young nodule lobes treated with osmium tetroxide. Microfiltration has also been used to separate vesicle clusters from crushed nodules followed by isolation in pour plates (Benson, 1982). Baker and O'Keefe (1984) reported the first isolation of *Frankia* directly from soil using a procedure in which samples were treated with 0.7% phenol just before placement in a sucrose density gradient and plating on a nitrogen-deficient selective medium. They reported that *Frankia* was more easily

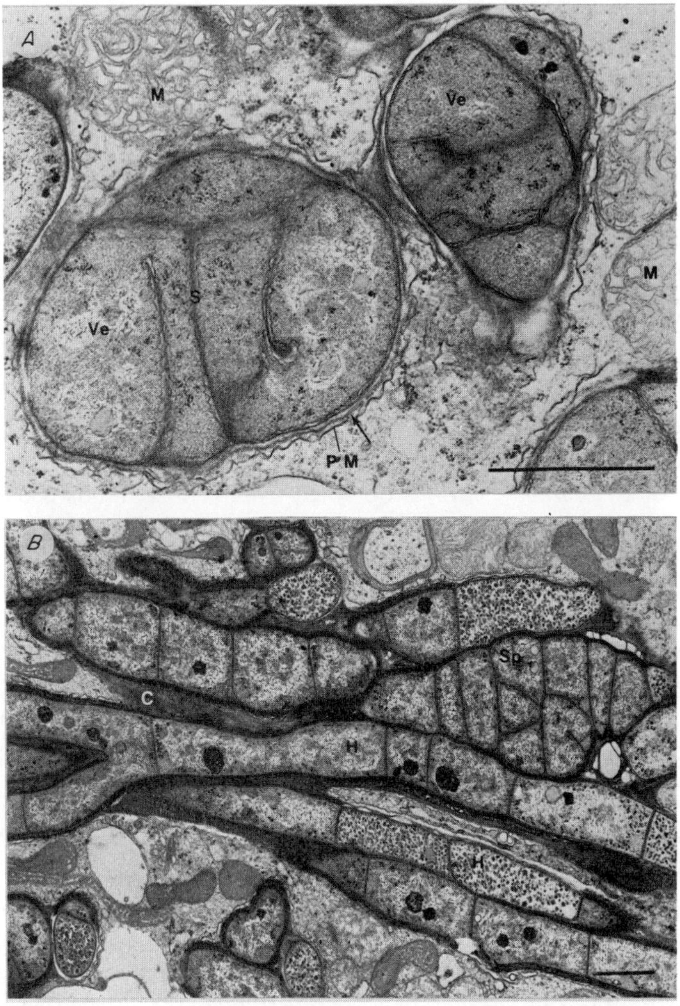

Fig. 3. Endophytic stages of *Frankia* sp. in a *Comptonia peregrina* nodule. Bars = 1.0 μm. (*A*) Vesicles (Ve) are the nitrogen-fixing stage of the endophyte and develop in the mature nodule. The vesicles are separated from the host plant cytoplasm by a capsule (arrow) and the host plasma membrane (P M). S = septum; M = host mitochondrion. (*B*) Colonizing hyphae (H) traverse the center of an infected host cell. A young sporangium (Sp) is also visible. Both hyphae and sporangia are surrounded by the host-derived capsule (C). TEM photomicrographs courtesy of Kate VandenBosch, Department of Biology, Texas A & M University.

obtained from subsurface soils (20–30 cm) than from surface soils (0–15 cm) and speculated that the higher organic matter content of the surface soils caused greater numbers of contaminating microorganisms than the subsurface soils. Interestingly, they also indicated that colonies of *Frankia* derived from the phenol-modified isolations exhibited "more robust growth" than colonies derived through the control procedure without phenol. Application of this phenol modification for isolation of *Frankia* directly from soil should facilitate studies of the ecology of the endophyte in its extranodular stage(s).

Few reports describe growth of *Frankia* in soil, and those that do are based on increases in the nodulation capacity of soils treated with nodule homogenates. Quispel (1955) reported an increase in the nodulation capacity of a homogenate of *Alnus glutinosa* upon incubation in a peat soil. Movement of the endophyte in soil was suggested by Rossi (1964) who used specially designed tubes filled with soil and inoculated at one end with a nodule homogenate. However, as pointed out by van Dijk (1979), avoidance of passive movement of the endophyte in such a system would be difficult. There is a strong need to better understand the behavior of *Frankia* in soil and to determine its capacity for host-independent growth and survival.

Long-term survival of *Frankia* in soil is suggested by studies which demonstrate the infectivity of soils from which the host has been absent for extensive periods. Wollum et al. (1968) found decreasing nodulation potential of soils for *Ceanothus velutinus* obtained from timber stands 200–350 years old. Soils from which the host had been absent for up to 100 years retained the ability to nodulate *Ceanothus*, suggesting long-term survival or the ability of the endophyte to grow in the absence of the host. Becking (1970) reported an absence of the alder endophyte in soils at the Agricultural University of Waginingen which had been free of alders for at least 43 years. Only plots receiving farmyard manure induced nodulation of alder test plants. He attributed the nodulation capacity of these soils to the presence of endophyte in the manure or to some substance in the manure which promoted the survival of the endophyte.

Van Dijk (1979) has shown that a clayey soil of near neutral pH retained a nearly constant level of infectivity for a period of at least a year, whereas the infectivity of a high organic matter, low pH soil declined to 20% that of the original level during a 30-week period. Similar findings were reported by Smolander et al. (1988) who observed a complete loss of nodulation capacity in an acid (pH 2.9) peat soil within 6 months of inoculation with a culture of isolate Ai1c from *Alnus incana*. In a separate study of the distribution of *Frankia* in acid forest soils devoid of actinorhizal plants a positive correlation was observed between pH and nodulation capacity of the soils (Smolander and Sundman, 1987). Interestingly, soils from stands of birch (*Betula* sp.) exhibited the highest nodulation capacity when a hybrid of *Alnus incana* and *Alnus glutinosa* was used as the test plant. These studies suggest a negative influence of low pH on long-term persistence of *Frankia* in soil. However, they also strongly suggest a long-term persistence of *Frankia* in soil and emphasize the need for more detailed study of the host-independent growth and survival of the organism. Perhaps the use of modern molecular techniques such as DNA probes can be brought to bear on this most interesting and elusive aspect of these symbioses.

While evidence for growth and survival of *Frankia* in soil and perhaps in non-host rhizospheres is very limited there are several reports of a host rhizosphere stimulation of *Frankia*. Becking (1975, 1977) and Lalonde (1977) provided ultrastructural evidence of growth of *Frankia* in the mucilaginous layer at the root surface of *Alnus glutinosa*. Such growth is a prerequisite to infection and nodule formation. Extranodular growth of *Frankia* has been observed on *in vitro*-propagated seedlings of *Casuarina equisetifolia* (Diem et al., 1982). Hyphae of the endophyte were observed emanating from the surface of nodules and these hyphae were shown to be capable of nodulation after they were scraped off the surface of the nodules. Whether such extranodular hyphae can serve as a source of inoculum in the soil awaits further study.

Diem et al. (1983) reported formation of vesicles in the rhizosphere of *C. equisetifolia* by an isolate (Cj 1–82) derived from a hybrid of *C. junghuhniana* and *C. equisetifolia*. The occurrence of vesicles in association with the external portions of root nodules is of considerable interest because they have not been observed in nitrogen-fixing nodules of *Casuarina* (Tyson and Silver, 1979). The question merits further investigation! From these studies it can be inferred that like other endosymbionts, *Frankia* must become established in the rhizosphere prior to the initiation of infection.

2.2. Abiotic Factors Affecting Growth of *Frankia* in Soil and the Rhizosphere

2.2.1. Moisture

The suggestions of growth and survival in soil and the evidence for rhizosphere growth imply that *Frankia* in its non-symbiotic state would be subject to numerous biotic and abiotic factors prior to and during the infection process. Amongst the abiotic factors, lack of moisture and low pH are known to reduce the growth and nodulation capacity of the endophyte (Cañizo et al., 1978; Shipton and Burgraaf, 1982). Shipton and Burgraaf (1982) studied isolates of *Frankia* from several genera of actinorhizal plants by using culture media with osmotic and matric-controlled water potentials. Their results indicated that growth of most of the isolates was reduced by 50% at water potentials between −5 and −8 bars and little growth was observed at −16 bars. Nitrogen fixation by an isolate from *Myrica* was greatest at −3 to −5 bars whereas an isolate from *Alnus glutinosa* showed greatest activity at −5 to −7.5 bars. Their isolates were more sensitive to a matric-controlled system (polyethylene glycol 4000) than to the osmotically controlled systems. It remains to be determined whether *Frankia* exhibits the well-known tolerance of desiccation shown by other actinomycetes. Certainly, this is an important trait for survival in soil in the absence of a host.

2.2.2. Temperature

Frankia exhibits a response to temperature characteristic of most mesophilic organisms (McNiel and Carpenter, 1980; Tjepkema et al., 1981; Shipton and

Burgraaf, 1983; Lechevalier et al., 1983; Winship and Tjepkema, 1985). The optimum temperature for growth *in vitro* was around 30 °C but an isolate from *Casuarina* (Cas JCT287) maintained growth at 35 °C better than isolates from *Hippophae* and *Myrica*. All grew between 14 °C and 40 °C but growth declined rapidly approaching 40 °C (Shipton and Burgraaf, 1983). The optimum temperature for nitrogenase induction *in vitro* for an isolate from *Alnus rubra* (ArI3) was 25–28 °C (Murry et al., 1984b). Similar temperatures have been reported for nitrogenase activity of nodules from *Elaeagnus, Shepherdia, Comptonia, Alnus* and *Hippophae* (McNeil and Carpenter, 1980). Nitrogenase activity of nodules of *E. angustifolia* was maximal at 40 °C. It has been suggested that temperature requirements of the endophyte are dependent on the climatic factors of the area from which the isolates are obtained. However, Moiroud et al. (1984) were unable to detect any degree of cold tolerance in isolates from cold-climate areas.

2.2.3. pH

In culture, *Frankia* grows best at near neutral pH. The pH values of most growth and isolation media are reported as 6.8–7.0 (Callaham et al., 1978; Tjepkema et al., 1980; Gauthier et al., 1981; Benson, 1982; Baker and O'Keefe, 1984; Murry et al., 1984a). Shipton and Burgraaf (1983) reported growth of several isolates in the pH range 5.6–8.9, with 6.0–8.0 being more favorable. There is at present little evidence to suggest that growth in soil would not be favored by similar pH ranges. Results of many nodulation studies (e.g. Cañizo et al., 1978) indicate a similar range of pH, though isolates from different species and different environments display unique ranges for nodulation. That survival of the organism may be influenced by pH was suggested in a study by van Dijk (1979) who reported that the infectivity of a clay soil (pH 6–7) remained relatively constant for one year following inoculation with a nodule homogenate, whereas an acid soil lost 80% of its infectivity within 30 weeks. These findings suggest that it may be necessary and desirable to select isolates for tolerance of various adverse environmental factors for use as inoculants of plants used in reclamation programs or amenity plantings (Bowen and Reddell, 1986).

2.2.4. Oxygen

Frankia is an aerobic bacterium capable of N_2 fixation under fully aerobic conditions (Murry et al., 1984a; Tjepkema et al., 1986). Nitrogenase activity of nodules of *Alnus rubra* was not affected by pO_2 ranging from 15 to 30 kPa O_2 (Winship and Tjepkema, 1985). Unlike most diazotrophs (especially free-living rhizobia) *Frankia* does not require reduced oxygen concentrations for free-living N_2 fixation. The ecological significance of this has not been fully explored but it is conceivable that free-living or root-associated (rhizosphere) N_2 fixation is within the capability of the organism and might be important for its survival in the non-symbiotic state. A striking finding regarding the oxygen tolerance of *Frankia* was that it could be "trained" to grow and fix N_2 at pO_2 up to 70 kPa (Parsons et al., 1987). The organism adjusted to the elevated pO_2 by increasing the thickness of the wall of the vesicle (the site of nitrogenase in the free-living organism). Thus it appears that the vesicle functions in a manner similar to that of the heterocyst in those N_2-fixing cyanobacteria that produce the specialized cells.

That the vesicle is the site of nitrogenase activity is further substantiated by the work of Murry et al. (1985). Working with an isolate from *Casuarina cunninghamiana* they showed that cultures grown at atmospheric oxygen concentration produced typical vesicles and nitrogen fixation was not sensitive to O_2, whereas cultures grown under microaerobic conditions did not produce abundant vesicles and nitrogen fixation was quite sensitive to inhibition by O_2. Noridge and Benson (1986) showed that nitrogenase activity was associated with isolated vesicles of *Frankia* sp. strain CpI1, providing conclusive evidence for the localization of nitrogenase in vesicles, and Meesters et al. (1987) demonstrated localization of nitrogenase in vesicles through the use of immunogold labeling of ultrathin cryosections of *Frankia* (EAN1pec) colonies.

2.2.5. Combined Nitrogen

Combined nitrogen affects growth of the endophyte as well as the nodulation process and nodule function. In the free-living state the formation of nitrogenase by *Frankia* is suppressed by ammonium, as is the production of vesicles (Murry et al., 1984b, 1985; Meesters et al., 1985; Tjepkema et al., 1986). Formation of nodules and nodule function are also adversely affected by combined N (MacConnell and Bond, 1957; Stewart and Bond, 1961; Huss-Danell et al., 1982; Granhall et al., 1983). Free-living frankiae can use amino acids, urea, adenine, uracil, ammonia and N_2 as sources of nitrogen for growth (see Tjepkema et al., 1986). Some of these compounds are undoubtedly available in the soil and rhizosphere and probably support growth of the endophyte, to some extent, in these environments.

2.3. Biotic Factors Influencing *Frankia* in Soils and Rhizospheres

While the autecology of free-living frankiae is largely unknown there are several reports suggesting that biotic factors may play a role in the survival and infection capability of some isolates. Among the biotic factors which have received some attention are the effects of other soil bacteria on the nodulation process (Knowlton et al., 1980; Knowlton and Dawson, 1983; Perinet and Lalonde, 1983) and the role of plant-derived phenolic compounds on the growth of *Frankia* and on the infection of alder (Dawson and Seymour, 1983; Perradin et al., 1983; Vogel and Dawson, 1985, 1986).

2.3.1. Effects of Other Bacteria

During attempts to reconstitute effective symbioses using pure cultures of *Frankia* it was observed that axenic plants nodulated very poorly or not at all whereas non-sterile plants nodulated freely (Knowlton et al., 1980). Subsequently it was shown that contaminating bacteria apparently facilitated nodulation by promoting root hair deformation on the host. Knowlton et al. (1980) identified species of *Pseudomonas* and a *Chromobacterium* sp. among the organisms which caused root hair deformation and promoted nodulation. They also studied several strains of *Pseudomonas cepacia* and *Rhizobium* strain 32H1 for their ability to cause root hair deformation and promote nodulation. All of the bacteria caused extensive root hair deformation, including *Rhizobium*, and none caused the formation of nodules on their own. *Pseudomonas cepacia* strain 85 caused nodulation of 100% of inoculated seedlings when administered with *Frankia*. Interestingly, *Rhizobium* administered with *Frankia* promoted nodulation of 30% of the inoculated seedlings. The conclusion was drawn that these bacteria served as "helper" bacteria by promoting root hair deformation and aiding the prolonged contact between the endophyte and the root, presumably by entrapment of the endophyte among the deformed root hairs.

In a subsequent study Knowlton and Dawson (1983) confirmed the ability of *Pseudomonas cepacia* strain 85 (PC85) to promote nodulation of *Alnus rubra*. They further demonstrated an interaction between the ability to promote nodulation and the pH of the plant growth medium. Nodulation was greater at pH 5.0–6.0 in the presence of PC85. They speculated that enhanced nodulation may have been due to the ability of the pseudomonad to neutralize the pH of the medium combined with its ability to cause deformation of root hairs. It must also be pointed out that *Frankia* can nodulate *in vitro*-propagated plantlets of *Alnus glutinosa* in the absence of any microbial contaminants or "helper" bacteria (Perinet and Lalonde, 1983). These studies indicate that there are distinct interactions of *Frankia* and its hosts with other soil microorganisms. The many questions regarding the effects of soil and rhizosphere bacteria on the extranodular growth and survival of *Frankia* and on the nodulation process provide fertile ground for further investigation. These studies will no doubt be important for implementing the practical uses of *Frankia* in the mass production of beneficial plant species (Perinet et al., 1985).

2.3.2. Effects of Plant-derived Phenolic Compounds

In its interaction with roots of host plants and perhaps with roots of non-hosts *Frankia* must encounter a broad range of plant-derived carbonaceous compounds. It can be expected that among these will be compounds detrimental or stimulatory to the growth of the endophyte. The fact that *Frankia* can grow in the rhizosphere (see earlier section) suggests that it finds substrates suitable for growth in this environment. On the other hand, it is also possible that the host may produce substances inhibitory to the endophyte or that products from the breakdown of lignaceous plant tissues in soil might be inhibitory.

Phenolic compounds are well known as inhibitors of a variety of microorganisms, including plant pathogens, and are thought to be a part of the defense mechanisms of plants against invading microorganisms. Several phenolic compounds have been studied with regard to their ability to inhibit growth of *Frankia in vitro* and at least one study has shown that nodulation of *Alnus glutinosa* in soil was reduced by a phenolic derivative, juglone, of walnut plants (Dawson and Seymour, 1983; Perradin et al., 1983; Vogel and Dawson, 1985, 1986). Ferulic, *o*-coumaric and *p*-coumaric acids (1 mM) were extremely effective in inhibiting growth of *Frankia*. Benzoic and *p*-hydroxybenzoic acids did not inhibit growth of *Frankia* but stimulated production of ''spherical septate vesicles'' on two strains of spore-negative *Frankia* (Perradin et al., 1983). Juglone (5-hydroxy-1,4-naphthoquinone) and caffeic acid were also inhibitory to growth of *Frankia*. Gentisic, *o*-hydroxyphenylacetic and vanillic acids were less inhibitory, and vanillic acid caused production of ''spherical structures unlike vesicles or sporangia in appearance'' (Vogel and Dawson, 1986). Extending these studies to the effects of phenolics on nodulation of *Alnus glutinosa* in soil, Vogel and Dawson showed that addition of aqueous suspensions of juglone (10^{-3} and 10^{-4} M) significantly decreased nodulation of black alder in a prairie soil inoculated with a *Frankia* isolate from red alder. Thus it is clear that plant-derived phenolics can affect the growth of *Frankia* as well as nodulation of at least one species of *Alnus*. The extent of these phenomena in soils and rhizospheres under natural conditions is unknown.

3. SOIL AND RHIZOSPHERE ASPECTS OF THE *RHIZOBIUM/BRADYRHIZOBIUM*-LEGUME SYMBIOSES

The *Rhizobium/Bradyrhizobium*-legume symbioses have been a subject of intensive investigation for nearly a century. The role of legumes in providing nitrogen for agricultural production as well as in other ecosystems is indisputable. The symbioses are especially important in the developing countries where lack of sufficient capital or infrastructure precludes the production and use of expensive nitrogenous fertilizers. Because of their utility in food, fiber and fuel production, and in the maintenance of productivity in non-agricultural ecosystems, the legume–rhizobia symbioses continue to be studied at all levels, ranging from the molecular to the field and ecosystem level. The extensive interest in these symbiotic associations has produced a tremendous volume of literature and many aspects of the subject have been reviewed frequently and in great detail. The information presented here will be largely restricted to the occurrence of rhizobia in the soil and rhizosphere and will not address post-infection events in the establishment of N_2-fixing root nodules.

With the continued intensive investigation of rhizobia and the application of physiological and molecular approaches has come the inevitable question as to the identity of the diverse groups of isolates from the many cross-inoculation groups of *Rhizobium*. It was obvious from numerous avenues of investigation that there was sufficient diversity across these groups to warrant creating a second genus to accommodate the so-called slow-growing rhizobia. Thus the genus *Bradyrhizobium* (*brady*: Gk. slow) was erected to include the slow-growing soybean rhizobia and some others. The slow-growing soybean rhizobia are now classified as *Bradyrhizobium japonicum*. These strains are differentiated from *Rhizobium* spp. by their longer generation times and the creation of alkaline conditions in culture media as well as other characteristics. It is safe to say that the taxonomy of rhizobia will likely undergo further revision as more information regarding the molecular biology of the bacteria accumulates. However, the ability of strains to nodulate only certain legume species or to nodulate a variety of species remains a valuable attribute in the classification of the bacteria.

3.1. Growth and Survival of Rhizobia in Soil

Study of the extranodular life of rhizobia has been impeded by lack of simple techniques for the isolation and identification of the bacteria. To date, there are still no effective selective culture media for the isolation of free-living rhizobia from soil and no simple tests to distinguish rhizobia from other bacteria. Isolation and identification of rhizobia usually involves some sort of plant-infection technique to recover the bacteria for a specific legume and the subsequent isolation of strains from nodules formed on the test plants. Because of these complications, knowledge regarding the ecology of rhizobia as free-living soil bacteria has been slow to accrue.

Study of the autecology of rhizobia depends on the ability to observe individual species or strains among the complex soil or rhizosphere bacterial populations (Schmidt, 1974; Bohlool and Schmidt, 1970, 1980). Therefore, investigations regarding the occurrence of a specific isolate in soil, indigenous or from an inoculant, necessarily rely on some marker such as intrinsic antibiotic resistance (Eaglesham, 1987) or the use of immunological techniques, including immunodiffusion (Parker et al., 1977), enzyme-linked immunosorbent assays (ELISA; Kishinevsky and Jones, 1986) and immunofluorescence (Schmidt et al., 1968). The use of strains marked with multiple antibiotic resistances is a relatively easy approach for studying the fate of inoculants. Bacteria are recovered on selective media containing the appropriate antibiotics after it has been established that the medium effectively controls the indigenous soil populations so that background resistance is minimal. Such methods can provide a means for the direct isolation of rhizobia from soil. Other methods usually capitalize on plant infection to retrieve rhizobia from the soil. Once they have been isolated from the nodule, a variety of immunological techniques can be used for the identification of the nodule occupants (Vincent, 1970; Wollum, 1986).

Immunofluorescence provides a means for the direct observation of rhizobia in soil or on roots. The bacteria can be recovered from soil suspensions through centrifugation and collection on membrane filters. The filters are then "stained" with the appropriate fluorescent antibody and numbers of rhizobia determined by viewing the treated filter with epifluorescence microscopy. Similarly, root pieces can be counterstained with a suitable reagent (e.g. a rhodamine-gelatin conjugate, or eriochrome black), treated with the fluorescent antibody and observed with epifluorescence (Bohlool and Schmidt, 1968, 1980). This technique provides the unique advantage of being able to observe the bacteria *in situ*. Simultaneous application of non-specific fluorochromes can enable the investigator to differentiate rhizobia from the rest of the bacterial population. It is of course a prerequisite to these studies that the immune serum be as free from cross-reactivity with non-specific strains as possible. Cross-reactive sera can frequently be rendered unreactive for non-specific strains by cross-absorbing the serum with the appropriate offending strain. Once the sera are confirmed specific, they are useful tools for labeling and detecting specific bacteria in a heterogeneous population.

At a higher level of organization, it is possible to use immunoelectron microscopy to observe rhizobia in the rhizosphere and in root tissues prior to and during invasion of the root. Such observations require the use of ferritin-labeled antibodies (Parker et al., 1977) or colloidal gold preparations (e.g. Bashan and Levanony, 1988).

Because the microscopy techniques mentioned above can be quite tedious and specific immune sera are not always readily available, many studies of the survival of rhizobia in soil have relied primarily on plant infection, most probable number techniques (Weaver and Frederick, 1982) or the use of antibiotic resistance. The volume of literature regarding the survival of rhizobia is great and will not be reviewed exhaustively. Lowendorf (1980) provides extensive information regarding many aspects of environmental factors influencing survival of rhizobia in soil.

Rhizobia can establish in the soil population following application of inoculants and can persist for several years following such applications (Parker et al., 1977). Persistence is enhanced by repeated growth of the host legume in inoculated fields. It is also possible that persistence and establishment can be aided by non-host plants (Robert and Schmidt, 1985). Some rhizobia also grow in unamended soil (Pena-Cabriales and Alexander, 1983a).

Recently, it has been determined that rhizobia can multiply in soil at the expense of added substrates (Pena-Cabriales and Alexander, 1983b; Viteri and Schmidt, 1987; Germida, 1988). Pena-Cabriales and Alexander (1983b) showed that numbers of *R. meliloti, R. japonicum* and *R. phaseoli* increased in response to the addition of mannitol (0.5 or 1.0% but not 0.01 or 0.1%). Increases in numbers declined within several days of the addition of the substrate, a phenomenon which the authors suggested was due to competition from other bacteria and predation by protozoa. The study involved the use of rhizobia resistant to streptomycin and erythromycin and greater proliferation of the bacteria was observed in antibiotic-treated soils also amended with mannitol. Viteri and Schmidt (1987) studied the response of indigenous *Bradyrhizobium japonicum* to readily available substrates. Using immunofluorescence assays they showed that isolates from serocluster 123 (a group of strains from serogroups 123, 127 and 129 interrelated through serology, bacteriophage sensitivity and DNA restriction digest analysis; Schmidt et al., 1986) and serogroups 110 and 138 (a serogroup designates a group of strains that have antigenic determinants in common but may be diverse with respect to other properties) increased (approx. 1 log) during the first 12 days of the incubation period in response to glucose, sucrose, arabinose, xylose or galactose. The bacteria reached maximum populations of about 10^6 within 36–48 days. The most rapid response was to arabinose, which caused generation times of about 3 days. The other sugars caused generation times ranging from 4 to 5.4 days. The study provided evidence that indigenous soil rhizobia could successfully compete with other soil bacteria for readily available carbohydrates in soil in the absence of a legume host or other rhizosphere.

Similar increases in indigenous populations of *R. leguminosarum* and *R. meliloti* in response to added substrates have been shown by Germida (1988). Using an indirect phage assay he showed that numbers of rhizobia increased in response to carbohydrates, organic acids and amino acids. All the organic substrates stimulated the rhizobia to some degree, but there were differences among strains depending on the substrate, its concentration and the soil type.

It is apparent from these studies that rhizobia can compete with other indigenous bacteria for readily available substrates. The suggestion is that some specialized mode of persistence is not necessary and that rhizobia probably function much like other heterotrophic bacteria in soil. The ability to compete for the substrates would almost certainly contribute to their ability to persist in soil until a host root becomes available. It is also worth noting that some rhizobia are capable of chemoautotrophic growth using hydrogen as an energy source (Hanus et al., 1979). The role of chemoautotrophy in survival and proliferation of rhizobia in soil has not been fully explored.

3.2. Growth of Rhizobia in the Rhizosphere

It is generally accepted that successful nodulation of the legume root is preceded by proliferation of an appropriate strain of *Rhizobium* in the rhizosphere/rhizoplane of the host (Dart, 1974; Dazzo, 1980; Vincent, 1980). How rhizobia arrive at the root surface has been a long-studied subject. It has been suggested that rhizobia migrate to roots through chemotactic responses (Currier and Strobel, 1976)—it is widely observed that rhizobia migrate toward root exudates. This chemotactic response is non-specific as many rhizobia will migrate to exudates of a wide array of legumes and to those of non-legumes as well

(Dowling and Broughton, 1986). However, it is generally regarded that chemotaxis plays only a minor role in the migration of rhizobia to roots and operates only on a microscale (Dazzo, 1980). Still, there are a few interesting reports describing enhanced nodulation by motile rhizobia versus non-motile mutants. Mellor et al. (1987) using Tn5 mutagenesis produced mutants defective in flagellar synthesis or function. The non-motile mutants produced as many nodules as wild-type *R. trifolii* TA1 when inoculated separately onto clover roots. Interestingly, when the wild-type and non-motile mutants were added as equal mixtures, the motile strain formed five times more nodules than the non-motile mutants, suggesting that motility was a factor in competition for nodule formation. Similar results have been shown for *R. meliloti* (Soby and Bergman, 1983) and *B. japonicum* strain 110 (Hunter and Fahring, 1980). Thus motility appears to be more important in competition for nodulation sites once the bacteria have arrived at the root (or vice versa) than in the actual migration of the bacteria to the root surface.

There are numerous reports of stimulation of specific rhizobia in the rhizosphere of the host legume (see Dazzo, 1980; Broughton, 1978). It is not entirely clear how the specific stimulation of the rhizobia is mediated but the quantity and the composition of root exudates undoubtedly play a role in the colonization of roots by specific rhizobia and other bacteria (Bowen and Rovira, 1976).

While it is clear that root exudates of hosts and non-hosts, including non-legumes, can stimulate rhizobia, such stimulation is not the sole basis of host specificity (Reyes and Schmidt, 1979; Vance, 1983; Moawad et al., 1984; Schmidt, 1988). Nor is sheer numerical dominance of a particular strain a guarantee that it will be the dominant organism occupying nodules (see Schmidt, 1988). The formation of effective root nodules following the arrival of an appropriate *Rhizobium* at the root surface is the result of an extremely complicated set of processes involving the host and the bacterium. Recognition is an important part of this process and probably holds the key to a complete understanding of the question of specificity. Considerable information has been amassed regarding the subject of recognition in plant–microbe associations (Dazzo, 1980; Lippincott and Lippincott, 1980; Pueppke, 1984; Halverson and Stacey, 1986). The subject of recognition is covered elsewhere in this volume (Chapter 6) and will not be reviewed here. However, it must be mentioned that much effort has been devoted to investigating the role of lectins in host specificity among *Rhizobium*–legume and other plant–microbe associations (Callow, 1977; Dazzo, 1980; Dazzo and Truchet, 1984; Pueppke, 1984). In conjunction with the role of lectins, considerable attention has been given to the role of extracellular polysaccharides produced by rhizobia (see Vance, 1983). It is likely that when the final chapter is written on specificity, both lectins and extracellular polysaccharides will have been shown to have critical roles in recognition and attachment of the bacteria to host plants.

3.3. Nitrogen Fixation by Free-living Rhizobia in Culture

In 1975, Keister reported the occurrence of nitrogen fixation (acetylene reduction) in pure cultures of *Rhizobium* (Keister, 1975). Since then numerous other investigators have confirmed this capability in many strains of *R. japonicum* and strains belonging to the so-called cowpea group or cowpea miscellany. The trait occurs mainly among slow-growing strains. Conditions for the occurrence of free-living N_2 fixation include low pO_2, a low level of combined nitrogen and a variety of carbon sources including mixtures of carbohydrates and organic acids (Krebs cycle intermediates) (Pankhurst, 1981; Tjepkema and Evans, 1975).

The occurrence of nitrogen fixation in free-living rhizobia poses some interesting questions. It suggests that given appropriate environmental conditions, some rhizobia might be able to fix nitrogen as free-living bacteria in soil or in the rhizosphere of legumes and non-legumes. A recent study by Ramaswamy and Bal (1987) suggested such a possibility. They reported that *Rhizobium* sp. 127E15 (a member of the cowpea group) fixed nitrogen in the rhizosphere of lima bean plants. There is room for question in this experiment since their measurement of rhizosphere N_2 fixation was based on nodulated plants from which "each and every visible nodule had been removed." Such an experiment does not rigorously exclude the possibility that unseen nodule primordia may have contributed to the observed acetylene reduction. It would be interesting to conduct experiments of this nature using non-nodulating isolines of legume species. There seems little reason not to believe that rhizobia could fix nitrogen, albeit in very small quantities, in the rhizosphere of a variety of plants under appropriate environmental conditions. Would this provide them with any advantages in terms of survival, etc.? The ecological aspects of free-living (associative?) nitrogen fixation by rhizobia need further study from the standpoint of the consequences for the bacterium and for plants colonized by the bacteria. Given the intense competition for substrate carbon in the rhizosphere, nitrogen fixation by rhizobia, if it occurs, would most likely be minimal.

3.4. Biotic Factors Influencing the Survival and Abundance of Rhizobia in Soils and Rhizospheres

3.4.1. Other Bacteria

Free-living rhizobia in soil or in the rhizosphere are subject to numerous biotic interactions with bacteriophages, other microorganisms and with the microfauna in soil. These interactions can be positive (i.e. growth promoting), negative or neutral. For example, Yahalom et al. (1987) demonstrated that *Azospirillum brasilense* Cd applied to *Macroptilium atropurpureum, Trifolium alexandrinum* and *Medicago polymorpha* 24 hours prior to inoculation with *Rhizobium* caused increased nodulation. They suggested that increased nodulation may have been due to increased production of epidermal cells that differentiate into root hairs. Increased nodule weight resulted from inoculation of soybean with *Pseudomonas fluorescens* and *Pseudomonas putida* when co-inoculated with *Bradyrhizobium japonicum* in a soil–perlite mix (Polonenko et al., 1987). Similar findings were observed when *Azotobacter* was co-inoculated with *B. japonicum* (El-Bahrawy, 1983). In contrast, dual inoculation with *Rhizobium* and *Azospirillum* reportedly reduced nodulation, yield and N content of peanut under field conditions (Raverkar and Konde, 1988) whereas inoculation with the single strains increased these parameters. The effects of co-inoculation of rhizosphere bacteria with rhizobia deserve further research.

3.4.2. Bacteriophages

Nodulation and performance of legumes has been reported to be adversely affected by bacteriophages of rhizobia (Dowling and Broughton, 1986; Hashem and Angle, 1988; Subba Rao, 1984). Recently, Hashem and Angle (1988) reported that a phage specific for *B. japonicum* USDA 117 reduced the number of USDA 117 in soil whereas it had no effect on USDA 110. Further, the phage reduced the numbers of USDA 117 when it was co-introduced with USDA 110. Nodulation and nitrogen fixation of soybean were reduced when the phage was added to a non-sterile, phage- and rhizobia-free soil amended with strain USDA 117. Similar findings have been reported for *Trifolium glomeratum* (Evans et al., 1979a, 1979b) and *Trifolium subterraneum* (Barnet, 1980). The ability of bacteriophages to selectively remove a strain of *Rhizobium* from a mixed inoculum raises some interesting possibilities. For example, Hashem and Angle (1988) speculated that phages might be used to selectively reduce nodule occupancy by an undesirable strain while concurrently increasing the occupancy by a desirable inoculant strain.

3.4.3. Vesicular-arbuscular Mycorrhizal (VAM) Fungi

Rhizobium–legume associations are generally positively affected by simultaneous infection with vesicular-arbuscular mycorrhizal fungi (Barea and Azcon-Aguilar, 1983). In these tripartite associations, VAM fungi often stimulate the production of greater numbers and weight of nodules with subsequent increases in nitrogen fixation and dry matter production (Barea and Azcon-Aguilar, 1983). The increased performance is generally attributed to enhanced supply of phosphorus in mycorrhizal plants. Not all interactions with soil fungi result in increased performance of the *Rhizobium*–legume symbiosis. *Trichoderma viride* has been reported to reduce numbers of *R. japonicum* as well as nodulation in a non-sterile soil (Angle et al., 1981). Nitrogen fixation was reduced with subsequent reductions in root and shoot weights.

3.4.4. Protozoa and Nematodes

Protozoa are active bacteriovores in the soil. Their effects on root colonization by rhizobia and other bacteria are not fully understood at present, though it is becoming more evident that they play a role in nitrogen mineralization through their "grazing" on soil bacteria (Clarholm, 1985a, 1985b) (see also Chapter 1). Numbers of rhizobia have been observed to decline in the presence of protozoa (Danso et al., 1975; Habte and Alexander, 1979; Ramirez and Alexander, 1981; Hossain and Alexander, 1984). A complete understanding of the impact of protozoa on the performance of *Rhizobium*–legume associations is still awaited.

Nematodes can also influence the performance of legume associations. Nematode–legume–*Rhizobium* interactions can be beneficial, detrimental or without effect. The subject has been reviewed by Huang (1987).

3.4.5. General Considerations

There is little doubt that rhizobia in soil and in the rhizospheres of legumes and non-legumes are subject to myriad interactions with other microorganisms and with microfaunal populations. The fate of rhizobia applied as inoculants is dependent on these interactions. As rhizosphere microbiology advances it may be possible to use these interactions to some advantage, or perhaps at least to reduce the negative effects of interactions with other microbes (Rovira, 1988). Studies cited above support the idea that applying other bacteria with rhizobia can be of benefit. This is an area that should not escape the attention of agricultural biotechnologists!

4. ASSOCIATIONS BETWEEN FREE-LIVING N_2-FIXING BACTERIA AND ROOTS OF NON-LEGUMES

4.1. "Hosts" and "Symbionts"

A vast array of non-leguminous plants, principally grasses, are colonized by one or more genera of free-living diazotrophs. Included among the list of grasses are economically important cereals such as wheat, corn, rice, sorghum and millet, which collectively make up most of the world's staple food crops. It is not suprising that there is strong interest in improving the abilities of such crops to obtain nutrients from natural sources, such as biological nitrogen fixation. Most if not all forage and turf grasses also harbor diazotrophs on their root systems, and sugarcane benefits from associations with a number of diazotrophic bacterial genera (Ruschel et al., 1978; Döbereiner and Pedrosa, 1987). It is probable that most plant species in natural environments are colonized to some degree by nitrogen-fixing bacteria and that at least a small portion of the nitrogen fixed by the bacteria eventually becomes available for plant growth.

The list of bacteria participating in associative symbioses with non-legumes continues to grow as new genera and species are described on a seemingly regular basis. The bacteria are predominantly heterotrophs, which have varied requirements for oxygen. Aerobes, microaerophiles, facultative anaerobes and anaerobes have all been reported as constituents of the diazotrophic microflora from roots of various plants. Among the aerobes and microaerophiles, species of *Azotobacter, Beijerinckia, Azospirillum, Herbaspirillum*, and *Derxia* are frequently reported as associates of a number of plant species. The facultative anaerobes are best represented by species of *Klebsiella, Enterobacter, Erwinia* and *Bacillus*. Species of *Clostridium* have been isolated from roots of grasses; however, their contribution to the nitrogen nutrition of plants remains largely unknown. It has been suggested that cellulolytic clostridia may be important in contributing nitrogen to soil through their activities in degrading cellulosic plant residues (Leschine et al., 1988).

Cavalcante and Döbereiner (1988) recently described a new genus of bacteria associated with the internal tissues of stems of sugarcane. The microaerobic organism, named *Saccharobacter*, shows some unusual properties, such as the ability to fix nitrogen at low pH (3.0). It is very likely that continued searching will uncover new species, some of which may prove useful in low-input, sustainable agricultural systems. It is these systems which are most likely to benefit from the activities of rhizosphere diazotrophs.

4.2. The Nature of N_2-fixing Rhizosphere Associations

The association of nitrogen-fixing bacteria with roots is not unlike that of other bacteria. The bacteria reside on or in the mucilaginous layer at the root surface and in the superficial tissues of the root, where they receive carbon in the form of exudates, secretions, lysates and sloughed cell debris (Rovira et al., 1979; Foster et al., 1983; Döbereiner and Pedrosa, 1987). The bacteria also occur abundantly in rhizosphere soils, where they may be enriched as much as 20-fold (Balandreau et al., 1978). The actual location of the bacteria and the abundance of the bacteria within various regions of the root are not fully resolved. The rhizosphere should be viewed as a continuum from the zone of soil where influence of the root is just detectable to the root epidermis or rhizoplane. The term "endorhizosphere" is often used to describe the intercellular colonization of the outer cell layers of the root. Diazotrophs can occur anywhere within this gradient, but proximity to the root almost certainly governs access to the supply of carbon necessary to drive the energy-intensive nitrogen fixation process.

Okon (1982) indicated that azospirilla occur mainly in the mucigel layer of roots of various crops. The bacterium also attaches to the root surface of *Panicum maximum* (guineagrass) and *Pennisetum americanum* (pearl millet) (Umali-Garcia et al., 1978, 1980) and may invade the middle lamella of cortical cells, the intercellular spaces between the cortex and endodermis, and dead cells of the cortex, endodermis and stele of *P. maximum* and *Digitaria decumbens* (digitgrass), and *Zea mays* (maize) (Döbereiner and Day, 1976; Umali-Garcia et al., 1978; Patriquin, 1982). *Azospirillum brasilense* strain Cd has been observed to enter wheat roots and reside in the intercellular spaces of the cortex as deep as the endodermis, but it did not enter the endodermis or the vascular system (Bashan and Levanony, 1988). The cells were confirmed as inoculated azospirilla through immunogold labeling techniques applied to ultrathin sections. Other bacteria in the rhizosphere were not labeled by the specific antibody.

Azospirillum degrades pectic substances *in vitro* (Tien et al., 1981; Umali-Garcia et al., 1980), but Myers and Hubbell (1987) could find no evidence to support the suggestion that polysaccharide components of cell walls of pearl millet might serve as substrates for growth following hydrolysis by the bacteria. Mono- and disaccharides supported growth of azospirilla but the polysaccharides did not. The role of polysaccharide-degrading enzymes in the invasiveness of azospirilla and other diazotrophs colonizing roots requires further clarification. It is clear, however, that if the diazotrophs can gain entry into the root without causing significant damage structurally or physiologically, they should have increased access to the carbon sources made available through the roots.

The question of specificity between the bacteria and plants involved in associative symbioses remains to be fully explored. Several specific associations have been reported. *Azotobacter paspali* associates specifically with *Paspalum notatum* cv batatais (Döbereiner, 1970) and *Beijerinckia* with sugarcane (Döbereiner, 1961; Döbereiner et al., 1972b). One species of *Bacillus* is reported to associate rather specifically with disomic substitution lines of wheat (Neal and Larson, 1976; Rennie and Larson, 1979). Baldani and Döbereiner (1980) have suggested that *Azospirillum lipoferum* associates preferentially with C4 grasses, whereas *A. brasilense* associates preferentially with C3 grasses. Differential responses with regard to nitrogen content of corn hybrids inoculated with strains of *Klebsiella, Escherichia, Derxia, Beijerinckia* and *Azospirillum* have also been reported (Saric et al., 1987). The subject of specificity in these and other rhizosphere associations (e.g. plant-growth-promoting rhizobacteria) is certainly deserving of further study. Successful manipulation of these associations will depend on finding appropriate bacteria for specific plants under specific environmental conditions.

4.3. Biotic Factors Affecting Colonization of Roots by N_2-fixing Bacteria

4.3.1. Competition

Capitalizing on the abilities of bacteria to stimulate plant growth, regardless of the mechanism of growth promotion, depends on successfully establishing the appropriate bacteria on the root. Despite decades of study on the use of nitrogen-fixing and other bacteria as inoculants no clear picture has emerged as to what specific properties make a bacterium a successful rhizosphere occupant. Assuming an inoculant bacterium can survive long enough in the soil to contact a root it faces intense competition for substrates, etc. to become established in the rhizosphere. Unless the rhizosphere is viewed as nitrogen-limiting for microbial growth, it is difficult to conceive that cells fixing N_2 would have an advantage over cells using combined nitrogen present in exudates or the soil solution. However, if nitrogen were to become exhausted in localized areas of abundant microbial growth, the ability to fix N_2 might be of advantage to the diazotroph. Most studies show that diazotrophs rarely if ever comprise a dominant fraction of the total rhizosphere population. While it is relatively easy to demonstrate a rhizosphere effect for nitrogen fixers, their numbers generally range between 1 and 10% of the total heterotrophic population (Gaskins et al., 1984/85; Baldani and Zuberer, unpublished). Thus, microbial competition appears to be a major factor in the establishment and maintenance of diazotrophs on roots. That competition is a factor in the establishment of azospirilla on roots of wheat was shown in a study by Bashan (1986) in which colonization by *Azospirillum brasilense* was enhanced following the application of four fungal and bacterial inhibitors to which the bacterium was resistant. Such treatment led to greater numbers of azospirilla in the rhizosphere and also increased plant dry weight and grain yield.

4.3.2. Predation

Predation is another factor which might influence the establishment of bacteria on roots. Few studies have been reported on the predation of specific free-living diazotrophs. Germida (1987) isolated bdellovibrios from Canadian soils and demonstrated that the *Bdellovibrio* spp. showed a preference for certain strains of *Azospirillum*. He suggested that predation might be a cause for reduced establishment of specific strains of azospirilla in the rhizosphere. Bacteriophages specific for *Azospirillum* have also been described but they do not appear to have a major effect on populations of the bacteria in soil (Germida, 1986). Protozoa are excellent consumers of bacteria in moist soils and rhizospheres and thus they might be a factor regulating the establishment and abundance of diazotrophs on roots (Alexander, 1977; Clarholm, 1985a, 1985b; Curl and Truelove, 1986). The effects of protozoa on abundance of free-living diazotrophs or other beneficial bacteria in the rhizosphere are largely unknown and the subject is worthy of further attention from microbial ecologists.

4.3.3. Antagonism

The role of antibiotic production by rhizosphere microorganisms in controlling the composition of the rhizosphere bacterial flora is not known. Actinomycetes and fungi occurring in rhizospheres of grasses are known to inhibit diazotrophic bacteria *in vitro* (Zuberer and Roth, 1982; Kulinska and Drozdowicz, 1983). Whether diazotrophs are inhibited in soil and the rhizosphere remains to be demonstrated. However, azospirilla resistant to low levels of streptomycin have been reported to occur in greater numbers in the rhizosphere of maize, wheat and beans (Döbereiner and Baldani, 1979; Baldani and Döbereiner, 1980). The authors suggested that the resistance conferred a competitive advantage upon those strains possessing the trait.

Azospirillum brasilense and *A. lipoferum* produce bacteriocins, which inhibit some enterobacteria, streptococci and other Gram-negative bacteria (Oliveira and Drozdowicz, 1981). The bacteriocins may play an additional role in the ecology of azospirilla in the rhizosphere.

4.3.4. The "Host" Plant

Certain strains of *Azospirillum* are known to establish better on some plants than others (see Boddey and Döbereiner, 1988). Further, there appear to be differences in the abilities of certain isolates to colonize the interiors of roots of various crops (Baldani et al., 1986). Such differences are undoubtedly attributable to the peculiarities of the host plant as well as the bacterial strain. Certainly, the quantity and the quality (composition) of the root-derived carbon sources play a role in apparent specificities observed. However, little information is presently available regarding the mechanism(s) involved in the observed instances of apparently specific associations. Developing this understanding is a challenging problem requiring a multidisciplinary approach. The payoff from learning what factors control the colonization of roots by specific bacteria will be great. Successful manipulation of rhizosphere associations demands that plant and bacterial factors contributing to useful associations be fully understood at the genetic and biochemical levels.

The examples above illustrate that microbial interactions in the rhizosphere are likely to be important in the establishment and persistence of desirable bacteria in the rhizosphere and on or in roots. Little is presently known about the actual impact of these factors under natural conditions but the current interest and activity in many aspects of rhizosphere microbiology will likely give rise to an expanded understanding of the complexities of these associations.

4.4. Abiotic Factors Affecting Free-living Diazotrophs

Numerous environmental factors, including availability of organic and inorganic nutrients, moisture, aeration, temperature and pH, affect survival and growth of bacteria in soil (Alexander, 1977; Balandreau et al., 1978; Lynch, 1983). Most relevant to diazotrophs are organic carbon supply, moisture and aeration. When they are dependent on N_2 for growth, diazotrophs must have access to abundant supplies of carbon for growth and energy production. Fixation of one molecule of N_2 requires approximately 16 molecules of ATP, thus considerable substrate must be available to N_2-fixing diazotrophs (Postgate, 1987). If sufficient combined nitrogen is present then dependence on nitrogenase is eliminated and presumably less substrate is required for growth. Few if any studies have demonstrated the abilities of free-living diazotrophs to utilize the commonly occurring substrates typical of the soil environment. The recent report of N_2-fixing cellulolytic clostridia (Leschine et al., 1988) suggests that these and other soil bacteria may be endowed with the capacity to hydrolyze plant-derived polysaccharides and to use these materials for nitrogen fixation. However, the failure of *Azospirillum* to utilize plant polysaccharides of *Pennisetum americanaum* (pearl millet) (Myers and Hubbell, 1987) provides evidence that this is not a universal property of diazotrophs.

4.4.1. Moisture

Moisture limits all microbial life in the soil, and on most non-irrigated land also limits plant production. Undoubtedly, growth and activities of bacteria in soil free of plants are limited when moisture is limiting (Harris, 1981). The availability of moisture to root-associated bacteria is probably adequate under conditions suitable for plant growth, especially where bacteria are found in the superficial cell layers of the root. Soil type affects the availability of soil water to roots and microbes by exerting matric effects. Thus fine-textured clay soils with many very small pores hold water with more tenacity than coarser soils with larger pores. In a study in our laboratory more water was necessary to support nitrogenase activity in a substrate-amended clay soil than in a silt loam (Zibilske and Zuberer, unpublished). It is apparent that some diazotrophs are able to withstand prolonged periods of drying because they can be isolated from air-dried soils which have been stored for long periods of time (Moreno et al., 1986). *Azotobacter* and *Azospirillum* are both able to form cysts or cyst-like cells, which probably aid their survival during periods of desiccation (Thompson and Skerman, 1979; Lamm and Neyra, 1981), and of course *Bacillus* and *Clostridium* produce endospores conferring excellent resistance in adverse environments.

4.4.2. Oxygen

Aeration probably exerts a greater effect on the growth of most diazotrophs under N_2-fixing conditions than it does when they are not fixing N_2. It might be expected that aerobic and microaerophilic bacteria would suffer under anoxic conditions; however, it must be pointed out that aerobic diazotrophs function best at reduced concentrations of oxygen due to the sensitivity of the nitrogenase complex to molecular O_2 (Postgate, 1987). The rhizosphere most likely contains numerous sites with lowered pO_2 providing favorable habitats for aerobic and microaerophilic diazotrophs (Zuberer and Alexander, 1986).

5. FACTORS AFFECTING NITROGEN FIXATION IN THE RHIZOSPHERE

It is generally accepted that nitrogen fixation occurs to some degree in association with roots of many plant species. With few exceptions (e.g. sugarcane and rice) rates of N_2 fixation are low. This does not mean that the contribution of fixed nitrogen is insignificant, rather it suggests that there are factors which hinder the expression of greater levels of nitrogenase activity in these associations. In undisturbed grasslands the contribution of fixed N to the ecosystem, though small, may be quite significant. However, in the production of agricultural crops, such as wheat and corn, rhizosphere nitrogen fixation provides only a small portion of the nitrogen needs of the crop. Why is the contribution of fixed N by biological nitrogen fixation in the rhizosphere not greater than the currently estimated 10–15% of the need?

5.1. Carbon Sources for Nitrogen Fixation in the Rhizosphere

The plant–microbe association is extremely complex but a few features can be described in terms of their effects on rates of N_2 fixation mediated by root-associated bacteria. Foremost among these is the provision of utilizable substrates in the form of exudates, secretions, lysates and sloughed cells (Rovira, 1969; Bowen and Rovira, 1976). The availability of carbon sources around and on plant roots is the driving force behind almost all rhizosphere phenomena, including nitrogen fixation. The subject of quantification of carbon loss from roots is covered elsewhere in this volume (Chapter 3) and will not be reviewed intensively for this discussion. It is necessary to point out that estimates of carbon transport to and loss from the below-ground portion of plants have been upwardly revised over the past two decades. The evidence suggests that 30–50% of the carbon fixed is translocated below ground, where most of it is used to create root biomass. Newman (1985) calculated—based on data of Whipps and Lynch (1983) and Whipps (1984)—that soluble exudates total 10–100 mg g^{-1} root and root cap plus mucigel comprise 20–50 mg g^{-1}. Together they contribute 3–15% as much organic matter as the roots themselves. Thus the great majority of carbon translocated to roots ends up in insoluble fractions, predominantly structural polysaccharides (cellulose, hemicelluloses, etc.).

The contribution of rhizosphere nitrogen fixers to the growth of the plant remains controversial. Much of the controversy stems from the energetic considerations of the nitrogen fixation process itself. It is generally agreed that diazotrophs fix 10–15 mg g^{-1} of substrate carbon consumed (Postgate, 1974a). If one applies this value to the quantities of exudates, etc. from solution-grown barley plants (0.2 mg sugar mg^{-1} plant dry weight) measured by Barber and Lynch (1977) and assumes all the carbon goes to diazotrophs fixing at the efficiency stated above, then about 2–3 μg N mg^{-1} plant dry weight would be fixed, or about 15% of the total plant N. This of course is an overestimate since competition for substrate would limit its flow to diazotrophs. There is much to be learned regarding the coupling of the flow of carbon to root-associated diazotrophs as well as other bacteria, and of the actual quantities of substrates available to rhizosphere microorganisms. These experiments should of necessity be conducted using soil-grown plants and this introduces a level of complexity which makes such experiments most difficult to interpret.

It is easy to demonstrate that nitrogen fixation in soil is limited by the supply of available carbon. Carbon-amended soils nearly always exhibit increased rates of nitrogen fixation under favorable environmental conditions (Brouzès et al., 1971; O'Toole and Knowles 1973; Zibilske and Zuberer, unpublished). Similarly, washed excised roots (Döbereiner, 1977; De-Polli et al., 1977; Haahtela et al., 1981) and soil–plant cores (Döbereiner et al., 1976; Zuberer and Silver, 1978) show enhanced nitrogen fixation when amended with sugars and organic acids. Alexander and Zuberer (1989) have recently shown that roots of intact maize plants at low pO_2 (2 kPa) exhibit higher rates of nitrogen fixation following addition of carbon to the nutrient solution or to the solid rooting medium supporting the plants. The increases in acetylene reduction were eliminated in the presence of chloramphenicol, suggesting that bacterial growth was associated with the increased rates (Alexander and Zuberer, 1989; Baldani and Zuberer, unpublished).

There is little doubt that root-associated nitrogen fixation is limited by the supply of available carbon in the rhizosphere. The limitation is exacerbated by the fact that there is intense competition for the limited quantity of substrates made available. Remember that diazotrophs rarely account for more than 10% of the rhizosphere bacterial population! More attention needs to be given to the utilization of the insoluble carbon sources in the root environment. It is possible that the actions of other bacteria and fungi might make available some of the carbon held in polymeric form. For example, it is known that fungi degrade the cellulosic component of wheat straw and facilitate nitrogen fixation (Lynch, 1985; Halsall and Gibson, 1985; Roper and Halsall, 1986). Lynch and Harper (1985) have proposed that wheat straw could be microbially "upgraded" using a consortium of microbes, including cellulolytic fungi (e.g. *Trichoderma* sp. parasitic on plant pathogenic fungi), an N_2-fixing *Clostridium butyricum*, and the bacterium *Enterobacter cloacae*, which produces an extracellular polysaccharide. They hypothesized that such a consortium might be useful in degrading straw and contributing to soil structure and friability due to the microbiological effects on aggregation. It was suggested that the diazotrophic clostridia could facilitate the fungal degradation of the celullosic component of straw by providing some nitrogen for the growth of the fungus. The cellulolytic N_2-fixing clostridia may be of significance in the utilization of moribund root tissues as well.

5.2. Other Environmental Factors Governing Nitrogen Fixation in the Rhizosphere

5.2.1. Oxygen

While available carbon is probably the principal limitation for root-associated nitrogen fixation, another factor of major importance is the level of oxygen in the microhabitat of the diazotroph. Oxygen irreversibly inactivates nitrogenase in most organisms and regulates its formation as well (Postgate, 1974b, 1987; Hill, 1988). Therefore, most diazotrophs have evolved mechanisms for coping with oxygen. The anaerobes and facultative anaerobes simply avoid the problem by fixing N_2 under anaerobic conditions or at very low pO_2 in the case of *Klebsiella* (Hill et al., 1984). Azotobacters deal with oxygen by respiring at very high rates, possibly to maintain a low pO_2 within the cell to protect nitrogenase (Robson and Postgate, 1980). Alternatively, when O_2 consumption cannot keep pace with the supply of oxygen nitrogenase can undergo a change in conformation which protects it from damage by oxygen (Robson and Postgate, 1980). *Azospirillum* has no such protective mechanisms and is restricted to fixing nitrogen most actively in microaerobic environments. The organism does not fix nitrogen in air (Day et al., 1975). Azospirilla exhibit maximal nitrogenase activity at pO_2 in the range 0.1–1.5 kPa (Day et al., 1975; Okon et al., 1977; Nelson and Knowles, 1978; Tibelius and Knowles, 1983). Likewise, the aerobic azotobacters fix nitrogen more efficiently at reduced pO_2 (Dalton and Postgate, 1969; Klugkist and Haaker, 1984). *Klebsiella* and *Enterobacter* also fix N_2 most rapidly at reduced pO_2 despite their reputations as facultatively anaerobic diazotrophs (Hill et al., 1984).

The data for most aerobic diazotrophs in culture clearly show that nitrogen fixation is most rapid under conditions where the pO_2 is much reduced below that of air (20 kPa). One might reasonably expect that nitrogen fixation in natural environments is also limited to some extent by the presence of oxygen. Döbereiner et al. (1973) showed that nitrogenase activity associated with washed, excised roots of *Paspalum notatum* exhibited a pattern similar to that of *Azotobacter* grown in culture (Drozd and Postgate, 1970) with a maximum at 4 kPa O_2. Other investigators have shown that root–soil cores express maximal rates of acetylene reduction when the oxygen concentration over the core is reduced (Tjepkema and Burris, 1976). Roots of intact maize and sorghum plants exposed to low pO_2 (0.5–2.5 kPa) for periods of 12–24 hours exhibit markedly increased rates of nitrogen fixation as compared with plants exposed to air (Zuberer and Alexander, 1986; Alexander et al., 1987). This was true of plants sparged in a hydroponic system or grown in a solid support medium (fritted clay). Nitrogenase activity as measured by acetylene reduction or incorporation of $^{15}N_2$ was greatest at pO_2 of 1.5–2.0 kPa. Thus the evidence accumulated through experiments using excised roots, soil–plant cores and intact plants indicates that rhizosphere nitrogen fixation is best described as a microaerophilic process. Anaerobes may contribute to some extent, but in many experiments conducted in our laboratory rates of acetylene reduction associated with intact plants were always greater under low pO_2 than under completely anaerobic conditions (cylinder nitrogen). We have also observed that nitrogenase activity was less sensitive to O_2 when the nutrient solution was supplemented with succinate (Alexander and Zuberer, 1989). Measurable acetylene reduction occurred at pO_2 of 5–10 kPa in carbon-amended systems.

5.2.2. Combined Nitrogen

Combined nitrogen (ammonium, nitrite, nitrate) may also govern the occurrence of nitrogen fixation in soil and in the rhizosphere. Low levels of ammonium repress the synthesis of nitrogenase so that in environments with sufficient available nitrogen the energetically costly nitrogen fixation process is well regulated. Nitrate also reduces or eliminates nitrogenase activity, though the mechanisms for the reduction of activity by nitrate are not well understood (Hom et al., 1980; Cejudo and Paneque, 1986).

Acetylene reduction in anaerobically incubated soils is eliminated or reduced at nitrogen concentrations ranging from 5 to 500 mg N kg^{-1} (Spiff and Odu, 1972; Knowles and Denike, 1974; Balandreau et al., 1975). The inhibition by nitrogen is proportional to the level of available carbon in the system (Knowles and Denike, 1974). Day et al. (1975) demonstrated that nitrogen fixation by soil cores containing *Paspalum notatum* was inhibited by 10 mg of NH_4^+- or NO_3^--N kg^{-1} soil. Acetylene reduction associated with washed roots is also inhibited by low levels of combined nitrogen (Neyra and Döbereiner, 1977; Trolldenier, 1977; van Berkum and Sloger, 1983; Yoch and Whiting, 1986). Addition of 4.2 mg NH_4^+-N l^{-1} rapidly reduced acetylene production associated with roots of intact maize plants in a hydroponic system at 2 kPa O_2 (Zuberer and Alexander, 1986). Nitrate (10–20 mg N l^{-1}) had similar effects. Nitrogen fixation resumed in the hydroponic system as the nitrogen was depleted through plant and microbial assimilation. It is probable that levels of combined nitrogen in the soil solution and in the exudates, etc. of plant roots control the rates of N_2 fixation associated with roots to some extent. In highly fertile soils it is likely that rhizosphere N_2 fixation is reduced or abolished until nutrient depletion zones around the roots are established.

Other fertility factors, such as phosphate supply and trace metal availability, may influence nitrogen fixation in these associations. However, the effects of these factors are virtually unexplored at present.

6. CONCLUSIONS

The phenomenon of nitrogen fixation in the rhizosphere has received considerable attention throughout the hundred years since the discovery of nitrogen-fixing bacteria. These efforts intensified as a result of worldwide oil shortages and the concomitant increases in the cost of nitrogenous fertilizers during the early 1970s. The need for fertilizers grows with the demands to feed an expanding global population. Thus interest in and utilization of N_2-fixing plant–microbe associations will remain high, particularly in less developed countries.

It is now widely recognized that rhizosphere nitrogen fixation contributes to plant growth, but the amounts of nitrogen fixed are low. Most studies indicate about 5–20 kg N ha^{-1} $year^{-1}$ are contributed through associative nitrogen fixation (Gibson et al., 1988). Sugarcane and rice are notable exceptions, receiving greater contributions of microbially fixed N.

Following an intensive period of research into the nature of grass–bacteria associations it has become apparent that nitrogen-fixing bacteria applied to plants affect plant growth through mechanisms other than nitrogen fixation. Chief among these mechanisms are: the production of plant growth regulators, such as indole acetic acid (Tien et al., 1979); alteration of the ability of roots to absorb nutrients (Lin et al., 1983; Boddey et al., 1986); and production of siderophores (Suslow, 1982) (see also Chapter 7). Indeed, some azospirilla are endowed with all of these capabilities. Thus a spin-off of research on associative N_2 fixation has been the increased understanding of ways in which root-associated bacteria may promote plant growth. The study of plant-growth-promoting rhizobacteria (so-called PGPRs) is becoming increasingly active, fueled by the increased understanding of mechanisms of growth promotion and the desire to develop useful systems and inoculants for enhanced plant nutrition and control of disease through modern biotechnological methods.

The rhizosphere *is* where the root meets the soil. Knowledge of how plants affect bacteria and how bacteria affect plants and their ability to derive sustenance from the soil will be critical for the development of improved plant–microbe associations and provision of food and fiber for a growing world population. The long-term goals of rhizosphere research include the maintenance of agricultural productivity with reduced need for chemical inputs and reduced negative impacts on the environment. These are truly worthwhile goals and it is certain that rhizosphere microbiologists will play an important part in their attainment.

REFERENCES

Alexander, D. B. and D. A. Zuberer (1989). Effects of organic substrates and chloramphenicol or nalidixic acid on acetylene reduction associated with roots of intact maize and sorghum plants. *Plant Soil*, **112:**61–67.

Alexander, D. B., D. A. Zuberer, and D. M. Vietor (1987). Nitrogen fixation (acetylene reduction) associated with roots of intact *Zea mays* in fritted clay at reduced oxygen tensions. *Soil Biol. Biochem*, **19:**1–6.

Alexander, M. (1977). *Introduction to Soil Microbiology*. John Wiley & Sons, New York, 467 pp.

Angle, J. S., B. K. Pugashetti, and G. H. Wagner (1981). Fungal effects on *Rhizobium japonicum*–soybean symbiosis. *Agron. J.*, **73:**301–306.

Baker, D. D., W. Newcomb, and J. G. Torrey (1980). Characterization of an ineffective actinorhizal microsymbiont, *Frankia* sp. EuIl (Actinomycetales). *Can. J. Microbiol.*, **26:**1072–1089.

Baker, D. D. and D. O'Keefe (1984). A modified sucrose fractionation procedure for the isolation of Frankiae from actinorhizal root nodules and soil samples. *Plant Soil*, **78:**23–28.

Baker, D. D., J. G. Torrey, and G. H. Kidd (1979). Isolation by sucrose-density fractionation and cultivation *in vitro* of actinomycetes from nitrogen-fixing roots. *Nature*, **281:**76–78.

Balandreau, J., G. Rinaudo, I. Fares-Hamad, and Y. Dommergues (1975). Nitrogen fixation in the rhizosphere of rice plants. In *Nitrogen Fixation by Free-living Micro-organisms* (W. D. P. Stewart, ed.). Cambridge University Press, London, pp. 57–70.

Balandreau, J., P. Ducerf, I. Hamad-Fares, P. Weinhard, G. Rinaudo, C. Millier, and Y. Dommergues (1978). Limiting factors in grass nitrogen fixation. In *Limitations and Potentials for Biological Nitrogen Fixation in the Tropics* (J. Döbereiner, R. H. Burris, and A. Hollaender, eds.). Plenum Press, New York, London, pp. 275–302.

Baldani, V. L. D. and J. Döbereiner (1980). Host plant specificity in the infection of cereals with *Azospirillum* spp. *Soil Biol. Biochem.*, **12:**433–439.

Baldani, V. L. D., M. A. B. Alvarez, J. I. Baldani, and J. Döbereiner (1986). Establishment of inoculated *Azospirillum* spp. in the rhizosphere and in roots of field-grown wheat and sorghum. *Plant Soil*, **90:**35–45.

Barber, D. A. and J. M. Lynch (1977). Microbial growth in the rhizosphere. *Soil Biol. Biochem.*, **9:**305–308.

Barea, J. and C. Azcon-Aguilar (1983). Mycorrhizas and their significance in nodulating nitrogen-fixing plants. *Adv. Agron.*, **36:**1–54.

Barnet, Y. M. (1980). The effect of rhizobiophages on populations of *Rhizobium trifolii* in the root zone of clover plants. *Can. J. Microbiol.*, **26:**572–576.

Bashan, Y. (1986). Enhancement of wheat root colonization and plant development by *Azospirillum brasilense* Cd. following temporary depression of rhizosphere microflora. *Appl. Environ. Microbiol.*, **51:**1067–1071.

Bashan, Y. and H. Levanony (1988). Interaction between *Azospirillum brasilense* Cd and wheat root cells during early stages of root colonization. In *Azospirillum IV: Genetics, Physiology, Ecology* (W. Klingmuller, ed.). Springer-Verlag, Berlin, pp. 166–173.

Becking, J. H. (1970). Plant–endophyte symbiosis in non-leguminous plants. *Plant and Soil*, **32:**611–654.

Becking, J. H. (1975). Root nodules in non-legumes. In *The Development and Function of Roots* (J. G. Torrey and D. T. Clarkson, eds.). Academic Press, London, pp. 507–566.

Becking, J. H. (1977). Endophyte and association establishment in non-leguminous nitrogen-fixing plants. In *Recent Developments in Nitrogen Fixation* (W. Newton, J. R. Postgate, and C. Rodriguez-Barrueco, eds.). Academic Press, New York, pp. 551–567.

Benson, D. (1982). Isolation of *Frankia* from alder actinorhizal root nodules. *Appl. Environ. Microbiol.*, **44:**461–465.

Benson, D. R. (1988). The genus *Frankia*: actinomycete symbionts of plants. *Microbiol. Sci.*, **5:**9–12.

Berry, A. and J. G. Torrey (1979). Isolation and characterization *in vivo* and *in vitro* of an actinomycetous endophyte form *Alnus rubra* Bong. In *Symbiotic Nitrogen Fixation in the Management of Temperate Forests* (J. C. Gordon, C. T. Wheeler, and D. A. Perry, eds.). Oregon State University, Corvallis, pp. 69–83.

Boddey, R. M. and J. Döbereiner (1988). Nitrogen fixation associated with grasses and cereals: recent results and perspectives for future research. *Plant Soil*, **108:**53–65.

Boddey, R. M., V. L. D. Baldani, J. I. Baldani, and J. Döbereiner (1986). Effect of inoculation of *Azospirillum* spp. on nitrogen accumulation by field-grown wheat. *Plant Soil*, **95:**109–121.

Bohlool, B. B. and E. L. Schmidt (1968). Nonspecific staining: its control in immunofluorescent examination of soil. *Science*, **162:**1012–1014.

Bohlool, B. B. and E. L. Schmidt (1970). Immunofluorescent detection of *Rhizobium japonicum* in soils. *Soil Sci.*, **110:**229–236.

Bohlool, B. B. and E. L. Schmidt (1980). The immunofluorescence approach in microbial ecology. *Adv. Microbial Ecol.*, **4:**203–241.

Bowen, G. D. and P. Reddell (1986). Nitrogen fixation in *Casuarinaceae. Proc. 18th IUFRO World Congress*, pp. 1–12.

Bowen, G. D. and A. D. Rovira (1976). Microbial colonization of plant roots. *Annu. Rev. Phytopathol.*, **14:**121–144.

Broughton, W. J. (1978). Control of specificity in legume–*Rhizobium* associations. *J. Appl. Bacteriol.*, **45:**165–194.

Brouzes, R., C. I. Mayfield, and R. Knowles (1971). Effect of oxygen partial pressure on nitrogen fixation and acetylene reduction in a sandy loam soil amended with glucose. *Plant Soil Special*: 481–494.

Callaham, D., P. Del Tredici, and J. G. Torrey (1978). Isolation and cultivation in vitro of the actinomycete causing root nodulation in *Comptonia. Science*, **199:**899–902.

Callow, J. A. (1977). Recognition, resistance and the role of plant lectins in host–parasite interactions. *Adv. Bot. Res.*, **4:**1–49.

Cañizo, A., C. Miguel, and C. Rodriguez-Barrueco (1978). The effect of pH on nodulation and growth of *Coriaria myrtifolia* L. *Plant Soil*, **49:**195–198.

Cavalcante, V. A. and J. Döbereiner (1988). A new acid-tolerant nitrogen-fixing bacterium associated with sugarcane. *Plant Soil*, **108:**23–31.

Cejudo, F. J. and A. Paneque (1986). Short-term nitrate (nitrite) inhibition of nitrogen fixation in *Azotobacter chroococcum. J. Bacteriol.*, **165:**240–243.

Clarholm, M. (1985a). Possible roles for roots, bacteria, protozoa and fungi in supplying nitrogen to plants. In *Ecological Interactions in Soil, Plants, Microbes and Animals* (A. H. Fitter, ed.). Blackwell Scientific Publications, Oxford, pp. 355–365.

Clarholm, M. (1985b). Interactions of bacteria, protozoa and plants leading to mineralization of soil nitrogen. *Soil Biol. Biochem.*, **17:**181–187.

Curl, E. A. and B. Truelove (1986). *The Rhizosphere*. Springer-Verlag, New York, 288 pp.

Currier, W. W. and G. A. Strobel (1976). Chemotaxis of *Rhizobium* sp. to plant root exudates. *Plant Physiol.*, **57:**820–823.

Dalton, H. and J. R. Postgate (1969). Effect of oxygen on growth of *Azotobacter chroococcum* in batch and continuous cultures. *J. Gen. Microbiol.*, **54:**463–473.

Danso, S. K. A., S. O. Keya, and M. Alexander (1975). Protozoa and decline of *Rhizobium* populations added to soil. *Can. J. Microbiol.*, **21:**884–889.

Dart, P. J. (1974). The infection process. In *The Biology of Nitrogen Fixation* (A. Quispel, ed.). North-Holland Publ. Co., Amsterdam, pp. 381–429.

Dawson, J. O. and P. E. Seymour (1983). Effects of juglone concentration on growth *in vitro* of *Frankia* ArI3 and *Rhizobium japonicum* strain 71. *J. Chem. Ecol.*, **9:**1175–1183.

Day, J. M., C. P. Neves, and J. Döbereiner (1975). Nitrogenase activity on the roots of tropical forage grasses. *Soil Biol. Biochem.*, **7:**107–112.

Dazzo, F. B. (1980). Adsorption of microorganisms to roots and other plant surfaces. In *Adsorption of Microorganisms to Surfaces* (G. Britton and K. C. Marshall, eds.). Wiley, New York, pp. 253–316.

Dazzo, F. B. and G. L. Truchet (1984). Attachment of nitrogen-fixing bacteria to roots of host plants. In *Current Developments in Biological Nitrogen Fixation* (N. S. Subba Rao, ed.). Edward Arnold, London, pp. 65–99.

De-Polli, H., E. Matsui, J. Döbereiner, and E. Salati (1977). Confirmation of nitrogen fixation in two tropical grasses by $^{15}N_2$ incorporation. *Soil Biol. Biochem.*, **9:**119–123.

Diem, H. G., D. Gauthier, and Y. R. Dommergues (1982). Extranodular growth of *Frankia* on *Casuarina equisetifolia. FEMS Microbiol. Lett.*, **15:**181–184.

Diem, H. G., D. Gauthier, and Y. Dommergues (1983). An effective strain of *Frankia* from *Casuarina* sp. *Can. J. Bot.*, **61:**2815–2821.

Döbereiner, J. (1961). Nitrogen-fixing bacteria of the genus *Beijerinckia* in the rhizosphere of sugar cane. *Plant Soil*, **15:**211–216.

Döbereiner, J. (1970). Further research on *Azotobacter paspali* and its variety specific occurrence in the rhizosphere of *Paspalum notatum* Flugge. *Zentralbl. Bakteriol. Parasitenkd.*, **124:**224–230.

Döbereiner, J. (1977). Physiological aspects of N_2 fixation in grass–bacteria associations. In *Recent Developments in Nitrogen Fixation* (W. Newton, J. R. Postgate, and C. Rodriguez-Barrueco, eds.). Academic Press, London, pp. 513–522.

Döbereiner, J. and V. L. D. Baldani (1979). Selective infection of maize roots by streptomycin-resistant *Azospirillum lipoferum* and other bacteria. *Can. J. Microbiol.*, **25:**1264–1269.

Döbereiner, J. and J. M. Day (1976). Associative symbiosis in tropical grasses: characterization of microorganisms and dinitrogen-fixing sites. In *Proceedings of the First International Symposium on Nitrogen Fixation* (W. E. Newton and C. Y. Nyman, eds.). Washington State University Press, Pullman, pp. 518–538.

Döbereiner, J. and F. Pedrosa (1987). *Nitrogen-Fixing Bacteria in Nonleguminous Crop Plants*. Science Tech Publishers, Madison, WI.

Döbereiner, J., J. M. Day, and P. J. Dart (1972a). Nitrogenase activity and oxygen sensitivity of the *Paspalum notatum-Azotobacter paspali* association. *J. Gen. Microbiol.*, **71**:103–116.

Döbereiner, J., J. M. Day, and P. J. Dart (1972b). Nitrogenase activity in the rhizosphere of sugar cane and some other tropical grasses. *Plant Soil*, **37**:191–196.

Döbereiner, J., J. M. Day, and P. J. Dart (1973). Rhizosphere associations between grasses and nitrogen-fixing bacteria: effect of O_2 on nitrogenase activity in the rhizosphere of *Paspalum notatum. Soil Biol. Biochem.*, **5**:157–159.

Döbereiner, J., I. E. Marriel, and M. Nery (1976). Ecological distribution of *Spirillum lipoferum* Beijerinck. *Can. J. Microbiol.* **22**:1464–1473.

Dowling, D. N. and W. J. Broughton (1986). Competition for nodulation of legumes. *Ann. Rev. Microbiol.*, **40**:131–157.

Drozd, J. W. and J. R. Postgate (1970). Effects of oxygen on acetylene reduction, cytochrome content and respiratory activity of *Azotobacter chroococcum. J. Gen. Microbiol.*, **63**:63–67.

Eaglesham, A. (1987). The use of intrinsic antibiotic resistance for *Rhizobium* study. In *Symbiotic Nitrogen Fixation Technology* (G. Elkan, ed.). Marcel Dekker Inc., New York, pp. 185–204.

El-Bahrawy, S. A. (1983). Associative effect of mixed cultures of *Azotobacter* and different rhizosphere fungi with *Rhizobium japonicum* on nodulation and symbiotic nitrogen fixation of soybean. *Zentralbl. Mikrobiol.*, **138**:443–449.

Evans, J., Y. M. Barnet, and J. M. Vincent (1979a). Effect of a bacteriophage on the colonization and nodulation of clover roots by a strain of *Rhizobium trifolii. Can. J. Microbiol.*, **25**:968–973.

Evans, J., Y. M. Barnet, and J. M. Vincent (1979b). Effect of a bacteriophage on colonization and nodulation of clover roots by paired strains of *Rhizobium trifolii. Can. J. Microbiol.*, **25**:974–978.

Fessenden, R. J. (1979). Use of actinorhizal plants for land reclamation and amenity planting in the U.S.A. and Canada. In *Symbiotic Nitrogen Fixation in the Management of Temperate Forests* (J. C. Gordon, C. T. Wheeler, and D. A. Perry, eds.). Oregon State University, Corvallis, Oregon, pp. 403–419.

Foster, R. C., A. D. Rovira, and T. W. Cock (1983). *Ultrastructure of the Root–Soil Interface.* American Phytopathological Society, St. Paul, Minn.

Gaskins, M. H., S. L. Albrecht, and D. H. Hubbell (1984/85). Rhizosphere bacteria and their use to increase plant productivity: a review. *Agriculture, Ecosystems and Environment*, **12**:99–116.

Gauthier, D., H. G. Diem, and Y. Dommergues (1981). In vitro nitrogen by two actinomycetes strains isolated from *Casuarina* nodules. *Appl. Environ. Microbiol.*, **41**:306–308.

Germida, J. J. (1986). Population dynamics of *Azospirillum brasilense* and its bacteriophage in soil. *Plant Soil.*, **90**:117–128.

Germida, J. J. (1987). Isolation of *Bdellovibrio* spp. that prey on *Azospirillum brasilense* in soil. *Can. J. Microbiol.*, **33**:459–461.

Germida, J. J. (1988). Growth of indigenous *Rhizobium leguminosarum* and *Rhizobium meliloti* in soils amended with organic nutrients. *Appl. Environ. Microbiol.*, **54**:257–263.

Gibson, A. H., M. M. Roper and D. M. Halsall (1988). Nitrogen fixation not associated with legumes. In *Advances in Nitrogen Cycling in Agricultural Systems* (J. R. Wilson, ed.). CAB International, Wallingford, UK.

Granhall, U., T. Ericsson, and M. Clarholm (1983). Dinitrogen fixation and nodulation by *Frankia* in *Alnus incana* as affected by inorganic nitrogen in pot experiments with peat. *Can. J. Bot.*, **61**:2956–2963.

Haahtela, K., T. Wartiovaara, V. Sundman, and J. Skujins (1981). Root-associated N_2 fixation (acetylene reduction) by Enterobacteriaceae and *Azospirillum* strains in cold-climate spodosols. *Appl. Environ. Microbiol.*, **41**:203–206.

Habte, M. and M. Alexander (1979). Further evidence for the regulation of bacterial populations in soil by protozoa. *Arch. Microbiol.*, **113**:181–183.

Halsall, D. M. and A. H. Gibson (1985). Cellulose decomposition and associated nitrogen fixation by mixed cultures of *Cellulomonas gelida* and *Azospirillum* species or *Bacillus macerans. Appl. Environ. Microbiol.*, **50**:1021–1026.

Halverson, L. J. and G. Stacey (1986). Signal exchange in plant–microbe interactions. *Microbiol. Rev.*, **50**:193–225.

Hanus, F. J., R. J. Maier, and H. J. Evans (1979). Autotrophic growth of H_2-uptake-positive strains of *Rhizobium japonicum* in an atmosphere supplied with hydrogen gas. *Proc. Nat. Acad. Sci. U.S.A.*, **76**:1788–1792.

Harris, R. F. (1981). Effect of water potential on microbial growth and activity. In *Water Potential Relations in Soil Microbiology* (D. M. Kral, ed.). Soil Science Society of America, Madison, Wisc., pp. 23–95.

Hashem, F. M. and J. S. Angle (1988). Rhizobiophage effects on *Bradyrhizobium japonicum*, nodulation and soybean growth. *Soil Biol. Biochem.*, **20**:69–73.

Hill, S. (1988). How is nitrogenase regulated by oxygen? *FEMS Microbiol. Rev.*, **54**:111–130.

Hill, S., G. L. Turner, and F. J. Bergersen (1984). Synthesis and activity of nitrogenase in *Klebsiella pneumoniae* exposed to low concentrations of oxygen. *J. Gen. Microbiol.*, **130**:1061–1067.

Hom, S. S. M., H. Hennecke, and K. T. Shanmugam (1980). Regulation of nitrogenase biosynthesis in *Klebsiella pneumoniae*: effect of nitrate. *J. Gen. Microbiol.*, **177**:169–179.

Hossain, A. K. M. and M. Alexander (1984). Enhancing soybean rhizosphere colonization by *Rhizobium japonicum. Appl. Environ. Microbiol.*, **48**:468–472.

Huang, J. (1987). Interactions of nematodes with rhizobia. In *Vistas on Nematology* (J. Veech and D. W. Dickson, eds.). Society of Nematologists, Hiatville, Md., pp. 301–306.

Hunter, W. J. and C. J. Fahring (1980). Movement by *Rhizobium* and nodulation of legumes. *Soil Biol. Biochem.*, **12**:537–542.

Huss-Danel, K., A. Sellstedt, A. Flower-Ellis and M. Sjöström (1982). Ammonium effects on function and structure of nitrogen-fixing root nodules of *Alnus incana* (L.) *Moench. Planta*, **156**:332–340.

Keister, D. L. (1975). Acetylene reduction by pure cultures of rhizobia. *J. Bacteriol.*, **123**:1265–1268.

Kishinevsky, B. D. and D. G. Jones (1986). Enzyme-linked immunosorbent assay (ELISA) for the detection and identification of *Rhizobium* strains. In *Symbiotic Nitrogen Fixation Technology* (G. H. Elkan, ed.). Marcel Dekker Inc., New York, pp. 157–184.

Klemmedson, J. O. (1979). Ecological importance of actinomycete-nodulated plants in the Western United States. *Bot. Gaz.*, **140**(Supp):s91–s96.

Klugkist, J. and H. Haaker (1984). Inhibition of nitrogenase activity by ammonium chloride in *Azotobacter vinelandii. J. Bacteriol.*, **157**:148–151.

Knowles, R. and D. Denike (1974). Effect of ammonium-, nitrite- and nitrate-nitrogen on anaerobic nitrogenase activity in soil. *Soil Biol. Biochem.*, **6**:353–358.

Knowlton, S., A. Berry, and J. G. Torrey (1980). Evidence that associated soil bacteria may influence root hair infection of actinorrhizal plants by *Frankia. Can. J. Microbiol.*, **26**:971–977.

Knowlton, S. and J. O. Dawson (1983). Effects of *Pseudomonas cepacia* and cultural factors on the nodulation of *Alnus rubra* roots by Frankia. *Can. J. Bot.*, **61**:2877–2882.

Kulinska, D. and A. Drozdowicz (1983). Occurrence of microorganisms antagonistic to *Azospirillum* spp. *Zentralbl. Mikrobiol.*, **138**:585–594.

Lalonde, M. (1977). Infection process of the *Alnus* root nodule symbiosis. In *Recent Developments in Nitrogen Fixation* (W. Newton, J. R. Postgate, and C. Rodriguez-Barrueco, eds.). Academic Press, New York, pp. 569–589.

Lalonde, M., H. E. Calvert, and S. Pine (1981). Isolation and use of *Frankia* strains in actinorhizae formation. In *Current Perspectives in Nitrogen Fixation* (A. H. Gibson and W. E. Newton, eds.). Australian Academy of Science, Canberra, pp. 296–299.

Lamm, R. B. and C. A. Neyra (1981). Characterization and cyst production of azospirilla isolated from selected grasses growing in New Jersey and New York. *Can. J. Microbiol.*, **27**:1320–1325.

Lechevalier, M. P., D. Baker, and F. Horriere (1983). Physiology, chemistry, serology, and infectivity of two *Frankia* isolates from *Alnus incana* subsp. *rugosa. Can. J. Bot.*, **61**:2826–2833.

Leschine, S. B., K. Holwell, and E. Canale-Parola (1988). Nitrogen fixation by anaerobic cellulolytic bacteria. *Science*, **242**:1157–1159.

Lin, W., Y. Okon and R. W. F. Hardy (1983). Enhanced mineral uptake by *Zea mays* and *Sorghum bicolor* roots inoculated with *Azospirillum brasilense. Appl. Environ. Microbiol.*, **45**:1775–1779.

Lippincott, J. A. and B. B. Lippincott (1980). Microbial adherence in plants. In *Bacterial Adherence* (E. H. Beachey, ed.). Chapman and Hall, London, pp. 377–395.

Lowendorf, H. S. (1980). Factors affecting survival of *Rhizobium* in soil. *Adv. Microb. Ecol.*, **4**:87–124.

Lynch, J. M. (1983). *Soil Biotechnology: Microbiological Factors in Crop Productivity*. Blackwell Scientific Publications, Oxford.

Lynch, J. M. (1985). Microbial saprophytic activity on straw and other residues: consequences for the plant. In *Ecological Interactions in Soils: Plants, Microbes and Animals* (A. H. Fitter, ed.). Blackwell Scientific Publications, Oxford, pp. 181–191.

Lynch, J. M. and S. H. T. Harper (1985). The microbial upgrading of straw for agricultural use. *Phil. Trans. R. Soc. Lond.*, **B310**:221–226.

MacConnell, J. T. and G. Bond (1957). A comparison of the effect of combined nitrogen on nodulation in non-legumes and legumes. *Plant Soil*, **8**:378–388.

McNiel, R. E. and P. L. Carpenter (1980). Effect of temperature on acetylene-reducing activity associated with root nodules on woody trees and shrubs. *J. Amer. Soc. Hort. Sci.*, **105**:543–546.

Meesters, T. M., S. Th. van Genesen, and A. D. L. Akkermans (1985). Growth, acetylene reduction activity and localization of nitrogenase in relation to vesicle formation in *Frankia* strains Ccl. 17 and Cpl. 2. *Arch. Microbiol.*, **143**:137–142.

Meesters, T. M., W. M. van Vliet, and A. D. L. Akkermans (1987). Nitrogenase is restricted to the vesicles in *Frankia* strain EAN1pec. *Physiol. Plantarum*, **70**:267–271.

Mellor, H. Y., A. R. Glenn, R. Arwas, and M. J. Dilworth (1987). Symbiotic and competitive properties of motility mutants of *Rhizobium trifoli* TA1. *Arch. Microbiol.*, **148**:34–39.

Moawad, H. A., W. R. Ellis, and E. L. Schmidt (1984). Rhizosphere response as a factor in competition among three serogroups of indigenous *Rhizobium japonicum* for nodulation of field-grown soybeans. *Appl. Environ. Microbiol.*, **47**:607–612.

Moiroud, A., M. Faure-Raynaud, and P. Simonet (1984). Influence de basses temperatures sur la croissance et la survie de souches pures de *Frankia* isolees de nodules d'Aulnes. *Plant Soil*, **78**:91–97.

Moreno, J., J. Gonsalez-Lopez and G. R. Vela (1986). Survival of *Azotobacter* spp. in dry soils. *Appl. Environ. Microbiol.*, **51**:123–125.

Murry, M., M. S. Fontaine, and J. D. Tjepkema (1984a). Oxygen protection of nitrogenase in *Frankia* sp. HFPArI3. *Arch. Microbiol.*, **139**:162–166.

Murry, M., M. S. Fontaine, and J. G. Torrey (1984b). Growth kinetics and nitrogenase induction in *Frankia* sp. HFPArI3 grown in batch cultures. *Plant Soil*, **78**:61–78.

Murry, M., Z. Zhongze, and J. G. Torrey (1985). Effect of O_2 on vesicle formation, acetylene reduction, and O_2-uptake kinetics in *Frankia* sp. HFPCcI3 isolated from *Casuarina cunninghamiana*. *Can. J. Microbiol.*, **31**:804–809.

Myers, M. L. and D. H. Hubbell (1987). Plant cell wall carbohydrates as substrates for *Azospirillum brasilense*. *Appl. Environ. Microbiol.*, **53**:2745–2748.

Neal, J. L. and R. I. Larson (1976). Acetylene reduction by bacteria isolated from the rhizosphere of wheat. *Soil Biol. Biochem.*, **8**:151–155.

Nelson, L. M. and R. Knowles (1978). Effect of oxygen and nitrate on nitrogen fixation and denitrification by *Azospirillum brasilense* grown in continuous culture. *Can. J. Microbiol.*, **24**:1395–1403.

Newman, E. I. (1985). The rhizosphere: carbon sources and microbial populations. In *Ecological Interactions in Soil: Plants, Microbes and Animals* (A. H. Fitter, ed.). Blackwell Scientific Publications, Oxford, pp. 107–121.

Neyra, C. A. and J. Döbereiner (1977). Nitrogen fixation in grasses. In *Advances in Agronomy Vol. 29* (N. C. Brady, ed.). Academic Press, New York, pp. 1–38.

Noridge, N. A. and D. R. Benson (1986). Isolation and nitrogen-fixation activity of *Frankia* sp. strain CpI1 vesicles. *J. Bacteriol.*, **166**:301–305.

Okon, Y. (1982). *Azospirillum* physiological properties, mode of association with roots and its application for the benefit of cereal and forage grass crops. *Israel J. Bot.*, **31**:214–220.

Okon, Y., J. P. Houchins, S. L. Albrecht, and R. H. Burris (1977). Growth of *Spirillum lipoferum* at constant partial pressures of oxygen, and the properties of its nitrogenase in cell-free extracts. *J. Gen. Microbiol.*, **98**:87–93.

Oliveira, R. G. B. and A. Drozdowicz (1981). Bacteriocins in the genus *Azospirillum. Rev. Microbiol.*, **12**:42–47.

O'Toole, P. and R. Knowles (1973). Efficiency of acetylene reduction (nitrogen fixation) in soil: effect of type and concentration of available carbohydrate. *Soil Biol. Biochem.*, **5**:789–797.

Pankhurst, C. E. (1981). Nutritional requirement for the expression of nitrogenase activity by *Rhizobium* sp. in agar cultures. *J. Appl. Bacteriol.*, **50**:45–54.

Parker, C. A., M. J. Trinick, and D. L. Chatel (1977). Rhizobia as soil and rhizosphere inhabitants. In *A Treatise on Dinitrogen Fixation. Section IV.* (R. W. F. Hardy and A. H. Gibson, eds.). John Wiley & Sons, New York, pp. 312–352.

Parsons, R., W. B. Silvester, S. Harris, W. T. M. Gruijters, and S. Bullivant (1987). *Frankia* vesicles provide inducible and absolute oxygen protection for nitrogenase. *Plant Physiol.*, **83**:728–731.

Patriquin, D. G. (1982). New developments in grass–bacteria associations. In *Advances in Agricultural Microbiology* (N. S. Subba Rao, ed.). Oxford & IBM Pub., London, pp. 139–190.

Pena-Cabriales, J. J. and M. Alexander (1983a). Growth of *Rhizobium* in unamended soil. *Soil Sci. Soc. Am. J.*, **47**:81–84.

Pena-Cabriales, J. J. and M. Alexander (1983b). Growth of *Rhizobium* in soil amended with organic matter. *Soil Sci. Soc. Am. J.*, **47**:241–245.

Perinet, P. and M. Lalonde (1983). Axenic nodulation of *in vitro* propagated *Alnus glutinosa* plantlets by *Frankia* strains. *Can. J. Bot.*, **61**:2883–2888.

Perinet, P., J. G. Brouillette, J. A. Fortin, and M. Lalonde (1985). Large scale inoculation of actinorhizal plants with *Frankia. Plant Soil*, **87**:175–183.

Perradin, Y., M. J. Mottet, and M. Lalonde (1983). Influence of phenolics on *in vitro* growth of *Frankia* strains. *Can. J. Bot.*, **61**:2807–2814.

Polonenko, D. R., F. M. Scher, J. W. Kloepper, C. A. Singleton, M. Laliberte, and I. Zaleska (1987). Effects of root colonizing bacteria on nodulation of soybean roots by *Bradyrhizobium japonicum. Can. J. Microbiol.*, **33**:498–503.

Postgate, J. R. (1974a). New advances and future potential in biological nitrogen fixation. *J. Appl. Bacteriol.*, **37**:185–202.

Postgate, J. (1974b). Prerequisites for biological nitrogen fixation in free-living heterotrophic bacteria. In *The Biology of Nitrogen Fixation* (A. Quispel, ed.). North-Holland, Amsterdam, pp. 674–686.

Postgate, J. R. (1987). *Nitrogen Fixation*, 2nd edn. Edward Arnold, London, 73 pp.

Pueppke, S. G. (1984). Adsorption of bacteria to plant surfaces. In *Plant–Microbe Interactions: Molecular and Genetic Perspectives* (T. Kosuge and W. E. Hester, eds.). Macmillan, New York, pp. 215–261.

Quispel, A. (1955). Symbiotic nitrogen fixation in non-legume plants. III. Experiments of the growth in vitro of the endophyte of *Alnus glutinosa. Acta Bot. Neerl.*, **4**:380–396.

Ramaswamy, P. and A. K. Bal (1987). Asymbiotic nitrogen fixation by *Rhizobium* sp. 127E15 in culture and in lima bean rhizosphere. *Curr. Microbiol.*, **15**:223–228.

Ramirez, C. and M. Alexander (1981). Increased bacterial colonization of the rhizosphere by controlling the soil protozoa in the *Rhizobium* legume system. In *Associative N_2 Fixation* (P. B. Vose and A. P. Ruschel, eds.). CRC Press, Boca Raton, FL, pp. 69–86.

Raverkar, K. P. and B. K. Konde (1988). Effect of *Rhizobium* and *Azospirillum lipoferum* inoculation on the nodulation, yield and nitrogen uptake of peanut cultivars. *Plant Soil*, **106**:249–252.

Rennie, R. J. and R. I. Larson (1979). Dinitrogen fixation associated with disomic chromosome substitution lines of spring wheat. *Can. J. Bot.*, **57**:2771–2775.

Reyes, V. G. and E. L. Schmidt (1979). Population densities of *Rhizobium japonicum* strain 123 estimated directly in soil and rhizospheres. *Appl. Environ. Microbiol.*, **37**:854–858.

Robert, F. M. and E. L. Schmidt (1985). Response of three indigenous serogroups of *Rhizobium japonicum* to the rhizosphere of pre-emergent seedlings of soybean. *Soil Biol. Biochem*, **17**:579–580.

Robson, R. L. and J. R. Postgate (1980). Oxygen and hydrogen in biological nitrogen fixation. *Ann. Rev. Microbiol.*, **34**:183.

Roper, M. and D. M. Halsall (1986). Use of products of straw decomposition by N_2-fixing (C_2H_2-reducing) populations of bacteria in three soils from wheat-growing areas. *Aust. J. Agric. Res.*, **37**:1–9.

Rossi, S. (1964). Propagation dans le sol de l'organisme causant nodosites dans les raciaires d'aune. *Ann. Inst. Pasteur*, **106**:505.

Rovira, A. D. (1969). Plant root exudates. *Bot. Rev.*, **35**:35–57.

Rovira, A. D. (1988). Ecology and management of the rhizosphere microflora. In *Microbiology in Action* (W. G. Murrell and I. R. Kennedy, eds.). John Wiley & Sons, New York, pp. 221–238.

Rovira, A. D., R. C. Foster, and J. K. Martin (1979). Note on terminology: origin, natures and nomenclature of the organic materials in the rhizosphere. In *The Soil–Root Interface* (J. C. Harley and R. S. Russell, eds.). Academic Press, London, pp. 1–4.

Ruschel, A. P., R. L. Victoria, E. Salati, and Y. Henis (1978). Nitrogen fixation in sugarcane (*Saccharum officinarum* L.). *Ecol. Bull.*, **26**:297–303.

Saric, M. R., Z. Saric, and M. Govedarica (1987). Specific relations between some strains of diazotrophs and corn hybrids. In *Genetic Aspects of Plant Mineral Nutrition* (H. W. Gabelman and B. C. Loughman, eds.). Martinus Nijhoff, Dordrecht, pp. 495–510.

Schmidt, E. L. (1974). Quantitative autecological study of microorganisms in soil by immunofluorescence. *Soil Sci.*, **118**:141–149.

Schmidt, E. L. (1988). Competition for legume nodule occupancy: a down-to-earth limitation on nitrogen fixation. In *World Crops: Cool Season Food Legumes* (R. J. Summerfield, ed.). Kluwer, pp. 663–674.

Schmidt, E. L., R. O. Bankole, and E. L. Bohlool (1968). Fluorescent antibody approach to the study of rhizobia in soil. *J. Bacteriol.*, **95**:1987–1992.

Schmidt, E. L., M. J. Zidwick, and H. M. Abebe (1986). *Bradyrhizobium japonicum* serocluster 123 and diversity among member isolates. *Appl. Environ. Microbiol.*, **51**:1212–1215.

Shipton, W. A. and A. J. P. Burgraaf (1982). *Frankia* growth and activity as influenced by water potential. *Plant Soil*, **69**:293–297.

Shipton, W. A. and A. J. P. Burgraaf (1983). Aspects of the cultural behaviour of *Frankia* and possible ecological implications. *Can. J. Bot.*, **61**:2783–2792.

Smolander, A. and V. Sundman (1987). *Frankia* in acid soils of forests devoid of actinorhizal plants. *Physiol. Plantarum*, **70**:297–303.

Smolander, A., C. van Dijk, and V. Sundman (1988). Survival of *Frankia* strains introduced into soil. *Plant Soil*, **106**:65–72.

Soby, S. and K. Bergman (1983). Motility and chemotaxis of *Rhizobium meliloti* in soil. *Appl. Environ. Microbiol.*, **46**:995–998.

Spiff, E. D. and C. T. I. Odu (1972). An assessment of non-symbiotic nitrogen fixation in some Nigerian soils by the acetylene reduction technique. *Soil Biol. Biochem.*, **4**:71–77.

Stewart, W. D. P. and G. Bond (1961). The effect of ammonium nitrogen on fixation of elemental nitrogen in *Alnus* and *Myrica*. *Plant Soil*, **14**:347–359.

Subba Rao, N. S. (1984). Interactions of nitrogen-fixing bacteria with other soil microorganisms. In *Current Developments in Biological Nitrogen Fixation* (N. S. Subba Rao, ed.). Edward Arnold, London, pp. 37–64.

Suslow, T. (1982) Role of root-colonizing bacteria in plant growth. In *Phytopathogenic Prokaryotes* (G. Lacy and H. Mount, eds.). Academic Press, New York.

Thompson, J. P. and V. D. B. Skerman (1979). *Azotobacteraceae: The Taxonomy and Ecology of the Aerobic Nitrogen-Fixing Bacteria*. Academic Press, New York, 417 pp.

Tibelius, K. H. and R. Knowles (1983). Effect of hydrogen and oxygen on uptake-hydrogenase activity in nitrogen-fixing and ammonium-grown *Azospirillum brasilense*. *Can. J. Microbiol.*, **29**:1119–1125.

Tien, T. M., M. H. Gaskins, and D. H. Hubbell (1979). Plant growth substances produced by *Azospirillum brasilense* and their effect on the growth of pearl millet (*Pennisetum americanum* L.) *Appl. Environ. Microbiol.*, **37**:1016–1024.

Tien, T. M., H. G. Diem, M. H. Gaskins, and D. H. Hubbell (1981). Polygalacturonic acid transeliminase production by *Azospirillum* species. *Can. J. Microbiol.*, **27**:426–431.

Tjepkema, J. and H. J. Evans (1975). Nitrogen fixation by free-living *Rhizobium* in a defined liquid medium. *Biochem. Biophys. Res. Commun.*, **65**:625–628.

Tjepkema, J. D. and R. H. Burris (1976). Nitrogenase activity associated with some Wisconsin prairie grasses. *Plant Soil*, **45**:81–94.

Tjepkema, J. D., W. Ormerod, and J. G. Torrey (1980). Vesicle formation and acetylene reduction in *Frankia* sp. CpI1 cultures in defined nutrient media. *Nature* (London), **287**:633–635.

Tjepkema, J. D., W. Ormerod, and J. G. Torrey (1981). Factors affecting vesicle formation and acetylene reduction (nitrogenase activity) in *Frankia* sp. CpI1. *Can. J. Microbiol.*, **27**:815–823.

Tjepkema, J. D., C. R. Schwintzer, and D. Benson (1986). Physiology of actinorhizal nodules. *Ann. Rev. Plant Physiol.*, **37**:209–233.

Torrey, J. G. and J. D. Tjepkema (1983). Introduction: International Conference on the biology of *Frankia. Can. J. Bot.*, **61**:2765.

Torrey, J. G., J. D. Tjepkema, G. L. Turner, F. J. Bergersen, and A. H. Gibson (1981). Dinitrogen fixation by cultures of *Frankia* sp CpI1 demonstrated by $^{15}N_2$ incorporation. *Plant Physiol.*, **68**:983–984.

Trolldenier, G. (1977). Influence of some environmental factors on nitrogen fixation in the rhizosphere of rice. *Plant Soil*, **47**:203–217.

Tyson, J. H. and W. S. Silver (1979). Relationship of ultrastructure to acetylene reduction (N_2 fixation) in root nodules of *Casuarina. Bot. Gaz.*, **140**(Supp.):s44–s48.

Umali-Garcia, M., D. H. Hubbell, and M. H. Gaskins (1978). Process of infection of *Panicum maximum* by *Spirillum lipoferum. Ecol. Bull.* (Stockholm), **26**:373–379.

Umali-Garcia, M., D. H. Hubbell, M. H. Gaskins, and F. B. Dazzo (1980). Association of *Azospirillum* with grass roots. *Appl. Environ. Microbiol.*, **39**:219–226.

van Berkum, P. and C. Sloger (1983). Interaction of combined nitrogen with the expression of root-associated nitrogenase activity in grasses and with the development of N_2 fixation in soybean (*Glycine max* L. Merr.). *Plant Physiol.*, **72**:741–745.

Vance, C. P. (1983). *Rhizobium* infection and nodulation: a beneficial plant disease. *Ann. Rev. Microbiol.*, **37**:399–424.

van Dijk, C. (1979). Endophyte distribution in the soil. In *Symbiotic Nitrogen Fixation in the Management of Temperate Forests* (J. C. Gordon, C. T. Wheeler, and D. A. Perry, eds.). Oregon State University, Corvallis, Oregon, pp. 84–94.

Vincent, J. M. (1970) *A Manual for the Practical Study of the Root Nodule Bacteria*. IBP Handbook No. 15. Blackwell Scientific Publications, Oxford.

Vincent, J. M. 1980. Factors controlling the legume-*Rhizobium* symbiosis. In *Nitrogen Fixation, Vol. II* (W. E. Newton and W. H. Orme-Johnson, eds.) University Park Press, Baltimore, pp. 103–129.

Viteri, S. E. and E. L. Schmidt (1987). Ecology of indigenous soil rhizobia: response of *Bradyrhizobium japonicum* to readily available substrates. *Appl. Environ. Microbiol.*, **53**:1872–1875.

Vogel, C. S. and J. O. Dawson (1985). Effect of juglone on growth *in vitro* of *Frankia* isolates and nodulation of *Alnus glutinosa* in soil. *Plant Soil*, **87**:79–89.

Vogel, C. S. and J. O. Dawson (1986). *In vitro* growth of five *Frankia* isolates in the presence of four phenolic acids and juglone. *Soil Biol. Biochem.*, **18**:227–231.

Weaver, R. W. and L. R. Frederick (1982). *Rhizobium*. In *Methods of Soil Analysis. Part 2. Chemical and Microbiological Properties* (A. L. Page, R. H. Miller, and D. R. Keeney, eds.). American Society of Agronomy, Madison, Wis., pp. 1043–1069.

Whipps, J. M. (1984). Environmental factors affecting the loss of carbon from the roots of wheat and barley seedlings. *J. Exp. Bot.*, **35**:767–773.

Whipps, J. M. and J. M. Lynch (1983). Substrate flow and utilization in the rhizosphere of cereals. *New Phytol.*, **95**:605–623.

Winship, L. J. and J. D. Tjepkema (1985). Nitrogen fixation and respiration by root nodules of *Alnus rubra* Bong.: effects of temperature and oxygen concentration. *Plant Soil*, **87**:91–107.

Wollum, A. G. (1986). Serological techniques for *Bradyrhizobium* and *Rhizobium* identification. In *Practical Symbiotic Nitrogen Fixation* (G. H. Elkan, ed.) Marcel Dekker, Inc., New York.

Wollum, A. G., C. T. Youngberg, and F. W. Chichester (1968). Relation of previous timber stand age to nodulation of *Ceanothus velutinus*. *Forest Science*, **14**:114–118.

Yahalom, E., Y. Okon, and A. Dovrat (1987). *Azospirillum* effects on susceptibility to *Rhizobium* nodulation and on nitrogen fixation of several forage legumes. *Can. J. Microbiol.*, **33**:510–514.

Yoch, D. C. and G. J. Whiting (1986). Evidence for NH_4^+ switch-off regulation of nitrogenase activity by bacteria in salt marsh sediments and roots of the grass *Spartina alterniflora*. *Appl. Environ. Microbiol.*, **51**:143–149.

Zuberer, D. A. and D. B. Alexander (1986). Effects of oxygen partial pressure and combined nitrogen on N_2-fixation (C_2H_2) associated with *Zea mays* and other gramineous species. *Plant Soil*, **90**:47–58.

Zuberer, D. A. and M. Roth (1982). *In vitro* inhibition of non-symbiotic nitrogen-fixing bacteria by rhizosphere actinomycetes associated with grasses. *Can. J. Microbiol.*, **29**:705–709.

Zuberer, D. A. and W. S. Silver (1978). Biological dinitrogen fixation (acetylene reduction) associated with Florida mangroves. *Appl. Environ. Microbiol.*, **35**:567–575.

13

SOIL ALGAE

B. METTING

Enviros Corporation, 600 Skyline Tower,
10900 NE 4th Street, Bellevue, WA 98004, USA.

Microscopic algae (microalgae) are ubiquitous components of soil surface microbial communities from high latitudes to the tropics. Algae can be isolated from dry, semi-arid, moist and flooded soils. With sufficient moisture, microbial communities on soil surfaces can be dominated by microalgal populations whose presence is often visible. The soil algal flora includes numerous taxa. Most common are the prokaryotic cyanobacteria (blue-green algae) and eukaryotic species belonging to the Chlorophyceae (green algae), Bacillariophyceae (diatoms), and Xanthophyceae (yellow-green algae). Euglenophyceae (eugleonids) and Rhodophyceae (red algae) are less often encountered (Metting, 1981).

Carbon fixation (CO_2-fixation), biological nitrogen fixation (N_2-fixation) and surface consolidation are the principal functional attributes of soil microalgae. Additional activities that might be important but for which only limited circumstantial evidence exists are solubilization of minerals, siderophore production, plant growth regulator (PGR) production, and uptake or partial degradation of soil pollutants. Soil algae are thus not important rhizosphere organisms *per se* but rather affect soil chemical and physical processes, which in turn influence root activity. While methods for studying soil algae are not well developed they nevertheless are receiving attention as N_2-fixing biofertilizers and soil conditioners in agriculture and as integral components of semi-arid rangeland ecosystems (Metting, 1988; Metting et al., 1988).

1. ECOLOGICAL ROLES OF SOIL ALGAE

1.1. Density and Primary Production

Primary production (CO_2-fixation) is the algal activity most important to soil microbial ecology as a source of organic matter and for provision of energy and reductant for N_2-fixation. Estimates of soil algal productivity, density (organisms per area) and standing crop (living biomass per area) vary widely and have most often been arrived at by indirect methods. The bulk of the literature on soil algae consists of floristic lists compiled either by microscopic examination or enrichment culture (Metting, 1981). Few lend insight into the functional dynamics of algal communities on soil.

Densities (colony-forming units) of microalgae on soil vary from none or a few propagules to values in excess of $10^9\ cm^{-2}$. Values ranging from 10^3 to 10^5 g^{-1} or cm^{-2} are most common. The only soils reported not to yield viable microalgae have been from the Atacama Desert (Chile) and samples from the deserts of western Australia and the Antarctic interior (Metting, 1981).

It is difficult to estimate standing biomass from density data for soil algae because many dominant taxa are filamentous or sarcinoid (packet or aggregate-forming). In particular, this has complicated the efforts to gauge the importance of algal N_2-fixation outside of the rice paddy because most of the important cyanobacterial species are filamentous. However, it seems probable that moist temperate soil with minimal shading will support from a few hundred to perhaps 1000 or more dry $kg\ ha^{-1}$ (Metting, 1981), whereas up to 24 tonnes (fresh weight ha^{-1} of cyanobacterial biomass has been reported from a flooded rice soil (Roger and Watanabe, 1986).

Two field and two laboratory studies include data relevant to estimating CO_2-fixation rate functions by microalgae on soil. Using an *in situ* $^{14}CO_2$ technique to study algal dynamics in a field cropped to a rye–sorghum rotation. Shimmel and Darley (1985) reported maximal production of 39 $mg\ C\ m^{-2}\ h^{-1}$ and estimated the total annual accrual to be 78 dry $g\ m^{-2}$. Also in the field, the single-celled green alga *Chlamydomonas sajao*, inoculated onto replicated plots at 5×10^3 cells cm^{-2}, achieved densities of $19 \times 10^6\ cm^{-2}$. Maximal accrual of biomass sampled over an 8-day period was 35 dry $mg\ cm^{-2}\ day^{-1}$ (Metting, 1986).

Estimates from the field agree reasonably well with laboratory data. Kroen (1984) inoculated *Chlamydomonas mexicana* onto sterile soil moistened with mineral medium and reported a maximum specific growth rate of 1.8 day^{-1}, slightly less than the 2.0 day^{-1} for the same alga in aqueous culture (Kroen and Rayburn, 1984). On non-sterile soil in the laboratory, *C. sajao* reproduced at rates ranging from 0.27 day^{-1} at -0.5 bar total water potential, to 0.68 day^{-1} at -0.1 bar, and 0.78 day^{-1} at -0.3 bar (Metting and Barry, 1986). Optimal soil water content probably varies among species, but too little evidence exists from which to generalize.

In contrast to these data, microalgal communities on marine sediments in marshes have been shown, with considerable variation, to produce up to 700 dry g m^{-2} yr^{-1} (Zedler, 1980; Colijn and de Jonge, 1984). In all likelihood, soil algae are (occasionally or often) more productive than estimated to date. Clearly, much remains to be learned about yield and production by soil algae in order to achieve a good understanding of their overall significance and indirect importance to root activity.

1.2. N_2-Fixation

Cyanobacterial N_2-fixation is the most thoroughly investigated subject regarding soil algae. Most research has been with either free-living cyanobacteria or the *Azolla/Anabaena azollae* symbiosis and rice cultivation. Utilization of the water fern *Azolla* has been developed as a reliable N source for rice in parts of China, Vietnam, the Philippines and West Africa. Depending on environmental conditions the fern is either incorporated as a green manure or grown as a companion crop with rice. Nitrogen fixation by the phycosymbiont *Anabaena* (thought by some researchers to be more properly categorized as *Nostoc*) is commonly sufficient for acceptable grain yields. Under ideal circumstances the symbiosis provides N in quantities as great as any reported for *Rhizobium*/legume symbioses in agriculture. The reader is referred to the following sources for more detail: Achtnich et al. (1986); Metting et al. (1988); Lumpkin and Plucknett (1982); Peters and Calvert (1983); and Roger and Watanabe (1986). Roger and Kulasooriya (1980) reviewed the literature on free-living cyanobacteria and rice. Cyanobacterial N_2-fixation and its relationship to fertility of arid and semi-arid soil crusts has also received attention (Metting et al., 1988; Skujiņš, 1984).

Nitrogen fixation rates and total nitrogen input by asymbiotic cyanobacteria vary widely within flooded paddies, on moist temperate soils, and in desert and semi-desert crusts (Tables 1 and 2). In rice paddy soils total annual input varies up to 90 kg N ha^{-1} yr^{-1}, with peak fixation rates of 0.5 kg N ha^{-1} day^{-1} (Roger and Kulasooriya, 1980; Stewart, 1980). By way of comparison, the *Azolla*

TABLE 1. Nitrogen fixation by soil Algae on non-flooded soils

Situation	*N_2-fixation (kg ha^{-1} yr^{-1})*	*Reference*
Savanna, Nigeria	3–9	Isichei (1980)
Algal crust, Sonoran Desert, USA	7–11	Mayland et al. (1966)
Algal crust, Sonoran Desert, USA	13–18	MacGregor and Johnson (1971)
Wheat field, England	13–28	Witty et al. (1979)
Various soils, Sweden	15–51	Henriksson (1971)

TABLE 2. N_2-fixation on flooded, uninoculated rice soils. Condensed from Roger and Kulasooriya (1980)

Country	*N_2-fixation (kg ha^{-1} crop^{-1})*
India	14–53
Japan	1–22
Philippines	9–63
Senegal	0–30
Mali	50–80

association has been reported by Chinese researchers to fix as much as 10.5 kg N ha^{-1} day^{-1} (Lumpkin and Plucknett, 1982).

Cyanobacteria on moist temperate soils are thought to fix as much as 28 kg N ha^{-1} yr^{-1}, with maximal short-term rates of 4.5 mg N m^{-2} h^{-1} (Henriksson, 1971; Henricksson et al., 1972; Witty et al., 1979). Algal and algal–lichen crusts dominated by cyanobacteria have been suggested as the major source of N in many deserts and steppes. Most research has been reported from the semi-arid American Great Basin desert, where maximal N_2-fixation rates by soil crusts of 30 mg N m^{-2} day^{-1} have been measured by ^{15}N methods (Skujiņš, 1984). Semi-arid and desert regions deserve more attention because cyanobacteria are probably the major avenue for N input in these ecosystems. Extrapolation to as much as 100 kg N ha^{-1} yr^{-1} has been reported (Rychert and Skujiņš, 1974), but in all likelihood this is seldom achieved in temperate ecosystems.

1.3. Consolidation of Soil Surfaces

Communities of eukaryotic microalgae and cyanobacteria consolidate surfaces against erosive forces whenever an adequate crust develops. Consolidation is effected by entanglement of soil particles by filamentous species and as a consequence of excretion of high molecular weight heteropolysaccharides (Metting, 1981).

Evidence has also been presented to suggest that algal crusts on dry soils might enhance seedling establishment, improve water retention, and provide nutrients from weathering of minerals. These phenomena have been well documented for decades in studies of soil colonization in primary and secondary plant community succession (Metting, 1981) and certainly have an impact on root biology, though this has not been subject to scientific scrutiny.

2. PHYSIOLOGICAL ECOLOGY OF SOIL ALGAE

2.1. Physical Factors

Sunlight and temperature, in addition to moisture (see Section 2.2), are the physical factors most important to metabolism, growth and reproduction of soil algae.

Irradiance, spectral quality and photoperiod all influence microalgae. Many algal communities function in full tropical sunlight, and a small number of studies have demonstrated a correlation between a successively more dense vascular plant canopy and decreasing algal density (e.g. Shimmel and Darley, 1985). On some rice soils, floristic succession has been shown to proceed from a community dominated by green microalgae and diatoms in full sunlight to one dominated by cyanobacteria under a mature canopy; cyanobacteria are better able to make use of long wavelengths by virtue of light-harvesting phycobiliprotein complexes (Reynaud and Roger, 1978). Laboratory studies have shown how spectral quality and/or photoperiod control or modify developmental cycles of both eukaryotic microalgae and cyanobacteria (Metting, 1981).

Soil algal metabolic activities and rates vary with temperature. Including those isolated from hot and cold deserts, most soil algae are mesophilic microorganisms, although respiration, photosynthesis and N_2-fixation occur over broad ranges. In common with other microorganisms, soil microalgae tolerate extreme temperatures either by spore formation or by physiological and biochemical modifications not manifested as distinctive cell types. Many cyanobacteria produce thick-walled akinetes (Whitton, 1987), while the zygotic product of gamete fusion in some green microalgae, e.g. *Chlamydomonas mexicana*, is a resting spore (Lewin, 1957).

The ability to photosynthesize or fix N_2 at high temperatures or at or below freezing varies among soil algae according to geographic origin. Accordingly, *Nostoc* and *Calothrix* species from tropical soils continue to metabolize at temperatures approaching 50 °C, albeit at reduced rates and efficiencies (Venkataraman, 1964), whereas *Nostoc commune* was shown to be able to fix N_2 near 0 °C during snowmelt episodes in a Canadian shortgrass prairie community (Coxson and Kershaw, 1983a, 1983b).

2.2. Water Requirement and Desiccation Response

Given sufficient sunlight, adequate moisture is the most important abiotic factor controlling algal activities on soil. This has been demonstrated many times from observation of naturally occurring soil algal communities (Metting, 1981) and in experimental work with application to agriculture (see Sections 3.1 and 3.2).

Photosynthesis, respiration and N_2-fixation by soil cyanobacteria have all been studied experimentally. Working with a desert crust of *Microcoleus* brought into the laboratory, Brock (1975) found it not to be particularly better adapted to low water potentials. Growth inhibition was noted at −7 bar total water potential; growth ceased at −18 bar. Photosynthesis was impaired at −18 and ceased at −28 bar. Scherer et al. (1984) studied the response of three terrestrial *Nostoc* species to desiccation and rewetting and found the surface-to-mass ratio of all three to be inversely correlated to speed of uptake and loss of water. Respiration started immediately after a 30-minute rewetting period, with maximum photosynthetic O_2 evolution in 6 and 8 hours, respectively, for *N. commune* and *N. flagelliforme*. Rewetted after 2 years, *N. commune* achieved maximal nitrogenase activity after 120–150 h. Kept dry for only 2 days, the same organism took 50 h to reach maximal activity. A lag of 4–5 h was noted prior to steady-state equilibria in intracellular pools of ATP (Scherer et al., 1986). Cryptogamic crusts in the Mojave Desert (USA) actively metabolized within 15–60 minutes following precipitation, while apparent N_2-fixation required a minimum moisture regime of -2 bar (Skujiņš, 1984).

Physiological and biochemical adjustments to desiccation have been extensively studied in *Nostoc commune*. In this cyanobacterium, water stress has been found to induce marked changes in the protein index (Potts, 1985), amounts of polysomes (Angeloni and Potts, 1986), and the qualitative composition of membrane fatty acids (Olie and Potts, 1986) and cellular carotenoids and phospholipids (Potts et al., 1987). The relationship among recovery of nitrogenase activity and stable intracellular ATP pools in the same series of experiments was influenced by CO_2 concentration as well as conditions and duration of desiccation and dry storage. Similar studies with eukaryotic microalgae from soil are needed.

2.3. Facultative Heterotrophy

Whether microalgae are active components of the microflora below the soil surface has been the subject of some contention since the classic work of Petersen (1935). Microalgae are easily isolated from soil collected throughout the profile and subsequently enriched in the light. At least seven studies from 1938 to 1972 reported microalgae in the rhizosphere of various plants, although most were microscopical observations and none offered evidence that the algae were metabolically active (Metting, 1981).

The systematics of eukaryotic green and yellow-green soil algae include as taxonomic criteria the utility of simple sugars and organic acids for photoheterotrophy or heterotrophic growth in dark axenic culture. Also of taxonomic consequence is production of extracellular gelatinase and amylase (Metting, 1981). Many cyanobacteria are also known to be facultative heterotrophs, some being able to synthesize and fix N_2 in the dark (Smith, 1982). However, it is doubtful that algae compete effectively with bacteria and fungi for organic C in the soil as it is rare to successfully select microalgae from soil by enrichment with simple organic compounds in the dark.

2.4. Soil Chemistry

Perhaps the most commonly cited relationship between algae and soil chemistry is that of floristic distribution and pH. Neutral to slightly alkaline soils support the greatest diversity of microalgae. Green and yellow-green species predominate on acid soils, whereas cyanobacteria and diatoms are more prevalent on alkaline soils.

Statistical interpretation of changing floristic diversity and fluctuating mineral concentrations in soils has been the focus of studies reviewed by Pipe and Shubert (1984). Hunt et al. (1979) measured algal density and diversity at monthly intervals, correlating these with pH and Mg for cyanobacteria and with Mg for eukaryotes. Starks and Shubert (1979, 1982) studied algal diversity and abundance on soils disturbed by mining. They found positive correlations during succession with levels of N, P_2O_4, Si, Mn, Al and Pb, while Cu, Li, Mo and Sr had negative correlations. Experimental studies in the laboratory or field have yet to be reported.

2.5. Pesticides and Soil Algae

Numerous laboratory and field observations and experiments have documented a wide variation in soil algae–pesticide interactions. These have been reviewed (Wright, 1978; Roger and Kulasooriya, 1980; Metting, 1981; Pipe and Shubert, 1984). Generally, herbicides that are metabolic poisons have detrimental effects, while those that are plant hormones or hormone analogs exhibit variable effects. Fungicides and insecticides are not usually detrimental, although effects are again variable. Pesticide effects are more pronounced in liquid culture than in soil. Microalgae have in some cases been shown to accumulate and partially degrade some pesticides (Wright, 1978).

3. APPLICATIONS

3.1. Cyanobacterial Biofertilizers

Algalization is the practice of inoculating rice soils with mixtures of free-living cyanobacteria (Venkataraman, 1972). Inconsistency of performance and the fact that algalization has not changed or grown in popular use where it has been championed in India for 20 years led Roger and Watanabe (1986) to label it an unproven techonology. Nevertheless, in many instances algalization trials have yielded statistically significant, positive crop responses (Roger and Kulasooriya, 1980).

Inoculation of temperate agricultural soils with cyanobacteria has been the subject of only three studies in the scientific literature. Witty (1979) inoculated irrigated soil cropped to wheat in England with a native *Nostoc ellipsosporum* and alien (to that soil) isolates of *N. punctiforme* and *Anabaena cylindrica.*

All three cyanobacteria successfully colonized the soil resulting in significantly earlier and greater N_2-fixation than that on non-inoculated soil. However, total fixation extrapolated from acetylene reduction assays was not agronomically significant, amounting to a maximum of only 4.75 kg N ha^{-1} for *N. ellipsosporum*, a quantity substantially less than that attributed to native cyanobacteria on nearby uncropped fields (28 kg N ha^{-1}: Witty et al., 1979). Competition from the homocystous filament *Lyngbya* was apparent, particularly under conditions of water stress. Following desiccation, only akinetes of the added species were recovered while *Lyngbya* became dominant.

Anderson et al. (1982) inoculated replicated plots sown to irrigated corn with a native *Anabaena* sp. and *Tolypothrix tenuis* (Venkataraman strain). Estimated by the most-probable-number method, populations of both added cyanobacteria increased by 2–3 orders of magnitude but nitrogenase activity was not significantly greater than that in control plots dominated by a native *Oscillatoria*. Following desiccation late in the cropping cycle, the heterocystous species disappeared. Upon rewetting, *Oscillatoria*, a non-heterocystous cyanobacterium, dominated all plots.

Reynaud and Metting (1988) screened 23 cyanobacteria from the USA (Washington State), Senegal, India and the Philippines for *in situ* colonization potential on a temperate soil irrigated for 1 week and allowed to dry for 2 weeks prior to sampling. Only 12 of the 23 strains could be recovered. Of these, the two recovered in greatest abundance, *Nostoc* sp. and *Calothrix* sp., were applied to 55 and 60 ha sprinkler-irrigated wheat fields (one species per field). Quantified by the method of Reynaud and Laloe (1985), *Nostoc* biomass increased 395-fold over the growing season to 79 dry kg ha^{-1} after inoculation at 0.2 kg ha^{-1}. Following harvest and desiccation of the soil, *Nostoc* biomass was reduced to 3 kg ha^{-1} but still represented nearly 64% of the total algal community. Although it could be recovered on each sampling date after inoculation at 0.1 dry kg ha^{-1}, *Calothrix* biomass peaked at 0.3 kg ha^{-1} during the growing season.

These few studies suggest that use of cyanobacterial biofertilizers on temperate agricultural land might only be successful under two situations. First, with low or minimum tillage operations permitting populations to achieve sufficiently great standing crops, such as for undisturbed soils at Broadbalk (Witty et al., 1979). Second, cyanobacterial biofertilizers might one day prove efficacious if concerted screening and selection for highly competitive, fast-growing strains identifies candidates amenable to genetic manipulation for permanent derepression of heterocyst differentiation and function. Wild-type cyanobacteria do not differentiate heterocysts or synthesize nitrogenase when sufficient amounts of combined N can be obtained from the environment. Because many intensively cultivated crops require quantities of N greatly in excess of what is biologically possible from cyanobacteria, competitive species will have to be engineered to fix N in the presence of fertilizer in order to make a contribution. In any case, technologies will be appropriate only for irrigated or flooded soils to which large application rates are commercially feasible.

3.2. Microalgal Soil Conditioners

Mucilaginous (palmelloid) green microalgae hold promise as biological soil conditioners of light, sprinkler-irrigated soil prone to crusting or wind erosion (Metting, 1988; Metting et al., 1988). In contrast to the heterocystous cyanobacteria referred to above (Section 3.1) species of *Asterococcus* and *Chlamydomonas* used as soil conditioners more consistently exhibit fast growth rates leading to successful colonization of moist soil (Metting, 1988).

Soil conditioning is any procedure or application of a product that improves the physical properties of soil. Desired effects include ease of cultivation, penetration or retention of water, better seedling emergence, reduced crusting, efficient gas exchange for healthy root development, and reduced erosion by water or wind. However, it is impractical, if not impossible, to modify soil and achieve all these ends in any one situation. Many of these benefits follow from genesis and stabilization of microaggregates and crumbs. Whereas mechanisms of aggregate formation are varied, most evidence suggests that microbial polysaccharides are responsible for aggregate stability (Burns and Davies, 1986; Lynch and Bragg, 1985).

Microalgal soil conditioners were first conceived in the mid-1970s when fresh mass cultures of non-palmelloid microalgae, such as *Chlorella* and *Scenedesmus*, were applied to thousands of hectares in the western USA. After replicated field trials demonstrated no benefit, the first commercial enterprise failed. Lewin (1977) first reported studies using palmelloid *Chlamydomonas*. Field observations suggested that massive inoculation of palmelloid microalgae through sprinkler systems could effectively influence aggregation (Metting and Rayburn, 1983). Field experimentation over a 3-year period (Table 3) has indicated that application of sufficiently large inocula onto constantly moist soil exposed to sunlight can significantly increase wet and dry aggregate stability (Metting, 1987).

Laboratory and field studies unequivocally demonstrate the requirements for adequate soil moisture and large enough rates of inoculation for successful colonization. Barclay and Lewin (1985) reported glasshouse experiments with *Chlamydomonas mexicana* and *C. sajao* in which the polysaccharide content of the upper 2 mm of soil was increased by 129%. While as much as 95% of the soil surface was colonized, growth and polysaccharide production was maximal only at 100% of field capacity.

Lewin (1956) showed that as much as 33% of C fixed by palmelloid *Chlamydomonas* species was excreted as polysaccharide material. Over a 3-week period on soil in the laboratory, *C. mexicana* accrued nearly 1600 μg polysaccharide (as glucose) cm^{-2} following inoculation at 10^3 cells cm^{-2} (Kroen, 1984). In the field, anthrone carbohydrate levels increased, although not significantly, throughout the season on plots inoculated with 5×10^{11} cells ha^{-1} (Metting, 1986).

TABLE 3. Increased aggregate stability by inoculation with *Chlamydomonas*. (After Metting, 1987)

	% *Wet aggregate stability*[a]		
Surface veneer	*At the end of: 1983*	*1984*	*1985*
Inoculated soil	41.1a	58.7a	65.8a
Uninoculated control	27.2a	45.8a	38.2a
First 30 cm of soil profile			
Inoculated soil	46.2	37.1	42.7b
Uninoculated control	27.6	31.7	29.8b
	% *Dry aggregate stability*		
Surface veneer	*At the end of: 1983*	*1984*	*1985*
Inoculated soil	42.0	31.1b	51.4c
Uninoculated control	43.0	27.8b	27.8c
First 30 cm of soil profile			
Inoculated soil	33.4b	51.6	37.8d
Uninoculated control	27.5b	37.9	14.2d

[a]Numbers in columns followed by the same letter were significantly different (5%).

Purified polysaccharide from *C. mexicana* and *C. sajao* added at a rate of 0.1% to soil in the laboratory significantly increased aggregation in soils of three textural classes by as much as 70% as estimated by slaking in water (Dow et al., 1976; Metting and Barry, 1986; Metting, 1988). As shown in Table 3, wet and dry aggregation (dry aggregation is estimated by rotary sieving of air-dried soil through nested cylindrical sieves) of soil particles was improved annually for 3 years for the surface veneer (Metting, 1986), and throughout the first 30 cm after 3 years (Metting, 1987).

3.3. Microalgae and Rangeland Management

Currently a topic of some controversy (Johansen, 1986), the role of algal crusts in rangelands is thought to be important in surface consolidation and has been correlated to better water retention and improved soil fertility. Controversial is the increasingly popular adoption of what has come to be known as the Savory method of short-duration intensive grazing of semi-arid and arid rangeland. Advocates of the Savory method argue that disruption of surface crusts enhances seedling establishment and results in greater secondary production (Savory, 1983). While this may be true for some situations in the short run, there is concern that this management technique will be detrimental to sustained production in fragile ecosystems as a consequence of elevated soil erosion. In particular, regions in Africa threatened by desertification might well be placed at greater risk to wind erosion by short-duration intensive grazing with subsequent loss of production for decades or even centuries.

In a series of glasshouse and field experiments, St. Clair et al. (1984, 1986) demonstrated the value of algal crusts in semi-arid ecosystems in the western USA. Seedling establishment in trampled versus undisturbed plots in the field showed a tendency toward increased establishment when crusts were present. Significant increases were reported from glasshouse experiments (St. Clair et al., 1984). Following disturbance by fire, St. Clair et al. (1986) inoculated study plots in the field with: (1) a soil crust slurry, (2) a subsoil slurry, and (3) water. Six months later, green microalgae and diatoms were not significantly different among treatments whereas significantly greater numbers of filamentous cyanobacteria were found on treated plots. They suggested that inoculation of small disturbed rangeland sites might prove feasible.

4. CONCLUSIONS

Eukaryotic microalgae and cyanobacteria are common inhabitants of soil surfaces. Often, they dominate the microbial communities and can significantly contribute to organic matter and nitrogen content while acting as agents for consolidation of the surface. Microalgal biofertilizer and soil conditioner technologies are under development. The chief advantage of microalgal products for agriculture is that the microbial habitat to be colonized (the soil surface) is readily accessible, although algae clearly have an ecological role even when enhancement is not feasible by inoculation. Additionally, these microorganisms need not compete for C or energy with the resident heterotrophic microflora.

Ecological limits to exploitation of soil algae are formidable, however. Successful colonization of soil surfaces leading to agronomically significant microalgal biomass within timeframes appropriate for a cropping cycle or growing season requires adequate sunlight, moisture and application rates. And finally, the value of algal crusts in semi-arid rangeland management is in need of further understanding in view of the fact that in natural semi-arid plant communities microalgae play an important role.

REFERENCES

Achtnich, W., A. M. Moawad, and C. S. Johal (1986). *Azolla*, a biofertilizer for rice. *Intl. J. Trop. Agric.*, **4**:188–211.

Anderson, D. B., K. R. Hanson, B. Metting, and P. M. Molton (1982). *Assessment of Blue-Green Algae as Living Sources of Fertilizer Nitrogen.* Pacific Northwest Laboratory Tech. Report No. 4187, Richland WA, USA, 41 pp.

Angeloni, S. V. and M. Potts (1986). Polysome turnover in immobilized cells of *Nostoc commune* (Cyanobacteria) exposed to water stress. *J. Bacteriol.*, **168**:1036–1039.

Barclay, W. R. and R. A. Lewin (1985). Microalgal polysaccharides as soil-conditioning agents. *Plant and Soil*, **88**:159–169.

Brock, T. D. (1975). Effect of water potential on a *Microcoleus* (Cyanophyceae) from a desert crust. *J. Phycol.*, **11**:316–320.

Burns, R. G. and J. A. Davies (1986). The microbiology of soil structure. *Biol. Agric. Hort.*, **3**:95–113.

Colijn, F. and V. N. de Jonge (1984). Primary production of microbenthos in the Ems-Dollard estuary. *Marine Ecol.-Prog. Ser.*, **14**:185–196.

Coxson, D. S. and K. A. Kershaw (1983a). Rehydration response of nitrogenase activity and carbon fixation in terrestrial *Nostoc commune* from *Stipa–Bouteloa* grassland. *Can. J. Bot.*, **61**:2658–2668.

Coxson, D. S. and K. A. Kershaw (1983b). The pattern of *in situ* nitrogenase activity and carbon fixation in terrestrial *Nostoc commune* from *Stipa–Bouteloa* grassland. *Can. J. Bot.*, **61**:2686–2693.

Dow, E. I., S. Roberts, T. A. Cline, and W. H. Weaver (1976). Effects of N, P, K and three soil amendment products on yield and nutrient uptake of five crops. *Circular 599*. College of Agric. Res. Center, Washington State Univ., Pullman, WA, USA, 16 pp.

Henriksson, E. (1971). Algal nitrogen fixation in temperate regions. In *Biological Nitrogen Fixation in Natural and Agricultural Habitats* (T. A. Lie and E. G. Mulder, eds.). *Plant and Soil* (Special Volume): 415–419.

Henriksson, L. E., P. H. Enckell, and E. Henriksson (1972). Determination of the nitrogen-fixing capacity of algae in soil. *Oikos*, **23**:420–423.

Hunt, M. E., G. L. Floyd, and B. B. Stout (1979). Soil algae in field and forest environments. *Ecology*, **60**:362–375.

Isichei, A. O. (1980). Nitrogen fixation by blue-green algal soil crusts in Nigerian savanna. In *Nitrogen Cycling in West African Ecosystems* (T. Rosswall, ed.). SCOPE/UNEP International Nitrogen Unit. Royal Swedish Acad. Sci., Stockholm, pp. 191–198.

Johansen, J. A. (1986). Soil algae and range management. *Appl. Phycol. Forum*, **3**:1–2.

Kroen, W. K. (1984). Growth and polysaccharide production by the green alga *Chlamydomonas mexicana* (Chlorophyceae) on soil. *J. Phycol.*, **20**:616–618.

Kroen, W. K. and W. R. Rayburn (1984). Influence of growth status and nutrients on extracellular polysaccharide synthesis by the soil alga *Chlamydomonas mexicana* (Chlorophyceae). *J. Phycol.*, **20**:353–357.

Lewin, R. A. (1956). Extracellular polysaccharides of green algae. *Can. J. Microbiol.*, **2**:665–672.

Lewin, R. A. (1957). Four new species of *Chlamydomonas*. *Can. J. Bot.*, **35**:321–326.

Lewin, R. A. (1977). The use of algae as soil conditioners. *Centros Invest. Baja Calif., Scripps Inst. Oceanogr.*, **3**:33–35.

Lumpkin, T. A. and D. L. Plucknett (1982). *Azolla as a Green Manure: Use and Management in Crop Production*. Westview Press, Boulder CO, USA, 230 pp.

Lynch, J. M. and E. Bragg (1985). Microorganisms and soil aggregate stability. *Adv. Soil Sci.*, **2**:133–171.

MacGregor, A. N. and D. E. Johnson (1971). Capacity of desert algal crusts to fix atmospheric nitrogen. *Soil Sci. Soc. Amer. Proc.*, **35**:843–844.

Mayland, H. F., T. M. McIntosh and W. H. Fuller (1966). Fixation of isotopic nitrogen on a semi-arid soil by algal crust organisms. *Soil Sci. Soc. Amer. Proc.*, **30**:56–60.

Metting, B. (1981). The systematics and ecology of soil algae. *Bot. Rev.*, **47**:195–312.

Metting, B. (1985). Soil microbiology and biotechnology. In *Biotechnology: Applications and Research* (P. A. Cheremisinoff and R. P. Ouellette, eds.). Technomic Publishers, Lancaster PA, USA, pp. 196–214.

Metting, B. (1986). Population dynamics of *Chlamydomonas sajao* and its influence on soil aggregate stabilization in the field. *Appl. Environ. Microbiol.*, **51**:1161–1164.

Metting, B. (1987). Dyanmics of wet and dry aggregate stability from a three-year microalgal soil conditioning experiment in the field. *Soil Sci.*, **143**:139–143.

Metting, B. (1988). Micro-algae in agriculture. In *Micro-algal Biotechnology* (M. A. Borowitzka and L. A. Borowitzka, eds.). Cambridge University Press, Cambridge, pp. 288–304.

Metting, B. (1988). Microalgal soil conditioners. In *Polysaccharides from Microalgae: A New Agro-industry*. (J. Ramus and S. Arad, eds.). Duke University Marine Lab., Beaufort, NC, pp. 56–61.

Metting, B. and A. D. Barry (1986). *Palmelloid Microalgae as Soil-Conditioning Agents.* U.S.D.A. Tech. Report No. 85-SBIR-8-0067 (unpublished).

Metting, B. and W. R. Rayburn (1983). The influence of a microalgal soil conditioner on selected Washington soils: an empirical study. *Soil Sci. Soc. Amer. J.*, **47**:682–685.

Metting, B., W. R. Rayburn, and P. A. Reynaud (1988). Algae and agriculture. In *Algae and Human Affairs* (C. A. Lembi and R. A. Waaland, eds.). Cambridge University Press, Cambridge, pp. 335–370.

Olie, J. J. and M. Potts (1986). Purification and biochemical analysis of the cytoplasmic membrane from the desiccation tolerant cyanobacterium *Nostoc commune* UTEX 584. *Appl. Environ. Microbiol.*, **52**:706–710.

Peters, G. A. and H. E. Calvert (1983). The *Azolla–Anabaena azollae* symbiosis. In *Algal Symbiosis* (L. J. Goff, ed.). Cambridge University Press, Cambridge, pp. 109–145.

Petersen, J. B. (1935). Studies on the biology and taxonomy of soil algae. *Dansk. Bot. Arkiv.*, **8**:1–180.

Pipe, A. E. and L. E. Shubert (1984). The use of algae as indicators of soil fertility. In *Algae as Ecological Indicators* (L. E. Shubert, ed.). Academic Press, New York, pp. 213–233.

Potts, M. (1985). Protein synthesis and proteolysis in immobilized cells of the cyanobacterium *Nostoc commune* UTEX 584 exposed to matric water stress. *J. Bacteriol.*, **164**:1025–1031.

Potts, M., J. J. Olie, J. S. Nickels, J. Parsons, and D. C. White (1987). Variation in phospholipid ester-linked fatty acids and carotenoids of desiccated *Nostoc commune* (Cyanobacteria) from different geographic locations. *Appl. Environ. Microbiol.*, **53**:4–9.

Reynaud, P. A. and F. Laloe (1985). La méthode des suspensions-dilutions adaptée à l'estimation des populations algales dans une rizière. *Rev. Ecol. Biol. Sol*, **22**:161–192.

Reynaud, P. A. and B. Metting (1988). Colonization potential of cyanobacteria on temperate irrigated soils in Washington state, USA. *Biol. Agric. Hort.*, **5**:197–208.

Reynaud, P. A. and P. A. Roger (1978). N_2-fixing algal biomass in Senegal rice fields. *Ecol. Bull. (Stockholm)*, **26**:148–157.

Roger, P. A. and S. A. Kulasooriya (1980). *Blue-Green Algae and Rice.* International Rice Research Institute, Los Baños, Philippines, 112 pp.

Roger, P. A. and I. Watanabe (1986). Technologies for utilizing biological nitrogen fixation in wetland rice: potentialities, current usage, and limiting factors. *Fertilizer Res.*, **9**:39–77.

Rychert, R. C. and J. Skujiņš (1974). Nitrogen fixation by blue-green algae–lichen crusts in the Great Basin Desert. *Soil Sci. Soc. Amer. Proc.*, **38**:768–771.

Savory, A. (1983). The Savory grazing method. *Rangelands*, **5**:155–159.

Scherer, S., A. Ernst, T.-W. Chen, and P. Böger (1984). Rewetting of drought-resistant blue-green algae: time course of water uptake and reappearance of respiration, photosynthesis, and nitrogen fixation. *Oekologia*, **62**:418–423.

Scherer, S., T.-W. Chen, and P. Böger (1986). Recovery of adenine-nucleotide pools in terrestrial blue-green algae after prolonged drought periods. *Oekologia*, **68**:585–588.

Shimmel, S. M. and W. M. Darley (1985). Productivity and density of soil algae in an agricultural system. *Ecology*, **66**:1439–1447.

Skujiņš, J. (1984) Microbial ecology of desert soils. *Adv. Microbial Ecol.*, **7**:49–91.

Smith, A. J. (1982). Modes of cyanobacterial carbon metabolism. In *The Biology of Cyanobacteria* (N. G. Carr and B. A. Whitton, eds.). Botanical Monographs Vol. 19, University of California Press, Berkeley, pp. 47–85.

Starks, T. L. and L. E. Shubert (1979). Algal colonization on a reclaimed surface mined area in western North Dakota. In *Ecology and Coal Resource Development* (M. K. Wali, ed.). Pergamon Press, New York, pp. 652–660.

Starks, T. L. and L. E. Shubert (1982). Colonization and succession of algae and soil–algal interactions associated with disturbance areas. *J. Phycol.*, **18**:99–107.

St. Clair, L. L., B. L. Webb, J. R. Johansen, and G. T. Nebeker (1984). Cryptogamic soil crusts: enhancement of seedling establishment in disturbed and undisturbed areas. *Reclamation Reveg. Res.*, **3**:129–136.

St. Clair, L. L., J. R. Johansen, and B. L. Webb (1986). Rapid stabilization of fire-disturbed sites using a soil crust slurry: inoculation studies. *Reclamation Reveg. Res.*, **4**:261–269.

Stewart, W. D. P. (1980). Systems involving blue-green algae. In *Methods for Evaluating Biological Nitrogen Fixation* (F. J. Bergersen, ed.). John Wiley & Sons, New York, pp. 583–635.

Venkataraman, G. S. (1964). Thermal resistance and viability of microalgae. *Phykos*, **3**:26–28.

Venkataraman, G. S. (1972). *Algal Biofertilizers and Rice Cultivation.* Today and Tomorrow's Printers and Publishers, New Delhi, 75 pp.

Whitton, B. A. (1987). Survival and dormancy of blue-green algae. In *Survival and Dormancy of Microorganisms* (Y. Henis, ed.). John Wiley & Sons, New York, pp. 109–167.

Witty, J. F. (1979). Algal nitrogen fixation on temperate arable fields. Algal inoculation experiments. *Plant and Soil*, **52**:165–183.

Witty, J. F., P. J. Keay, P. J. Frogatt, and P. J. Dart (1979). Algal nitrogen fixation on temperate arable soils. The Broadbalk experiment. *Plant and Soil*, **52**:151–164.

Wright, S. J. L. (1978). Interactions of pesticides with micro-algae. In *Pesticide Microbiology* (S. J. L. Wright, ed.). Academic Press, London, pp. 535–602.

Zedler, J. B. (1980). Algal mat productivity: comparisons in a salt marsh. *Estuaries*, **3**:122–131.

14

FAUNA–MICROFLORA INTERACTIONS

E. A. CURL

Department of Plant Pathology, 139 Funchess Hall, Auburn University, Alabama 36849, USA

J. D. HARPER*

Department of Entomology, Auburn University, Alabama 36849, USA

Complex terrestrial ecosystems are created largely by the interactive processes involving protozoa, nematodes and microarthropods in concert with components of the microflora, all responding to the ambient abiotic environment. The relatively large populations and small body size of the micro- and mesoscopic animals, along with their roles in microbial grazing and in nutrient transformations, qualify them for participation in critical rhizosphere phenomena affecting root growth and plant health. By definition of the rhizosphere (Curl and Truelove, 1986) or ectorhizosphere (*sensu* Lynch, 1982), larger animals such as the macroarthropods are not as likely to be directly influenced by environmental changes occurring along the root–soil interface, since they are more transient and not confined to this specific ecological niche.

While many small soil animals are unspecialized feeders, many are microphagous and dependent to a large extent on microflora food sources. These microbial populations are largely dependent on carbon sources derived from rhizodeposition. Consequently, populations of fungi and bacteria are usually larger in the rhizosphere than distant from roots, and microphagous animals tend to congregate in that region.

The interaction of small soil invertebrates and the microflora in organic matter decomposition and nutrient cycling has been extensively documented (Reichle,

*Present address: Department of Entomology, Box 7613, North Carolina State University, Raleigh, NC 27695, USA

1977; Stout, 1980; Freckman, 1982; Peterson and Luxton, 1982; Coleman et al., 1983, 1984a, 1984b, 1988; Anderson et al., 1984; Seastedt, 1984). While fungi and bacteria are the primary decomposers, microphagous protozoa, nematodes and microarthropods affect the dynamics of the microflora community and directly or indirectly regulate the decomposition process. At the same time, these microbe-feeding animals influence the quantitative and qualitative nature and the behavior of rhizosphere fungi and bacteria that are known to either promote plant growth and health or incite disease.

In this chapter we consider a neglected and poorly understood topic, namely how the activities of soil invertebrates might influence plant pathogen behavior and plant disease. The general relation of small fauna to plant disease, including inoculum transmission, was discussed by Beute and Benson (1979). Here, we attempt to create a new awareness of certain invertebrate activities which may contribute to disease suppression and enhance crop productivity; some background information on the subject has been provided (Curl, 1988). At the same time, equal awareness must be focused upon animal activities that might favor disease.

1. RHIZOSPHERE FAUNA

Before considering our specific topic of disease-related interactions it may be useful to review briefly the functional groups of small invertebrate animals commonly inhabiting the rhizosphere. For convenience these animal groups are assigned to the arbitrary categories of microfauna (free-living protozoa and nematodes) with body widths of < 100 mm and mesofauna (microarthropods) with body widths of < 1–2 mm (Swift et al., 1979). The macrofauna (< 10–20 mm), though important in organic matter fragmentation and decomposition, are largely inhabitants of surface litter rather than of microniches of the rhizosphere. Thus, the micro- and mesoscopic forms, as they feed upon bacteria and fungi at the root surface, offer the greatest potential for altering certain natural rhizosphere processes affecting plant growth, health, and vigor. Except for plant parasitic nematodes, relatively little study has been directed to fauna populations specifically in the rhizosphere. Indeed, the common definition of rhizosphere (Curl and Truelove, 1986) infers a root-system zone of influence primarily affecting activities of the microflora.

1.1. Protozoa

Free-living members of the Protozoa are represented by flagellates, ciliates and amoebae. Laybourn-Parry (1984) has described the classification and functional biology of various groups of the Protozoa. Together they are the most abundant of the micro- and mesofauna, followed by nematodes and the microarthropods. Precise comparisons of their populations with the microflora or with other small animals is difficult due to great differences in methods of isolation and enumeration. For the purpose of this discussion we are primarily concerned with a few genera of the Amoebida which have been studied for implication in the suppression of plant-pathogenic fungi through mycophagous activities. Naked amoebae, estimated at 2×10^3 to $17\times10^5\,g^{-1}$ dry weight of soil, constitute the dominant group of soil protozoa (Heal, 1971; Chakraborty and Old, 1985; Old and Chakraborty, 1986). Protozoa, largely amoebae, in an arable field soil have been estimated at 5 g dry wt m^{-2} (Schnurer et al., 1982). Elliott and Coleman (1977) found that 95% of protozoa in a shortgrass prairie following irrigation were naked amoebae. In field experiments Clarholm (1981) observed that peaks of amoebae quickly followed peaks in numbers and biomass of bacteria immediately after rainfall. Interrelationships of amoebae and bacteria leading to acceleration of nutrient mineralization have been extensively researched and discussed (for example, see Elliott et al., 1979; Clarholm, 1981; Coleman et al., 1983, 1984b, 1988). In a suitable physical environment, populations of soil amoebae are largely determined by the availability of food, this being provided mostly by bacteria and to some extent by fungi and algae (Heal, 1970). Since populations of the microflora are greater in the rhizosphere–rhizoplane than distant from roots, it is reasonable to expect higher populations of microphagous amoebae as well within the zone of root influence.

1.2. Nematodes

Rhizosphere and root-surface influence on plant parasitic nematodes is well known, largely relating to egg-hatching reactions to root exudates and the attraction or congregation of larvae around roots in the region of cell elongation. The many ecological phenomena that influence populations and behavior of plant-parasitic nematodes (Norton, 1978; Maggenti, 1981) and their relation to the rhizosphere (Curl and Truelove, 1986) have been discussed. We are concerned here primarily with free-living forms, particularly those that feed upon fungi. Many species of non-parasitic nematodes are associated with plant roots, feeding superficially on the microflora along the rhizoplane. Population densities of free-living nematodes vary widely with season and different kinds of plant cover; estimates for agroecosystems range from 1×10^4 to $18\times10^5\,m^{-2}$ (Stinner and Crossley, 1977, 1982). Hendrix et al. (1986) provided summary estimates showing total numbers and biomass of non-parasitic nematodes in conventional-tillage (3008 m^{-2}/344 mg dry wt m^{-2}) and no-tillage (2437 m^{-2}/252 mg dry wt m^{-2}) agroecosystems. Bacteria-feeding species are abundantly represented in such populations, and mycophagous species of *Aphelenchoides, Aphelenchus, Ditylenchus, Neotylenchus,* and other genera are common (Nicholas, 1984).

These nematodes, like the Protozoa, are significantly involved in organic matter decomposition and mineralization, principally through their grazing activities on the microflora (Wasilewska and Bienkowski, 1985; Elliott et al., 1988). Bacterial-feeding nematodes may significantly stimulate bacterial populations in the rhizosphere and increase nutrient availability and uptake

(Coleman et al., 1984b). It follows that intensified microphagous activity in the rhizosphere, while affecting the mineralization of nutrients, may also directly or indirectly influence the competitive behavior of root-infecting fungi and their microbial antagonists.

1.3. Microarthropods

Like other soil invertebrates, populations of microarthropods vary with season, soil chemical and physical factors, kind of vegetative cover, and food availability. Relative numbers of microarthropods per square meter of soil in a European grassland have been estimated at 120 000 for the Acarina (mites) and 40 000 for Collembola (springtails) (MacFadyen, 1957; Kevan, 1965). In Michigan grasslands (Mallow et al., 1985), mite populations reached 80 000 m^{-2} and collembolan populations 18 000 m^{-2} in July. In conventional agroecosystems, the biomass of mites was estimated to be 118 mg and collembolans 17 mg dry wt m^{-2}; biomass figures for both were somewhat higher in no-tillage systems (Hendrix et al., 1986). Though individual species may deviate from the whole-population trend, numbers of both mites and collembolans are generally much higher (depending on season) in grasslands with dense fibrous root systems than in cultivated lands.

The Cryptostigmata or oribatid mites are predominantly soil-dwelling saprophages which feed upon decomposing litter (macrophytic feeders) or on bacteria, fungi and algae (microphytic feeders). Panphytophages are active in either of these categories. The oribatids have exhibited extremely varied feeding preferences, accepting certain foods and rejecting others; they have been most commonly reared on brewer's yeast and other living fungi, dried mushrooms, chopped lichen, and artificial diets such as dextrose and casein (see Butcher et al. (1971) for more on the bioecology of the Acarina). The soil-dwelling Cryptostigmata are most likely to impact upon other biotic components of the rhizosphere whereas the Mesostigmata, Prostigmata and Astigmata are primarily inhabitants of surface litter, compost and dung. The oribatids can be expected to congregate in niches of the rhizosphere where a preferred food supply is available.

Collembola, the most abundant order of the primitive soil-dwelling Apterygota are, like the Acarina, of major importance in organic matter decomposition and nutrient cycling (Edwards et al., 1970; Anderson and Ineson, 1984; Coleman et al., 1984a). Currently, these predominantly microphagous animals also are being recognized for their interactions with mycorrhizal fungi (Warnock et al., 1982; Finlay, 1985; Shaw, 1985; Cromack et al., 1988; Rabatin and Stinner, 1988) and plant pathogens (Curl et al., 1988). The Arthropleona, with elongate, visibly six-segmented abdomens, are typically soil-dwelling and well represented in agricultural soils to 15 cm depth, often following avenues provided by growing roots. Wiggins et al. (1979) found a direct relationship between increased cotton-root density and populations of *Proisotoma minuta* Tullberg and *Onychiurus encarpatus* Denis; further, these species were consistently more abundant in the rhizosphere than in root-free soil. Other common genera in agricultural, forest and grassland soils are *Tullbergia, Friesea, Folsomia* and *Isotoma*.

Populations of collembolans respond positively to both organic and inorganic fertilization (Edwards and Lofty, 1969; Wiggins et al., 1979), and have been known to create problems in greenhouse experiments by feeding upon pathogen inoculum introduced into non-sterilized high-organic plant-growth media. Certain species of *Onychiurus* have been cited as plant pests in situations of low preferred-food availability in fields of sugar beet (Heijbroek, 1972; Dunning and Baker, 1977; Ulber, 1980; Brown, 1985) and several other crops (Ulber, 1980; Getzin, 1985).

2. RHIZOSPHERE FLORA

The quantitative and qualitative nature of the microflora around roots has, unlike that of the fauna, received an enormous amount of research attention since 1904, when Professor Lorenz Hiltner (Hiltner, 1904) coined the term "rhizosphere". Populations and activities of the rhizosphere microflora (Curl and Truelove, 1986) and their influence on crop productivity (Whipps and Lynch, 1986) have been thoroughly documented. Therefore, it will suffice at this point to look at specific groups of bacteria and fungi which relate prominently to plant health and which may be affected by invertebrate animal activities.

Microbial populations, stimulated by rhizodeposition, are almost invariably higher in the rhizosphere than in root-free soil (Fig. 1). Four groups of organisms,

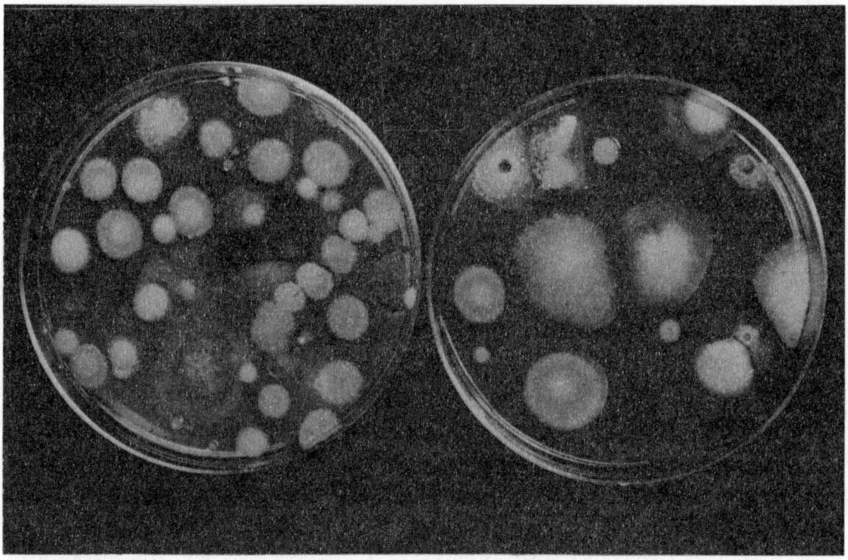

Fig. 1. Dilution plates of fungal colonies from rhizosphere soil (left) and non-rhizosphere soil (right) of wheat, illustrating the greater food potential for mycophagous fauna around roots.

all directly or indirectly related to plant growth and health, are potentially subject to alteration by the activities of microbe-feeding invertebrates. These groups are the plant-growth-promoting rhizobacteria (PGPR), root pathogens, biocontrol organisms and symbiotic organisms.

Populations of bacteria in field soils may exceed 100 million g^{-1} as estimated by soil dilution and plate counts. From such counts the rhizosphere/non-rhizosphere ratio (R/S) values are most frequently in the range 2–20, but ratios over 100 have been recorded. The ratio biomass fungi/biomass bacteria in an arable soil was found to be 5–10 (Shields et al., 1973), and the ratio percent root surface covered by fungi/percent cover by bacteria has been estimated ranging from 0.28 to 14.0 for *Lolium perenne* (Newman et al., 1981). Most prevalent in the rhizosphere are the PGPR (Kloepper et al., 1988), largely composed of *Pseudomonas* spp. Plant growth can be enhanced by *P. fluorescens* and *P. putida* through several mechanisms (Chapter 7).

Soil actinomycetes, predominantly *Streptomyces* spp., are usually second to bacteria in populations, with common R/S ratios of 5 to 10, based on plate counts. Actinomycetes are best known for antibiotic production and the inhibition of plant pathogens and other soil microorganisms. Rouatt et al. (1951) found higher relative numbers of antibacterial actinomycetes in the rhizospheres of several crop plants than in root-free soil.

In general populations of fungi are lower than those of bacteria and actinomycetes, though 100 000 colony-forming units g^{-1} dry wt of soil is not uncommon (Rouatt et al., 1960). The R/S ratios derived from plate counts may range from 3 : 1 to >100 : 1 but most frequently are 10 : 1 to 20 : 1 for crop plants. By the use of selective inhibitors and measurement of CO_2 production, the metabolically active biomass ratio of fungi/bacteria has been found to range from 1.5 to 9 in agricultural, grassland and forest soils, and from 0.13 to 1.5 in rhizospheres of crop plants and prairie grasses (Anderson and Domsch, 1975, 1980; Vancura and Kunc, 1977; Nakas and Klein, 1980). Thus, whereas fungi in bulk soil usually exceed bacteria in biomass, either group in the rhizosphere may have the larger biomass (Newman, 1985). Saprophytic fungi as well as bacteria in the rhizosphere create a competitive deterrent to pathogen colonization of the rhizoplane. Some saprophytes, such as species of *Trichoderma* and *Gliocladium*, produce toxins or become directly parasitic on root pathogens. Other common fungi in the rhizosphere include species of *Penicillium, Aspergillus, Fusarium, Cladosporium, Cephalosporium*, the Mucorales and nematode-trapping Hyphomycetes.

Garrett (1970) categorized root-infecting fungi as either specialized root-inhabiting fungi characterized by a declining or weakly competitive saprophytic phase after death of the host plant, or soil-inhabiting pathogens characterized by a strong competitive ability (expanding saprophytic phase in absence of the host plant). Most highly specialized are the symbiotic fungi represented by ectomycorrhizal species of Basidiomycetes in the genera *Pisolithus*, *Laccaria*, *Boletus*, *Thelephora*, *Suillus*, *Amanita*, *Lactarius*, *Corticium*, and *Rhizopogon*. Endomycorrhizal species of the Endogonaceae are found in the genera

Gigaspora, Glomus, Acaulospora and *Sclerocystis* (see also Chapter 11). Specialized root pathogens are represented by such common species as *Plasmodiophora brassicae* Woron., *Phymatotrichum omnivorum* (Sheer) Duggar, *Gaeumannomyces graminis* (Sacc.) Arx et Olivier, *Verticillium dahliae* Kleb. and the vascular wilt fusaria. Root-infecting fungi with high competitive saprophytic ability include *Pythium* spp., *Rhizoctonia solani* Kühn, *Aphanomyces euteiches* Drechs., *Fusarium solani* f. sp. *phaseoli* (Burk.) Snyd. et Han., and *Cochliobolus sativus* (Ito et Kuirg.) Drechs. ex Dastur.

These fungi collectively produce a variety of propagules (conidia, chlamydospores, sporangia, zoospores, sclerotia, rhizomorphs) which serve as inocula and/or survival structures. They provide opportunities for mycophagous invertebrate animals to influence pathogen behavior or survival, either in the passive state of the fungus or during root colonization and infection.

3. ROLE OF FAUNA IN PLANT DISEASE SUPPRESSION

Great differences in the natural disease-suppressive capacity of cropped soils have been observed since the early 1900s but the mechanisms involved are not entirely clear (Baker and Cook, 1974; Schneider, 1982; see also Chapter 15). While most efforts to explain the phenomenon relate to microbial antagonisms, soil mycostasis and physico-chemical soil properties, it is essential that we adopt a holistic approach (Lyda, 1982) to such studies, knowing that no single biological phenomenon in the rhizosphere is likely to determine the capacity of a soil-borne pathogen to establish a disease relationship. Invertebrate animals that consume or inactivate pathogen inocula have only recently received serious attention as disease suppressants. Mycophagous amoebae, nematodes and

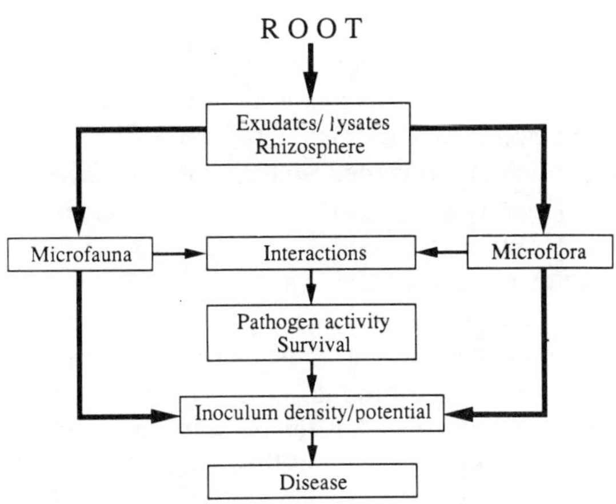

Fig. 2. Schematic representation of rhizosphere influence on the soil microfauna and microflora, ultimately affecting root pathogen activities and disease potential.

microarthropods are the targets of relatively new research in this regard. Fig. 2 shows a schematic representation of the interrelationship of microfauna and flora in the rhizosphere and potential impact on pathogen activities and disease. Biotic relationships are further influenced by the ambient chemical and physical environment around roots.

3.1. Mycophagous Amoebae in the Rhizosphere

High populations of amoebae in plant rhizospheres are reflected in reports of R/S ratios ranging from 2 : 1 for total protozoa to 20 : 1 for specific mycophagous amoebae (Rouatt et al., 1960; Geltzer, 1963; Darbyshire and Greaves, 1967, 1973; Napolitano, 1983; Chakraborty et al., 1985). Old and Chakraborty (1986) have reviewed the biology and significance of mycophagous soil amoebae in relation to suppression of soil-borne plant pathogens and disease. Therefore, it is only necessary here to summarize certain key points and provide update information.

Among the genera of amoebae known to be fungal feeders are *Arachnula, Cashia, Gephyramoeba, Mayorella, Saccamoeba, Thecamoeba, Theratromyxa* and *Vampyrella*. A wide range of fungal species, and in many cases bacteria and algae, serve as food for amoebae; food selectivity or preference is common. Feeding activity results in either perforation lysis, in which cell walls are penetrated, leaving uniform holes 0.2–6.0 μm in diameter, or general lysis, in which fungal propagules taken into digestive vacuoles are ruptured following lysis of the cell wall. Different fungal species have been reported to be resistant to attack by specific amoebae (Esser et al., 1975; Old and Darbyshire, 1978; Chakraborty et al., 1983). However, the great nutritional diversity of amoebae and the ability of certain species to destroy both pigmented and hyaline propagules suggest the improbability that any fungal species is totally resistant.

Recognition that mycophagous amoebae may destroy considerable fungal biomass, particularly in the rhizosphere, prompted serious investigations to determine the potential impact of such activities on inoculum density–disease relationships. Anderson and Patrick (1978) demonstrated that vampyrellid amoebae isolated from field soil in Ontario perforated and fed on chlamydospores and mycelium of *Thielaviopsis basicola* (Berk. et Br.) Ferr. and conidia of *Cochliobolus sativus*, resulting in reduced populations of viable propagules of both fungi. Preliminary greenhouse tests showed that populations of both fungi could be reduced by adding cysts of the amoebae to soil. The amoebae feed on pigmented spores, as well as hyaline propagules, thus suggesting a potential for destroying spores generally believed resistant to many other biotic factors (Old and Patrick, 1979). During this period, the effectiveness of mycophagous amoebae as biological control agents under field conditions was postulated but not tested.

Chakraborty et al. (1983) and Chakraborty (1985) found that the population of viable propagules of *Gaeumannomyces graminis* var. *tritici* Walker (take-all disease pathogen) declined more rapidly in take-all suppressive soil than in non-suppressive soil. Mycophagous amoebae were strongly implicated when higher numbers were found associated with the fungus in suppressive soils than in disease-conducive soils. Subsequently, Chakraborty and Warcup (1985) reported the suppression of *G. graminis* var. *tritici* and take-all disease of wheat in pot bioassays using mycophagous amoebae. *Gephyramoeba* sp., *Saccamoeba* sp. and *Thecamoeba granifera* Greef. subsp. *minor* Pussard, when added to sterilized non-suppressive soil infested with the pathogen, reduced disease severity and increased plant height and shoot dry weight. Mixed populations of the amoeba species also provided significant disease control. Thus, it is possible that inoculum-destroying amoebae may contribute to the natural suppressive-soil phenomenon in wheat fields.

3.2. Nematode–Fungus Relationships

Fungus-feeding nematodes are ubiquitous in soil; many belong to the genera *Aphelenchus, Aphelenchoides, Ditylenchus, Neotylenchus* and *Paraphelenchus* (Nicholas, 1984). Some species are primarily bacterial feeders, omnivores or predators. A wide variety of fungi, both saprophytes and plant pathogens, serve as suitable food sources for mycophagous nematodes, thus suggesting unspecialized feeding habits. Consideration of mycophagous nematodes for biological suppression of fungal pathogens and root diseases has involved primarily species of *Aphelenchus* and *Aphelenchoides*.

Aphelenchus avenae Bastian is commonly found associated with roots, tubers and rhizomes, but most nematologists consider the species to be mycophagous or saprophagous. Mankau and Mankau (1963), in laboratory cultures, demonstrated highly favorable effects of *Rhizoctonia solani, Periconia* sp., *Verticillium alboatrum* (Reinke et Berth.), *Fusarium solani* (Martius) Sacc. and *Thielaviopsis basicola* on reproduction and development of *A. avenae.* Contrastingly, saprophytic soil fungi such as *Aspergillus terreus* Thom, *Penicillium* sp., *Rhizopus nigricans* Ehrenberg and *Trichoderma viride* Pers. ex Fries., were not favorable food sources. Pythiaceous fungi, including plant pathogens *Pythium ultimum* Trow, *Phytophthora parasitica* Dastur., *P. citrophthora* (R. et E. Smith) Leonian and *P. cryptogea* Pethy et Laff., also were poor food bases for *A. avenae* and apparently were toxic. Studies in France (Cayrol and Combettes, 1977) showed that neither *A. avenae* nor *Aphelenchoides composticola* Frank prevented growth of *P. ultimum* in Petri dishes but both were aggressive on cultures of *R. solani.* Townshend (1964) observed that *A. avenae* could feed and reproduce on 54 of 59 fungal species tested. The nematode can readily pierce cell walls of gill tissue in basidiomycetes and ingest the cytoplasm (Walker, 1984).

Aphelenchoides spp. are largely polyphagous, feeding and reproducing on a variety of phytopathogenic fungi. Their populations were predominant around lower stems of barley and wheat in Germany (Rossner and Urland, 1983). Certain species can be reared in large populations on specific fungi, such as *Agaricus campestris* Link ex Fries, for experimental purposes (Rossner and Nagel, 1984).

Effective control of several root-infecting fungi in artificial greenhouse environments has been achieved by using relatively large populations of *A. avenae*. For control of root rot (*Pythium arrhenomanes* Drechs.) of corn, 125 000 nematodes per pot (15 cm diameter) of autoclaved soil were required (Rhoades and Linford, 1959). The nematode only invaded roots that had been invaded by *Pythium* and did not invade healthy roots where *Pythium* was absent. *Aphelenchus avenuae* populations of 100 000 per pot (13 cm diameter) of soil provided excellent control of *Rhizoctonia solani* on bean (Barker, 1964), whereas 10 000 nematodes were relatively ineffective. Barnes et al. (1981) controlled *R. solani* and *Fusarium solani* on alfalfa seedlings by infesting pots of soil with 250 000 to 1 million nematodes. While these large populations would not be expected to occur commonly around roots, such artificial tests do suggest a potential role for mycophagous nematodes in the natural reduction of inoculum density of certain pathogens.

Aphelenchoides cibolensis Riffle and *A. composticola* suppressed the development of *Armillaria mellea* (Vahl. ex Fries) Karst. and root-rot disease of *Pinus ponderosa* Laws in flasks of agar–sand cultures (Riffle, 1973). However, the fungus was not eradicated, suggesting that the nematode may be more effective in preventing development of the fungus initially than in preventing invasion of root tissue after the pathogen becomes established at the rhizoplane. In greenhouse pot trials with *Aphelenchoides hamatus* Thorne et Malek, damage to wheat caused by *Fusarium culmorum* (W. G. Smith) Sacc. was significantly reduced (Rossner and Nagel, 1984). Thus, while extensive field tests have not yet been conducted, the potential role of these relatively abundant pathogens seems evident.

3.3. Role of Microarthropods in Plant Pathogen Suppression

Prominent among the Acarina are non-predatory forms that feed primarily as mycophages and saprophages. The ability of some mite species to feed on higher plants probably evolved from early development of mycophagous habits (Krantz and Lindquist, 1979). In managed agricultural systems, the phytophagous economic pest species have received most attention, while the bulk of literature on the fungus-feeding acarinids has concentrated on their role in organic matter decomposition and nutrient cycling in forest and pasture ecosystems. Whereas the possibility that soil mites may transport pathogen propagules and promote plant disease is recognized (Beute and Benson, 1979), little effort has been made to determine the potential of mycophagous species to suppress pathogen activities. Many of the Cryptostigmata, or oribatid mites, prefer the pigmented fungi (Mitchell and Parkinson, 1976), which otherwise are usually highly resistant to stress conditions in field soil. The capacity of mites to alter the inoculum density of dark pigmented pathogens, such as *Rhizoctonia solani, Thielaviopsis basicola* and *Cochliobolus sativus*, and reduce disease potential is worthy of more investigation.

3.3.1. Collembolan Grazing Activities

Selective grazing pressure by collembolans on the microflora of soil/litter systems and roots could be significant, depending upon the animal populations, growth rate of the grazed organisms, and suitability of the fungi or bacteria as food sources; some produce toxic metabolites (Visser, 1985). We have observed in laboratory feeding tests with *Proisotoma minuta* and *Onychiurus encarpatus*, the most prevalent collembolan species in the cotton rhizosphere in Alabama, that fewer than 100 insects per 30 mm-diameter colony of *Rhizoctonia solani* were unable to totally destroy the pathogen (unpublished data); the rapid growth rate of the fungus partially negated the grazing effect of the animals. A population of 200 insects, however, rapidly consumed a young colony leaving only mycelial aggregations or multiple strands. The effect of grazing on respiration of a fungus also may be linked with the nutrient concentration of the growth medium (Hanlon, 1981). Grazing by *Folsomia candida* Willem. on cultures of the fungus *Botrytis cinerea* Pers. ex Fries, grown on vermiculite granules at low nutrient levels derived from Czapek-Dox solutions, resulted in a significant decline in fungal activity. Conversely, grazing on cultures grown at higher nutrient concentrations stimulated fungal respiration.

While collembolans generally are unspecialized feeders, the capacity of some species to destroy or restrict growth of certain fungi and inflict little damage on others is clearly evident. Mixed populations (200 animals) of *P. minuta* and *O. encarpatus* totally prevented growth from mycelial disks of the soil-borne plant pathogens *R. solani, Macrophomina phaseolina* (Tassi) Goid., *Verticillium dahliae*, and *Fusarium oxysporum* f. sp. *vasinfectum* (Atk.) Snyd. et Hans. on Czapeks agar, but at the same time had little effect on the pathogen *Sclerotium rolfsii* Sacc. or the popular biocontrol agent *Trichoderma harzianum* Rifai (Curl et al., 1985a). In further laboratory tests, Lartey et al. (1989) found that the two collembolan species grazed preferentially on *R. solani* in the presence of *T. harzianum* and two other biocontrol agents, *Gliocladium virens* Miller, Giddens et Foster and *Laetisaria arvalis* Burdsall (Table 1). It was discovered that high mortality rates among the insects occurred on cultures of *T. harzianum* and *G. virens* at the time of fungal sporulation, suggesting either toxic effects of fungal metabolites or direct parasitism of insects by the fungi. Cultures of *L. arvalis* were not lethal to the insects, but egg production was reduced, compared to high egg populations of insects grazing on *R. solani*.

3.3.2. Collembolan Relation to Fungal Propagules

Fungal spores and hyphal fragments, particularly of the dematiaceous hyphomycetes, are commonly found in the gut contents of collembolans (Poole, 1959). This suggests the probability that ingested fungal propagules, perhaps still viable, could be deposited in fecal pellets at the root–soil interface, or that the inoculum density of a root pathogen might be reduced and affect disease incidence.

TABLE 1. Attraction of mycophagous insects to colonies of *R. solani* and three biocontrol fungal agents when the insects were applied to a water agar surface simultaneously with the fungi. From Lartey et al. (1989)

	Number of insects attracted to:		
Fungal agent	*Fungal agent*[a]	*R. solani*[a]	χ^{2b}
	Proisotoma minuta		
G. virens	196	333	17.74***
T. harzianum	67	384	222.81***
L. arvalis	82	307	136.16***
	Onychiurus encarpatus		
G. virens	84	360	171.57***
T. harzianum	81	350	167.89***
L. arvalis	150	340	73.76***

[a]Values represent total counts for eight replications; only insects in contact with a food source were included in the analysis.
[b]*** represents $P<0.001$.

The collembolans, *P. minuta* and *O. encarpatus*, from the rhizosphere of cotton, were used in a series of laboratory tests to determine feeding effects on pathogen propagules (Curl et al., 1985a). The insects were attracted to washed chlamydospores of the cotton-wilt pathogen, *Fusarium oxysporum* f. sp. *vasinfectum* on a water–agar surface; spores ingested and redeposited in fecal pellets were reduced in germinability. When chlamydospores in water suspension were placed on Millipore membrane filters and buried in either sterilized or non-sterilized sandy loam artificially populated with collembolans (1600 kg^{-1} soil), the spore population after 10 hours was significantly reduced. Germination of chlamydospores that were ingested and redeposited in fecal pellets also was greatly reduced.

The collembolans were not able to infest or destroy mature macrosclerotia of *Sclerotium rolfsii* or the microsclerotia of *Verticillium dahliae* and *Macrophomina phaseolina*, thus they did not reduce sclerotial populations. However, the germination rate of sclerotia in the presence of the insects was significantly reduced (Fig. 3); the highly efficient grazing of germ tubes by the animals essentially rendered the sclerotia ineffective as viable propagules.

3.3.3. Modification of Fungal Populations in the Rhizosphere

Collembolans can alter the quantitative nature of fungi in the rhizosphere (Curl et al., 1983). A mixed population of *P. minuta* and *O. encarpatus* (500/140 g soil) was applied to a natural sandy loam in large glass tubes with cotton seedlings. Six days later, soil dilution and plating procedures revealed a change in the R/S ratio of total colony-forming units of fungi per gram of soil from

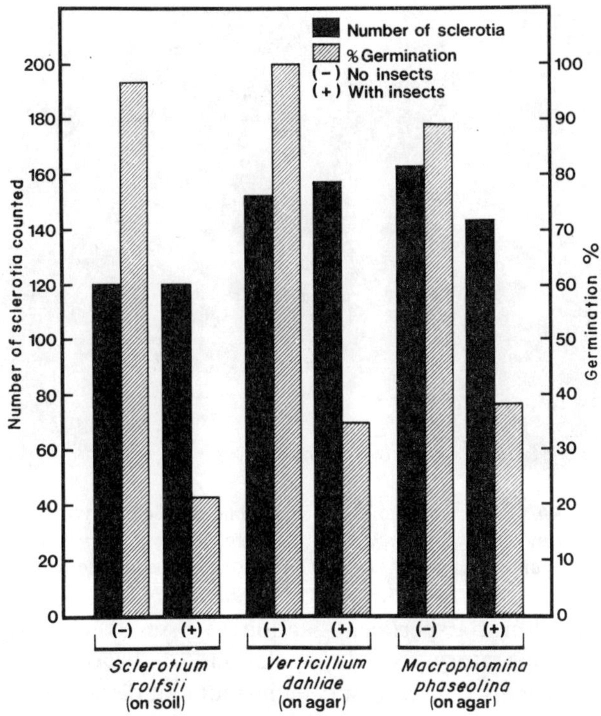

Fig. 3. Effects of collembolans on the sclerotia of three plant pathogenic fungi.

R/S = 2.40 in the absence of insects to R/S = 0.41 in soil with insects. The R_0/R_c ratio (R_0 = no insects, R_c = collembolans present) for rhizosphere soil was 6.94, indicating a significantly lower fungal population in soil with insects present.

Specific tests with *Fusarium oxysporum* f. sp. *vasinfectum* further demonstrated the impact of the collembolan species on inoculum density. Chlamydospores of the pathogen were applied to sterile hypocotyls of pregerminated cotton seed before transplanting to tubes of sterilized soil with or without insect populations. A significant decrease in the viable spore populations of rhizosphere soil had occurred 48 hours after the soil was infested with collembolans. The relation of this activity to the incidence of fusarium wilt in cotton has not been determined.

3.3.4. Biological Control of Root Disease

The capacity of the Collembola to consume relatively large amounts of fungal mycelia and suppress pathogen activities led to further investigations at the

Fig. 4. Control of *Rhizoctonia solani* and cotton seedling disease by mycophagous collembolans (*Proisotoma minuta* and *Onychiurus encarpatus*, mixed populations). (–) = no insects in soil infested with *R. solani* inoculum at 0.1 g kg^{-1}, (+) = 1700 insects kg^{-1} added to pathogen-infested soil.

Alabama Agricultural Experiment Station to assess the biological control capabilities of laboratory reared populations of *Proisotoma minuta* and *Onychiurus encarpatus*. The earliest demonstration of disease suppression (Curl, 1979) was performed with cotton seedlings in large glass tubes (35 mm × 20 cm) of sterilized sandy loam infested with *Rhizoctonia solani* mycelium at 0.05 kg g^{-1} (wet wt) and mixed populations of the two insect species at either 1600 kg^{-1} soil or 4800 kg^{-1}. Surface sterilized cotton seeds were germinated on a nutrient agar and uncontaminated seedlings transplanted to the tubes. Assessment 12 days after planting showed that seedling root disease was significantly reduced by the collembolan populations, the higher number of insects being most effective. Plants in sterilized soil containing only Collembola were stimulated in growth, suggesting that plant-growth-promoting bacteria from the insect gut may have colonized the rhizosphere.

Subsequent greenhouse tests (Curl et al., 1985b) showed that collembolan populations of 1000 to 2000 kg^{-1} soil were about equally effective for disease suppression where the inoculum density of *R. solani* was less than 0.5 g kg^{-1} soil (Fig. 4). Lartey et al. (1986) discovered that disease control was further enhanced when either of the three biocontrol fungi, *T. harzianum, G. virens* or *L. arvalis* was added to soil along with insect populations. These fungal agents are not preferred food for the two collembolan species (Lartey et al., 1989).

4. CONCLUSIONS AND PROJECTED EMPHASIS

The rhizosphere harbors a complex community of interacting microflora and small invertebrate fauna, responding to root exudates, lysates and sloughed root cells, and directly or indirectly affecting the growth and productivity of plants. Protozoa, free-living nematodes, and microarthropods have long been recognized for their role in organic matter decomposition and nutrient cycling, whereas their activities in the rhizosphere relating to root pathogen behavior and survival have received relatively little attention. However, recent investigations have provided strong evidence that certain species of these predominantly mycophagous animals can consume pathogen propagules in sufficient quantity to significantly reduce inoculum density and disease potential, thus effecting a degree of biological control.

This kind of faunal activity suggests the need for a new look at the mechanisms that have been proposed to explain the phenomenon of disease-suppressive versus conducive soils (Schneider, 1982). Where disease-conducive soils, once heavily infested with specific pathogen inocula, gradually become suppressive, the potential contributions of micro- and mesoscopic fauna to this process should be considered along with the popular theory of microbial antagonism (competition and antibiosis). This seems particularly logical in fields subjected to cultural practices (rotations, turning under crop refuse, minimal tillage) which increase the organic matter content of soils, this condition being highly favorable for small fauna activity.

While natural disease suppression by any biotic agent is viewed as highly desirable, it is recognized at the same time that small invertebrates, particularly the microarthropods with bristled bodies and gut contents of ingested microbial components, may transport undesirable organisms from the edaphosphere (non-rhizosphere soil) into the rhizosphere, creating adverse conditions for plant growth. Further, current efforts to genetically transform and manage bacterial and fungal agents for disease control must take into account the small-animal relationship, whether compatible with or adverse to specific biocontrol agents to be established on seeds and roots.

Finally, we know little about the interactions of fungus-feeding microarthropods and other small animals with the ecto- and endomycorrhizal fungi. Few plants in natural ecosystems are without mycorrhizas, and it is now recognized that the rhizodeposition chemistry of roots with the fungal symbionts may create a mycorrhizosphere which determines the composition of the microflora and in turn affects animal species diversity (see Fogel, 1988). Depending upon the food preference of specific animals within an ecological niche, herein lies the potential for mycophagous species to reduce mycorrhizal competition from other fungi or to interfere with establishment of the symbiotic association. The end result could affect not only nutrient cycling but also the incidence of root disease, since the ectomycorrhizal mantle or sheath over young feeder roots is believed to function in a protective capacity (see Schenck, 1981).

A thorough understanding of specific fauna–microflora relationships in the rhizosphere and their consequences to plant health, growth and function could lead to management strategies beneficial to crop production.

REFERENCES

Anderson, J. P. E. and K. H. Domsch (1975). Measurement of bacterial and fungal contributions to respiration of selected agricultural and forest soils. *Can. J. Microbiol.*, **21**:314–322.

Anderson, J. P. E. and K. H. Domsch (1980). Quantities of plant nutrients in the microbial biomass of selected soils. *Soil Science,* **130**:211–216.

Anderson, J. M. and P. Ineson (1984). Interactions between microorganisms and soil invertebrates in nutrient flux pathways of forest ecosystems. In *Invertebrate–Microbial Interactions* (J. M. Anderson, A. D. M. Rayner and D. W. H. Walton, eds.). Cambridge University Press, Cambridge, pp. 59–88.

Anderson, T. R. and Z. A. Patrick (1978). Mycophagous amoeboid organisms from soil that perforate spores of *Thielaviopsis basicola* and *Cochliobolus sativus. Phytopathology*, **68**:1618–1626.

Anderson, J. M., A. D. M. Rayner, and D. W. H. Walton (eds.) (1984). *Invertebrate–Microbial Interactions.* Cambridge University Press, Cambridge, London, New York, 349 pp.

Baker, K. F. and R. J. Cook (1974). *Biological Control of Plant Pathogens.* W. H. Freeman & Co., San Francisco, 433 pp.

Barker, K. R. (1964). On the disease reduction and reproduction of the nematode *Aphelenchus avenae* on isolates of *Rhizoctonia solani. Plant Dis. Rep.*, **48**:428–432.

Barnes, G. L., C. C. Russell, W. D. Foster, and R. W. McNew (1981). *Aphelenchus avenae*, a potential biological control agent for root rot fungi. *Plant Dis.*, **65**:423–424.

Beute, M. K. and D. M. Benson (1979). Relation of small soil fauna to plant disease. *Annu. Rev. Phytopathol.*, **17**:485–502.

Brown, R. A. (1985). Effects of some root-grazing arthropods on the growth of sugar-beet. In *Ecological Interactions in Soil: Plants, Microbes, and Animals* (A. H. Fitter, D. Atkinson, D. J. Read, and M. B. Usher, eds.) Blackwell Scientific Publications, Oxford, pp. 285–295.

Butcher, J. W., R. Snider, and R. J. Snider (1971). Bioecology of edaphic Collembola and Acarina. *Annu. Rev. Entomol.*, **16**:249–288.

Cayrol, J. C. and S. Combettes (1977). Etudes preliminaires sur les possibilities d'utilisation des nematodes mycophages comme agents de lutte biologique contre les "fontes de semis". *Rev. Hortic*, No. 180, p. 33.

Chakraborty, S. (1985). Survival of wheat take-all fungus in suppressive and non-suppressive soils. *Pedobiologia*, **28**:13–18.

Chakraborty, S. and K. M. Old (1985). Mycophagous amoebas from arable, pasture, and forest soils. In *Ecology and Management of Soilborne Plant Pathogens* (C. A. Parker, A. D. Rovira, K. J. Moore, P. T. W. Wong, and J. F. Kollmorgan, eds.). American Phytopathological Society, St. Paul, Minn., pp. 14–16.

Chakraborty, S. and J. H. Warcup (1985). Reduction of take-all by mycophagous amoebas in pot bioassays. In *Ecology and Management of Soilborne Plant Pathogens* (C. A. Parker, A. D. Rovira, K. J. Moore, P. T. W. Wong, and J. F. Kollmorgan, eds.). American Phytopathological Society, St. Paul, Minn., pp. 107–109.

Chakraborty, S., K. M. Old, and J. H. Warcup (1983). Amoebae from a take-all suppressive soil which feed on *Gaeumannomyces graminis tritici* and other soil fungi. *Soil Biol. Biochem.*, **15**:17–24.

Chakraborty, S., C. Theodorou, and G. D. Bowen (1985). The reduction of root colonization by mycorrhizal fungi by mycophagous amoebae. *Can. J. Microbiol.*, **31**:295–297.

Clarholm, M. (1981). Protozoan grazing of bacteria in soil—impact and importance. *Microb. Ecol.*, **7**:343–350.

Coleman, D. C., C. P. P. Reid, and C. V. Cole (1983). Biological strategies of nutrient cycling in soil systems. *Adv. Ecol. Res.*, **13**:1–55.

Coleman, D. C., C. V. Cole, and E. T. Elliott (1984a). Decomposition, organic matter turnover, and nutrient dynamics in agroecosystems. In *Agricultural Ecosystems, Unifying Concepts* (R. Lowrance, B. R. Stinner, and G. J. House, eds.). John Wiley & Sons, New York, pp. 83–104.

Coleman, D. C., R. E. Ingham, J. F. McClellan, and J. A. Trofymow (1984b). Soil nutrient transformations in the rhizosphere via animal–microbial interactions. In *Invertebrate–Microbial Interactions* (J. M. Anderson, A. D. M. Rayner, and D. W. H. Walton, eds.). Cambridge University Press, Cambridge, pp. 35–58.

Coleman, D. C., D. A. Crossley Jr., M. H. Beare, and P. F. Hendrix (1988). Interactions of organisms at root/soil and litter/soil interfaces in terrestrial ecosystems. *Agriculture, Ecosystems and Environment*, **24**:117–134.

Cromack, K. Jr., B. L. Fichter, A. M. Moldenke, J. A. Entry, and E. R. Ingham (1988). Interactions between soil animals and ectomycorrhizal fungal mats. *Agriculture, Ecosystems, and Environment*, **24**:161–168.

Curl, E. A. (1979). Effects of mycophagous Collembola on *Rhizoctonia solani* and cotton seedling disease. In *Soil-borne Plant Pathogens* (B. Schippers and W. Gams, eds.). Academic Press, London, New York, pp. 253–269.

Curl, E. A. (1988). The role of soil microfauna in plant-disease suppression. *CRC Critical Reviews in Plant Sciences*, **7**:175–196.

Curl, E. A. and B. Truelove (1986). *The Rhizosphere*. Springer-Verlag, Berlin, 288 pp.

Curl, E. A., R. T. Gudauskas, J. D. Harper, and C. M. Peterson (1983). Modification of the rhizosphere fungus flora by mycophagous insects. *Abstracts, 3rd International Mycological Congress*, Mycological Society of Japan, Tokyo, p. 425.

Curl, E. A., R. T. Gudauskas, J. D. Harper, and C. M. Peterson (1985a). Effects of soil insects on populations and germination of fungal propagules. In *Ecology and Management of Soilborne Plant Pathogens* (C. A. Parker, A. D. Rovira, K. J. Moore, P. T. W. Wong, and J. F. Kollmorgen, eds.). American Phytopathological Society, St. Paul, Minn. pp. 20–23.

Curl, E. A., J. D. Harper, C. M. Peterson, and R. T. Gudauskas (1985b). Relationships of mycophagous Collembola and *Rhizoctonia solani* populations in biocontrol. *Phytopathology*, **75**:1360.

Curl, E. A., R. Lartey, and C. M. Peterson (1988). Interactions between root pathogens and soil microarthropods. *Agric. Ecosystems and Environment*, **24**:249–261.

Darbyshire, J. F. and M. P. Greaves (1967). Protozoa and bacteria in the rhizosphere of *Sinapsis alba* L., *Trifolium repens* L., and *Lolium perenne* L. *Can. J. Microbiol.*, **13**:1057–1068.

Darbyshire, J. F. and M. P. Greaves (1973). Bacteria and protozoa in the rhizosphere. *Pestic. Sci.*, **4**:349–360.

Dunning, R. A. and A. N. Baker (1977). Some sugar beet cultural practices in relation to incidence and damage by soil-inhabiting pests. *Ann. Appl. Biol.*, **87**:528–532.

Edwards, C. A. and J. R. Lofty (1969). The influence of agricultural practice on soil micro-arthropod populations. In *The Soil Ecosystem* (J. G. Sheals, ed.). Systematics Association Publication No. 8, London, pp. 237–247.

Edwards, C. A., D. E. Reichle, and D. A. Crossley Jr. (1970). The role of invertebrates in turnover of organic matter and nutrients. In *Ecological Studies 1: Analysis of Temperate Forest Ecosystems* (D. E. Reichle, ed.), Springer-Verlag, Berlin, New York, pp. 147–172.

Elliott, E. T. and D. C. Coleman (1977). Soil protozoan dynamics in a shortgrass prairie. *Soil Biol. Biochem.*, **9**:113–118.

Elliott, E. T., D. C. Coleman, and C. V. Cole (1979). The influence of amoebae on the uptake of nitrogen by plants in gnotobiotic soil. In *The Soil–Root Interface* (J. L. Harley and R. Scott Russell, eds.). Academic Press, London, pp. 221–229.

Elliott, E. T., D. C. Coleman, R. E. Ingham, and J. A. Trofymow (1984). Carbon and energy flow through microflora and microfauna in the soil subsystem of terrestrial ecosystems. In *Current Perspectives in Microbial Ecology* (M. J. Klug and C. A. Reddy, eds.), Am. Soc. Microbiol., Washington, D. C., pp. 424–433.

Elliott, E. T., H. W. Hunt, and D. E. Walter (1988). Detrital foodweb interactions in North American grassland ecosystems. *Agriculture, Ecosystems and Environment*, **24**:41–56.

Esser, R. P., W. H. Ridings, and E. K. Sobers (1975). Ingestion of fungus spores by protozoa. *Soil Sci. Soc. Fl. Pro.*, **34**:206–208.

Finlay, R. D. (1985). Interactions between soil microarthropods and endomycorrhizal relationships of higher plants. In *Ecological Interactions in Soil: Plants, Microbes and Animals* (A. H. Fitter,

D. Atkinson, D. J. Read, and M. B. Usher, eds.). Blackwell Scientific Publications, Oxford, pp. 319–329.

Fogel, R. (1988). Interactions among soil biota in coniferous ecosystems. *Agriculture, Ecosystems, and Environment*, **24**:69–85.

Freckman, D. W. (ed.) (1982). *Nematodes in Soil Ecosystems*. University of Texas Press, Austin, 206 pp.

Garrett, S. D. (1970). *Pathogenic Root-Infecting Fungi*. Cambridge University Press, Cambridge, 294 pp.

Geltzer, J. G. (1963). On the behavior of soil amoebae in the rhizosphere of plants. *Pedobiologia*, **2**:249–251.

Getzin, L. W. (1985). Chemical control of the springtail *Onychiurus pseudarmatus* (Collembola: Onychiuridae). *J. Econ. Entomol.*, **78**:1337–1340.

Hanlon, R. D. G. (1981). Influence of grazing by Collembola on the activity of senescent fungal colonies grown on media of different nutrient concentration. *Oikos*, **36**:362–367.

Heal, O. W. (1970). Methods of study of soil protozoa. In *Methods of Study in Soil Ecology* (J. Phillipson, ed.), UNESCO, Paris, pp. 119–126.

Heal, O. W. (1971). Protozoa. In *Methods of Study in Quantitative Soil Ecology: Population Production and Energy Flow* (J. Phillipson, ed.) Blackwell Scientific Publications, Oxford, pp. 51–71.

Heijbroek, W. (1972). De mogelijkheden voor de bestrijding van de belangrijkste voorjaarsplagen. III. De Springstaart *Onychiurus armatus* Tullb. *Med. Inst. voor Rat Suikerproductie*, **1**:1–48.

Hendrix, P. F., R. W. Parmelee, D. A. Crossley Jr., D. C. Coleman, E. P. Odum, and P. M. Groffman (1986). Detritus food webs in conventional and no-tillage agroecosystems. *BioScience*, **36**:374–380.

Hiltner, L. (1904). Über neuere Erfahrungen und Probleme auf dem Gebiet der Bodenbakteriologie und unter besonderer Berücksichtigung der Grundungung und Brache. *Arb. Dtch. Landwirt. Ges.*, **98**:59–78.

Kevan, D. K. McE (1965). The soil fauna—its nature and biology. In *Ecology of Soil-Borne Plant Pathogens* (K. F. Baker and W. C. Snyder, eds.). University of California Press, Berkeley, pp. 33–51.

Kloepper, J. W., R. Lifshitz, and M. N. Schroth (1988). *Pseudomonas* inoculants to benefit plant production. *ISI Atlas of Science: Animal and Plant Sciences/1988*, pp. 60–64.

Krantz, G. W. and E. E. Lindquist (1979). Evolution of phytophagous mites (Acari). *Annu. Rev. Entomol.*, **24**:121–158.

Lartey, R. T., E. A. Curl, and C. M. Peterson (1986). Compared biological control of *Rhizoctonia solani* by fungal agents and mycophagous Collembola. *Phytopathology*, **76**:1104.

Lartey, R. T., E. A. Curl, C. M. Peterson, and J. D. Harper (1989). Mycophagous grazing and food preference of *Proisotoma minuta* (Collembola: Isotomidae) and *Onychiurus encarpatus* (Collembola: Onychiuridae). *Environ. Ent.*, **18**:334–337.

Laybourn-Parry, J. (1984). *A Functional Biology of Free-Living Protozoa*. University of California Press, Berkeley, 218 pp.

Lyda, S. D. (1982). Physical and chemical properties of suppressive soils. In *Suppressive Soils and Plant Disease* (R. W. Schneider, ed.). American Phytopathological Society, St. Paul, Minn, pp. 9–22.

Lynch, J. M. (1982). Interactions between bacteria and plants. In *Bacteria and Plants* (M. E. Rhodes-Roberts and F. A. Skinner, eds.). Academic Press, London, pp. 1–23.

Macfadyen, A. (1957). *Animal Ecology: Aims and Methods*. Pitman, London, 264 pp.

Maggenti, A. (1981). *General Nematology*. Springer-Verlag, Berlin, New York, 306 pp.

Mallow, D., R. J. Snider, and L. S. Robertson (1985). Effects of different management practices on Collembola and Acarina in corn production systems. II. The effects of moldboard plowing and atrazine. *Pedobiologia*, **28**:115–131.

Mankau, R. and S. K. Mankau (1963). The role of mycophagous nematodes in the soil. I. The relationships of *Aphelenchus avenae* to phytopathogenic fungi. In *Soil Fauna, Soil Microflora and Their Relationships* (J. Doeskin and J. Van Der Drift, eds.), North-Holland, Amsterdam, pp. 271–280.

Mitchell, M. J. and D. Parkinson (1976). Fungal feeding of oribatid mites (Acari: Cryptostigmata) in an aspen woodland soil. *Ecology*, **57**:302–312.

Nakas, J. P. and D. A. Klein (1980). Mineralization capacity of bacteria and fungi from the rhizosphere-rhizoplane of a semiarid grassland. *Appl. Environ. Microbiol.*, **39**:113–117.

Napolitano, J. J. (1983). Presence of amoebae in the rhizosphere of beach grass. *J. Protozool.*, **30**:540–541.

Newman, E. I. (1985). The rhizosphere: carbon sources and microbial populations. In *Ecological Interactions in Soil: Plants, Microbes, and Animals* (A. H. Fitter, D. Atkinson, D. J. Read, and M. B. Usher, eds.). Blackwell Scientific Publications Oxford, pp. 107–121.

Newman, E. I., A. J. Heap, and R. A. Lawley (1981). Abundance of mycorrhizas and root-surface micro-organisms of *Plantago lanceolata* in relation to soil and vegetation: a multi-variate approach. *New Phytologist*, **89**:95–108.

Nicholas, W. L. (1984). *The Biology of Free-Living Nematodes*, 2nd edn. Clarendon Press, Oxford, 235 pp.

Norton, D. C. (1978). *Ecology of Plant Parasitic Nematodes*. Wiley, New York, 257 pp.

Old, K. M. and S. Chakraborty (1986). Mycophagous soil amoebae: their biology and significance in the ecology of soil-borne plant pathogens. *Progress in Protistology*, **1**:163–194.

Old, K. M. and J. F. Darbyshire (1978). Soil fungi as food for giant amoebae. *Soil Biol. Biochem.*, **10**:93–100.

Old, K. M. and Z. A. Patrick (1979). Giant soil amoebae, potential biocontrol agents. In *Soil-Borne Plant Pathogens* (B. Schippers and E. Gams, eds.). Academic Press, London, pp. 617–628.

Peterson, H. and M. Luxton (1982). A comparative analysis of soil fauna populations and their role in decomposition processes. *Oikos*, **39**:287–388.

Poole, T. B. (1959). Studies on the food of Collembola in a Douglas fir plantation. *Proc. Zool. Soc. Lond.*, **132**:71–82.

Rabatin, S. C. and B. R. Stinner (1988). Indirect effects of interactions between VAM fungi and soil-inhabiting invertebrates on plant processes. *Agriculture, Ecosystems and Environment*, **24**:135–146.

Reichle, D. E. (1977). The role of soil invertebrates in nutrient cycling. In *Soil Organisms as Components of Ecosystems* (U. Lohm and T. Peterson, eds.). *Ecol. Bull. Stockholm*, **25**:145–156.

Rhoades, H. L. and M. B. Linford (1959). Control of *Pythium* root rot by the nematode *Aphelenchus avenae. Plant Dis. Rep.*, **43**:323–328.

Riffle, J. W. (1973). Effect of two mycophagous nematodes on *Armillaria mellea* root rot of *Pinus ponderosa* seedlings. *Plant Dis. Rep.*, **57**:355–357.

Rossner, J. and K. Urland (1983). Mycophage nematoden der Gattung *Aphelenchoides* an der Halmbasis von Getreidepflanzen und ihre Wirkung Gegen Fusskrankheitserreger von Getreide. *Nematologica*, **29**:454–462.

Rossner, J. and S. Nagel (1984). Untersuchungen zur Okologie und vermehrung des mycohagen nematoden *Aphelenchoides hamatus. Nematologica*, **30**:90–98.

Rouatt, J. W., M. Lechevalier, and S. A. Waksman (1951). Distribution of antagonistic properties among actinomycetes isolated from different soils. *Antibiot. Chemother.*, **1**:185–192.

Rouatt, J. W., H. Katznelson, and T. M. B. Payne (1960). Statistical evaluation of the rhizosphere effect. *Soil Sci. Soc. Am. Proc.*, **24**:271–273.

Schenck, N. A. (1981). Can mycorrhizae control root disease? *Plant Dis.*, **65**:230–234.

Schneider, R. W. (ed.) (1982). *Suppressive Soils and Plant Disease*. American Phytopathological Society, St. Paul, Minn., 88 pp.

Schnurer, J., M. Clarholm, and T. Rosswall (1982). Microorganisms. In *Ecology of Arable Land—The Role of Organisms in Nitrogen Cycling* (T. Rosswall, ed.), Progress Report 1981, Swedish University of Agricultural Sciences, Uppsala, pp. 77–88.

Seastedt, T. R. (1984). The role of microarthropods in decomposition and mineralization processes. *Annu. Rev. Entomol.*, **29**:25–46.

Shaw, P. J. A. (1985). Grazing preferences of *Onychiurus armatus* (Insecta: Collembola) for mycorrhizal and saprophytic fungi of pine plantations. In *Ecological Interactions in Soil: Plants, Microbes, and Animals* (A. H. Fitter, D. Atkinson, D. J. Read, and M. B. Usher, eds.). Blackwell Scientific Publications, Oxford, pp. 333–336.

Shields, J. A., E. A. Paul, and W. E. Lowe (1973). Turnover of microbial tissue in soil under field conditions. *Soil Biol. Biochem.*, **5**:753–764.

Stinner, B. R. and D. A. Crossley Jr. (1977). The influence of invertebrates on root growth of crops with minimal and zero cultivation. In *Soil Organisms as Components of Ecosystems*. (U. Lohm and T. Peterson, eds.). Swedish Natural Science Research Council, Stockholm, Ecology Bull. No. 25, pp. 348–356.

Stinner, B. R. and D. A. Crossley Jr (1982). Nematodes in no-tillage agroecosystems. In *Nematodes in Soil Ecosystems* (D. W. Freckman, ed.). University of Texas Press, Austin, pp. 14–28.

Stout, J. D. (1980). The role of protozoa in nutrient cycling and energy flow. *Adv. Microbiol. Ecol.*, **4**:1–50.

Swift, M. J., O. W. Heal, and J. M. Anderson (1979). *Decomposition in Terrestrial Ecosystems*. Blackwell Scientific Publications, Oxford, 372 pp.

Townshend, J. L. (1964). Fungus hosts of *Aphelenchus avenae* Bastian, 1865, and *Bursaphelenchus fungivorus* Franklin and Hooper, 1962, and their attractiveness to these nematode species. *Can. J. Microbiol.*, **10**:727–737.

Ulber, Von B. (1980). Untersuchungen zur Nahrungswahl von *Onychiurus fimatus* Gisin (Onychiuridae, Collembola) einer Aufgangsschadling der Zucherrube. *Z. Ange. Entomol.*, **90**:333–346.

Vancura, V. and F. Kunc (1977). The effect of streptomycin and actidione on respiration in the rhizosphere and non-rhizosphere soil. *Zentralblatt für Bakteriologie II*, **132**:472–478.

Visser, S. (1985). Role of the soil invertebrates in determining the composition of soil microbial communities. In *Ecological Interactions in Soil: Plants, Microbes, and Animals* (A. H. Fitter, D. Atkinson, D. J. Read, and M. B. Usher, eds.). Blackwell Scientific Publications, Oxford, pp. 297–317.

Walker, G. E. (1984). Ecology of the mycophagous nematode *Aphelenchus avenae* in wheat-field and pine-forest soils. *Plant Soil*, **78**:417–428.

Warnock, A. J., A. H. Fitter, and M. B. Usher (1982). The influence of a springtail (*Folsomia candida*) on the mycorrhizal association of leek (*Allium porrum*) and the vesicular-arbuscular mycorrhizal endophyte, *Glomus fasciculatus. New Phytol.*, **90**:285–292.

Wasilewska, L. and P. Bienkowski (1985). Experimental study on the occurrence and activity of soil nematodes in decomposition of plant material. *Pedobiologia*, **28**:41–57.

Whipps, J. M. and J. M. Lynch (1986). The influence of the rhizosphere on crop productivity. *Adv. Microbiol. Ecol.*, **9**:187–244.

Wiggins, E. A. and E. A. Curl (1979). Interactions of Collembola and microflora in cotton rhizosphere. *Phytopathology*, **69**:244–249.

Wiggins, E. A., E. A. Curl, and J. D. Harper (1979). Effects of soil fertility and cotton rhizosphere on populations of Collembola. *Pedobiologia*, **19**:75–82.

15

THE IMPACT OF CROPPING SYSTEMS ON RHIZOSPHERE ORGANISMS AFFECTING PLANT HEALTH

A. D. ROVIRA

CSIRO Division of Soils, Private Bag No. 2, Glen Osmond, South Australia 5064

L. F. ELLIOTT

USDA-ARS Cotton Research Station, 17053 Shafter Avenue, Shafter, CA 93263, USA

R. J. COOK

USDA-ARS, Johnson Hall, Washington State Univeristy, Pullman, WA, 99164-6421, USA

1. INTRODUCTION

The term "cropping system" is used in this paper to mean the hierarchy of primary and secondary factors of practices that make up the total system used to grow crops in a given field. The factors of cropping systems thought to be most relevant to our subject and discussed in this paper are: crop rotation (including cultivar rotation); method or intensity of tillage; trash retention; planting date; row spacing (or seeding rate); soil water; liming; herbicide use; and fertilizer practices.

Our collective interests in rhizosphere colonization as influenced by cropping systems are mainly in relation to root health and crop yield. Each specific cropping system presents unique ecological niches available for occupancy by microorganisms, some beneficial and some detrimental to plant growth. Since no two farms and probably no two fields are managed in exactly the same way, farms and fields present a range of habitats and ecological niches. Organisms best adapted to those habitats and niches developed as a result of the farming system will gradually replace those not so well adapted. The appearance of a root disease in one field but not in a neighbouring field reflects differences in cropping systems imposed over one or more years (Cook, 1984).

The cropping system offers a means to manage the populations of both root pathogens (including the so-called "minor" pathogens, also referred to as deleterious microorganisms) and indigenous populations of beneficial rhizosphere colonists antagonistic to root pathogens (Schippers et al., 1987). The cropping system is also relevant to any programme aimed at biological control of root pathogens using introduced rhizosphere colonists, whether as a vehicle for the introduction or establishment of an introduced rhizosphere colonist, or as a means to influence the rhizosphere environment and the competition from indigenous rhizosphere colonists. Aspects of the influence of cropping systems on rhizosphere colonization by organisms affecting plant health and crop yield are presented in this review.

It is important when considering the impact of cropping systems on rhizosphere organisms to remember that the primary inoculum for these rhizosphere colonists originates from cells or colonies in the soil with which the root makes contact, or in the case of fungi the propagules close enough to roots to respond to water-soluble and volatile substances released by the root. Most of these microbial cells and propagules are associated with particulate organic matter that originates from previous crops or pastures. In cultivated soil, this organic matter is initially particulate and mixed through the soil, and, as the organic matter decomposes, colonies of bacteria or fungi develop in microniches in and between soil aggregates (Hattori, 1973; Foster et al., 1983; Foster, 1988). In the no-till system, organic matter from the top residues accumulates at the surface so that, apart from that which is buried or incorporated by the action of soil animals, or the readily decomposable carbon which leaches to a maximum depth of 4 cm (Cochran et al., 1980), the bulk of the particulate organic matter in the soil originates from roots of previous crops. Although more organic matter is incorporated into cultivated soil than in no-till systems, the flush in decomposition which occurs with each cultivation (Rovira and Greacen, 1957) results in a lower organic matter content in cropping systems using cultivation.

These are important factors when we consider the biology of the rhizosphere of plants in the "real world" of agriculture, which, by definition, must involve a range of cropping, fertilization and cultivation strategies and often the use of chemicals to control pests, weeds and diseases.

Studies of root diseases caused by soil-borne pathogens and of beneficial, detrimental and symbiotic associations are, in reality, examples of applied rhizosphere biology, because before an organism becomes pathogenic, symbiotic or mycorrhizal it must first pass through or proliferate in the rhizosphere, and hence these organisms will be influenced by roots directly via exudates, lysates and root cell debris and indirectly via the rhizosphere microflora.

Crop rotations and tillage systems are used worldwide to maintain or increase crop and pasture production. In view of this, our examples of the effects of cropping systems on the rhizosphere are taken mainly from rotations involving pasture–cereals, continuous cereals, cereal–fallow and grain legume–cereals as used in southern Australia and the Inland Pacific Northwest (IPNW) (parts of Idaho, Oregon and Washington) of the USA. Average winter temperatures are lower in the IPNW than in southern Australia, but both areas are characterized by a Mediterranean-type climate with hot dry summers and most of the effective precipitation in autumn, winter and spring. Both areas are also characterized by a growing interest in change, from intensive wheat (short rotations or monoculture) grown with intensive tillage (conventional tillage) to more diversified, and sometimes longer rotations with less or no tillage (conservation tillage). Conservation tillage cropping systems are defined as any ''tillage sequences which reduce loss of soil and water relative to conventional tillage'' (Papendick and Elliott, 1983). Definitions and descriptions of the various forms of conservation tillage in the USA are covered by Sprague and Triplett (1986) and in Australia by Cornish and Pratley (1987).

Selected studies from other environments which demonstrate the effect of cropping systems on rhizosphere organisms affecting plant health will also be discussed in this paper.

2. EFFECT OF SOIL FUMIGATION

Soil fumigation studies have demonstrated that the controlling influence of the rotation effect involves the mix of detrimental and beneficial microorganisms in the rhizosphere. The studies in southern Australia with methyl bromide or chloropicrin and selective fungicides and nematicides demonstrated that root pathogenic fungi and nematodes imposed constraints upon crop production (Rovira, 1976; Meagher et al., 1978; Rovira and Ridge, 1978; Rovira and Simon, 1985). In some of these trials the release of available N as NH_4^+-N and the death of nitrifying bacteria could not be ruled out as cause for part of the crop response, but the inclusion of fertilizer-N treatments demonstrated that the major benefit of soil fumigation was due to destruction of root pathogens.

Following the lead of Wilhelm (1965) in California for vegetable crops and strawberries, and the Australian work on wheat, soil fumigation was used in the IPNW to (a) reveal the significance of ''soil microbiological limitations'' to wheat production in different rotations, and (b) provide a means through selective treatment of soil to narrow the search for causal agents responsible for any root damage in these cropping systems. Some 50 experiments were carried out over a 10-year period throughout north-eastern Oregon, eastern

Washington, and northern Idaho (Cook and Haglund, 1982; Cook et al., 1987; Cook, unpublished).

The yield responses for areas with 400–450 mm or more of available water annually averaged 30–50% in fields cropped every year to wheat, 20–25% in fields cropped to wheat every second year and 5–10% in fields cropped every third year to wheat (3-year rotations) (Cook and Haglund, 1982; Cook et al., 1987). In drier areas the wheat also exhibited the typical increased growth response to fumigation, but yields were no greater because water, and not root health, was the yield-limiting factor. The fumigation response was as great for wheat after barley as for wheat after wheat, which was not surprising if root pathogens are involved—barley and wheat serve as hosts for the same root pathogens. On the other hand, the rotation effect was the same whether potato, lentils, alfalfa or peas was used as the break crop. Evidence that the increased growth response from fumigation did not result from nitrogen released from the microbial biomass killed by the fumigants was obtained experimentally by soil treatment with 1,3-dichloropropane (Telone) and the same fumigant amended with 15–17% chloropirin (Telone C). Both fumigants at 240 l ha^{-1} (product) resulted in equivalent releases of nitrogen as ammonium and both delayed nitrification, but only Telone C, which destroyed root pathogens, increased growth and yield of wheat.

3. INFLUENCE OF CROP ROTATIONS

3.1. Root Pathogens

In southern Australia the major agricultural enterprises are based upon a cereal–sheep system. Cereals are generally grown in rotation with annual self-sown pastures made up of grasses and either subterranean clover (*Trifolium subterraneum* L.) in neutral to acid soils or annual medics (*Medicago* spp.) in neutral to alkaline soils. The major root pathogens in this farming system in southern Australia are *Gaeumannomyces graminis* var. *tritici*, which causes take-all, *Rhizoctonia solani*, which causes rhizoctonia root rot (bare or purple patch), and *Heterodera avenae*, which is the cereal cyst nematode/eelworm. The root pathogens of lesser importance are common root rot (*Bipolaris sorokiniana*), crown rot (*Fusarium graminearum*) and pythium root rot (*Pythium* sp.).

In the IPNW the major influences on the colonization of the wheat rhizosphere appear to be crop rotation, i.e. whether wheat or barley were last grown in the field (no break crop), 2 years previously (a single break crop or 2-year rotation) or 3 years previously or longer (3+ year rotation), and the method of crop residue management. It has been documented for many crops, but especially for wheat and potatoes, that productivity of the crop on any given field is inversely related to the frequency with which that crop was grown in fields within the range of a 3- to 6-year rotation. Not only diseasc appears to be responsible for the declining yields. Replant problems associated with crop monoculture exist and are accepted including: autotoxicity, allelopathy and depletion of soil nutrients. The aetiology of these problems will not be resolved simply by documenting symptoms, such as yellow leaves, stunting, reduced branching or tillering, or nutrient-deficiency symptoms. These are the symptoms of poor root health but do not give a clue as to the cause(s). An accurate diagnosis also requires more than simply completing Koch's postulates under controlled conditions, since many organisms isolated from roots or the rhizosphere might cause injury under specific laboratory conditions but have little to do with problems in the field.

The sugar cane industry in northern Australia is based on long-term monoculture in which many fields have grown nothing but sugar cane for over 100 years. It is of concern to this industry that yields are declining, despite introduction of new varieties, new insecticides and better fertilizer practices. There is little available suitable new land for expanding the industry or changing from monoculture, but where experimental plantings are made into virgin rainforest or grassland soils, yields are higher than on the commercial cane fields. This led to an investigation of the reasons for "New Land–Old Land" differences and also of a problem termed PRS (poor root syndrome) in which root systems are discoloured and show considerable rotting; it was thought that PRS and "New Land–Old Land" differences could be related. By using soil fumigation (Croft et al., 1984) and solarization (Reghenzani, 1988) it was shown that the yield decline was due to biological causes and not to chemical or physical changes in the soil. This research established that the build-up of *Pythium graminicola* (Croft and Magarey, 1984) and a previous undescribed oomycete (*Pachymetra chaunorhiza*), found only in old sugar cane soils of Queensland (Dick et al., 1989), is responsible for poor root systems and yield decline. Magarey (personal communication) has shown that this new root pathogen is favoured by certain cultivars of sugar cane and after four years the numbers of *Pachymetra* spores/g under these varieties were 4 times greater than under resistant varieties. This problem provides a classic example of how monoculture can build up organisms which are specific for that crop which can be accentuated by different cultivars.

3.1.1. Pythium Root Rot

Differential soil fumigation in the IPNW has helped unravel some of the causes of poor wheat growth. A partial microbiological analysis of the soils fumigated with either Telone or Telone C revealed:

1. No effect on the *Fusarium* population of soil, thereby ruling out pathogens of this genus as responsible for the injury eliminated by soil fumigation.
2. Elimination of ectoparasitic nematodes (mainly *Pratylenchus* species) by both fumigants, thereby ruling out control of these root parasites as factors in the response.

3. Elimination of *Pythium* species by Telone C but not Telone, thereby leading to further study of this ubiquitous fungus genus well known to cause root disease on many crops (Wilhelm, 1965; Yarwood, 1966).

Dilution-plate counts of the *Pythium* populations revealed that most fields in the study area contained 300–400 propagules per gram of soil in the top 10 cm. No field contained less than 100 propagules g^{-1}, and some contained more than 1000 propagules g^{-1} soil. Correlation coefficients of -0.79 to -0.94 were obtained between the *Pythium* population in plots in the autumn at the time of sowing and heights of adult plants in the same plots 9 months later (Cook et al., 1987). Heat treatment of the soil also reduced or effectively eliminated the *Pythium* population and produced the typical increased growth response. This was accomplished in the field equally well by either solarization beneath clear plastic (Katan, 1980) or by burning a thick layer of wheat straw (20 t ha^{-1}) on the surface of moist soil. In greenhouse studies, the *Pythium* population was reduced and wheat seedling growth coordinately increased by moist heat treatment of the soil at only 42 °C for 10 minutes, the threshold heat treatment.

Subsequent studies revealed that the *Pythium* species began colonizing wheat seed in the first 24 hours after planting and established in the embryo region before the seed even germinated (Hering et al., 1987). Commonly 50–60% and often up to 80–90% of the wheat seedlings emerge in these soils with *Pythium* species already established in the scutellum and region of root shoot transition following colonization of the embryo (Fukui, 1988; Cook, unpublished). Although non-lethal, these infections are not harmless and cause a stunting effect on the seedlings.

Treatment of seed or soil with metalaxyl, a fungicide specific for control of *Pythium* and closely related oomycetous fungi, protected against embryo colonization (and to some extent also against root colonization) and increased growth in both field and greenhouse experiments. These results provide further evidence for the importance of *Pythium* root rot of wheat in the IPNW and corroborate data from soil fumigation studies pointing to the importance of this group of root pathogens.

Of several *Pythium* species isolated from wheat roots or wheat-field soils (Chamswarng, 1984), two which stand out as both the most common and the most pathogenic to wheat are *P. ultimum* and *P. irregulare* (Chamswarng and Cook, 1985; Ingram, Cook and Chamswarng, 1990). At any given inoculum density, *P. ultimum* is more efficient than *P. irregulare* as an embryo colonist, whereas *P. irregulare* may be more important as a root colonist (Ingram and Cook, submitted).

Since wheat under dryland conditions in this region may be grown in rotation with barley, peas or lentils, and since the magnitude of the growth response to fumigation was inversely related to the frequency of wheat in the cropping system, it became important to know the relative susceptibility of each crop to each *Pythium* species. At 5 °C, the soil temperature during establishment of fall-sown wheat or spring-sown barley, peas or lentils, *P. irregulare* was the most pathogenic to barley while *P. ultimum* was the most pathogenic to pea and wheat (Ingram, Cook and Chamswarng, 1990). Bratoloveanu (1985) also found for soils and cropping systems in South Australia that *P. irregulare* was the most pathogenic species on barley. Thus, it would appear that the common wheat/barley/pea (or lentil) rotation in the IPNW would not control and may even favour *Pythium* root rot of wheat. On the other hand, while the 5–10% yield increase in response to fumigation in fields cropped in these 3-year rotations is best explained by control of *Pythium* damage, the 20–25% yield increases in fields cropped in 2-year rotations and 30–50% yield increases in continuous wheat, must involve control of pathogenic rhizosphere and/or root colonists other than *Pythium* species.

In southern Australia, Pankhurst and McDonald (1988) reported that rotation and season influence the levels of spores of *Pythium* spp. in soil. At time of sowing wheat higher numbers of *Pythium* occurred following annual pasture than wheat or lupins. These higher inoculum densities are thought to be linked with higher amounts of plant residues returned to the soil with pasture compared with wheat or lupins. Pankhurst (personal communication) has also demonstrated that infection of the wheat seed embryo occurs within 48 hours of sowing in soil following pasture where inoculum densities are greatest. The major species at a site with 500 mm rainfall p.a. were *P. ultimum* and *P. graminearum*. The effects of these *Pythium* populations on the growth and yield of wheat are still to be assessed.

3.1.2. Take-all

Take-all is a major root disease of wheat and barley worldwide (Garrett, 1942), including southern Australia (Butler, 1961; Walker, 1975; Rovira et al., 1990) and in the IPNW (Sprague, 1950). In Australia, the disease is important in all regions below 32°S with dominant winter rainfall and hot dry summers and the cereal–sheep farming system described earlier in this chapter.

Rotations have a large effect on take-all because the host range of *Gaeumannomyces graminis* var. *tritici* is confined to grasses, wheat, barley and triticale which maintain or increase the numbers of propagules through the soil. The importance of grasses in pasture in maintaining or increasing inoculum of *G. graminis* var. *tritici* in soil has been known for decades, but it was quantified when MacNish and Nicholas (1987) showed that as the proportion of grass in pasture increased the levels of take-all in the following wheat crop increased.

The concepts developed by Gilligan (1979, 1980a, 1980b) and Gilligan and Simons (1987) in the models of infection of wheat roots by *G. graminis* have relevance to the effects of cropping strategies on take-all (Chapter 8). In these models of the "pathozone" (the furthest distance from the host of inoculum from which infection occurred), account is taken of propagule number, proportion of propagules capable of causing infection when placed on roots, the decline in infection efficiency of propagules with increasing distance from the host roots and the energy reserves of the propagule. All these factors are affected by both rotation and cultivation. Rotation influences the carry-over from infected roots and crowns of grasses and propagule size—the longer the periods between wheat crops, the smaller the size of particles and debris containing *G. graminis* var. *tritici* (Wilkinson et al., 1985a). Cultivation fragments the propagules and reduces their infectivity (Moore and Cook, 1984; Wilkinson et al., 1985b) and mixes them through soil resulting in shorter distances between propagules and roots of the following wheat crop (Wilkinson et al., 1985b).

Roget and Rovira (1990) have demonstrated that, in a particular cropping system and environment, soil moisture conditions favourable to the growth of the take-all fungus on and inside roots influence the amount of the inoculum carried over from a grass-medic pasture to the following wheat crop, and that the effect of this inoculum on yield is greater when the September rainfall is high.

In the IPNW direct inspection of washed roots recovered from wheat fields has confirmed that infections by *G. graminis* are common even in the traditional dryland areas (250–450 mm) where typical take-all (blackened stem bases with white heads) does not occur. Obviously the brief "window" of moist soil in these dry areas is sufficient for maintenance of this rhizosphere colonist and root pathogen. Economically damaging take-all is limited to areas with 400–450 mm of precipitation annually and to the irrigated areas (Cook, 1986). These are also the areas with available water sufficient to permit annual cropping, i.e. no fallow.

Typical of most wheat-growing areas of the world where soil conditions are suitable for take-all (Cook, 1981b), the disease in the IPNW is most severe with consecutive crops of wheat or wheat after barley, and is mild or non-existent where wheat follows non-host crops in which grass weeds have been controlled. However, while a single year's break to a non-host crop before returning to wheat provides significant control of take-all, some low but significant incidence of disease occurs under irrigation and in the high-rainfall areas (500–600 mm) of eastern Washington and northern Idaho in fields cropped every other year to wheat (Cook, unpublished). This is shown by yield data for irrigated spring wheat at Lind, Washington, in an experiment carried out on a site cropped 12 consecutive years to winter wheat prior to a 4-year rotation study. Yield of spring wheat was 4.2 t ha^{-1} in the 16th year when spring wheat was grown in years 13, 14, 15 and 16, while yield was 6.1 t ha^{-1} in the 16th year where potatoes had been grown in years 14 and 15; a 1-year break to potatoes (in year 15) did not control take-all completely and spring wheat yielded 5.0 t ha^{-1}.

However, infections by the take-all fungus are common (5–10% of young plants with one or more root lesions) in wheat fields following only a 1-year break to lentil or pea in the dryland eastern Washington and northern Idaho. The significance of this low but real incidence of infection is not known, but it must be taken into account when attempting to explain the additional yield response of wheat to soil fumigation in this area in fields in a 2-year rotation compared with those in a 3-year rotation, especially considering the efficient means of secondary spread of this pathogen from root to root and plant to plant during the course of a growing season. Take-all in this dryland area is better described as chronic rather than acute (Cook, 1984), being associated with slight stunting and yellow leaves but rarely white heads and rarely in patches. In Australia, this subclinical form of take-all causes losses estimated at $200–300 million p.a. (Rovira et al., 1990) in most seasons. The appearance of white heads requires a high incidence of infected plants (>50%) and high spring rainfall, and in these conditions yield losses of 60–75% occur (Roget and Rovira, 1990).

There is now evidence in Australia of suppression to take-all developing with continuous wheat. Rovira and Wildermuth (1981) reported that a field in South Australia with 13 successive wheat crops was suppressive when *G. graminis* inoculum was added to the soil, but this suppressiveness was lost following 2 years of severe drought when wheat was not grown, indicating the need for a wheat rhizosphere to maintain the suppressive microorganisms. A similar situation occurred in a soil from Victoria which had been cropped continuously for 70 years to wheat but lost its suppressiveness after a severe drought.

MacNish (1988) found that the incidence of take-all in sandy soils declined with continuous wheat; this decline was hastened by the application of ammonium fertilizers but was lost when lime was added to the soil. Simon et al. (1987, 1988a) established that biological suppression of the saprophytic growth of *G. graminis* was involved in the decline, and Simon and Sivasithamparam (1988a) went on to demonstrate that *Trichoderma koningii* was involved in the suppression. The complexity of the mechanisms involved in suppression of take-all was demonstrated by the interactions which Simon and Sivasithamparam (1988b) found between *T. koningii*, soil bacteria and *G. graminis*.

The long-term rotation × tillage trial conducted at Avon, South Australia, demonstrates the need for field trials to continue over several years in order to understand the mechanisms responsible for changes in amounts of root disease. Roget and Rovira (1990) related the incidence of take-all in relation to rainfall in wheat in two rotations, namely pasture–wheat and continuous wheat, and found a high incidence of take-all for the first 2 years, followed by a decline. This decline occurred in both pasture–wheat and continuous wheat and is attributed to rainfall rather than microbially induced suppression. Following a year of average rainfall, the incidence of take-all was 70–80% in wheat following pasture with conventional cultivation and direct drilling, low (25%) in the continuous wheat with cultivation, but high (60%) in continuous wheat sown by direct drilling (Rovira and Roget, unpublished). The speculative interpretation could be that the mixing of soil with cultivation distributes the biocontrol agents through the soil and improves the chances of these organisms colonizing wheat roots.

3.1.3. Rhizoctonia Root Rot

Rhizoctonia "bare patch" or "purple patch" of cereals is caused by *Rhizoctonia solani*, which has a wide host range and so the disease affects many plant species. As a consequence of this wide host range and a high competitive saprophytic ability in colonizing particulate soil organic matter (Neate, 1987) one would not expect rotations to exert the same effects upon this disease as they do on take-all. This pathogen has been known to cause yield losses in cereals in limited areas of southern Australia for many years (Samuel and Garrett, 1932), but since the introduction of direct drilling or no-till farming it has caused widespread yield losses.

While rotation had little effect on root damage caused by *Rhizoctonia* in direct drilled wheat, the area of crop lost through "bare patches" was greater after grass-medic pasture than in wheat after grass-free medic pasture or peas (Rovira, 1986). This ameliorating effect of medic pasture and peas could be due to less inoculum or higher levels of mineral-N producing more vigorous wheat seedlings that are better able to combat the root pruning effects of the pathogen. Also, Neate (1987) showed that residues of grasses or cereals carry more *Rhizoctonia* than residues from annual *Medicago* spp.

Rhizoctonia root rot was first reported in the IPNW on wheat and barley by Weller et al. (1986) and occurred in patches similar to those described in south Australia by Samuel and Garrett (1932). Moreover, *R. solani* AG8 (Neate and Warcup, 1985), the same anastomosis group (AG) of *R. solani* responsible for bare patch of wheat in Australia, was the most common isolate of *R. solani* from the diseased wheat and barley plants in the IPNW and caused typical symptoms of the disease on both wheat and barley when tested at 10 °C (Ogoshi et al., 1990). *R. oryzae*, the second most frequently isolated *Rhizoctonia* species, caused symptoms of the disease at 20 °C but was only mildly pathogenic or non-pathogenic at 10 °C (Ogoshi, Cook and Bassett, submitted). The results suggest that both species play a role in the disease, depending on the soil temperature.

After learning to recognize the disease by the typically reddish brown to brown, pruned-off ends of both the laterals and main root axes, this disease is now widely recognized in its more chronic form (pruned roots but no patches of stunted or dead plants) on the majority of wheat plants examined from random fields of wheat or barley in the IPNW (Cook, unpublished). Thus, some credit for the increased growth response to soil fumigation in the respective rotations of this region might also belong to control of chronic rhizoctonia root rot.

Root ratings of wheat plants from adjoining fields cropped the previous year to spring wheat and lentils, respectively, revealed more severe rhizoctonia root rot in sites where wheat was grown the previous year (no rotation) than where peas or lentils were grown the previous year. Glasshouse tests revealed that isolates of *R. solani* AG8 from either wheat or barley are highly pathogenic to peas and lentils as well as to wheat and barley. This observation matches the experience in southern Australia, where rhizoctonia root rot is less severe after the medic or peas than after the wheat or grass-medic, in spite of the susceptibility of the medic and peas to the pathogen. The benefits of a legume preceding wheat could relate to the much smaller amount of carry-over inoculum from plants with a tap root system and relatively few laterals in the top 10–15 cm of soil (where this pathogen is active) compared to plants with a fibrous root system.

In Texas, USA, the incidence of rhizoctonia root rot of sugar beet was 65% following wheat and sorghum, 35% after alfalfa (lucerne), and 15% after cotton, sunflower and fallow. These results similarly show that, despite the broad host range of *R. solani*, root rot caused by this pathogen is subject to amelioration by crop rotation. The possibility exists that the strain of *R. solani* responsible for root rot of sugar beet (AG2.2) does not have the wide host range of the *R. solani* (AG8) responsible for root rot of wheat in Australia and the IPNW. Rush and Winter (1988) hypothesized that the amount of residue returned or available in the soil after cotton, sunflower and fallow was less than after wheat, sorghum and alfalfa, resulting in less substrate for *R. solani* in its saprophytic phase in soil. These differences could also be explained by less root mass available for colonization by *R. solani* and therefore less carry-over of inoculum. Much emphasis has been given in the past to the role of saprophytism in the epidemiology of rhizoctonia, but possibly the roots of living plants have a more important role in sustaining this pathogen than dead plant material.

3.1.4. Common Root Rot

Common root rot of cereals is caused by *Bipolaris sorokiniana* (Sacc.) Shoen. It causes yield losses of wheat in Canada (Tinline et al., 1975), Australia (Wildermuth, 1990) and Brazil (Diehl et al., 1982), and is influenced by crop rotation. Wildermuth (1990) found that after 2 years of wheat the number of spores in soil had increased by 98% and with barley by 206%, while following safflower, chick peas, snail medic and cocksfoot the numbers decreased by 36–46%. These results are typical of those expected from rotation with a non-host crop, as the conidia of this pathogen in soil have a finite life expectancy, owing to many kinds of physical and biological stresses. The numbers measured at any one time are the net change resulting from a rate of decline and the rate of increase on a host whilst in the absence of a host there is only decline.

3.1.5. Cereal Cyst Nematode

Cereal cyst nematode (CCN) (*Heterodera avenae*) is an obligate parasite requiring roots of hosts for multiplication. Australian research has shown that the rate of build-up of this pathogen is affected by crop rotation (Meagher and Rooney, 1966; Meagher and Brown, 1974; Rovira and Simon, 1982). Roots of susceptible cereals will increase egg levels by 25–100% in one season—the magnitude of the increase is greater in sandy soils than clay loams. Rotation with non-host plants, e.g. pasture or grain legumes, reduces the egg levels by 60–80% in a single season.

In England, Kerry (1981) found that after several successive wheat or barley crops the populations of CCN declined due to the build-up of the fungus *Nematophthora gynophila*, which is an obligate parasite on cyst nematodes. This fungus has a motile zoospore stage which enables it to move through wet soil to its nematode host and colonize the developing female inside the root.

Although CCN is a major problem of wheat in south-eastern Australia, *Nematophthora* has never been isolated, even from fields where wheat had been grown for several successive seasons. However, even if this fungus existed in Australia it is unlikely that it would be an effective biocontrol agent because the soil is seldom wet enough for the zoospores to move. In addition, biological control of CCN by an obligate parasite such as *N. gynophila* would be difficult to utilize in Australia because of the relatively low population of nematodes required for economic damage, namely 2 eggs g^{-1} soil in Australia (Meagher and Brown, 1974) compared with a threshold of 10 eggs g^{-1} soil in England (Kerry, unpublished).

The use of a fungus such as *Verticillium chlamydosporium*, which grows on roots and in the rhizosphere and maintains its population in the absence of the nematode, offers advantages over an obligate fungus such as *Nematophthora*, which depends upon high levels of females of CCN on roots for growth and maintenance of its population. Thus, the filamentous fungus is a more likely candidate for biological control in Australia because it grows through soil of low moisture content, grows in the absence of the nematode and some isolates have good "rhizosphere competence" (Kerry et al., 1984).

3.1.6. Root Lesion Nematodes

The root lesion nematode, *Pratylenchus thornei*, has been shown by Thompson (1990) to be increased by certain rotations to levels which cause severe losses to successive wheat crops. This nematode enjoys a wide host range but can be reduced by either long (>6 months) fallow or a cropping frequency of host crops (e.g. wheat, maize, linseed, chickpeas and soybean) of at most one year in three, using sorghum, sunflower or barley (non-hosts) as break crops. By contrast, *P. neglectus* (*P. minyus*) is wisespread at high levels in the Australian wheat belt, without causing yield losses, except in association with *R. solani*, and apparently is not affected by rotation.

3.2. Deleterious Rhizosphere Microorganisms

Deleterious rhizosphere microorganisms (DRMO) have been defined as minor pathogens that affect plants by their metabolites without parasitizing plant tissues (Schippers et al., 1987); they include both bacteria and fungi (Suslow and Schroth, 1982). Pathogenicity is difficult to demonstrate because their effect on plants is to retard root and/or shoot growth with few other symptoms. The effects of these microorganisms result from their extracellular metabolites or toxins (Fredrickson and Elliott, 1985a, 1985b; Schippers et al., 1987; Bolton and Elliott, 1989). There is evidence that DRMO have host specificity for wheat, citrus and downy brome (*Bromus tectorum* L.) (Gardner et al., 1984; Fredrickson and Elliott, 1985a, 1985b; Elliott and Kennedy, 1988).

Evidence is accumulating that DRMO are affected by crop rotation (Schippers et al., 1987; Kaufman and Elliott, unpublished data). Schippers et al. (1987) found that yields of wheat and especially potatoes were sensitive to the frequency of the rotation, but were unable to attribute the low yields to known soil-borne pathogens and postulated that the yield reductions resulted from non-spore-forming DRMO. Studies in the IPNW on DRMO pseudomonads indicated that as the frequency of winter wheat in the rotation increased, the populations of DRMO (pseudomonads) increased on the wheat rhizoplane (Kaufman and Elliott, unpublished). Other causes of the plant growth retardation cannot be ruled out, and the direct effect of the DRMO cannot be established in natural soil with current technology.

3.3. Vesicular-arbuscular Mycorrhizal Fungi

One of the few examples of vesicular-arbuscular mycorrhizas (VAM) being important in wheat production and linked with rotations has been demonstrated by Thompson (1987) in southern Queensland, where a problem with wheat called "long fallow disorder" was associated with a decline in the population of VAM in soil. After a 10-month fallow the numbers of VAM propagules dropped and crops showed symptoms of Zn and P deficiencies. Chickpea, grain sorghum, sudan grass, sunflower and soybean are similarly affected by declines in populations of VAM following long fallows. Glasshouse experiments revealed that the disorder was corrected by the introduction of VAM fungi. Thompson (1987) and Thompson et al. (1986) showed that the practical solution was to avoid clean fallows of 12 months or more and to grow crop plants the roots of which maintained or built up the populations of VAM.

3.4. Disease Suppression

3.4.1. Take-all Decline

"Take-all decline" (TAD) describes the phenomenon whereby the disease increases in severity with consecutive crops of wheat up to a maximum and then declines with continued monoculture of wheat due to the soil becoming suppressive (Shipton, 1975). If a break crop, such as potatoes or oats, is grown in the field the soil loses its suppressive properties and the cycle must be repeated (Cook, 1981a). Unravelling the mechanisms of take-all decline could point to a practical means by which rhizosphere microorganisms can be managed for biological control of *G. graminis* and possibly other root pathogens. Indeed, the microorganisms from roots of wheat growing in soil from fields that have undergone take-all decline are showing promise as biocontrol agents of take-all when introduced into the rhizosphere of wheat (Weller and Cook, 1983; Cook and Weller, 1987; Cook et al. 1988; Weller, 1988).

The first breakthrough pointing to a microbiological explanation for take-all decline in the IPNW occurred 20 years ago when soil suppressiveness was successfully transferred from one field to another by introducing suppressive soil into conducive soil at about 0.5% (w/w) to a depth of 15 cm (Shipton et al., 1973; Baker and Cook, 1974). The small amount of soil required to produce the effect and the need to grow wheat a second year to establish suppression demonstrated the biological nature of the factor and the prospect of transferring the effect into new fields where it could be expected to multiply.

Using the "pot assay" of Shipton et al. (1973) for suppressiveness to take-all, Cook and Rovira (1976) showed for soils from long-term wheat fields in the IPNW and Victoria, Australia, that the introduction of 1% (w/w) of these soils to a methyl bromide fumigated soil gave significant protection to 4-week-old wheat seedlings against take-all. The work indicated that the suppression occurred on or within the wheat roots, a lead supported by an earlier observation of Vojinovic (1972) that the suppressive microorganisms were in the lesions caused by the take-all fungus. Using a "cone assay" that permits more precise control of conditions for infection, Cook et al. (1986) showed that neither the minimum size of the inoculum particles (fragments of oat grains colonized by the pathogen) nor the infection efficiency (number of lesions per milligram of inoculum particles) was affected by the known suppressiveness of the soils. These results provide indirect evidence that suppression occurs on or within the root after infection because, if suppression occurred in the bulk soil, i.e. on or within the inoculum particles, there would have been more infection with larger particles and/or fewer infections per unit mass of inoculum. However, Wildermuth et al. (1979) proposed that suppression of *G. graminis* var. *tritici* occurred in the soil or the rhizosphere rather than on the root surface, because they observed colonization and lysis of hyphae emerging from oat inoculum in suppressive soil but not in conducive soil.

Extensive studies over the past 15 years and ones continuing at present all point to a major role of root-colonizing, antibiotic-producing, fluorescent *Pseudomonas* species in the suppression of take-all in fields cropped for many years to wheat. The roots of wheat, either alone or when infected by the take-all fungus, somehow favour a change in the population of rhizobacteria towards both a higher total population of fluorescent pseudomonads and a higher proportion of pseudomonads inhibitory to *G. graminis*. Studies of pseudomonads as rhizosphere colonists became possible with the development by Sands and Rovira (1970) of a selective medium for their isolation and enumeration. Using this medium, combined with a standard test for antibiosis by pure cultures of random isolates from dilution plates, Smiley (1978a, 1978b) obtained the first evidence for such qualitative and quantitative changes towards greater inhibitory activity in the rhizosphere of ammonium-fed wheat. Smiley's work explained the microbiological basis for greater suppression of take-all in the rhizosphere of wheat fertilized with ammonium than that fertilized with nitrate nitrogen. Cook and Rovira (1976) found that eight bacterial cultures out of 200 tested protected wheat against take-all in fumigated soil. All eight cultures came from roots of wheat growing in suppressive soil and seven were pseudomonads.

Rovira and Campbell (1974, 1975) observed with scanning electron microscopy that as lesions caused by *G. graminis* developed on roots, the mucilage covering the roots disintegrated and rod-shaped bacteria proliferated. This sequence of events in the roots and rhizosphere of plants infected with *G. graminis* was confirmed by transmission electron microscopy (Foster et al., 1983). Rovira and Wildermuth (1981) advanced the hypothesis that enrichment of suppressive bacteria occurred on the lesions in roots through increased exudation, which selectively stimulated fluorescent pseudomonads of which a high proportion are suppressive to the take-all fungus. These bacteria survive on and inside roots and crowns of wheat in which the pathogen over-summers, and in the following growing season these suppressive pseudomonads are strategically placed to colonize new wheat roots and suppress the growth of *G. graminis* in the rhizosphere and so reduce its ectotrophic growth on roots (Wildermuth and Rovira, 1977; Wildermuth et al., 1985).

Weller (in Cook and Baker, 1983) and Weller et al. (1988) have presented the strongest evidence to date for a role of pseudomonads in take-all decline by showing that the shift in the rhizosphere populations of wheat towards both higher total numbers and a higher percentage inhibitory to *G. graminis* was unique to soils from fields that had been cropped many years to wheat. Strains effective against take-all can be found in the rhizosphere of wheat growing in soils previously cropped only occasionally or not at all to wheat, but the chances of finding effective strains are 25 times better from soils showing take-all decline.

The rise and decline of take-all along with a coordinated drop and increase in yields of winter wheat was recorded in a plot of irrigated wheat near Lind, Washington, during a 20-year period, from the 1967–68 crop year (Cook, 1989) through the 1986–87 crop year. Soil fumigation was used to provide an indication of the yield potential each year in the absence of take-all. Best agronomic practices such as planting date, fertilization, tillage and weed control were used at each site. Take-all increased while yields decreased in each of the first 7 years. Thereafter, disease declined while yields increased to a peak in the 15th year. The low yield in the 7th year was only about 50% of the fumigated check, and the high yield in the 15th year was within 95% of the fumigated check. Strain 2–79 of *P. fluorescens* (Weller and Cook, 1983) was isolated from the rhizosphere of wheat growing in soil taken from this plot in 1979, the 10th year of the study and the 3rd year after onset of take-all decline. This strain is now widely studied because of its biocontrol activity against take-all (Weller and Cook, 1983), and is effective because of its ability to produce a phenazine antibiotic (Brisbane and Rovira, 1988; Brisbane et al., 1987; Thomashow and Weller, 1988a, 1988b).

The involvement of phenazine-l-carboxylic acid (PCA) in suppression of *G. graminis* by strain 2-79 was demonstrated by Thomashow and Weller (1988a) when they showed that Tn-5 transposon mutants lacking the ability to produce PCA were less protective against *G. graminis* than the parent culture; this activity was restored by re-inserting genes responsible for PCA production. Subsequently, Thomashow and Weller (1988b) reported that PCA could be extracted from sterile and non-sterile rhizosphere soil of wheat treated with strain 2-79. Using the same pseudomonad, Brisbane and Rovira (1988) demonstrated that PCA was produced under axenic conditions in sand in which wheat treated with strain 2-79 was growing. These results provide conclusive evidence that production of an antibiotic occurs in the rhizosphere of wheat and is involved in biological control of take-all. An interesting aspect of the role of PCA in suppression is the effect of pH on the ionization and biological activity of this antibiotic, the activity of which is greatest at pH 5.5 (Brisbane et al., 1987); thus it should be more effective in soils of low pH or in the rhizospheres of ammonium-fed plants due to the lower rhizosphere pH. This highlights the importance of edaphic factors in biological control.

Simon et al. (1987, 1988a) used growth of *G. graminis* on membrane filters to demonstrate that "pathogen suppression" (as distinct from "disease suppression") occurred in bulk soil in which decline of take-all had occurred following successive wheat crops to which ammonium fertilizer had been applied. Simon and Sivasithamparam (1988a) attributed this pathogen suppression to a build-up of *Trichoderma koningii* as the soils became more acidic. Subsequently, Simon (1989) suppressed take-all in unsterilized soil by mixing propagules of *T. koningii* through the soil prior to planting wheat. The most suppressive strain of *T. koningii* against take-all in soil produces volatile pyrone compounds which are active against a range of fungi *in vitro* (Simon et al., 1988b); it is not yet known whether these volatile compounds are responsible for the activity of *T. koningii* in soil. Claydon et al. (1987) isolated two strains of *Trichoderma harzianum* which produced identical volatile pyrones with activity against a wide range of fungi, including *G. graminis*, and also reduced the rate of damping off in lettuce.

3.4.2. Antagonistic Interactions Between Populations of Root Pathogens

Some disease suppression apparently results from the competitive or other antagonistic effects of interacting populations of root pathogens in the rhizosphere. Such an effect can be inferred from the data for disease and yield of wheat in the 20-year continuous wheat plot at Lind (Cook, 1989). After a peak yield with essentially no evidence of take-all in the 15th year (1982), yields began inexplicably to decline with further wheat cropping, yet take-all remained obscure. A diagnosis had to wait until 1984 when rhizoctonia root rot was recognized for the first time in the IPNW, including this plot (Weller et al., 1986). Yields continued to decline in the plots while rhizoctonia root rot increased to severe proportions, exhibiting bare patches. Obviously the suppressiveness of the soil to take-all was ineffective against rhizoctonia root rot. It may be that rhizoctonia root rot could not develop until take-all came under control.

Garren (in Baker and Snyder, 1965) observed in 1961 that application of PCNB to soil significantly increased pod rot of peanuts, while application of Dexon to the same field controlled root rot. The following year in another field, he got the opposite result—Dexon increased root rot and PCNB controlled this disease. PCNB is effective against *Rhizoctonia* while Dexon is effective against *Pythium*. Similar results have been obtained over several years in eastern Washington with wheat; fungicides, such as Rhizolex and PCNB that control *Rhizoctonia* but not *Pythium*, favour root disease and lower wheat yields but not if metalaxyl is also added and not if the chemicals are applied to wheat planted in fumigated (methyl bromide) soil (Cook and Ogg, unpublished). Conversely, metalaxyl by itself often results in significantly lower wheat yields and symptoms of more not less root disease, but it is not detrimental when combined with PCNB or soil fumigation. Phytotoxicity of the chemicals to wheat can be ruled out on the basis of no harmful effects in soil made "pathogen-free" by soil fumigation. While definitive experiments are needed, it may be that the *Pythium* and *Rhizoctonia* species pathogenic to wheat are somehow antagonistic to each other in the rhizosphere of wheat, and possibly also in the rhizosphere of other susceptible crops grown in the eastern Washington dryland agroecosystem. It would appear better to leave these two populations of root pathogens to coexist in their natural balance rather than to control only one component of the mixture and upset this balance.

Strains of *Phialophora* species taxonomically related to the take-all fungus have been shown to protect wheat against take-all (Scott, 1970; Deacon, 1974) in fields where wheat follows grass pasture favouring high populations of *Phialophora*-like fungi. Perhaps a similar phenomenon occurs between species of *Pythium* or *Rhizoctonia*. Ingram, Cook and Chamswarng (1990) showed that while the common but non-pathogenic *P. heterothallicum* and *P. torulosum* were only slightly less efficient that *P. ultimum* and more efficient than *P. irregulare* as colonists of wheat embryos, they had no pathogenic effect on emerging seedlings of wheat when established in embryos; conceivably if established in embryos first they could provide protection against a subsequent challenge by either *P ultimum* or *P irregulare*.

Wheat roots are also colonized by many multinucleate and binucleate *Rhizoctonia* species in addition to the pathogenic *R. solani* AG8 and *R. oryzae* (Ogoshi et al., 1989). The binucleate *Rhizoctonia* strains in particular have shown evidence of ability to protect roots of other plants against pathogenic, multinucleate strains (Cardoso and Echandi, 1987). Presumably the various species and subspecies of *Pythium* and *Rhizoctonia* are adapted differentially to wheat, barley, peas, lentils and other crops and, therefore, can be expected to fluctuate in relative numbers and activity in response to each of these crops. The occurrence of a certain mixture of root diseases with one crop rotation (or lack of rotation) and a different mixture in response to a different rotation could be the result of a declining population of non-pathogenic strains having some ability to protect that crop, as well as selective enrichment of strains having the ability of attack that crop.

4. INFLUENCE OF TRASH RETENTION

The retention of maximal trash—the residue left in the field after harvest of the crop—has long been recognized as valuable both for control of soil erosion and conservation of water. The U.S. Department of Agriculture recognizes "conservation tillage" as any method of residue management that provides at least 30% cover of the soil at the time of sowing (Papendick and Miller, 1977). The influence of tillage on rhizosphere colonization is covered in a later section of this chapter. In this section, we address the influence of the trash specifically.

Following the introduction of stubble-mulch farming into North America in the 1930s it was soon recognized that wheat grew relatively poorly and yielded less in trashy soils compared with clean soils, i.e. ones where the crop residue was buried by mould-board plough. Wheat plants were stunted, had yellow leaves and tillered poorly in the mulch-tilled compared with clean-tilled fields or plots. The problem was greatest in the wet years or wet areas within the fields; Zing and Whitfield (1957) showed for 108 separate comparisons at eight locations in the Great Plains and IPNW that the wetter the area, the greater the yield advantage of clean-tilled plots over plots with maximal trash left on or near the soil surface.

4.1. Evidence that Yield Depressions Associated with Retention of Trash are caused by Microorganisms in the Soil

Early research on the aetiology of the yield depression of wheat associated with retention of trash focused on wheat straw rather than the soil, since the problem was associated with straw regardless of site, and since immobilization of nutrients such as nitrogen (Ferguson and Garby, 1964), phosphorus, and sulphur could not account for the effect (McCalla and Army, 1961). This led to extensive studies of possible phytotoxins produced during decomposition of the straw (Norstadt and McCalla, 1968; Kimber, 1967, 1973a, 1973b; Elliott et al., 1984; Cochran et al., 1977). Establishing the role of a phytotoxin as the cause of any given symptom or combination of symptoms requires that the toxin be identified, be shown to occur *in vivo* at concentrations sufficient to cause the damage, and be shown to cause the exact symptoms of the syndrome observed in nature (see Chapter 7).

Harris and Kimber (1983) examined evidence that phenolic acids produced from decomposing straw may inhibit plant growth. Norstadt and McCalla (1968) considered that the major factor associated with high levels of wheat stubbles on the soil under wet conditions was the production of patulin by *Penicillium urticae*. Subsequently, in England, Lynch (1977, 1978) and Lynch et al. (1981) attributed the poor seedling emergence and growth in crops direct drilled into wheat stubble to the production of acetic acid during anaerobic decomposition of the straw. In the field these substances probably occur in high concentrations on the surfaces of crop residues (Lynch et al., 1980).

Studies at sites throughout the intermediate to high-rainfall areas (400–600 mm p.a.) of the IPNW consistently revealed that leaving all straw on the soil surface (where it makes contact with the soil), with no soil disturbance (no-till), presented no problem for wheat growth if the soil was treated with methyl bromide in advance of sowing (Cook et al., 1980; Moore and Cook, 1984; Douglas et al., unpublished). On the other hand, wheat growth and yield was significantly lower in these same experiments in non-fumigated plots. Moreover, removal of the straw resulted in markedly improved wheat growth and yield, even without fumigation, and return of wheat straw to the surface of burned plots resulted in the same poor growth with yellow leaves and low yields recorded in the undisturbed plots. Clearly the problem is caused by microorganisms, the population or activity of which is eliminated by soil fumigation. Straw left on the soil surface plays a role in the aetiology and/or epidemiology of the problem.

Evidence that the causal organisms are in the soil (probably old roots and past rhizosphere soil) and not in the straw was obtained by deep-placement of chloropicrin in the same band as the nitrogen fertilizer (anhydrous ammonia) at the time of direct drilling into plots with heavy wheat straw residue on the soil surface and also in adjacent replicated plots with the straw removed by burning (Cook and Haglund, unpublished). The rate of chloropicrin was only $10\,l\,ha^{-1}$ injected 12–15 cm deep to avoid damage to the wheat seeds planted 5 cm deep and 7.5 cm to the left and right side of each nitrogen/fumigant placement. Adult plants of wheat in plots with surface residues were of uneven height, tillered less and were a lighter green compared to the wheat in adjacent plots without the straw residue. Deep-band treatment with chloropicrin corrected most of this problem and resulted in about $600\,kg\,ha^{-1}$ higher yield. In contrast, yields were high whether or not fumigant was injected in those plots where the residue was burned. The fumigant would have diffused slowly at biocidal concentrations through the soil but would not have attained or maintained effective concentrations at the soil–straw interface and in the 2- to 5-cm thick layer of straw on the soil surface (Cook and Haglund, unpublished).

Greenhouse studies have confirmed that the responsible organisms are in the soil rather than the straw (Cook, Chamswarng and Tang, in press). Fresh straw had no effect on seedling emergence or vigour when added untreated (natural) to soil that had been heated with moist heat to 60 °C for 30 minutes. It is likely, therefore, that the deleterious activity arises from rapid colonization of the straw as a substrate.

4.2. Influence of Trash on Wheat Root Pathogens

Trash left after harvest of the wheat affects both the inoculum potential and the soil environment of the root pathogens of wheat. In the IPNW these effects are mostly favourable to the epidemiology of take-all and pythium and rhizoctonia root rots and unfavourable to fusarium root and crown rot caused by *Fusarium culmorum*.

Trash affects the inoculum potential of the root pathogens by serving both as a primary and secondary source of energy (foodbase). *Pythium* species, for example, are capable of colonizing wheat straw and chaff through saprophytic activity (Cook, Chamswarng and Tang, 1990). The injurious effect of fresh chaff on growth of wheat seedlings in natural soil was virtually non-existent if metalaxyl was added to the soil; however, it occurred in the heat-treated soil if either *P. ultimum* or *P. irregulare* was re-established as oospores in the soil (Cook, Chamswarng, and Tang, 1990; Ingram, 1987; Ingram and Cook, unpublished). *Rhizoctonia solani* is also well known as a saprophytic colonist of plant residue and presumably can also use wheat straw as a secondary source of energy, along with roots colonized through parasitism—its primary means of obtaining energy.

In a long-term tillage experiment in northern Idaho the population of *Pythium* species in the top 10 cm of soil was highest in plots managed as minimum-till for the previous 7–8 years, lowest in plots clean-tilled (residue buried by mould-board plough in the autumn after each harvest), and intermediate in plots direct drilled during the same period (Cook, Chamswarng and Tang, 1990). At 15–30 cm deep, on the other hand, the population of *Pythium* species was highest where residue had always been buried by mould-board plough. Any given population of *Pythium* species reflects the past history of activity of this fungus genus, both as a saprophyte in fresh crop residue and as a parasite of plant roots. The occurrence of highest populations in the surface 10 cm where the crop residue was repeatedly mixed into the surface soil, and highest below this depth where residue was repeatedly buried, is evidence that this fungus does, indeed, multiply in the field on or within the trash. The intermediate population in direct-drilled plots could reflect past activity of this pathogen as a root parasite since all trash was left on the soil surface and presumably out of reach of the *Pythium* inoculum in the soil.

More importantly than energy sources, trash on the soil surface functions as a barrier to drying of the top soil where these pathogens are active. In work by Douglas, Elliott, and Cook (unpublished), removal of the straw by burning resulted in a 50% reduction in the incidence of take-all (defined as all seminal roots black, with infection moving up the subcrown internode), and returning the straw as a layer on burned soil resulted in an even higher incidence of take-all than occurred in the undisturbed checks. Pythium and rhizoctonia root rots are favoured by wet soil and hence can be expected to be more severe in soil covered with straw than in completely bare soil exposed to drying by the sun and wind. Fusarium root and crown rot, on the other hand, is favoured by dry soil and plant water stress (Papendick and Cook, 1974) so, not surprisingly, leaving maximal straw on the soil surface did not result in more of this disease. In Nebraska, fusarium stalk rot of sorghum caused by *F. moniliforme* is severe in fields made bare of residue of the previous wheat crop and virtually only mild in fields left covered by the trash of previous wheat crops where up to 10–12 cm of additional water is available to the sorghum crop (Doupnik and Boosalis, 1980).

Kollmorgen et al. (1987) found that with a mulch equivalent to 4 t ha^{-1} over a 36-week period, survival of *G. graminis* in crowns was 21% compared with

51% with no mulch, probably because the soil would remain wetter for longer under the mulch, increasing decomposition of crowns and colonization by competing microorganisms.

4.3. Deleterious Rhizosphere Microorganisms

Residue placement has physical, chemical and biological implications on rhizosphere colonization. Crop residues serve as a secondary foodbase or energy source (assuming living hosts or roots are the primary food sources) for potential rhizosphere colonists, whether they be pathogens or non-pathogens. Whether the residues are placed on the surface or buried influences their rate of degradation because of water and temperature relationships (Stroo et al., 1989), which affects microbial colonization and substrate availability, and in turn influences rhizosphere colonization. When crop residues are managed on the soil surface, readily available carbon from the residue has limited penetration as compared with incorporated residues (Cochran et al., 1980). Surface residues reflect heat, form a thermal blanket if thick enough, retard water loss and improve water infiltration as compared with incorporated residues (Papendick et al., 1973).

Cherrington and Elliott (1987) studied the incidence of deleterious rhizosphere microorganisms (DRMO) on the rhizoplane of winter wheat, peas, lentils, winter barley and the weed, downy brome, and found that while pseudomonads occurred on the roots of all the plants, no-till seeded winter wheat and winter barley and natural stands of downy brome had the largest proportion of pseudomonads capable of inhibiting root growth. When pseudomonad isolates specific against winter wheat were tested, the organisms extensively colonized barley, oat, lentils, peas and wheat roots, but winter wheat root growth was inhibited to the greatest degree (Fredrickson et al., 1987).

More non-fluorescent pseudomonads inhibitory to winter wheat root growth were isolated from the rhizoplane of the direct-drilled winter wheat roots growing in plots with residues than from the rhizosphere of winter wheat growing in plots where residues had been burned (Elliott and Lynch, 1984). The inhibitory pseudomonads are usually non-fluorescent and not detectable on the rhizoplane of winter wheat until late winter or early spring. These bacteria grow well at 5 °C, colonize wheat roots, and inhibit wheat root growth by the production of a toxin; the extent of inhibition is influenced by soil type, and the organisms and toxins are fairly specific against winter wheat (Elliott and Lynch, 1984, 1985; Fredrickson and Elliott, 1985a, 1985b; Fredrickson et al., 1987). The toxin is of low molecular weight, polar, pH- and temperature-labile and produced during the late exponential growth phase (Fredrickson and Elliott, 1985b; Bolton and Elliott, 1989; Bolton et al., 1989a, 1989b).

Evidence is accumulating that these detrimental organisms are strongly associated with the previous crop residue, especially when the residue remains on the soil surface. When non-sterile winter wheat straw was inoculated with a rifampicin-marked inhibitory pseudomonad at 1/1000th the population of the native bacterial flora at 5 and 15 °C, the organism multiplied rapidly to almost the same numbers as the native bacterial flora. Fairly high proportions of the introduced rifampicin-resistant mutant were also obtained on pea residue, winter barley straw and chaff, and winter wheat chaff (Fredrickson et al., 1987).

The aggressive growth by these organisms on plant roots and residues indicates that they compete well for available substrate. Rhizosphere competency of Tn5 TOX minus mutants of the organisms was as good as the wild-type in the rhizosphere of winter wheat grown either axenically or in non-sterile soil (Bolton et al., 1989b), which implies that toxin production has little or no role in the ability of these organisms to grow in the rhizosphere (Kennedy, Bolton, and Elliott, unpublished data).

Pseudomonas spp. are usually not tolerant to water stress (Harris, 1981), but the inhibitory pseudomonads appear unusually tolerant to water stress as shown in studies in which the introduced inhibitory pseudomonad became a dominant proportion of total bacterial flora on wheat straw at water potentials of -0.9 MPa. At lower water content the numbers declined but when the straw was wetted they increased rapidly. When the organism was applied to barley straw in the fall in the field and the plots were no-till sown to winter wheat, the introduced inhibitory pseudomonad competed successfully with the native bacterial flora and comprised 5–10% of the total rhizoplane bacterial flora in the following spring. During the study the surface residues showed about a 10-fold greater number of the inhibitory pseudomonads than did the incorporated residues, indicating that these organisms may be favoured by no-till seeding (Stroo et al., 1988). Winter wheat root growth was inhibited in the inoculated direct-drilled plots when inhibitory bacteria had been applied to the surface before sowing. Obviously the environmental changes in soil caused by residue management will greatly influence the importance of these organisms.

When non-fluorescent pseudomonads were isolated from winter wheat roots it was found that significantly greater numbers were TOX^+ when incubated at 5 °C in a medium containing nitrate. Possibly nitrification of aqua ammonia fertilizer and the overwinter cold induces toxin expression by the inhibitory pseudomonads (Kaufman and Elliott, unpublished).

Weeds have been responsible for the failure of many conservation tillage cropping systems. One weed that has been particularly troublesome and for which there is no effective control in the IPNW is downy brome (*Bromus tectorum*). Laboratory studies showed that pseudomonads inhibitory to downy brome and not winter wheat could be isolated from the rhizoplanes of both plants growing in the field (Cherrington and Elliott, 1987). When such strains were applied to no-till seeded winter wheat in farmers' fields by spraying onto the crop residues in late autumn, there was a significant growth reduction of downy brome and increased yields of winter wheat (Elliott and Kennedy, unpublished).

4.4. Disease Suppression

Whether the pathogens such as *Pythium, Rhizoctonia* and *Fusarium* can use trash to increase their inoculum potential through saprophytism depends on whether the trash is available for colonization or is already colonized by other microorganisms (Yarwood, 1966; Garrett, 1970; Hancock, 1977). *F. culmorum* is highly aggressive as a saprophytic colonist of wheat straw (Butler, 1953) but is pre-empted from doing so in the dryland, trashy-fallow management system of the IPNW by the common fungi in the straw. Chaff is not suitable as an energy source for *Pythium* species (Rush et al., 1986) and is non-injurious to wheat seedlings when added to natural soil (Cook et al., 1989). Thus, practices that leave most of the wheat residue on the soil surface and temporarily out of reach of pathogen in the soil can provide disease suppression by effectively eliminating the residue as an energy source for these pathogens.

5. INFLUENCE OF TILLAGE

Cultivation of soil prior to planting crops is so universal that it is the accepted conventional wisdom that it must be useful. For thousands of years civilizations throughout the world have depended upon arable agriculture for survival, but the decline of most of these civilizations after 1000 to 1500 years has been associated with declining soil fertility and land degradation. Cultivation was an essential part of the agriculture of these civilizations—the Romans and Greeks advocated up to 12 cultivations to destroy weeds and provide good soil–seed contact. This custom persists to the present day in most countries, despite the fact that each time a soil is cultivated, microbial activity is stimulated and organic matter, valuable in so many ways, is lost. With the development of herbicides for controlling weeds before and after seeding the need for multiple cultivations is removed. Some weed seeds germinate better after soil disturbance and are difficult to control in crops, but in such cases reduced cultivation may suffice. However, one benefit of cultivation is the fragmentation and destruction of plant residues harbouring plant pathogens. Except for weed control, and a few benefits such as the protective effects of a dust mulch again evaporative loss of water during summer fallow in the low-rainfall IPNW (Papendick et al., 1973; Davidson and Santleman, 1973) and the better root penetration when special tools are used to break up a plough-pan or loosen compact soil (Allmaras et al., 1988), there is little experimental evidence that tillage is agronomically necessary for crop production.

5.1. Root Pathogens

Most research intended to compare the effects of tillage and no-tillage treatments on crop growth is done without disease-free checks, and therefore responses to the treatments, whether positive or negative, cannot be interpreted. Even loosening the soil by subsoil tillage may be beneficial to the crop because of root-disease control, as shown by Burke et al. (1972) for bean cultivars susceptible to root rot caused by *Fusarium solani* f. sp. *phaseoli*; cultivars with the greatest resistance to this disease showed the least response to subsoiling. Since tillage has direct and multiple effects on the environment of root pathogens, tillage can also be expected to affect the epidemiology of diseases caused by these pathogens.

Studies with wheat in the high-production areas of the IPNW (400–450 mm or more annual precipitation and cropped every year), using either soil fumigation or crop rotation to rid the soil of root pathogens and soil-inhabiting pests (e.g. wireworms), have consistently shown no agronomic benefit of one kind of tillage (or no tillage) over another. If anything, yields are highest with no-tillage (direct drilling), presumably because more water is available to the crop when the soil is left undisturbed and protected from drying by surface mulch. In our experience, supported by numerous reports in the literature, yield depressions for wheat in response to minimum or no tillage are invariably experiments carried out in combination with wheat after wheat (no break crop) or wheat after barley, or in fields where grass weeds and volunteer (self-sown) wheat and barley were allowed to grow until a few days before sowing. These situations present unsanitary soil conditions for wheat since root pathogens are given every opportunity to increase their populations or inoculum potential before the next crop is sown.

In some areas (e.g. Georgia, USA) take-all was less severe with direct drilling than with conventional tillage (Rothrock, 1987). However, take-all (Moore and Cook, 1984), rhizoctonia root rot (Weller et al., 1986) and pythium root rot (Cook et al., 1980) all have been shown experimentally to be favoured in the IPNW by direct drilling. This increased incidence and severity of these diseases can be explained by the generally wetter top soil and additional energy sources associated with maximum retention of trash on or near the soil surface. However, there is also evidence that these root pathogens benefit when their soil habitat is disturbed minimally or not at all.

Moore and Cook (1984) showed that total mass of tiller bases and associated tiller roots infested with the take-all fungus was up to 10 times greater where the soil was left undisturbed (no tillage) following harvest of a wheat crop severely affected by take-all than where the soil was repeatedly disturbed by tillage. Each tillage operation reduces the size of fragments of wheat residue infested by *G. graminis* and exposes new surfaces of the infested residue to colonization by potential competitors. The take-all pathogen is the pioneer colonist of these fragments of host residue, having become established in them while the host was alive, but it is subject to displacement in the residue by other microorganisms and apparently the process is accelerated by tillage.

In Australia the effects of cultivation of take-all have been variable. At the Avon (S.A.) site with 350 mm average rainfall and calcareous sandy loam it was found that when wheat followed peas and medicago pasture there was more take-all in 7 out of 8 years with direct drilling; this is attributed to incomplete grass control in the peas and medic, which were also sown by direct drilling. When wheat followed grass–medicago pasture there was more take-all in 2 out of 4 years with direct drilling. Similar results were obtained at another field site with a different soil type and rainfall of 495 mm per annum—yields of direct-drilled wheat following lupins were similar to those obtained with cultivation if free of grass but not so when there were grassy weeds in the lupins (Rovira, unpublished).

Kollmorgen et al. (1987) showed that when wheat was sown into fallow soil which had been kept free of weeds, either with herbicides or with cultivation, the disease incidence was similar but relatively high (28–40%). In other experiments where high infection was established in wheat the year before the tillage trial by adding *G. graminis* propagules to soil, again there was also no effect of tillage. This confirms the finding of Moore and Cook (1984) in the IPNW: where the amount of introduced inoculum was the same, the severity of take-all was not affected by tillage. Kollmorgen et al. (1987) found that after 36 weeks (including summer) there was a 75% survival of crowns on the surface compared with 31% and 17% when buried at 5 and 10 cm depths respectively. In view of this one would expect a greater attrition of take-all propagules with cultivation through their burial and fragmentation, as reported by Moore and Cook (1984). These results of Kollmorgen et al. (1987) and Kollmorgen and Walsgott (1984) suggest that under Australian conditions more take-all might be expected if a blade plough was used for weed control and the crop direct drilled.

In Australia, the effects of tillage on rhizoctonia root rot are consistent throughout the four southern mainland states. This pathogen limits wheat growth and yield so severely that it is the major factor preventing the acceptance of direct drilling or no-till farming. Over a 3-year field trial in a calcareous sandy loam the average root rot rating on a 0–5 scale was 0.75 in cultivated soil compared with 2.5 in direct-drilled wheat, across four different rotations. In sandy soils "bare patches" with poor seedling growth occur at root damage ratings >2.0. The method of direct drilling (and hence degree of soil disturbance) affects the severity of rhizoctonia root damage and area of crop lost. With barley, direct drilling with 15 cm wide shares giving full disturbance resulted in less root damage and higher yield than sowing with the "Sirodrill" (Rovira, 1986), which gave minimum soil disturbance.

MacNish (1984) demonstrated that soil disturbance *per se* reduced the effects of rhizoctonia in pots and subsequently reported field responses through cultivation (MacNish, 1985). This direct effect due to soil disturbance is probably due to a severance of hyphae from the particulate organic matter colonized by *Rhizoctonia* to form the propagules. These propagules are concentrated in the surface 5 cm (Neate, 1987) and with direct drilling the seed is placed in the zone with the greatest concentration of propagules.

The original concept for no-till farming in southern Australia was based upon the "graze–spray–seed" principle, namely following the autumn rains, grasses, clovers and medics germinate and are grazed by the sheep until just before spraying with desiccant herbicide and sowing. The worst rhizoctonia "bare patch" disease occurred when farmers adopted this practice. Roget et al. (1987) demonstrated that the roots of ryegrass and barley grass were colonized by *Rhizoctonia* and unless the grasses were sprayed early enough, these infected roots provided a highly virulent inoculum into which cereals and grain legumes were sown. With grass plants numbering $>5000\,m^{-2}$ and each grass seedling having three roots, each 5–10 cm long after 4 weeks, it is clear why root disease is higher with direct drilling. The "graze–spray–seed" technique has now been replaced in commercial practice by a chemical fallow of 3 to 4 weeks before seeding.

In contrast with results on the effects of cultivation on take-all and *Rhizoctonia* there is the finding by Roget and Rovira (1985) that the severity and carry-over of cereal cyst nematode is less in direct-drilled wheat than when the soil is cultivated. They found that direct-drilled wheat suffered half the root damage of wheat sown into cultivated soil, and there was a lower population of white females (hence less carry-over of eggs) with direct drilling; these effects on the nematode resulted in a substantial yield increase. Subsequent experiments (Rovira and Roget, unpublished) have demonstrated that the above effect is due to soil disturbance, either mixing cysts, eggs and nematodes through soil and/or improvement of physical conditions to aid movement of nematodes through soil.

Pankhurst and McDonald (1988) found populations of up to 450 spores g^{-1} of *Pythium* spp. in no-till plots, which was two to four times the population of cultivated plots; this points to a potential problem in wet periods after sowing wheat by direct drilling. As stated earlier (see Section 3.1.1), the significance of these population differences on yield has yet to be assessed. The major manifestations of *Pythium* damage to wheat are reduced root lengths to explore soil for nutrients and reduced top growth affecting the potential productivity of the crop (Cook et al., 1987).

5.2. Detrimental Rhizobacteria

Tillage and residue management affect rhizosphere colonization mainly through effects on the physical and chemical properties of the soil and on the availability of food for the microflora. Soil in the rooting zone when either tilled to leave maximum amounts of crop residue on the soil surface (conservation tillage) or no-till seeded through the previous crop residue, is commonly wetter, slower to cool down in the autumn, and slower to warm up in the spring than soil tilled to bury the maximum amounts of crop residue (clean tillage). Size of soil aggregates, soil drainage, soil aeration, and distribution of soil organic matter and presumably nutrients are affected also (Griffith et al., 1983; Triplett, 1983). These physical and chemical factors affected by tillage are important parameters of the soil environment and could affect rhizosphere colonization.

No-till seeded wheat plants which appear healthy in late autumn can appear unthrifty in early spring and improve as the season gets warmer; this may correlate with the fact that detrimental rhizobacteria are not usually detectable on the wheat rhizoplane in the autumn and that they grow well at 5 °C. Field trials in which selected deleterious pseudomonads were coated onto wheat seed resulted in overwinter stand losses up to 50% and yield reductions of 30% (Bolton et al., unpublished).

In many of the higher rainfall (400–600 mm per annum) areas of New South Wales, farmers have moved away from direct drilling of wheat because of a disorder termed "poor early growth". Chan et al. (1987) examined the physical and biological properties of these hard-setting duplex soils in which "poor early growth" was a problem. Plant measurements indicated that the poor plant growth in no-till, with or without fallow, was due to poor growth after germination rather than poor seedling emergence. Soil physical differences (bulk density on and between drill rows and shear strength) could not account for the condition, although the soil bulk density was considerably higher in the no-till than tilled plots. However, densities in the top 150 mm beneath the drill row were highest for cultivated, and lowest with the nil fallow treatments respectively. When the bacterial isolates of the rhizospheres of roots from the three treatments were assessed for their ability to inhibit root growth, populations were higher in the rhizosphere of plants direct drilled into soil without chemical fallow than where soil had been cultivated. Roberts (unpublished) showed that most of the root-inhibiting bacteria were pseudomonads. Chan et al. (1987) conclude that the cause may be physical or biological but make no reference to root pathogens, such as *Pythium* and *Rhizoctonia*, which can cause poor growth in seedlings of direct-drilled wheat. Masle and Passioura (1987) demonstrated under controlled environment conditions that leaf area, shoot and root dry weights were negatively correlated with soil strength even with adequate nutrients and water and no root diseases. Root growth was affected less than top growth, and the authors suggest that with increasing resistance to penetration, the supply of plant growth promoters (auxins, gibberellins and cytokinins) to the tops is reduced. Such changes in the physiology of plants due to soil compaction would also affect the supply of assimilates to the roots, root exudation and longevity of root cells, all of which affect the rhizosphere microflora.

At this stage, we must consider that the relative importance of direct hormonal effects, detrimental rhizosphere microorganisms and root pathogens in the oft-reported poorer growth of direct-drilled crops is open to question and requires further study. Certainly, it is unlikely that the same mechanism, or even a single factor, will operate in all situations considering the wide range of soil and environmental conditions. The development of new no-till drilling technology has alleviated some growth problems associated with this seeding method. The ability to seed through heavy residues, place fertilizer with precision and push trash away from the seed row has improved stand and visible plant health (Cook, unpublished).

5.3. Disease Suppression

Under IPNW conditions, suppression of take-all did not develop in direct-drilled fields and hence expression of take-all decline is dependent on some form of tillage. At Lind, Washington, a plot cropped continuously to irrigated wheat which had achieved maximum suppression of take-all beginning in the 15th year suffered devastating take-all in direct-drilled subplots with the 17th, 18th and 19th wheat crops. Moreover, in spite of the increasing rhizoctonia root rot at this site favoured by direct drilling, take-all was the dominant root disease in each of these 3 years in the direct-drilled subplots. Take-all remained suppressed in the tilled subplots and, there, rhizoctonia root rot was the main yield-limiting factor.

Results of Rovira and Cook (Cook, 1981b), using the pot assay to measure suppressiveness of soils to take-all, show that expression of suppression is sensitive to changes in soil porosity associated with compaction. Addition of 1% (w/w) of suppressive soil (from "Quincy field") to a fumigated soil from South Australia, potted dry and then watered, made the soil suppressive. However, take-all on the seedlings was severe when a portion of the total water was added in advance of potting and the remainder then added from the top. It was necessary to compress this premoistened soil to fit the predetermined amount into each pot. While this treatment and the treatment potted air-dry and then watered provided the same bulk density and total water content, and hence the same total air-filled porosity, they presented two different physical environments for root–pathogen–antagonist interactions. The total porosity of soil potted dry would have been made up of a range of large and small pores; water moved rapidly as a wetting front through this medium, indicating good drainage characteristics and hence good aeration of the wetted soil. In contrast, this same total porosity of the soil potted wet and compressed into the pots would have been made up of more uniformly sized pores. This medium had poor drainage characteristics: water moved slowly as a wetting front through this soil and presumably aeration was relatively poor. The root-colonizing fluorescent pseudomonads thought to be responsible for inhibition of *G. graminis* in the rhizosphere are strictly aerobic and their inhibitory activity depends on production of antibiotics (e.g. phenazines). Loosening the soil by tillage may be sufficient to tip the balance in favour of greater numbers of these bacteria, or greater or more rapid antibiotic production. A further factor may be that cultivation spreads the microorganisms through the soil to improve the opportunity for them to colonize roots contacting microcolonies in the soil. Once colonization of the rhizosphere has occurred, the biological control agents are strategically placed to inhibit the take-all fungus, either as it grows through the rhizosphere or as it grows on or in the roots.

6. INFLUENCE OF SOWING DATE, ROW SPACING AND SEEDING RATE

6.1. Root Pathogens

Time of sowing, plant density and the spatial arrangement of plants in the field can be highly influential singly and collectively by controlling: (1) the time of the onset of a root disease epidemic; (2) the rate of development of the epidemic; or (3) the ultimate intensity of the epidemic. We will illustrate these effects by using wheat root pathogens as examples.

The time of infection of winter wheat by *Pythium* species can be delayed in the high-rainfall areas of the IPNW (>400–450 mm precipitation) by sowing early in the autumn, e.g. mid to late September, when conditions are sufficiently moist for germination but too dry for *Pythium* infection. The lower limit for wheat seed germination in these soils is −0.4 to −0.5 MPa matric potential, but embryo infections by *Pythium* seldom occur if the Palouse and Thatuna silt loams are drier than −0.05 to −0.2 MPa matric potential (Hering et al., 1987; Fukui, 1988). However, such early seeding will not allow the plants to escape root infections, which occur as soils become wetter, and have less opportunity to drain once the normal rainy season begins (late October and November). Nevertheless, helping the plants escape colonization by *Pythium* until past the three-leaf stage can be significant in terms of tillering and hence yield potential. Unfortunately, early seeding is not practical in fields where wheat follows wheat or barley because the soils usually are too dry for seed germination until October, and hence the practice is limited to rotations where wheat follows peas or lentils.

Taylor et al. (1983) reported that late sowing of wheat after wheat or barley delays the onset of infection by *G. graminis* and limits the rate of disease development, especially when combined with tillage. The onset of take-all is delayed largely by the fact that there is a lower inoculum potential of the pathogen—the longer time between crops forces the pathogen into a saprophytic existence, and in this state it is vulnerable to displacement in the crop residue by the competing soil microbiota. Late-sown plants tend to "catch-up" with plants started earlier in the season, requiring less real time to complete development and hence allowing less time for *G. graminis* to develop on and inside roots. Unfortunately, plants started late also produce fewer tillers and therefore have a lower yield potential than those started at a date more ideal for plant development.

In Australia, early sown crops similarly are more prone to take-all than late-sown crops. This was illustrated at the CSIRO experimental site at Avon in 1988 where three adjoining experiments in one field with a uniform level of take-all inoculum were sown on June 2nd, 7th and 15th. The percentage of wheat plants showing lesions caused by the take-all fungus 12 weeks after seeding were 60%, 38% and 16% respectively. In another experiment in 1988 at another site, wheat sown early (May 26th) had 52% plants with take-all lesions compared with 26% incidence in an adjacent crop sown 16 days later. This reflects two factors—with delayed sowing the *G. graminis* propagules (grass roots and crowns) in soil are exposed for longer to colonization and displacement by competitors while the time for the pathogen to grow in the rhizosphere and on and inside wheat roots is reduced. Cotterill and Sivasithamparam (1988) also report reduced incidence of take-all with delayed sowing in Western Australia.

These examples with *Pythium* and *G. graminis* illustrate how the saprophytic and parasitic phases of root diseases are affected by the length of time between host crops and soil moisture conditions at the time of sowing. It is relatively simple to identify such effects with pathogens that can be identified, but it is feasible that many saprophytic bacteria and fungi which colonize the rhizosphere would be similarly affected, especially if they are dependent upon organic substrates from living roots for growth.

Sowing rate is another factor which will affect the colonization of roots by pathogens. Adams and Colquhoun (1935) found that in a clay loam, spacing of plants 5, 10 and 20 cm apart led to infection spreading 51, 41 and 20 cm along rows from planting to maturity. The disease starts as primary infections, which lead to secondary spread of the pathogen from root to root, and plant to plant. Obviously, the greater distance that the fungus needs to grow through soil the more prone it is to attack by suppressive organisms and the less infective the propagules become. An indication of the influence of seeding density on the spread of *G. graminis* through soil from host to host comes from Wildermuth et al. (1984) who reported that *G. graminis* could colonize roots 9 to 14 mm from propagules in natural soils but grew only 2 mm through soil in the absence of wheat roots. Thus, it is not suprising that Clark (1942) and Chambers (1962) found that high sowing rates favour the spread of take-all. These findings are supported in practice—wheat crops sown at low rates (e.g. 35 kg ha^{-1}) develop less take-all under continuous wheat than crops sown at higher rates (60 kg ha^{-1}) (Manley, personal communication).

An important factor in the growth of *G. graminis* through soil to infect roots which is related to sowing rates is the energy base of the propagule. Wilkinson et al. (1985a) demonstrated that the larger the propagules from ground-colonized oat grain the greater the distance the pathogen grew through soil to infect wheat roots. Gilligan and Simons (1987) obtained similar results with both *G. graminis* and *R. solani* when they found that high-energy propagules (colonized millet seed) infected roots from a greater distance than did low-energy propagules (colonized sand grains).

As an alternative to different rates of seeding, plants can be arranged in rows of different spacings. Whilst pathogens such as *G. graminis* can grow several centimetres through soil as saprophytes (Grose et al., 1984; Glenn and Parker, 1988), field trials have shown little or no spread between rows 15 cm apart (Ryder, personal communication). In eastern Washington, the patches of take-all are more circular in fields planted with a narrow row spacing (15–25 cm), and more oblong with a wide spacing (40 cm), the long direction of the patch coinciding with the direction of sowing. While this observation confirms the epidemiological significance of a higher plant density and crowding, wide row spacings are not practical agronomically in these high-production systems because of the lower yield potential of the crop.

Paired-row spacings, i.e. rows in pairs spaced 12–15 cm apart with 35–40 cm between pairs of rows, can provide the advantage of optimal plant density for maximum yield potential while simultaneously limiting plant-to-plant spread of the pathogen. This method of sowing was made popular during recent years in the IPNW by the introduction of the "Yielder" no-till drill (Papendick et al., 1984) in which fertilizer is placed in a band 2–4 cm deeper than the seed between each pair of rows. Working with an experimental drill that plants wheat either in paired rows (17.5/42.5 cm) or uniformly spaced rows (30 cm), at Lind, Washington, during 1987 and 1988, Cook (unpublished) found that where root diseases were controlled or minimized by either crop rotation or planting date, yields of winter wheat were identical between paired and uniform row spacing. However, when take-all and rhizoctonia were present, paired row spacing resulted in less take-all and less rhizoctonia root rot in winter wheat but more rhizoctonia root rot in no-till spring barley. These data indicate no agronomic advantage of one method of sowing over the other if root pathogens are controlled, but suggest further that spacing, soil disturbance by the fertilizer shank at the time of sowing, or possibly placement of the fertilizer can be manipulated to advantage in fields where root pathogens are not controlled by other methods.

7. INFLUENCE OF FERTILIZERS

7.1. Root Pathogens

Fertilizers applied for crop production might also serve as a source of mineral nutrition for pathogens during their saprophytic existence in soil and during prepenetration growth *en route* to infection, but there is no evidence that these rhizosphere and root colonists benefit directly from the fertilizers either before or after establishment inside the host root. Survival structures such as oospores, chlamydospores and sclerotia contain the right combinations of energy reserves and nutrients to support both long-term survival and the growth necessary to reach a host root. Possibly an external source of nitrogen (if limiting) or some other essential mineral nutrient could extend longevity, efficiency of infection, or distance of prepenetration growth to initiate infection. Studies of such subtle but potentially significant effects of fertilizers on the biology of root pathogens and epidemiology of root diseases should be carried out.

Survival of the take-all fungus in host residues is enhanced by an external source of nitrogen (Garrett, 1940, 1976); these studies provide the first, and still the best, examples of a benefit of fertilizers to survival of a root pathogen. This classic work of Garrett showed that the ability of *G. graminis* to use the cellulose during its saprophytic survival in wheat tiller bases was limited by the amount of nitrogen in this host residue. As a consequence, *G. graminis* was vulnerable to displacement by competitors better adapted to a high C:N ratio. The addition of nitrogen allowed the pathogen as the pioneer colonist of the residue to maintain metabolism and keep possession *sensu* Bruehl (1975). The advantage to the pathogen is a greater energy supply for infection of the next host, i.e. a greater inoculum potential, but a disadvantage could be the more rapid exhaustion of the foodbase and hence more rapid disappearance of the fungus from the soil. To exploit this phenomenon in consecutive crops of wheat, Garrett (1956, 1970) recommended practices that kept the nitrogen status of the topsoil low during the period of saprophytic existence of the pathogen between wheat crops and high only as necessary to meet the nitrogen needs of the next crop. Today, the nitrogen can be injected below the zone of infected soil and thereby ‘‘feed’’ the crop but not the pathogen.

Garrett’s work has important implications in the cereal–pasture farming system of southern Australia, in which the aim is to use the legume–*Rhizobium* association to build up soil nitrogen sufficiently to supply most of the needs of the cereal crops; such levels would aid the survival of *G. graminis* in roots and crowns of grasses. To counter this effect it is necessary to develop appropriate management strategies to reduce the survival of *G. graminis*, e.g. use of selective herbicides to kill grasses early enough in the season to restrict growth of the pathogen in the grass roots and to ensure sufficient time for decomposition of these roots.

The availability of an external nitrogen supply can be beneficial to *Fusarium solani* f. sp. *phaseoli,* both for chlamydospore germination and prepenetration growth. During saprophytic growth in the rhizosphere, this pathogen, and possibly other *Fusarium* pathogens, depend on at least a portion of their nitrogen from the soil solution (Cook and Snyder, 1965). Snyder et al. (1959) showed that this dependency could be exploited by incorporating barley straw, sawdust or other material of high C:N ratio into the infested layer of soil prior to planting beans. The disease-controlling effect of such materials was offset if fertilizer nitrogen was also added. Both percentage chlamydospore germination and length of the germ tubes in response to an artificial supply of sugars and amino acids (to mimic root exudates) were significantly less in soil with barley straw, but not if additional nitrogen was added (Cook, 1962).

Probably the greatest effects of fertilizer on root pathogens are the indirect effects—those mediated through influence on the soil or rhizosphere pH, antagonism from the associated soil microbiota, or changes in host plant resistance. This is covered in Section 8.

Wheat is more susceptible to take-all if nutritionally deficient in any of several macro or minor nutrients, including P (Clark, 1942; Graham and Menge, 1982), Mg, K, Mn, Cu, Zn or Fe (Reis et al., 1982). Most Australian wheat-growing soils were low in P prior to the introduction of superphosphate, and in those days the common symptoms of the disease was seedling blight—hence the invention of the term ‘‘take-all’’ in Australia as early as the 1850s. With the improvement of P nutrition came the introduction of pasture legumes to provide an input of N into the soils; concomitantly, the field symptoms of the disease became the haying off of crops by anthesis due to a slower development of the disease in which the fungus and polyphenols block the water-conducting vessels in the plant (Foster et al., 1983) and induce water stress.

In the IPNW, P can become more limiting and take-all more severe with intensive cropping to wheat, but if take-all is controlled by soil fumigation the crop may not respond to P (Reis et al., 1982). Similar results were obtained with Zn and a mixture of Zn, Mn, Cu and Fe at another site in the IPNW, namely less take-all and higher yields with these trace nutrients in unfumigated soil but no response in fumigated soil.

The mechanisms by which these nutrients are thought to affect the take-all fungus are many and varied. P, Mg, K, Mn, Cn, Zn and Fe are involved in basal cell maintenance and metabolism and, as pointed out by Cook and Baker (1983), each affects the intrinsic susceptibility of wheat and hence the rate of disease development (Reis et al., 1982, 1983). Those nutrients critical to formation of new tissues, e.g. N, S and Ca, help the plant to outgrow the disease but do not affect intrinsic susceptibility. Graham and Menge (1982) postulated that P deficiency increased root exudation, which favoured the growth of *G. graminis* in the rhizosphere.

Graham and Rovira (1984) hypothesized that Mn-deficient wheat plants were more susceptible to infection and spread of *G. graminis* inside the roots—this may be connected with the role of Mn in lignification as a response by roots to infection. Field research by Wilhelm et al. (1988) showed that application of Mn, either to the soil or as a seed dressing, reduced take-all in wheat and increased grain yield—an important management strategy for reducing the impact of the disease in soils low in Mn.

7.2. Disease Suppression

The pattern of greater and lesser severity of take-all in response to fertilization of wheat with nitrate and ammonium respectively, might be explained by the respectively higher and lower rhizosphere pH values being unfavourable to take-all (Smiley and Cook, 1973) (but compare views in Chapter 9). This is a third-level interaction, namely antagonistic microorganisms are more abundant or effective at the lower pH in the rhizosphere induced by ammonium-N. Smiley (1978b) obtained further evidence for this explanation by showing that fluorescent pseudomonads inhibitory to *G. graminis* occurred at a higher frequency in the rhizosphere of wheat fertilized with ammonium-N than with nitrate-N. Using *P. fluorescens* strain 2-79 distinguishable from native bacteria by resistance to rifampicin and nalidixic acid in two acid soils (pH 5.0) unamended or amended with lime to pH 7.0 and with nitrogen either as nitrate alone or as ammonium plus soil nitrate, Howie (1985) found that populations of *P. fluorescens* 2-79 in the rhizosphere were higher in each of the two soils at rhizosphere pH values in the range of 6.0 to 6.5 than at 7.0 or 5.0. There was no response to lime or ammonium except as mediated through rhizosphere pH. Populations of 2-79 were depressed at pH 5.0, but a different mechanism of take-all suppression involving competition from saprophytic fungi in the foodbase occupied by the take-all pathogen is thought to account for the low incidence of take-all in soils of pH 5.0–5.5 and below (Cook and Weller, 1987). The studies by MacNish (1988) and Simon et al. (1988a) on the effects of nitrogenous fertilizers and liming in continuous wheat showed the importance of soil fungi, especially *Trichoderma* spp., in the suppression of growth in *G. graminis* through soil at the low soil pH developed from the use of ammonium-based fertilizer.

8. INFLUENCE OF pH AND LIME

For centuries it has been traditional to correct soil acidity by applying lime to increase productivity. Such applications can reduce plant toxicity from excess manganese and aluminium, hasten the decomposition of plant residues and mineralization of nitrogen, and increase earthworm activities.

8.1. Root Pathogens

The increase in incidence and severity of take-all with the addition of lime to acid soils has been shown in Australia (Butler, 1961; Garrett, 1970), and in the highly acidic, ferruginous soils of southern Brazil, where lime is added at $10\,t\,ha^{-1}$ to raise the pH from 4.7 to 6.7 to ensure nitrogen-fixation by the soybean–*Rhizobium* symbiosis (Anon. 1977; Reis et al., 1982).

Reis et al. (1983) conducted a series of sand culture experiments to ascertain whether the effect of liming on take-all was due to the added calcium *per se* or the pH change. They found that increasing the amount of calcium without a change in pH had no effect on the disease, whereas increasing the pH from 4.5 to 5.5 and above with sodium hydroxide increased disease incidence and severity. Considering the large areas of calcareous sandy soils of pH 8 or more used for wheat in Victoria, South Australia and Western Australia, the finding of Reis et al. (1983) that the severity of the disease steadily increased with each pH unit to 8.5 has important implications. It means that measures to control take-all, e.g. by eliminating grasses, will need to be more stringent in such soils than in soils with a neutral to acidic reaction.

Many soils in southern Australia have become acidic through the build-up of soil organic matter, through the fixation of nitrogen by clovers in improved pastures fertilized with superphosphate, or through leaching of nitrate-N (Helyar, 1976; Cregan et al., 1979; Coventry et al., 1987). Although liming acidic soils improves pasture production and nitrogen fixation and reduces concentrations of manganese and aluminium to non-toxic levels, it has often not resulted in expected increases in cereal production. Murray et al. (1987) applied lime at 0–5 t ha^{-1} at four sites in New South Wales, grew successive wheat crops without further applications of lime and for 3 years monitored root disease and soil pH. They found that the amount of take-all remained low (5%) at pH values (in 1:2.5 soil:0.01 M $CaCl_2$) between pH 4.1 (pre-lime) and 4.7 (after 2 t lime ha^{-1}) but increased to 10–36% at a pH raised to 5.4 with 5 t ha^{-1} of lime. For each 1% higher incidence of plants with blackened basal stems, they recorded 0.76% lower yield and found a correlation of −0.80 between yield and incidence of take-all. In north-eastern Victoria, Coventry and Kollmorgen (1987) found an association between the application of lime to subterranean clover pastures and the incidence of take-all in the first and second wheat crops sown after 3 years of pasture. Coventry et al. (1989) subsequently found that the yield losses in wheat associated with applications of lime to pasture could be reversed by using superphosphate amended with the fungicide flutriafol to control take-all. Field trials in Western Australia on the effect of form on nitrogen fertilizer with and without a single application of lime to raise the pH from 5.0 to 6.0 (1:2.5 soil/0.1 M $CaCl_2$) showed that the incidence of take-all increased from 17% to 28% over the 5 years following liming; without liming the low incidence of take-all achieved with continuous wheat and ammonium fertilizer was maintained (MacNish, 1988).

Fertilizing plants with predominantly ammonium-N results in rhizosphere pH values lower than the bulk soil, while fertilization with nitrate-N results in rhizosphere pH values higher than the bulk soil. Root pathogens most active or damaging in acid soils are favoured by fertilization of their host plant with ammonium-N, and those most active or damaging in neutral or alkaline soils, e.g. take-all, are favoured by fertilization with nitrate-N and are suppressed with the ammonium-N (Smiley, 1975).

In addition to temporary effects mediated through changes in rhizosphere pH, long-term use of nitrogen fertilizers has a permanent acidifying effect on the topsoil. Soils in the IPNW have dropped two or more pH units since anhydrous ammonia came into use nearly 40 years ago. The top 15–20 cm of some soils in the Palouse region of south-eastern Washington and adjacent Idaho have pH values as low as 4.5. Acid soils are highly favourable to cephalosporium stripe of winter wheat (Bruehl and Lai, 1968), a vascular wilt-type disease caused by *Cephalosporium graminearum*. The widespread occurrence of acid soils has made it imperative that growers do not grow winter wheat in the same field more than every third year (3-year rotation), or if a 2-year rotation is used, that they avoid highly susceptible cultivars.

Infection of wheat by *Pythium* has been shown to be influenced by soil pH by Fukui (1988) in a study with 25 soils collected from random wheat fields in the Palouse region and maintained at −0.01 MPa. He found that clay content and soil pH were important in determining embryo colonization: when a slightly alkaline soil was acidified, embryo colonization increased, and when the pH of an acid soil was raised, embryo colonization decreased. The soils of this region have up to 30–40% clay content, but the many years of fertilization with anhydrous ammonia, and resultant acidification, possibly have made them even more favourable to *Pythium* activity. This change can explain, at least partially, why *Pythium* damage to wheat has become such a yield-limiting factor in the region in only the past 10–20 years (Cook and Haglund, 1982; Cook et al., 1987).

8.2. Disease Suppression

Expression of the suppressiveness responsible for take-all decline is strongly influenced by soil pH, being most rapid or effective at pH values below neutral and retarded in alkaline soils. The relatively slow onset of take-all decline in the plot of irrigated wheat at Lind—15 years to maximum expression—possibly relates to the fact that the soil at this site was pH 7.5 when the experiment was started in 1967–68 and gradually dropped to about pH 6.0 by the 15th year (Cook, unpublished). In contrast to the Lind site, take-all decline occurred in western Washington in soil at pH 5.5–6.0 by the third year after two severe outbreaks, but took 6 years longer where the soil pH had been elevated by liming (Cook and Baker, 1983).

The complexity of the pH/lime effects on take-all were also demonstrated by Christensen and Brett (1985) who found that liming to raise soil pH from 5.5 to 6.6 increased the nitrification rate from NH_4Cl and $(NH_4)_2SO_4$ fertilizer. With this speeding up of nitrification following liming the ratio of NH_4^+–N to NO_3^-–N fell to below 3 and take-all became more severe. Their studies, which showed that a NH_4^+–N:NO_3^-–N ratio >3 reduced take-all, is consistent with the findings of Smiley and Cook (1973) and Smiley (1978a, 1978b) that NH_4^+–N fed roots had a lower pH and less disease than nitrate-fed roots.

These strong effects of rhizosphere pH on the suppression of take-all via the populations of pseudomonads may be linked with the pseudomonads' ability to produce antibiotics and the activities of antibiotics at different rhizosphere pHs. If, for example, the suppression is due to the production of phenazine-1-carboxylic acid, as found for *P. fluorescens* strain 2-79, then the activity of the antibiotic against the take-all fungus would be over 100 times greater at pH 5.2 than at pH 7.2 (Brisbane and Rovira, 1988).

On the sites used by MacNish (1988), Simon et al. (1988a) demonstrated that the liming reduced the ability of ammonium sulphate-treated soil to suppress the saprophytic growth of *G. graminis*. The activity of *Trichoderma* spp. in the ammonium sulphate-treated soil was reduced with lime treatment, which suggests that the build-up and activity of *Trichoderma* spp. is involved in the suppression in these soils where the pH is lowered.

8.3. pH Measurement

Whilst on the topic of soil pH and its importance in the ecology and epidemiology of take-all, we should stress that most plant pathologists and soil microbiologists often make no reference to the method of preparing the soil suspension for pH measurements. This makes it difficult to compare results from different laboratories. Readers are referred to the paper by Smiley and Cook (1972) entitled "Use and abuse of the soil pH measurement", in which they report that in one soil, the pH ranged from 6.4 to 8.5 depending upon the method used. In South Australia differences of one pH unit can be obtained for a single soil sample depending whether soil suspensions are made in water or in $CaCl_2$ (Rovira and Simon, unpublished).

9. INFLUENCE OF SOIL WATER

9.1. Root Pathogens

There has been a great amount of research conducted on this topic and the reader is referred to two reviews (Cook and Papendick, 1972; Cook and Duniway, 1981), and the book by Cook and Baker (1983).

The growth rate of *G. graminis* on agar in which the osmotic potential has been controlled to different levels shows that maximum growth rate occurs at −0.12 to −1.12 MPa (Cook et al., 1972; Grose et al., 1984). Other studies conducted with soil using the membrane soil sandwich method (Grose et al., 1984) and growth along a bisected length of autoclaved straw (Cook et al., 1972) have shown that maximum growth occurred at the wettest end of the moisture curve. The sensitivity of *G. graminis* to soil water content indicates that the growth of this pathogen in the rhizosphere and as a pathogen inside the root will be markedly affected by soil water.

Under the dryland farming conditions of southern Australia excess water is seldom a problem so the proliferation and survival of *G. graminis* generally occurs at water contents below field capacity. This has been demonstrated by Roget and Rovira (1990) in long-term field trials which show the relationship between disease incidence and the rainfall in the previous season, and also show that the September rainfall affects the yield loss suffered by the wheat crop for any given level of disease. These results reflect the need of the take-all fungus for sufficient soil water to grow through soil, colonize the rhizosphere, and infect and colonize roots sufficiently to form propagules that remain viable through to the following growing season and develop sufficiently in the following wheat crop to reduce yield.

These results are consistent with findings in both Australia and the USA that take-all can cause catastrophic losses to irrigated wheat or crops receiving supplementary spring irrigation.

9.2. Deleterious Rhizosphere Bacteria

The detrimental rhizosphere pseudomonads grow over a wide range of water potentials in the laboratory and hence would be expected to operate over a similar range in the field. When pseudomonads detrimental against winter wheat were sprayed onto standing barley residues in a warm, dry autumn, there were high numbers on the rhizoplane of winter wheat sown into the barley residues without tillage, indicating a spread from surface straw down into the profile with water movement (Stroo et al., 1988).

10. INFLUENCE OF HERBICIDES

Herbicides are an integral part of modern agriculture, making up 50% of all pesticides used and with world sales valued at $7100 million (Strohbusch, 1974; Sommerville, 1987). Despite their widespread use as a part of cropping management, scant attention is paid to the impact of herbicides on the rhizosphere microflora, whether they be pathogens or non-pathogens.

10.1. Root Pathogens

In a recent review of this topic, Altman and Rovira (1989) list many instances in which herbicides have increased the damage from root diseases. One example is the carbamate herbicide (Cycloate) which predisposed sugar beet to increased rhizoctonia disease (Altman and Campbell, 1977) and increased damage due to the beet cyst nematode (*Heterodera schachtii*) (Abivardi and Altman, 1978). Another example is cotton seedling blight caused by *Rhizoctonia solani*, which is increased following treatment of soil with the preplant herbicide, trifluralin (Chandler and Santelmann, 1968).

In Australia chlorsulfuron (Glean®), applied to a calcareous sandy loam of pH 8.3 for weed control in wheat, increased root disease caused by *R. solani* in a following barley crop 1 year later and reduced yield by up to 1 t ha^{-1} (Rovira and McDonald, 1986). Glasshouse experiments demonstrated that Glean applied at the equivalent of 5 g ha^{-1} (one-third the recommended rate) increased root disease caused by *R. solani*. No effect of Glean was detected on the early infection of wheat seedling roots by the take-all fungus.

The report by Trottman and Thompson (1978) that the herbicides mecoprop, ioxynil, dicamba and a dicamba/2,3,6-TBA-mecoprop-MCPA mixture all increased take-all lesions on roots by an average of 60%, with dicamba giving the greatest increase, has not been followed up despite its important implications, considering the global significance of take-all and the widespread use of herbicides on wheat.

A graphic example of the long-term effects of herbicides was demonstrated in South Australia in 1988. In 1985 trial plots were sprayed with a dicamba-MCPA mixture, while the surrounding farmer's crop received MCPA only; in that year some whiteheads appeared in the plots but none appeared in the surrounding crops. The whole field was under annual grass—*Medicago* spp.

pasture in 1986, 1987—and sown to wheat in 1988. In 1988 the area which had been in plots (and received dicamba) in 1985 was decimated with take-all and yielded 0.39 t ha^{-1} while the surrounding crop yielded 2.23 t ha^{-1}. Dicamba would not persist in the soil over the 3 years, so it is probable that in 1985 the herbicide increased take-all infection of the roots and crowns of the wheat in the plots. This higher infestation with the take-all fungus would be maintained by the grasses in the pasture over two seasons and result in a heavy infection in the wheat planted in 1988.

Glyphosate (Roundup®) is used worldwide to kill weeds prior to planting crops by direct-drill or no-till, and in most cases there are no obvious ill effects through increased root diseases. However, in Western Australia extensive crop losses occurred following the use of glyphosate immediately before the direct drilling of cereals; these losses were largely attributed to a seedling root rot caused by *Pythium*, which built up to high levels on the pasture residues after glyphosate use (Blowse, 1987).

Research in eastern Washington and north-eastern Oregon has shown the same effect for rhizoctonia root rot, an effect dubbed "Roundup injury". The pathogen increased its inoculum potential on the roots of volunteer cereals and grass weeds under stress from the herbicides and was therefore more destructive on the next crop sown into this dead and dying mass of plants (Ogg, Cook and Smiley, unpublished). The problem was largely avoided by spraying the weeds and volunteers at least 2 weeks before planting rather than a few days before as had been practised hitherto. A similar strategy of "chemical fallowing" with herbicide 3 to 4 weeks before direct drilling wheat or barley to reduce rhizoctonia root rot has been developed in South Australia (Roget et al., 1987).

Little is known of the mechanisms by which the herbicides influence the rhizosphere microflora, symbiotic nitrogen fixers, mycorrhizas or root pathogens. They may affect the exudation of organic substrates from roots into the rhizosphere: Altman and Campbell (1977) reported increased glucose exudate from the hypocotyl of sugar beet in the presence of Ro-neet (S-ethyl-N-ethyl-thiocyclohexane carbamate), which would stimulate the rhizosphere microflora and cause lower calcium levels in the plant tissues which may reduce resistance to root pathogens.

11. CONCLUSIONS

In this chapter we have brought together some examples of how cropping systems influence rhizosphere organisms that affect plant health and crop production. There is a vast literature on this topic and hence we have restricted much of this review to results of our own research in the USA and Australia. Nevertheless, this research is sufficiently broad to cover many crop and soil management systems associated with broad-acre agriculture. Most of our examples originate from field observations and we consider it important that this more applied aspect of rhizosphere microbiology be explored further.

With the tremendous demands for increased world food production likely in the future, there is the need to develop cropping systems that, while achieving maximum productivity, improve soil properties and prevent land degradation. Root health is an essential component of efficient cropping systems, and an understanding of the microbiology of the rhizosphere under different management systems will enable the development of strategies that will meet the above criteria of maximum productivity and improvement of the biological, physical and chemical properties of the soil.

REFERENCES

Abivardi, C. and J. Altman (1978). Effects of cycloate on development of *Heterodera schachtii* and growth of three *Beta* species. *J. Nematol.*, **10**:90–94.

Adams, D. B. and T. T. Colquhoun (1935). The spread of take-all through soil. *J. Aust. Inst. Agric. Sci.*, **1**:172–174.

Allmaras, R. R., J. M. Kraft, and D. E. Miller (1988). Effects of soil compaction and incorporated crop residue on root health. *Ann. Rev. Phytopathol.*, **26**:219–243.

Altman, J. and C. L. Campbell (1977). Pesticide plant disease interactions: effect of cycloate on sugar beet damping-off induced by *Rhizoctonia solani. Phytopathology*, **67**:1163–1165.

Altman, J. and A. D. Rovira (1989). Herbicide–pathogen interactions in soil-borne root diseases. *Can. J. Pl. Path.* (in press).

Anon (1977). Biennial Report from Wheat Research Centre, Passo Fundo, 1976–77.

Baker, K. F. and R. J. Cook (1974). *Biological Control of Plant Pathogens*. W. H. Freeman and Co., San Francisco, 433 pp. (Book reprinted in 1982, American Phytopathological Society, St Paul, MN.)

Baker, K. F. and W. C. Snyder (eds.) (1965). *Ecology of Soil-Borne Plant Pathogens. Prelude to Biological Control*. University of California Press, Berkeley, 571 pp.

Blowse, W. M. (1987). Effect of rye grass residues, knockdown herbicides, and fungicides on the emergence of barley in sandy soils. *Aust. J. Exp. Agric.* **27**:785–790.

Bolton Jr., H. and L. F. Elliott (1989). Toxin production by a rhizobacterial *Pseudomonas* sp. that inhibits wheat root growth. *Plant and Soil,* **114**:269–278.

Bolton Jr., H., L. F. Elliott, S. Gurusiddaiah, and J. K. Fredrickson (1989). Characterization of a toxin produced by a rhizobacterial *Pseudomonas* sp. that inhibits wheat growth. *Plant and Soil*, **114**:279–287.

Bolton Jr., H., J. K. Fredrickson, A. C. Kennedy, H. F. Stroo, S. W. Li, and L. F. Elliott (1990). Generation and characterization of Tn5 TOX^- mutants of winter wheat (*Triticum aestivum* L.) root inhibitory rhizobacteria. *Soil Biol. Biochem*. (submitted).

Bratoloveanu, J. I. (1985). Species of *Pythium* associated with barley in South Australia. Ph.D. Dissertation, University of Adelaide, South Australia, 158 pp.

Brisbane, P. G. and A. D. Rovira (1988). Mechanisms of inhibition of *Gaeumannomyces graminis* var. *tritici* by fluorescent pseudomonads. *Plant Path.*, **37**:104–111.

Brisbane, P. G., L. J. Janik, M. E. Tate, and R. F. O. Warren (1987). Revised structure for the phenazine antibiotic from *Pseudomonas fluorescens* (NRRL B-15132). *Antimicrobial Agents and Chemotherapy,* **31**:1967–1971.

Bruehl, G. W. (1975). Systems and mechanisms of residue possession by pioneer fungal colonists. In *Biology and Control of Soil-borne Plant Pathogens* (G. W. Bruehl, ed.). American Phytopathological Society, St Paul, MN, pp. 77–83.

Bruehl, G. W. and P. Lai (1968). The probable significance of saprophytic colonization of wheat straw in the field by *Cephalosporium graminearum. Phytopathology*, **58**:464–466.

Burke, D. W., D. E. Miller, L. D. Holmes, and A. W. Barker (1972). Counteracting bean root rot by loosening the soil. *Phytopathology*, **62**:306–309.

Butler, F. C. (1953). Saprophytic behaviour of some cereal root rot fungi. I. Saprophytic colonization of wheat straw. *Ann. Appl. Biol.*, **40**:284–297.

Butler, F. C. (1961). Root and foot rot diseases of wheat. *Sci. Bull. Dept Agric. NSW*, **77**:1–98.

Cardoso, J. E. and E. Echandi (1987). Nature of protection of bean seedlings from Rhizoctonia root rot by a binucleate *Rhizoctonia*-like fungus. *Phytopathology,* **77**:1548–1551.

Chambers, S. C. (1962). Root diseases in wheat on clover-ley—factors under investigation. 3. The effect of seeding rate. *J. Agric. West. Aust.,* **3**:719–720.

Chamswarng, C. (1984). Etiology and epidemiology of *Pythium* root rot of wheat. Ph.D. Dissertation, Washington State University, Pullman, 161 pp.

Chamswarng, C. and R. J. Cook (1985). Identification and comparative pathogenicity of *Pythium* species from wheat roots and wheat-field soils in the Pacific Northwest. *Phytopathology,* **75**:821–827.

Chan, K. Y., J. A. Mead, and W. P. Roberts (1987). Poor early growth of wheat under direct drilling. *Aust. J. Agric. Res.,* **38**:791–800.

Chandler, J. M. and P. W. Santelmann (1968). Interactions of four herbicides with *Rhizoctonia solani* on seedling cotton. *Weed Sci.,* **16**:453–456.

Cherrington, C. A. and L. F. Elliott (1987). Incidence of inhibitory pseudomonads in the Pacific Northwest. *Plant and Soil,* **101**:159–165.

Christensen, N. W. and M. Brett (1985). Chloride and liming effects on soil nitrogen form and take-all of wheat. *Agron. J.,* **77**:157–163.

Clark, F. E. (1942). Experiments towards the control of the take-all disease of wheat and the Phymatotrichum root rot of cotton. *USDA Tech. Bull. No. 835*, 1–27.

Claydon, N., M. Allan, J. R. Hanson, and A. G. Avent (1987). Antifungal alkyl pyrones of *Trichoderma harzianum. Trans. Br. Mycol. Soc.*, **88**:503–513.

Cochran, V. L., L. F. Elliott, and R. I. Papendick (1977). The production of phytotoxins from surface crop residues. *Soil Sci. Soc. Am. J.*, **41**:903–908.

Cochran, V. L., L. F. Elliott, and R. I. Papendick (1980). Carbon and nitrogen movement from surface-applied wheat (*Triticum aestivum*) straw. *Soil Sci. Soc. Amer. J.*, **44**:978–982.

Cook, R. J. (1962). Influence of barley straw on the early stages of pathogenesis in Fusarium root rot of bean. *Phytopathology,* **52**:728.

Cook, R. J. (1981a). The influence of rotation crops on take-all decline phenomenon. *Phytopathology*, **71**:189–192.

Cook, R. J. (1981b). The effect of soil reaction and physical conditions. In *Biology and Control of Take-all* (M. J. C. Asher and P. J. Shipton, eds.). Academic Press, London, pp. 343–352.

Cook, R. J. (1984). Root health: importance and relation to farming systems. In *Organic Farming: Current Technology and Its Role in a Sustainable Agriculture* (D. F. Bezdicek and J. F. Power, eds.). Amer. Soc. Agron. Special Publication 26, 192 pp.

Cook, R. J. (1986). Wheat management systems in the Pacific Northwest. *Plant Disease*, **70**:894–898.

Cook, R. J. (1989). Biological control and holistic plant care in agriculture. *Am. J. Altern. Agric.*, **3**:51–62.

Cook, R. J. and K. F. Baker (1983). *The Nature and Practice of Biological Control of Plant Pathogens*. American Phytopathological Society, St Paul, MN, 539 pp.

Cook, R. J. and J. M. Duniway (1981). Water relations in the life cycles of soil-borne plant diseases. In *Water Potential Relations in Soil Microbiology* (J. F. Parr, ed.). Amer. Soc. Agron. Special Publication, Madison, Wisconsin.

Cook, R. J. and W. A. Haglund (1982). *Pythium Root Rot: A Barrier to Yield of Pacific Northwest Wheat*. Washington State College of Agriculture Res. Bull. No. XB0913, 20 pp.

Cook, R. J. and R. I. Papendick (1972). Influence of water potential of soils and plants in root disease, *Ann. Rev. Phytopathology,* **10**:349–372.

Cook, R. J. and A. D. Rovira (1976). The role of bacteria in the biological control of *Gaeumannomyces graminis* by suppressive soils. *Soil Biol. and Biochem.*, **8**:269–273.

Cook, R. J. and W. C. Snyder (1965). Influence of host exudates on growth and survival of germlings of *Fusarium solani* f. *phaseoli* in soil. *Phytopathology*, **55**:1021–1025.
Cook, R. J. and D. M. Weller (1987). Management of take-all in consecutive crops of wheat or barley. In *Innovative Approaches to Plant Disease Control* (I. Chet, ed.). John Wiley & Sons, New York, pp. 41–76.
Cook, R. J., R. I. Papendick, and D. M. Griffin (1972). Growth of two root-rot fungi as affected by osmotic and matric water potential. *Soil Sci. Soc. Amer. Proc.*, **36**:78–82.
Cook, R. J., J. W. Sitton, and J. T. Waldher (1980). Evidence for *Pythium* as a pathogen of direct-drilled wheat in the Pacific Northwest. *Plant Disease*, **64**:102–103.
Cook, R. J., H. T. Wilkinson, and J. R. Alldredge (1986). Evidence that microorganisms in suppressive soil associated with wheat take-all decline do not limit the number of lesions produced by *Gaeumannomyces graminis* var. *tritici*. *Phytopathology*, **76**:342–345.
Cook, R. J., J. W. Sitton, and W. A. Haglund (1987). Increased growth and yield responses of wheat to reduction in the *Pythium* populations by soil treatments. *Phytopathology*, **77**:1192–1198.
Cook, R. J., L. S. Thomashow, and D. M. Weller (1988). Fungal–bacterial interactions in the root region, with special reference to antagonism of wheat root-infecting fungi by rhizobacteria. In *Proceedings of the Fourth International Symposium on Microbial Ecology, Ljubljana, Yugoslavia, 24–29 August, 1986*, pp. 401–405.
Cook, R. J., C. Chamswarng, and W-N. Tang (1990). Influence of wheat chaff and tillage on *Pythium* populations in soil and *Pythium* damage to wheat. *Soil Biol. Biochem.*, **22**:(in press).
Cornish, P. S. and J. E. Pratley (1987). *Tillage. New Directions in Australian Agriculture*. Inkata Press, Melbourne, Sydney, 448 pp.
Cotterill, P. J. and K. Sivasithamparam (1988). Effect of sowing date on take-all of wheat in Western Australia. *Phytophylactica*, **20**:11–14.
Coventry, D. R. and J. F. Kollmorgen (1987). An association between lime application and the incidence of take-all symptoms on wheat on an acid soil in north-eastern Victoria. *Aust. J. Exp. Agric.*, **27**:695–699.
Coventry, D. R., T. G. Reeves, H. D. Brooke, and W. J. Slattery (1987). Increasing wheat yields in north-eastern Victoria by liming and deep ripping. *Aust. J. Exp. Agric.*, **27**:33–35.
Coventry, D. R., H. D. Brooke, J. F. Kollmorgen, and D. J. Ballinger (1989). Increases in wheat yield on limed soil after reduction of take-all by fungicide application and crop rotation. *Aust. J. Exp. Agric.*, **29**:85–89.
Cregan, P. D., J. A. Sykes, and A. J. Dymock (1979). Pasture improvement and soil acidification. *Agric. Gaz. of N.S.W.*, **90(5)**:33–35.
Croft, B. J. and R. C. Magarey (1984). Pathogenic fungi associated with northern poor root syndrome of sugar cane. *Proc. Aust. Soc. Sugar Cane Technol. 1984 Conf.*, 55–61.
Croft, B. J., J. R. Reghenzani, and A. P. Hurney (1984). Northern poor root syndrome of sugar cane—studies on soil transmission and the effects of various fungicidal, nutritional and agronomic treatments. *Proc. Aust. Soc. Sugar Cane Technol. 1984 Conf.*, 69–77.
Davidson, J. M. and P. W. Santleman (1973). *An Evaluation of Various Tillage Systems for Wheat*. Oklahoma Agric. Exp. Stn Bull. B711, 17 pp.
Deacon, J. W. (1974). Interactions between varieties of *Gaeumannomyces graminis* var. *tritici* and *Phialophora radicicola* on roots, stem bases, and rhizomes of the Gramineae. *Plant Pathol.*, **25**:85–92.
Dick, M. W., B. J. Croft, R. C. Magarey, A. W. A. M. de Cock, and G. Clark, (1989). A new genus of the *Verrucalvaceae (Oomycetes)*. *Bot. J. Linn. Soc.* **99**:97–113.
Diehl, J. A., R. D. Tinline, R. A. Kochlann, J. Shipton, and A. D. Rovira (1982). The effect of fallow periods on common root rot of wheat in Rio Grande du Sol., Brazil. *Phytopathol.*, **72**:1297–1301.
Doupnik, B., Jr. and M. G. Boosalis (1980). Ecofallow—a reduced tillage system—and plant diseases. *Plant Disease*, **64**:31–35.
Elliott, L. F. and A. C. Kennedy (1988). Method for screening bacteria and application thereof for field control of the weed downy brome. Patent application 6/2/1988, filed 6/6/1988 under serial number 7-207592.

Elliott, L. F. and J. M. Lynch (1984). Pseudomonads as a factor in the growth of winter wheat (*Triticum aestivum* L.). *Soil Biol. Biochem.*, **16**:68–71.

Elliott, L. F. and J. M. Lynch (1985). Plant growth-inhibitory pseudomonads colonizing winter wheat (*Triticum aestivum* L.) roots. *Plant and Soil*, **84**:57–65.

Elliott, L. F., R. I. Papendick, and V. L. Cochran (1984). Phytotoxicity and microbial effects on cereal growth. In *Proceedings of the Sixth Manitoba-North Dakota Tillage Workshop, Bismark, North Dakota, January, 1984*. North Dakota State University Press, Fargo, ND, pp. 40–47.

Ferguson, W. S. and B. J. Gorby (1964). Effect of straw on availability of nitrogen to cereal crops. *Can. J. Soil Sci.*, **44**:286–291.

Foster, R. C. (1988). Microenvironments of soil microorganisms. *Biol. Fertil. Soils*, **6**:189–203.

Foster, R. C., A. D. Rovira, and T. W. Cook (1983). *Ultrastructure of the Root–Soil Interface*. American Phytopathological Society, St Paul, MN, 157 pp.

Fredrickson, J. K. and L. F. Elliott (1985a). Colonization of winter wheat roots by inhibitory rhizobacteria. *Soil Sci. Soc. Amer. J.*, **49**:1172–1177.

Fredrickson, J. K. and L. F. Elliott (1985b). Effects on winter wheat seedling growth by toxin-producing rhizobacteria. *Plant and Soil*, **83**:399–409.

Fredrickson, J. K., L. F. Elliott, and J. C. Engibaus (1987). Crop residues as substrate for host-specific inhibitory pseudomonads. *Soil Biol. Biochem.*, **19**:127–134.

Fukui, R. (1988). Factors affecting the inoculum density–embryo infection relationship for *Pythium* in germinating seeds of wheat. M.S. Thesis, Washington University, Pullman, 104 pp.

Gardner, J. M., J. L. Chandler, and A. W. Feldman (1984). Growth promotion and inhibition by antibiotic-producing fluorescent pseudomonads on citrus roots. *Plant and Soil*, **77**:103–113.

Garrett, S. D. (1940). Soil conditions and the take-all disease of wheat. V. Further experiments on the survival of *Ophiobolus graminis* in infected wheat stubble buried in soil. *Ann. Appl. Biol.*, **27**:199–204.

Garrett, S. D. (1942). The take-all disease of cereals. *Tech. Comm. Bur. Soil Sci. Harpenden* **41**:40 pp.

Garrett, S. D. (1956). *Biology of Root-Infecting Fungi*. Cambridge University Press, London, 294 pp.

Garrett, S. D. (1970). *Pathogenic Root-Infecting Fungi*. Cambridge University Press, London, 294 pp.

Garrett, S. D. (1976). Influence of nitrogen on cellulolysis rate and saprophytic survival in soil of some cereal root-rot fungi. *Soil Biol. Biochem.*, **8**:229–294.

Glenn, O. F. and C. A. Parker (1988). Growth and infectivity of *Gaeumannomyces graminis* var. *tritica* in soil. *Soil Biol. Biochem.* **20**:525–576.

Gilligan, C. A. (1979). Modeling rhizosphere infections. *Phytopath.*, **69**:782–784.

Gilligan, C. A. (1980a). Dynamics of root colonization by the take-all fungus, *Gaeumannomyces graminis*. *Soil Biol. Biochem.*, **12**:507–512.

Gilligan, C. A. (1980b). Zone of potential infection between host roots and inoculum units of *Gaeumannomyces graminis*. *Soil Biol. Biochem.*, **12B**:513–514.

Gilligan, C. A. and S. A. Simons (1987). Inoculum efficiency and pathozone width for two host–parasite systems. *New Phytol.*, **107**:549–566.

Graham, J. H. and Menge, J. A. (1982). Influence of vesicular-arbuscular mycorrhizae and soil phosphorus on take-all disease. *Phytopath.*, **72**:95–98.

Graham, R. D. and A. D. Rovira (1984). A role for manganese in the resistance of wheat to take-all. *Pl. Soil*, **78**:441–448.

Griffith, D. R., J. V. Mannering, and J. E. Box (1983). Soil moisture management with reduced tillage. In *No-Tillage and Surface-Tillage Agriculture: The Tillage Revolution*.(M. A. Sprague and G. B. Triplett, eds.). John Wiley & Sons, New York, pp. 19–57.

Grose, M. J., C. A. Parker, and K. Sivasithamparam (1984). *Soil Biol. Biochem*, **16**:211–216.

Hancock, J. G. (1977). Factors affecting soil populations of *Pythium ultimum* in the San Joaquin Valley of California. *Hilgardia*, **45**:107–122.

Harris, J. R. and Kimber, R. W. L. (1983). Phytotoxins in soil. In *Soils: An Australian Viewpoint*. CSIRO/Academic Press, Melbourne, pp. 741–757.

Harris, R. F. (1981). Effect of water potential on microbial growth and activity. In *Water Potential Relations in Soil Microbiology* (J. F. Parr, W. R. Gardner, and L. F. Elliott, eds.). Soil Science Society of America Special Publication No. 9, American Society of Agronomy, Madison, WI, pp. 23–96.

Hattori, T. (1973). *Microbial Life in the Soil. An Introduction*. Marcel Dekker Inc., New York, 427 pp.

Helyar, K. R. (1976). Nitrogen cycling and soil acidification. *J. Aust. Inst. Agric. Sci.*, **42**:217–221.

Hering, T. F., R. J. Cook, and W. H. Tang (1987). Infection of wheat embryos by *Pythium* species during seed germination and the influence of seed age and soil matric potential. *Phytopathology*, **77**:1104–1108.

Howie, W. J. (1985). Factors affecting colonization of wheat roots and suppression of take-all by pseudomonads antagonistic to *Gaeumannomyces graminis* var. *tritici*. Ph.D. Thesis, Washington State University, Pullman, 82 pp.

Ingram, D. M. (1987). Pathogenicity of *Pythium* species on major crops in the Palouse. Ph.D. Thesis, Washington State University, Pullman, 75 pp.

Ingram, D. M., R. J. Cook, and C. Chamswarng (1990). Pathogenicity of four *Pythium* species to wheat, barley, peas, and lentils. *Plant Pathology* (in press).

Katan, J. (1980). Solar pasteurization of soils for disease control: status and prospects. *Plant Disease*, **64**:450–454.

Kerry, B. R. (1981). Fungal parasites: a weapon against cyst nematodes. *Plant Disease*, **65**:390–393.

Kerry, B. R., A. Simon, and A. D. Rovira (1984). Observations on the introduction of *Verticillium chlamydosporium* and other parasitic fungi into soil for the control of cereal cyst nematode, *Heterodera avenae*. *Ann. Appl. Biol.*, **105**:509–516.

Kimber, R. W. L. (1967). Phytotoxicity from plant residues. I. The influence of rotted wheat straw on seedling growth. *Aust. J. Agric. Res.*, **18**:361–374.

Kimber, R. W. L. (1973a). Phytotoxicity from plant residues. II. The effect of time of rotting of straw from some grasses and legumes on the growth of wheat seedlings. *Plant Soil*, **38**:347–361.

Kimber, R. W. L. (1973b). Phytotoxocity from plant residues. III. The relative effect of toxins and nitrogen immobilization on the germination and growth of wheat. *Plant Soil*, **38**:543–555.

Kollmorgen, J. F. and D. N. Walsgott (1984). Saprophytic survival of *Gaeumannomyces graminis* var. *tritici* at various depths in soil. *Trans. Br. Mycol. Soc.*, **82**:346–348.

Kollmorgen, J. F., P. E. Ridge, and R. F. de Boer (1987). Effects of tillage and straw mulches on take-all in the northern Wimmera of Victoria. *Aust. J. Exp. Agric.*, **27**:419–423.

Lynch, J. M. (1977). Phytotoxicity of acetic acid produced in the anaerobic decomposition of wheat straw. *J. Applied Bact.*, **42**:81–87.

Lynch, J. M. (1978). Production and phytotoxicity and acetic acid in anaerobic soils containing plant residues. *Soil Biol. Biochem.*, **10**:131–135.

Lynch, J. M., K. B. Gunn, and L. M. Pantling (1980). On the concentration of acetic acid in straw and soil. *Pl. Soil*, **56**:93–98.

Lynch, J. M., F. B. Ellis, S. H. T. Harper, and D. G. Christian (1981). The effect of straw on the establishment and growth of winter cereals. *Agriculture and Environment*, **5**:321–328.

MacNish, G. C. (1984). The use of undisturbed soil cores to study methods of controlling Rhizoctonia patch of cereals. *Plant Pathol.*, **33**:355–359.

MacNish, G. C. (1985). Methods of reducing Rhizoctonia patch of cereals in Western Australia. *Plant Pathol.*, **34**:175–181.

MacNish, G. C. (1988). Changes in take-all (*Gaeumannomyces graminis* var. *tritici*), rhizoctonia root rot (*Rhizoctonia solani*) and soil pH in continuous wheat with annual applications of nitrogenous fertilizer in Western Australia. *Aust. J. Exp. Agric.*, **28**:333–341.

MacNish, G. C. and D. A. Nicholas (1987). Some effects of field history on the relationship between grass production in subterranean clover pasture, grain yield and take-all (*Gaeumannomyces graminis* var. *tritici*) in a subsequent crop of wheat at Bannister, Western Australia. *Aust. J. Agric. Res.*, **38**:1011–1018.

Masle, J. and J. B. Passioura (1987). The effects of soil strength on the growth of young wheat plants. *Aust. J. Pl. Physiol.*, **14**:643–647.

McCalla, T. M. and T. J. Army (1961). Stubble-mulch farming. *Adv. Agron.*, **13**:125–196.

Meagher, J. W. and R. H. Brown (1974). Microplot experiments on the effect of plant hosts on populations of the cereal cyst nematode (*Heterodera avenae*) and on the subsequent yield of wheat. *Nematologica,* **20**:337–346.

Meagher, J. W. and D. R. Rooney (1966). The effect of crop rotations in the Victorian Wimmera on the cereal cyst nematode (*Heterodera avenae*), nitrogen fertility and wheat yield. *Aust. J. Exp. Agric. Anim. Husb.,* **6**:425–431.

Meagher, J. W., R. A. Brown, and A. D. Rovira (1978). The effects of cereal cyst nematode (*Heterodera avenae*) and *Rhizoctonia solani* on the growth and yield of wheat. *Aust. J. Agr. Res.,* **29**:1127–1137.

Moore, K. J. and R. J. Cook (1984). Increased take-all of wheat with direct drilling in the Pacific Northwest. *Phytopathology,* **74**:1044–1049.

Murray, G. M., B. J. Scott, Z. Hochman, and B. J. Butler (1987). Failure of liming to increase grain yield of wheat and triticale in acid soils may be due to the associated increase in the incidence of take-all (*Gaeumannomyces graminis* var. *tritici*). *Aust. J. Exp. Agric.,* **27**:411–417.

Neate, S. M. (1987). Plant debris in soil as a source of inoculum of *Rhizoctonia* in wheat. *Trans. Br. Mycol. Soc.,* **88**:157–162.

Neate, S. M. and J. H. Warcup (1985). Anastomosis grouping of some isolates of *Thanatephorus cucumeris* in South Australia. *Trans. Br. Mycol. Soc.,* **85**:615–620.

Norstadt, F. A. and T. M. McCalla (1968). Microbially induced phytotoxicity in stubble-mulched soil. *Soil Sci. Soc. Am. Proc.,* **32**:241–245.

Ogoshi, A., R. J. Cook, and E. N. Bassett (1990). Rhizoctonia species and anastomosis groups causing root rot of wheat and barley in the Pacific North-West. *Phytopathology,* **80**:(in press).

Pankhurst, C. E. and H. J. McDonald (1988). Influence of tillage and crop rotation on the ecology of *Pythium* spp. in a South Australian wheat-growing soil. *Proc. 5th ICPP, Kyoto, Japan,* Abstracts V. 1–6, p. 163.

Papendick, R. I. and R. J. Cook (1974). Plant water stress and development of *Fusarium* root rot in wheat subjected to different cultural practices. *Phytopathology,* **64**:358–363.

Papendick, R. I. and L. F. Elliott (1983). Soil physical factors that affect plant health. In *Challenging Problems in Plant Health* (T. Kommendahl and P. H. Williams, eds.). American Phytopathological Society, St Paul, MN, pp. 168–180.

Papendick, R. I. and D. E. Miller (1977). Conservation tillage in the Pacific Northwest. *Journal of Soil and Water Conservation,* **32**:49–56.

Papendick, R. I., M. J. Lindstrom, and V. L. Cochran (1973). Soil mulch effects on seedbed temperature and water during fallow in eastern Washington. *Proc. Soil Sci. Soc. Amer.,* **37**:307–314.

Papendick, R. I., L. F. Elliott, V. L. Cochran, and L. A. Morrow (1984). *Paired-Row Concept for No-Till Planting.* Agronomy Abstracts, American Society of Agronomy, Madison, WI, p. 252.

Reghenzani, J. R. (1988). Northern sugar cane response to soil solarization. *Proc. Aust. Soc. Sugar Cane Technol. 1988 Conf.*, 163–169.

Reis, E. M., R. J. Cook, and B. L. McNeal (1982). Effect of mineral nutrition on take-all of wheat. *Phytopathology,* **72**:224–229.

Reis, E. M., R. J. Cook, and B. L. McNeal (1983). Elevated pH and associated reduced trace-nutrient availability as factors contributing to take-all upon soil liming. *Phytopathology,* **73**:411–413.

Roget, D. K. and A. D. Rovira (1985). Effect of tillage on *Heterodera avenae*. In *Ecology and Management of Soil-borne Plant Pathogens* (C. A. Parker, A. D. Rovira, K. J. Moore, P. T. W. Wong, and J. F. Kollmorgen, eds.). American Phytopathological Society, St Paul, pp. 252–254.

Roget, D. K. and A. D. Rovira (1990). The relationships between incidence of take-all infection of wheat, rainfall and yield. *Aust. J. Exp. Agric.* (in press).

Roget, D. K., N. R. Venn, and A. D. Rovira (1987). Reduction of Rhizoctonia root rot of direct drilled wheat by short term chemical fallow. *Aust. J. Exp. Agric.,* **27**:425–430.

Rothrock, C. S. (1987). Take-all of wheat as affected by tillage and wheat-soybean double cropping. *Soil Biol. Biochem.*, **19**:307–311.

Rovira, A. D. (1976). Studies on soil fumigation. I. Effects of ammonium, nitrate and phosphate in soil and on the growth, nutrition and yield of wheat. *Soil Biol. Biochem.*, **8**:241–247.

Rovira, A. D. (1986). Influence of crop rotation and tillage on Rhizoctonia bare patch of wheat. *Phytopath.*, **76**:669–673.

Rovira, A. D. and R. Campbell (1974). Scanning electron microscopy of microorganisms on the roots of wheat. *Microb. Ecol.*, **1**:15–23.

Rovira, A. D. and R. Campbell (1975). A scanning electron microscope study of interactions between micro-organisms and *Gaeumannomyces graminis* (Syn. *Ophiobolus graminis*) on wheat roots. *Microb. Ecol.*, **2**:177–185.

Rovira, A. D. and E. L. Greacen (1957). The effect of aggregate disruption on the activity of microorganisms in the soil. *Aust. J. Agric. Res.*, **8**:659–673.

Rovira, A. D. and H. McDonald (1986). Effects of the herbicide chlorsulfuron on Rhizoctonia bare patch and take-all of barley and wheat. *Plant Disease*, **70**:879–882.

Rovira, A. D. and E. H. Ridge (1978). The effect of methyl bromide and chloropicrin on some chemical and biological properties of soil and on the growth and nutrition of wheat. In *Soil Disinfestation* (D. Mulder, ed.). Elsevier Scientific Publishing Company, Amsterdam, pp. 231–250.

Rovira, A. D. and A. Simon (1982). Integrated control of *Heterodera avenae*. *Eur. Pl. Prot. Org.*, **12**:517–523.

Rovira, A. D. and A. Simon (1985). Growth, nutrition and yield of wheat in calcareous sandy loams of South Australia: effects of soil fumigation, fungicide, nematicide and nitrogen fertilizer. *Soil Biol. Biochem.*, **17**:279–284.

Rovira, A. D. and G. B. Wildermuth (1981). The nature and mechanism of suppression. In *The Biology and Control of Take-all* (P. J. Shipton and M. J. Asher, eds.). Academic Press, London, pp. 385–415.

Rovira, A. D., J. F. Kollmorgen, G. C. MacNish, K. Sivasithamparam, and P. T. Wong (1990). Take-all in Australia. In *Soil-borne Diseases of Wheat in the Australian Environment* (L. W. Burgess, ed.). Inkata Press, Melbourne, Sydney.

Rush, C. M. and S. R. Winter (1988). The influence of crop rotation on the rhizoctonia root rot progression on sugar beet. Poster presented at the International Conference on Dryland Farming Tour of the Texas Agricultural Experiment Station, Bushland, Texas, August 1988.

Rush, C. M., R. E. Ramig, and J. M. Kraft (1986). Effects of wheat chaff and tillage on inoculum density of *Pythium ultimum* in the Pacific Northwest. *Phytopathology*, **76**:1330–1332.

Samuel, G. and S. D. Garrett (1932). *Rhizoctonia solani* on cereals in South Australia. *Phytopathology*, **22**:827–836.

Sands, D. C. and A. D. Rovira (1970). Isolation of fluorescent pseudomonads with a selective medium. *Appl. Microbiol.*, **20**:313–514.

Schippers, R., A. W. Bakker, and P. A. H. M. Bakker (1987). Interactions of deleterious and beneficial rhizosphere microorganisms and the effect of cropping practices. *Ann. Rev. Phytopathol.*, **25**:339–358.

Scott, P. R. (1970). *Phialophora radicicola*, an avirulent parasite of wheat and grass roots. *Trans. Brit. Mycol. Soc.*, **55**:163–167.

Shipton, P. J. (1975). Take-all decline during cereal monoculture. In *Biology and Control of Soil-borne Plant Pathogens* (G. W. Bruehl, ed.). American Phytopathological Society, St Paul, MN, pp. 137–144.

Shipton, P. J., R. J. Cook, and J. W. Sitton (1973). Occurrence and transfer of a biological factor in soil that suppresses take-all in wheat in eastern Washington. *Phytopathology*, **63**:511–517.

Simon, A. (1989). Biological control of take-all of wheat by *Trichoderma koningii* under controlled environmental conditions. *Soil Biol. Biochem.*, **21**:323–326.

Simon, A. and K. Sivasithamparam (1988a). The soil environment and the suppression of saprophytic growth of *Gaeumannomyces graminis* var. *tritici*. *Can. J. Microbiol.*, **34**:865–870.

Simon, A. and K. Sivasithamparam (1988b). Interactions among *Gaeumannomyces graminis* var. *tritici*, *Trichoderma koningii* and soil bacteria. *Can. J. Microbiol.*, **34**:871–876.

Simon, A., K. Sivasithamparam, and G. C. MacNish (1987). Biological suppression of the saprophytic growth of *Gaeumannomyces graminis* var. *tritici* in soil. *Can. J. Microbiol.*, **33:**515–519.

Simon, A., K. Sivasithamparam, and G. C. MacNish (1988a). Effect of application to soil of nitrogenous fertilizers and lime on biological suppression of *Gaeumannomyces graminis* var. *tritici*. *Tr. Brit. Mycol. Soc.*, **91:**287–294.

Simon, A., R. W. Dunlop, E. C. Ghisalberti, and K. Sivasithamparam (1988b). *Trichoderma koningii* produces a pyrone compound with antibiotic properties. *Soil Biol. Biochem.*, **20:**263–264.

Smiley, R. W. (1975). Forms of nitrogen and the pH in the root zone and their importance to root infections. In *Biology and Control of Soil-borne Plant Pathogens* (G. W. Bruehl, ed.). American Phytopathological Society, St Paul, MN, pp. 55–62.

Smiley, R. W. (1978a). Antagonists of *Gaeumannomyces graminis* from the rhizoplane of wheat in soils fertilized with ammonium- or nitrate-nitrogen. *Soil Biol. Biochem.*, **10:**169–174.

Smiley, R. W. (1978b). Colonization of wheat roots by *Gaeumannomyces graminis* inhibited by specific soils, microorganisms and ammonium-nitrogen. *Soil Biol. Biochem.*, **10:**175–179.

Smiley, R. W. and R. J. Cook (1972). Use and abuse of soil pH measurement. *Phytopathology,* **42:**193–194.

Smiley, R. W. and R. J. Cook (1973). Relationship between take-all of wheat and rhizosphere pH in soils fertilized with ammonium vs. nitrate-nitrogen. *Phytopathology,* **73:**882–890.

Snyder, W. C., M. N. Schroth, and T. Christou (1959). Effect of plant residue on root rot of bean. *Phytopathology,* **49:**755–756.

Sommerville, L. (1987). Perspectives on side effects testing. In *Pesticide Effects on Soil Microflora* (L. Sommerville and M. P. Greaves, eds.). Taylor and Francis, London. pp. 5–13.

Sprague, M. A. and G. B. Triplett (1986). *No-tillage and Surface Tillage Agriculture. The Tillage Revolution.* John Wiley & Sons, New York.

Sprague, R. (1950). *Disease of Cereals and Grasses in North America.* Ronald Press, New York, 538 pp.

Strohbusch, D. F. (1974). *Changes in Agriculture and their Influence on the use of Agrochemicals.* Adv. Div. Plastic, Univ. Manchester, Sept 1974, Bull. 8/74 BASF.

Stroo, H. F., L. F. Elliott, and R. I. Papendick (1988). Growth, survival and toxin production of root-inhibitory pseudomonads on crop residues. *Soil Biol. Biochem.,* **20:**201–207.

Stroo, H. F., K. E. Bristow, L. F. Elliott, R. I. Papendick, and G. S. Campbell (1989). Predicting rates of wheat residue decomposition. *Soil Sci. Soc. Amer. J.,* **53:**91–99.

Suslow, T. V. and M. N. Schroth (1982). Role of deleterious rhizobacteria as minor pathogens in reducing crop growth. *Phytopathology,* **72:**111–115.

Taylor, R. G., T. L. Jackson, R. L. Powelson, and N. W. Christensen (1983). Chloride, nitrogen form, lime, and planting date effects on take-all root rot of winter wheat. *Plant Disease,* **67:**1116–1120.

Thomashow, L. S. and D. M. Weller (1988a). Role of a phenazine antibiotic from *Pseudomonas fluorescens* in biological control of *Gaeumannomyces graminis* var. *tritici. J. Bact.,* **170:**3499–3508.

Thomashow, L. S. and D. M. Weller (1988b). Role of antibiotics in root disease suppression by fluorescent pseudomonads. *Abstr. 5th ICPP Congress, Kyoto, Japan,* **II.7-4,** 94.

Thompson, J. P. (1987). Decline of vesicular-arbuscular mycorrhizae in long fallow disorder of field crops and its expression in phosphorus deficiency of sunflower. *Aust. J. Agric. Res.,* **38:**847–867.

Thompson, J. P. (1990). Root lesion nematodes. In *Soil-borne Diseases of Wheat in the Australian Environment* (L. W. Burgess, ed.). Inkata Press, Melbourne, Sydney (in press).

Thompson, J. P., T. G. Clewett, M. Fiske, and P. Lister (1986). Role of vesicular-arbuscular mycorrhizae (VAM) in long-fallow disorder. In *Biennial Report of Queensland Wheat Research Institute 1982–1984*, pp. 35–36.

Tinline, R. D., R. J. Ledingham, and B. J. Sallans (1975). Appraisal of loss from common root rot in wheat. In *Biology and Control of Soil-borne Plant Pathogens* (G. Bruehl, ed.). American Phytopathological Society, St Paul, pp. 22–27.

Triplett, G. B. (1983). Crop management practices for surface-tillage systems. In *No-Tillage and Surface-Tillage Agriculture: The Tillage Revolution* (M. A. Sprague and G. B. Triplett, eds.). John Wiley & Sons, New York, pp. 149–182.

Trottman, D. R. and W. Thompson (1978). The influence of herbicides on the incidence of take-all disease (*Gaeumannomyces graminis*) on the roots of winter wheat. *Proc. 1978 Brit. Crop Prot. Conf.—Weeds,* pp. 609–615.

Vojinovic, Z. D. (1972). *Antagonists From Soil and Rhizosphere to Phytopathogens.* Final Tech. Rep., Inst. Soil Sci., Beograd, Yugoslavia, 130 pp.

Walker, J. (1975). Take-all disease of Gramineae: A review of recent work. *CMI Rev. Pl. Path.,* **54:**113–144.

Weller, D. M. (1988). Biological control of soil-borne plant pathogens in the rhizosphere with bacteria. *Ann. Rev. Phytopathol.,* **26:**379–407.

Weller, D. M. and R. J. Cook (1983). Suppression of take-all of wheat by seed treatments with fluorescent pseudomonads. *Phytopathology,* **73:**463–469.

Weller, D. M., R. J. Cook, G. MacNish, E. N. Bassett, R. L. Powelson, and R. R. Petersen (1986). Rhizoctonia bare patch of small grains favoured by reduced tillage in the Pacific Northwest. *Plant Disease,* **70:**70–73.

Weller, D. M., W. J. Howie, and R. J. Cook (1988). Relationship between in vitro inhibition of *Gaeumannomyces graminis* var. *tritici* and suppression of take-all of wheat by fluorescent pseudomonads. *Phytopathology,* **78:**1094–1100.

Wildermuth, G. B. (1990). Common root rot. In *Soil-borne Diseases of Wheat in the Australian Environment* (L. W. Burgess, ed.). Inkata Press, Melbourne, Sydney (in press).

Wildermuth, G. B. and A. D. Rovira (1977). Hyphal density as a measure of suppression of *Gaeumannomyces graminis* var. *tritici* on wheat roots. *Soil Biol. Biochem.,* **9:**203–205.

Wildermuth, G. B., A. D. Rovira, and J. H. Warcup (1979). Mechanisms and site of suppression of *Gaeumannomyces graminis* var. *tritici* in soil. In *Soil-borne Plant Pathogens* (B. Schippers and W. Gams, eds.). Academic Press, London, pp. 157–164.

Wildermuth, G. B., J. H. Warcup, and A. D. Rovira (1984). Growth of *Gaeumannomyces graminis* var. *tritici* in soil in the presence and absence of wheat roots. *Trans. Br. Mycol. Soc.,* **82:**435–444.

Wildermuth, G. B., A. D. Rovira, and J. H. Warcup (1985). Mechanism and site of suppression of *Gaeumannomyces graminis* var. *tritici* in soil. *Trans. Br. Mycol. Soc.,* **84:**3–10.

Wilhelm, S. (1965). *Pythium ultimum* and the soil fumigation growth response. *Phytopathology,* **55:**1016–1020.

Wilhelm, N. S., R. D. Graham, and A. D. Rovira (1988). Application of different sources of manganese sulphate decreases take-all (*Gaeumannomyces graminis* var. *tritici*) of wheat grown in a manganese deficient soil. *Aust. J. Agric. Res.,* **39:**1–10.

Wilkinson, H. T., J. R. Alldredge, and R. J. Cook (1985a). Relation of inoculum size and concentration to infection of wheat roots by *Gaeumannomyces graminis* var. *tritici. Phytopathology,* **75:**98–103.

Wilkinson, H. T., J. R. Alldredge, and R. J. Cook (1985b). Estimated distances for infection of wheat roots by *Gaeumannomyces graminis* var. *tritici* in soils suppressive and conducive to take-all. *Phytopath,* **75:**557–559.

Yarwood, C. E. (1966). Detection of *Pythium* in soil. *Pl. Dis. Reptr.,* **50:**791–792.

Zing, A. W. and C. J. Whitfield (1957). A summary of research with stubble-mulch farming in the western states. *U.S. Dept Agric. Tech. Bull.,* **1166:**1–56.

APPENDIX: COMMON AND BOTANICAL NAMES OF SOME PLANT SPECIES CITED IN THIS BOOK

Common name	Botanical name
Almond	*Prunus amygdalus*
Arolla pine (Swiss stone pine)	*Pinus cembra*
Australian pine	*Casuarina cunninghamiana*
Avocado	*Persia gratissima*
Bahia grass	*Paspalum notatum*
Barley	*Hordeum vulgare*
Beach or shore pine	*Pinus contorta*
Bean	*Phaseolus vulgaris*
Bermuda grass	*Cynodon dactylon*
Birdsfoot trefoil	*Lotus corniculatus*
Blue grama grass	*Bouteloua gracilis*
Buckthorn	*Hippophae* spp
Buffaloberry	*Shepherdia* spp
Carnation	*Dianthus carophyllus*
Celery	*Apium graveolens*
Cherry	*Prunus avium*
Chickpea	*Cicer arietinum*
Citrange	*Poncirus trifoliata*
Clover	*Trifolium* spp
Cocksfoot	*Dactylis glomerata*
Common alder	*Alnus glutinosa*
Cotton	*Gossypium*
Cowpea	*Vigna sinensis*; *V. radiata*
Cucumber	*Cucumis sativus*
Digitgrass	*Digitaria decumbens*
Douglas fir	*Pseudotsuga taxifolia*

Downy brome	*Bromus tectorum*
Eucalyptus	*Eucalyptus* spp
Faba bean	*Vicia faba*
Fimbriate medick	*Medicago polymorpha*
Flax	*Linum* spp
Galleta (warm season grass)	*Hilaria jamesii*
Grain sorghum	*Sorghum vulgare*
Grey alder	*Alnus incana*
Guineagrass	*Panicum maxicum*
Heather	*Calluna* spp
Leek	*Allium porrum*
Lima bean	*Phaseolus lunatus*
Loblolly pine	*Pinus taeda*
Lodgepole pine	*Pinus contorta* (= *P. murrayana*)
Lucerne (alfalfa)	*Medicago sativa*
Lupin	*Lupinus* sp
Maize	*Zea mays*
Muskmelon	*Cucumis melo*
Mustard	*Sinapsis alba*
Norway spruce	*Picea excelsa* (= *P. abies*)
Oak	*Quercus* spp.
Oats	*Avena sativa*
Onion	*Allium cepa*
Orange	*Citrus sinensis*
Orchid	*Goodyera repens* Bv
Oregon or red alder	*Alnus rubra*
Peach	*Prunus persica*
Pearl millet	*Pennisetum americanum*
Pea	*Pisum* spp. (*Pisum sativum*)
Peanut	*Arachis hypogaea*
Pepper (green or cayenne)	*Capsicum annum* (= *C. frutescens*)
Pigeon pea	*Cajanus cajan*
Potato	*Solanum tuberosum*
Prairie grass (short)	*Bouteloua gracilis* + *Buchloe dactyloides*
Radish	*Raphanus sativus*
Ragwort	*Senecio* spp
Rape	*Brassica napus*
Red maple	*Acer* spp
Red clover	*Trifolium pratense*
Rhododendron	*Rhododendron* spp
Rice	*Oryza sativa*
Rockrose	*Helianthemum* spp
Rose	*Rosa* spp
Russian olive or silver berry	*Elaeagnus* spp
Rye	*Secale cereale*
Ryegrass	*Lolium perenne*
Safflower	*Carthamus tinctoria*
Sagebrush	*Artemisia tridentata*
Scented vernal grass	*Anthoxanthum odoratum*
Scots pine	*Pinus sylvestris*
Small buffalo grass	*Panicum coloratum*
Smooth brome	*Bromus inermis*
Snail medic	*Medicago scutellata*
Sour orange	*Citrus aurantium*

Soybean	*Glycine max*
Strawberry	*Fragaria xananassa*
Subterranean clover	*Trifolium subterranean*
Sudan grass	*Sorghum vulgare*
Sugar beet	*Beta vulgaris*
Sunflower	*Helianthus annus*
Swamp or bull oak	*Casuarina equisetifolia*
Sweet potato	*Ipomoea batatas*
Sweet fern	*Comptonia peregrina*
Swiss chard	*Beta vulgaris*
Tobacco	*Nicotiana tabacum*
Tomato	*Lycopersicon esculentum*
Trefoil	*Lotus* spp
Vetch	*Lathyrus odoratus*
Walnut	*Juglans regia*
Water melon	*Citrullus vulgaris*
Wheat	*Triticum aestivum*
White birch	*Betula verulosa*
Yellow poplar	*Liriodendron tulipifera*
Yellow vetchling	*Lathyrus aphaca*
Zinnia	*Zinnia* spp

INDEX

Index compiled by Liza Weinkove